Lattice Boltzmann Method: *Theory and Applications*

格子 Boltzmann 方法的理论及应用

何雅玲 李庆 王勇 童自翔 编著

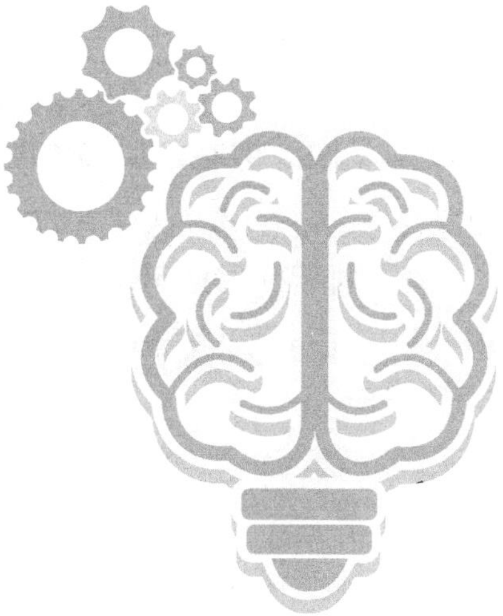

中国教育出版传媒集团
高等教育出版社 · 北京

内容提要

格子 Boltzmann 方法自 20 世纪 80 年代被提出后，受到了国内外众多学者的关注，在理论和应用研究等方面都迅速发展，并逐渐成为相关领域国际研究的热点之一。本书结合作者近二十年在相关科研实践中的体会和所积累的经验，通过采取"基础理论＋应用基础模型＋应用案例"的递进模式，系统地介绍了格子 Boltzmann 方法的基本概念、基础理论、应用基础模型、边界处理、网格划分、数值实施以及应用等，并配以典型的算例和程序代码，以帮助读者快速掌握该方法。在呈现方式上，本书增加了可扫码查看的数字资料，激发读者的学习兴趣，增强学习效果。

本书适用于高等学校研究生，以及科研单位的工程技术人员和研究人员，可作为能源动力、机械、数学、物理及生物工程等大类专业的计算流体力学与计算传热学课程的教材或参考用书，也可供相关技术人员参考。

图书在版编目（CIP）数据

格子 Boltzmann 方法的理论及应用 ／ 何雅玲等编著. -- 北京：高等教育出版社，2023.2（2024.5 重印）
ISBN 978-7-04-059357-0

Ⅰ. ①格⋯ Ⅱ. ①何⋯ Ⅲ. ①流体动力学-计算方法-教材 Ⅳ. ①O351.2

中国版本图书馆 CIP 数据核字（2022）第 159347 号

Gezi Boltzmann Fangfa de Lilun ji Yingyong

| 策划编辑 | 宋　晓 | 责任编辑 | 宋　晓 | 封面设计 | 王　琰 | 版式设计 | 杜微言 |
| 责任绘图 | 黄云燕 | 责任校对 | 刘丽娴 | 责任印制 | 高　峰 | | |

出版发行	高等教育出版社	网　　址	http://www.hep.edu.cn
社　　址	北京市西城区德外大街 4 号		http://www.hep.com.cn
邮政编码	100120	网上订购	http://www.hepmall.com.cn
印　　刷	固安县铭成印刷有限公司		http://www.hepmall.com
开　　本	787 mm×1092 mm　1/16		http://www.hepmall.cn
印　　张	19		
字　　数	460 千字	版　　次	2023 年 2 月第 1 版
购书热线	010-58581118	印　　次	2024 年 5 月第 2 次印刷
咨询电话	400-810-0598	定　　价	40.00 元

本书如有缺页、倒页、脱页等质量问题，请到所购图书销售部门联系调换
版权所有　侵权必究
物　料　号　59357-00

序

我的老师陶文铨院士所作此篇序言，原载于2009年出版的《格子Boltzmann方法的理论及应用》，现在读来仍然能感受到先生的谆谆鼓励和殷殷期望，感人至深，为我们的编著增辉添彩。现谨把先生此文也作为本书序，我们认为具有特别意义，既能回望我们一路走来的研究历程，也赋予我们源源不断的前行力量。——何雅玲注

在我的案头上放着一卷由何雅玲教授编著的《格子Boltzmann方法的理论及应用》的书稿，这就使我不禁想起：十年前，冒着冬日的严寒，笔者和雅玲教授为了一篇我国学者关于格子气研究的较早的文献遍访某大城市而一无所获的情形。

我们的研究团队，在十余年前开始接触到关于格子气以及格子Boltzmann方法在流动与换热中应用的文献，从一开始我们就被这种方法的独特优点所吸引：从微观的动力学观点出发而可以获得宏观的流场与温度场的参数。这以后，我们就从实践中学习、钻研这种方法，努力收集并阅读有关参考文献，本文一开始提到的情形就是一个例子。雅玲教授在指导研究生的过程中，对格子Boltzmann方法的基本理论，例如BGK近似、Navier-Stokes方程与Boltzmann方程的关系等都进行了认真的推演。正是有这样的一个过程，使得该书具备了一个十分有利于读者的特点：作者从基本理论出发，根据他们自己的理解、掌握，直到有所发展、有所创新的经历来撰写，给该书增添了针对读者的困惑和难点而循循善诱、娓娓道来的亲切气氛，有利于学习者的理解和掌握。

雅玲教授是我国工程热物理领域的知名学者，在新型低温制冷机、新型换热器、电子器件冷却、新能源的利用等方面的研究，都取得了很好的成绩。据笔者所知，她研究格子Boltzmann方法的基本目的之一是想利用此法的优点来揭示新型低温制冷机的基本物理过程——交变流动与换热的基本规律。当将格子Boltzmann方法应用到这类新型低温制冷机工作过程的模拟时，因为可压缩交变流动的复杂性，她(他)们发现文献中现有的格子Boltzmann方法还存在很大的不足。正是从一开始该书作者就是带着问题来研究格子Boltzmann方法，而不是急于简单地应用现有程序去求解问题、发表文章，他们从最基本的知识理论开始钻研，步步深入，使得该书作者具备了坚实的理论功底，进而能够有所创新。近几年来雅玲教授和她所指导的研究生连续在国际著名期刊上发表文章，对将格子Boltzmann方法应用于可压缩与交变流动等方面提出了自己的数值方法与格式，有的论文是由国际期刊的主编在未经外审条件下就直接接受的，可见研究结果的学术水平已经得到国际刊物的认可。因为这本书稿不仅有关于格子Boltzmann方法基本思想的介绍，同时还融入了作者自己的科研成果，所以该书既是一本凝结了作者心血的专著，同时也可以作为有关教师和科研人员学习格子Boltzmann方法的有益的参考书。

笔者从事传热与流动过程的数值模拟的学习、研究已经整整30年了。笔者的深刻体会之一就是:数值模拟方法的学习与研究不能只停留在理解数值方法的思想上,更重要的是应用该方法进行数值计算的实践。笔者高兴地看到,雅玲教授在这一点上与笔者很有共识,即不仅要理解、掌握数值方法的基本思想,而且一定要通过程序的编制学会如何在工程实际和理论研究中实现这种方法,并反思总结达到“明齐全而析其微”的境界。该书不仅在数值方法的介绍上颇具特色,而且书中还附有作者自己开发的用格子Boltzmann方法求解流场的程序,并介绍了编程思想和方法。尽管这些程序可能还需要改进,但是对初学者而言,有了这样一个范本,就会对数值方法有更深刻的理解和认识,也给读者提供了进一步发展的基石。

2000年,笔者在由科学出版社出版的《计算传热学的近代进展》一书中用一节的篇幅向我国广大读者介绍了格子Boltzmann方法,并且引用了20世纪80年代后半期格子气方法刚诞生时美国《华盛顿邮报》记者的热情报道。近十年来,我国学者在格子Boltzmann方法的学习、研究方面表现出了极大的热情,迫切需要一本从基础入手,包含自己研究成果又有程序可以参照的著作。该书正是满足了这样的要求。笔者相信,该书的出版一定会对推动我国在应用与研究格子Boltzmann方法方面起到积极的作用。

“读书心细丝抽茧,炼句功深石补天。”在此也祝愿雅玲教授在以后的科研中,继续勤奋探索求新,争取有更多的佳作问世。

陶文铨
中国科学院院士
西安交通大学教授
2008年10月于西安

前言

随着航空航天、空间技术、国防军工、生物医学、互联信息、低碳能源、智能制造等领域现代科学技术的高速发展，流动与传热现象的研究范围也在不断向极端条件、多物理场、多因素耦合等方面延拓，人们对流动与传热的机理认知有了越来越深入的需求，而这单靠以宏观尺度为基础的实验研究和理论分析是无法完成的，需要从微观和跨尺度的介观的角度才能深入揭示。

在这样的应用背景下，格子 Boltzmann 方法(lattice Boltzmann method)自 20 世纪 80 年代被提出后，受到了国内外众多学者的关注，在理论和应用研究等方面都迅速发展，并逐渐成为相关领域国际研究的热点之一。与传统模拟方法不同，格子 Boltzmann 方法基于分子动理论，在宏观上是离散方法，在微观上是连续方法，因而被称为介观模拟方法。该方法具有清晰的粒子运动图像和介观物理背景，能够非常方便地从底层刻画流体内部以及流体与固体之间的相互作用，从而在复杂流动与传热问题的模拟方面具有显著的优势。在许多传统模拟方法难以胜任的领域，如微尺度流动与换热、多孔介质流动、多相/多组分流动、生物流体、晶体生长等，格子 Boltzmann 方法都取得了成功的应用并揭示了多种复杂现象的机理，推动了相关学科的发展。所以，格子 Boltzmann 方法不仅仅是一种数值模拟方法，而且是一项重要的科学研究手段。此外，格子 Boltzmann 方法还具有天生的并行特性，以及边界条件处理简单、程序易于实现等优点。随着计算机技术的进一步发展，以及计算方法的逐渐丰富，格子 Boltzmann 方法将会取得更多成果，并为科技发展发挥重要作用。

作者团队从事格子 Boltzmann 方法的研究与应用研究已有 20 余年，经过不断努力取得了一些成果。其间在与不少同行交流时，我们深切感受到该领域迫切需要一本入门著作，能细致明了地展示格子 Boltzmann 方法的基本概念、基础理论以及发展现状，并适当配以具体的算例以帮助读者快速入门。因此，我们于 2009 年编写了《格子 Boltzmann 方法的理论及应用》一书，由科学出版社出版。该书不仅包含了我们在格子 Boltzmann 方法领域所取得的部分研究成果，以及在相关科研实践中积累的经验，而且还尽可能全面地介绍了格子 Boltzmann 方法当时的发展现状，并系统阐明该方法的流体力学与统计力学背景知识、模型推导、边界处理、数值实施等问题，以便读者了解相关基础知识，拓展全局视角，更好地掌握格子 Boltzmann 方法并灵活运用于科研工作。该书出版后，受到了广大同行与读者的喜爱与欢迎，据了解，许多初学者纷纷采用该书作为学习格子 Boltzmann 方法的参考用书。13 年过去了，格子 Boltzmann 方法无论是在基础理论还是在模型构建与应用方面都有了很大的进步，许多读者提出希望能够在该著作基础上了解更新的理论与应用研究成果。考虑到读者的需求，以及格子 Boltzmann 方法的发展，特别是随着格子 Boltzmann 方法在多相流、多孔介质流动与传热、相变传热等领域的广泛应用，基于对相关内容的反复推敲、精心编著，在高等教育出版社的帮助下，我们推出了新版的《格子 Boltzmann 方法的理论及应用》。

本书采取了“基础理论 + 应用基础模型 + 应用案例”递进模式逐渐展开，与 2009 年的著作相比，新增了多相流、多孔介质流动与传热、相变传热等方面应用基础模型的相关内容；此外，根据格子 Boltzmann 方法的最新发展，对基础理论的相关章节增加了多松弛格子 Boltzmann 模型、格子 Boltzmann 方法作用力格式等方面的介绍；同时，为了帮助读者更好地体会与理解格子 Boltzmann 方法“简单算法模拟复杂问题”的独特魅力，本书还增加了有关沸腾传热和多孔介质流动与传热的应用举例，并且提供了关于相分离格子 Boltzmann 模拟的程序代码。

本书共分 10 章。第 1 章为绪论，介绍了格子 Boltzmann 方法的发展历程；第 2 章总结了从连续介质层次出发的连续方程、动量方程和包括焓方程和熵方程在内的多种形式能量方程，这些流体力学基本方程是传统计算流体力学的基础，也是格子 Boltzmann 方法所应恢复的宏观方程；第 3 章从介观的气体动理论出发，系统地阐述了包括多松弛模型在内的格子 Boltzmann 方法基础理论；第 4 章对单相流动与传热的格子 Boltzmann 模型进行了介绍；在第 5~7 章中，基于格子 Boltzmann 方法在多相流、相变传热以及多孔介质流动与传热方面的基础模型和应用进展，对气液两相流格子 Boltzmann 模型、流固两相流模型及固液相变格子 Boltzmann 模型、多孔介质流动与传热的格子 Boltzmann 模型分别予以专题介绍；第 8 章针对模型的边界条件，从最基本的反弹处理和镜面反射处理，到适用于运动边界和复杂曲线边界的处理格式，以及复杂边界质量修正和温度边界处理格式等，都做了介绍；第 9 章介绍了格子 Boltzmann 方法的网格处理技术，涉及均分网格、多块网格、多重网格、结构化网格以及非结构化网格等技术；第 10 章给出了格子 Boltzmann 方法的部分算例以及在工程热物理领域的部分应用，特别增加了池沸腾中气泡成核与生长以及泡沫金属内相变材料熔化问题的应用案例。书末，还附有读者学习过程中可能用到的数学基础，格子单位与物理单位的转换，并公布了作者自己编写的单相流动模型和伪势多相模型的程序代码，以帮助初学者更快速地上手。

我们希望能够为读者提供一本从内容到形式都更为优良的书籍，并通过涵盖基础理论，到应用基础模型，再到应用案例的写作和组织模式，使读者能够更好地了解、更容易掌握格子 Boltzmann 方法，并在此基础上能够通过自编程序或者使用该方法的开源代码与商用软件，更好地开展相关的科学研究。

本书可供高等学校研究生和科研单位的技术人员和研究人员使用，可作为数学、物理、机械、能源动力和生物工程等大类专业的计算流体力学与计算传热学课程的教材或参考用书。对于那些对格子 Boltzmann 方法感兴趣而又尚未进入该领域的人员，本书可作为入门的学习用书，通过本书的学习，能对格子 Boltzmann 方法有基本的了解，能完成基本的数值模拟，从而为独立开展相应的研究工作打下较好的基础；对于那些已经初步了解格子 Boltzmann 方法但又尚存疑惑的人员，本书可以起到答疑解惑的作用；对于已经掌握格子 Boltzmann 方法的研究人员来说，本书可作为深入学习的工具书。

本书得到了我的老师陶文铨院士的热切关心，先生在数值计算方面给予我多方面的指导和帮助，一直支持我们进行格子 Boltzmann 方法的研究，先生在 2008 年欣然为我们的《格子 Boltzmann 方法的理论及应用》撰写序言，现在读来，仍然能感受到先生的谆谆鼓励和殷殷期望，感人至深，为我们的编著增辉添彩，在此向陶先生致以最衷心的感谢和祝福！也谨把先生此篇文章作为本书的序，它将赋予我们源源不断的前行力量。

本人指导的一些博士生，在不同程度上对格子 Boltzmann 方法进行了研究和应用，特别是现在任教于中南大学的李庆教授，持之以恒地开展这方面工作，取得了可喜的成绩，他为本书付出了辛勤努力，做了大量工作；刘清博士论文中的一些工作体现在第 7 章的部分内容中。在此一并致谢！

英国皇家工程院院士、伦敦大学学院（University College London）Kai H. Luo 教授在百忙之中对本书仔细审阅，给予了中肯的评价和宝贵的建议；高等教育出版社的宋晓编辑为本书的出版研赘斧正，做了大量工作。他们给予了大力支持，我们表示衷心的感谢！

本书所涉及的研究工作，得到了国家自然科学基金委创新群体项目以及科学中心项目（Nos. 51721004、51888103、51822606）的资助，在此表示感谢！

本书虽数易其稿，但因格子 Boltzmann 方法的内容博大精深，而我们学识有限，纵心有所向，然力有未逮，书中不足和疏漏之处在所难免，热诚欢迎广大读者提出宝贵意见！

中国科学院院士
西安交通大学教授
2022 年 3 月于西安

目录

第1章 绪　论

格子 Boltzmann 方法是 20 世纪 80 年代末开始发展起来的一种数值模拟方法。该方法是介于微观分子动力学模拟方法和宏观传统数值模拟方法之间的一种介观层次的数值模拟方法。从描述流体运动的方式来说,微观分子动力学模拟着眼于每个流体分子的动力学行为(图 1.1),通过刻画每个分子的具体运动并采用统计方法来描述流体的运动情况,而宏观方法则将流体看作连续介质,它着眼于流体质点(流体质点连续布满整个物理空间),并通过一组偏微分方程来直接描述流体的运动。格子 Boltzmann 方法则着眼于流体粒子的分布函数,根据相应的控制方程来描述分布函数的时空演化,流体的宏观运动信息则通过宏观物理量与分布函数之间的关系来获得。与传统数值模拟方法相比,格子 Boltzmann 方法的介观物理背景和清晰的粒子运动图像使其能够非常方便地从底层刻画流体内部以及流体与固体之间的相互作用,从而在多相/多组分流动、多孔介质流动等复杂流动的模拟方面具有显著的优势。

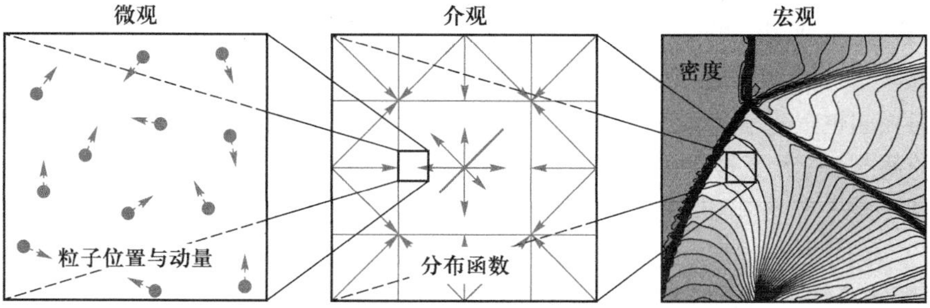

图 1.1　描述流体运动的三个层次

作为一门多学科的交叉产物,格子 Boltzmann 方法涉及统计物理、流体力学、传热学以及计算数学等诸多学科。为了让读者了解并掌握该方法,本书介绍了与格子 Boltzmann 方法相关的统计物理基础知识,并对该方法在理论建模、边界实施、网格划分和实际应用等方面的基础理论以及相关进展都做了一定的介绍。这些内容,包含了我们从事相关研究实践所取得的成果和积累的经验;同时,也是对格子 Boltzmann 方法相关知识的一次梳理。希望透过本书,能够促进同行交流,并进一步传播该方法。

格子 Boltzmann 方法作为一种介观数值模拟方法,属于计算流体力学与计算传热学的一个分支,作为基础知识的储备,本章先简要介绍计算流体力学与计算传热学,并引出格子 Boltzmann 方法。随后,重点阐述格子 Boltzmann 方法的发展历程,并给出本书各章节框架以及各章的主要内容。

1.1 计算流体力学与计算传热学

流体力学与传热学都是人类社会从古老的实践活动中逐渐形成学科。无论是我国古代大禹治水，还是古罗马人大规模供水管道系统的建设，无不体现了古人对流体力学的认知水平。流体力学学科的形成可追溯到古希腊时期的阿基米德时代[1]。阿基米德创立了包括物体浮力定律和浮体稳定性在内的液体平衡理论，奠定了流体静力学的基础。另外，虽然人类从“钻木取火”的时代就开始利用传热学的规律，但作为独立的学科，传热学形成较晚。一般认为，传热学产生于18世纪30年代的工业革命时期[2]。1822年，傅里叶发表了他的著名论著“热的解析理论”[3]，成功地完成了创建导热理论的任务，为从理论上和实验上正确理解和定量研究传热学问题奠定了基础。随后，对流换热理论随着工业革命的兴起而逐步发展起来。

傅里叶的人物小传

虽然历经数个世纪的发展，流体力学与传热学已成为理论体系日趋完善的基础学科，但它们的发展却从未停止。**计算流体力学**(computational fluid dynamics)[4-5]与**计算传热学或数值传热学**(numerical heat transfer)[6-8]的出现，就是这两门基础学科发展史中的大事。

20世纪，随着计算机技术的出现和发展，数值计算逐渐成为与理论分析、实验研究相并列的三大重要科学研究手段之一。通过数值计算，可以对许多无法或很难进行理论分析的复杂问题进行求解，扩大了研究范围。同时，数值计算还可以在一定程度上替代实验研究，从而节约实验成本，加快研究进度。数值计算方法在流体力学与传热学方面的应用，促使了计算流体力学与计算传热学的出现，通过运用数值计算方法求解一组非线性的质量、动量、能量或组分的偏微分方程(如Euler方程组、Navier-Stokes方程组等)，可以获得流体流动、传热、传质等现象的细节，从而揭示相关的物理规律。计算流体力学与计算传热学的兴起也促进了理论分析与实验研究的发展，三者相辅相成。尤其是对湍流等复杂问题的研究，三者更是缺一不可。同时，在工业应用中，计算流体力学与计算传热学已被大规模应用于产品设计、生产以及销售等诸多环节，它们的出现推动了社会水平的进步，丰富了人类对世界的认知。相应的，工业应用的新成果，如并行计算机、矢量计算机的出现，又推动计算流体力学与计算传热学的发展；工业应用中出现的新问题，也会促使它们的进一步发展与完善。

从方法论的角度，对流体流动与传热的描述可以从**宏观**(macroscopic)、**介观**(mesoscopic)和**微观**(microscopic)三个层次进行，参见图1.1。相应的，数值模拟方法也可以分为宏观方法、介观方法和微观方法三类。

马蹄涡效应

在宏观层次，流体被假设为连续介质。流体运动满足质量守恒、动量守恒以及能量守恒，并由Euler方程组或Navier-Stokes方程组进行描述。在数值计算中，通过各种离散方法，将非线性的Euler方程组或Navier-Stokes方程组离散成各种代数方程组，再由计算机求解。现有的大多数数值模拟方法都属于宏观方法，例如有限差分法(finite difference method)、有限容积法(finite volume method)、有限元法(finite element method)、边界元法(boundary element method)、谱方法(spectral method)等。这些方法发展较为成熟，已被广泛应用于各种宏观物理问题的研究，并揭示了一些新的物理现象，如槽道湍流中的倒马蹄涡就是通过直接数值模拟发现而后由实验予以证实[9-10]。同时，这些宏观方法在多个工业领域也得到了广泛的应用,例如叶轮机械的型线设计、换热器的优化计算、建筑物室内流场与温度场的模拟、飞行器升阻力的测算等。一些基于宏观方法的商业软件,如PHOENICS、FLUENT、STAR-CD、CFX 等,也应运而生,并被应用于解决多种工程实际问题。

在微观层面,流体由大量分子组成,分子的运动特性受到分子间相互作用力以及外加作用力的影响。任何体系的宏观特性,在微观都表现为分子的不规则热运动。因而,可以通过模拟每个分子的运动,再进行统计平均,从而获得流体流动与传热的宏观规律。1957年,Alder 和 Wainwright 基于硬球模型,采用分子动力学研究了气体和液体的状态方程[11],开创了分子动力学模拟的先例。分子动力学模拟的步骤比较直观,它的计算对象为指定空间内的模型分子体系,按照演化规律进行计算,通过对结构空间内的分子体系采样,可以获得分子的详细轨道图景以及系统的各种物理量分布[12]。在分子动力学模拟中,分子的动力学行为通常遵循经典运动方程。分子动力学模拟的程序较复杂,计算量非常大,对内存要求高。由于受到计算机速度和内存的限制,分子动力学模拟的空间尺度和时间尺度都很有限。随着计算机技术的飞速发展,计算机的运算速度越来越快,再加上多体势函数的提出和发展,分子动力学方法已在流体热力性质与输运性质[13-14]、相界面及相变[15-18]、微纳尺度流动[19-20]等研究方面,取得了令人瞩目的成绩。

在介观层次,一方面,可以通过数值求解气体动理论的基本方程——Boltzmann 方程得到分布函数的时空演化规律,然后根据宏观量与分布函数之间的关系获得宏观流动与传热信息。另一方面,也可以将流体看作是由一系列的流体粒子组成,这些粒子比分子要大得多(粗粒化),但在宏观上又无限小;考虑到单个分子的运动细节并不影响流体的宏观特性,因而可以构造出符合一定物理规律的演化机制,让这些流体粒子进行演化,从而获得与物理规律相符的模拟结果。常见的介观方法有格子气自动机(lattice gas automata)、格子 Boltzmann 方法、气体动理论格式(gas-kinetic scheme)、直接模拟蒙特卡罗方法(direct simulation Monte Carlo method)等。其中,格子气自动机和格子 Boltzmann 方法属于全离散的数值模拟方法,即流体被离散成流体粒子,粒子的速度空间被离散成有限个速度的集合,物理空间被离散成一个个规则的格子,时间则被离散成一个时间序列。按照简化的动理学模型,流体粒子被约束在有限的格线上运动,宏观的密度、速度、温度等物理量需要对这些粒子的相关特性值做平均获得。与宏观数值模拟方法相比,格子 Boltzmann 方法具有粒子运动图像清晰、边界条件处理简单、程序易于实施、天生并行等优点。

由于出发点不同,三个层次的方法有着不同的适用范围。当连续介质假设不成立时,如高空非常稀薄的气体流动、微纳尺度气体流动等,宏观方法不再适用。微观方法的出发点是流体分子最基本的运动规律,原理上它既适用于微纳尺度也适用于宏观尺度,但是由于需要跟踪每个分子的具体运动,受到计算条件的限制,微观方法存在时间与空间尺度上的局限性。需要强调的是,这三个层次并非完全割裂的,而是相互关联的。例如,基于统计物理,可以从微观跨入介观;基于 Chapman-Enskog 多尺度展开[21],可以从介观的 Boltzmann 方程恢复到宏观 Euler 方程组或 Navier-Stokes 方程组等。

1.2 格子 Boltzmann 方法发展历程

格子 Boltzmann 方法自诞生至今已取得了长足的发展,被誉为现代流体力学的一场变革。

1.2.1 孕育

元胞自动机

最早的格子 Boltzmann 模型并不是基于气体动理论建立的，而是直接从格子气自动机发展而来。格子气自动机是**元胞自动机**（cellular automata）在流体力学中的应用[22]。元胞自动机是一个时间和空间离散的数学模型，根据若干简单的局域规则运行，通过计算机模拟相关演化过程来获得所需的解，曾被广泛应用于细胞生长、分形结构、城市交通流等复杂问题的研究。20 世纪 60 年代，Broadwell[23] 率先提出了离散速度模型，用于研究流体中的激波结构。但在他们的模型中，时间和空间仍然保持连续。

20 世纪 70 年代，为了研究流体的输运性质，Hardy 等[24-25] 提出了第一个完全离散的模型。在该模型中，除了流体被离散成一系列粒子以外，时间和空间也进行了相应的离散，且空间离散采用正方形格子。根据作者的名称缩写，该模型被命名为 **HPP 模型**。这是历史上的第一个格子气自动机模型，它采用四个离散速度（图 1.2），每个格点处的粒子分布状况用一个布尔变量 $n_\alpha(\boldsymbol{r},t)$ 来表示，相应的时空演化方程为

$$n_\alpha(\boldsymbol{r}+\boldsymbol{e}_\alpha,t+\delta_t)=n_\alpha(\boldsymbol{r},t)+\Omega_\alpha[n_\alpha(\boldsymbol{r},t)], \quad (\alpha=1,\cdots,4) \tag{1.1}$$

式中：$\boldsymbol{r}$ 为格点的空间位置；$\boldsymbol{e}_\alpha$ 为离散速度；δ_t 为时间步长；Ω_α 为碰撞算子。粒子在格点上的碰撞规则为：当两个相反的粒子到达同一格点时，在另外两个方向没有粒子的情况下，发生对头碰撞；在其他情况下，粒子不改变原有速度。

HPP 模型粒子碰撞演示

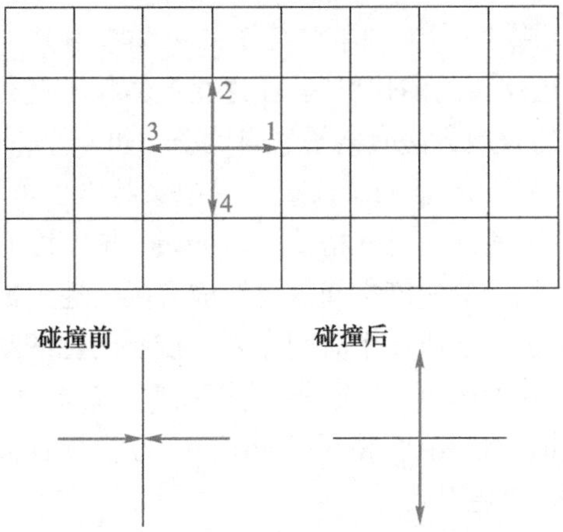

图 1.2 HPP 模型及粒子碰撞规则

由于正方形格子缺乏足够的对称性，HPP 模型的应力张量不满足各向同性，且存在伪随机守恒量，这就导致 HPP 模型既不能恢复 Navier-Stokes 方程的非线性项，也不能恢复相应的耗散项。因而，在很长一段时期，HPP 模型并未引起研究者们的重视。1986 年，Frisch 等[26] 提出了具有足够对称性的二维正六边形的格子气自动机模型，即 **FHP 模型**。同年，d'Humières 等[27] 建立了四维面心立方（face-centered-hyper-cubic，FCHC）模型及其在三维空间的投影。FHP 模型和 FCHC 模型成功克服了 HPP 模型格子对称性不足的缺点，分别可以恢复二维和三维的 Navier-Stokes 方程。

在格子气自动机中，流体粒子位于离散的格子节点上，并沿着格线迁移。所有粒子按照一定的规则同步地进行碰撞和迁移。由于粒子的演化只涉及相邻节点，因而格子气自

动机可以非常方便地采用区域分裂方法进行并行运算。在固体边界上,格子气自动机只需对边界节点上的粒子进行反弹或反射处理,边界处理非常简单,适合多孔介质等复杂几何结构的模拟。此外,由于采用 1 或 0 来描述格点上粒子的有无,也即进行布尔运算,因此格子气自动机可以无条件稳定运行。然而这种方法也存在以下一些缺点:(1)由格子气自动机演化方程所导出的动量方程不满足伽利略不变性(Galilean invariance);(2)状态方程不仅依赖于密度和温度,还与宏观速度有关;(3)由于采用布尔运算,因而局部量往往存在数值噪声,需要对时间和空间做平均,这样就增加了计算量;(4)碰撞算子具有指数复杂性,对计算量和存储量有较高要求。

为了克服格子气自动机的上述缺点并保留其优点,格子 Boltzmann 方法应运而生。因而,我们将这一段格子气自动机的发展过程,称为格子 Boltzmann 方法的孕育期。

1.2.2 诞生

为了消除格子气自动机的统计噪声,1988 年,McNamara 和 Zanetti[28]提出把格子气自动机中的布尔运算变成实数运算,即格点上的粒子数不再用 0 或 1 来表征,而是采用实数来表示系综平均后的**局部粒子分布函数**(local particle distribution function)。这是最早的格子 Boltzmann 模型,它的出现开启了格子 Boltzmann 方法的历史大门。

与格子气自动机相比,McNamara 和 Zanetti 的格子 Boltzmann 模型消除了系统的大部分统计噪声。由于粒子分布函数的介观特性,该模型继承了格子气自动机的许多优点。然而,它保留了格子气自动机的碰撞项,虽然碰撞过程直观明了,但当粒子数增加时,碰撞算子的选择会变得非常困难,所以该模型的碰撞项仍然具有指数复杂性。

玻尔兹曼的人物小传

1989 年,Higuera 和 Jimenez[29]提出了一种简化的碰撞模型:通过引入平衡态分布函数,将碰撞算子线性化,碰撞算子由一个碰撞矩阵来代替,矩阵中各元素的值满足质量守恒及动量守恒(如果是热流体模型,还应当满足能量守恒)。该模型不再需要碰撞规则表,并且忽略各个粒子间的碰撞细节,相比于多粒子碰撞模型,容易构造。但是当粒子种类增加时,矩阵变得很大,矩阵元素的定义及碰撞演算变得很复杂。同年,Higuera 等[30]进一步提出了强化碰撞算子的方法,以增强模型的数值稳定性。这两类模型被称为**矩阵模型**(matrix model)。矩阵模型克服了指数复杂性,使得计算量和存储量都大大降低。经历了上述几类模型,格子 Boltzmann 方法消除了统计噪声,克服了碰撞算子的指数复杂性,但是由于依然使用 Fermi-Dirac 平衡态分布函数,格子气自动机的其他缺点仍然存在。

1991~1992 年,Qian 等[31]、Chen 等[32]、Koelman[33]分别独立提出在格子 Boltzmann 方法中采用**单松弛碰撞算子**(single-relaxation-time),即用同一个松弛时间来控制不同方向分布函数靠近其平衡态的快慢,从而简化了碰撞算子。因为这种方法源自 1954 年 Bhatnagar、Gross 和 Krook 为了简化 Boltzmann 方程中的碰撞积分项而提出的碰撞间隔理论[34],所以这类格子 Boltzmann 模型通常也被称为**格子 BGK 模型**(BGK 为三人名字的首字母)。与前面的模型不同,当粒子种类数增加时,单松弛碰撞算子本身不会变得复杂。此外,格子 BGK 模型摒弃了 Fermi-Dirac 平衡态分布函数,转而采用 Maxwell 分布函数的截断形式。具体的平衡态分布函数形式与流场的维数以及离散速度模型有关。通过选取适当的平衡态分布函数,可以保证从格子 Boltzmann 方程所导出的宏观方程具有伽利略不变性,且状态方程与宏观速度无关。

格子 BGK 模型的出现使得格子 Boltzmann 方法克服了格子气自动机的一系列缺点,

从格子气自动机中脱胎而出，展现出了全新的生命力，也因此吸引了国内外学者的广泛关注，并成为国际研究热点之一。

1.2.3 基础理论体系的构筑期

近30年来，格子Boltzmann方法的发展成长大体上可以分为两个阶段，即基础理论体系的构筑期(1992—2006年)和深入发展期(2006年至今)。

1992—2006年，格子Boltzmann方法经历了几次大的发展。首先，以He和Luo[35-38]为代表的一些学者分别从不同角度严格地证明了格子Boltzmann方程可以从气体动理论中的连续Boltzmann-BGK方程导出。这一发现，一方面，建立了格子Boltzmann方法与气体动理论之间的联系，有助于人们更加深刻地理解格子Boltzmann方法的物理内涵和介观背景；另一方面，它为构建格子Boltzmann模型提供了新的思路，即可以直接从相关的动理学理论出发构建相应的格子Boltzmann模型，后来的许多格子Boltzmann模型就是在这一思路的指导下产生的。例如，1998年He等[39]从连续Boltzmann-BGK方程出发，通过引入一个新的分布函数——热力学能分布函数，构建了一个包含密度分布函数和热力学能分布函数的双分布函数格子Boltzmann模型。在这一模型中，流场仍然由密度分布函数来求解，而温度场则由热力学能分布函数求解。由于热力学能分布函数的演化方程是根据连续Boltzmann-BGK方程推导出来的，因此具有清晰的物理背景。双分布函数格子Boltzmann方法作为一种具备较完备理论体系的方法是始于He等[39]的热力学能分布函数模型，它的出现推动了格子Boltzmann方法在传热问题中的广泛应用。

随后，Lallemand和Luo[40]对d'Humiéres在一次会议上所提出的广义格子Boltzmann模型进行了系统的理论分析。他们发现，与单松弛格子Boltzmann模型相比，广义格子Boltzmann模型在数值稳定性和参数选取等方面具有明显的优势。广义格子Boltzmann模型和单松弛格子Boltzmann模型的主要区别在于广义格子Boltzmann模型的碰撞算子采用多个松弛时间，因而也被称为**多松弛碰撞算子**(multiple-relaxation-time)。Lallemand和Luo的研究工作使得长期被人们忽视的多松弛格子Boltzmann模型开始受到了重视。在这一理论的指导下，一些新的格子Boltzmann模型迅速出现[41-48]，并在许多领域得到应用。由于多松弛格子Boltzmann模型具有良好的数值稳定性，因而对格子Boltzmann方法在高雷诺数、高瑞利数流动与传热问题中的应用具有重要的意义。

此外，Shan等[49]和Philippi等[50]分别独立研究了构造高阶格子Boltzmann模型的系统化理论方法，其根本目标是从连续的Maxwell分布函数推导出高阶格子Boltzmann模型所需的平衡态分布函数和离散速度模型。在Shan等[49]的方法中，离散速度通常与格子不一致，于是分布函数的演化就会偏离格子，在计算时需要进行特殊处理。Philippi等[50]的方法正好相反，他们是预先给定与规则格子一致的基本格子矢量，从而能够保证粒子在格线上迁移。但这样带来的结果是离散速度的构造丧失了灵活性，且在模型阶数相同的情况下，离散速度的个数往往要多于前者。此外，Chikatamarla和Karlin[51-52]也提出了类似的方法。这些系统化方法的出现，为构建高阶格子Boltzmann模型指明了方向。

与格子BGK模型的出现一样，上述几次大的发展在格子Boltzmann方法的发展历程中也具有重要的意义。通过建立格子Boltzmann方法与气体动理论的联系，在理论上巩固了其介观方法的地位。前面曾提到，介观方法主要分为两类，一类是通过数值求解Boltzmann方程，得到分布函数的时空演化规律；另一类则是通过粗粒化构造介观层面的流体粒子，并设计出符合一定物理规律的演化机制，让这些流体粒子进行演化（例如，流体粒子在格子上碰撞迁移）。与第一类介观方法相比，第二类介观方法的优势在于其清晰的粒子运动图像和简单的算法。格子 Boltzmann 方法是基于第二类介观方法孕育而成的，但随后格子 Boltzmann-BGK 方程被证明是连续 Boltzmann-BGK 方程的一种特殊离散形式，这也就使得格子 Boltzmann 方法具有其他任何介观方法都不具备的“双重属性”。

这一时期，格子 Boltzmann 方法在应用基础模型、边界处理格式、作用力格式、网格技术等方面也取得了重要进展和突破。应用基础模型方面以多相流模型为例，Gunstensen 等[53]提出了第一个多相流格子 Boltzmann 模型，他们用不同颜色的分布函数来区分不同相态的流体，相与相之间的相互作用通过颜色梯度来实现，并根据它来调整粒子的运动趋势，实现流体的分离和混合。1993 年，Shan 和 Chen[54-55]提出用一种伪势相互作用力来反映相与相之间的相互作用。与颜色梯度模型相比，伪势模型能够更直接地描述流体粒子之间的相互作用。随后，Shan 和 Doolen[56]对伪势模型中平衡态分布函数的速度以及混合流体的速度进行了重新定义。Swift 等[57]则直接从自由能泛函出发，构造了与热力学理论一致的多相流格子 Boltzmann 模型。其基本思想是根据自由能泛函来设计模型的平衡态分布函数，通过引入热力学压力张量来描述相分离和表面张力效应。但当界面附近存在较大的密度梯度时，Swift 等[57]的自由能模型不满足伽利略不变性，从而会导致一些非物理现象。1998 年，He 等[58]从包含作用力项的连续 Boltzmann-BGK 方程出发构建了一个基于动理论的多相流模型。由于界面附近的密度梯度通常很大，在计算有效作用力时会产生数值不稳定。为了解决这一问题，He 等[59]引入了一个指标函数来跟踪相界面，建立了格子 Boltzmann 方法中最早的相场模型。至此，形成了以**颜色梯度模型**（color-gradient model）、**伪势模型**（pseudopotential model）、**自由能模型**（free-energy model）和**相场模型**（phase-field model）为主体的多相流格子 Boltzmann 方法体系。

伪势模型模拟相分离

边界处理格式方面，格子 Boltzmann 方法逐渐发展出了适合不同边界条件、不同模型的边界处理格式。这些边界处理格式可分为启发式格式、动力学格式、外推格式以及复杂边界处理格式等。不同的边界处理格式对数值模拟的精度、稳定性以及计算效率会产生较大的影响。网格技术方面，标准格子 Boltzmann 方法通常是基于均匀对称的网格。由于均匀网格在计算效率等方面存在不足，从而促使了各种非均匀网格、多块网格、多重网格等网格技术的出现。这些网格技术可以在一定程度上提高格子 Boltzmann 方法对不规则计算区域的适应性以及计算效率，从而拓展了格子 Boltzmann 方法的应用范围，推动格子 Boltzmann 方法走向更多的实际应用场景。

1.2.4 深入发展期

经历了前一时期的发展，格子 Boltzmann 方法形成了相对完备的基础理论体系，但仍然存在着两类突出的问题。一方面，对于某些流动与传热问题，仍然缺乏合适的基本模型；另一方面，早期模型存在一些缺陷或不足。例如，伪势多相模型长期受困于热力学不一致性（相平衡时两相共存密度与热力学 Maxwell 等面积法则的结果不一致）、低密度比、表面张力不能自由调节等缺陷，从而限制了它的应用范围。为了解决这两类突出问题，国内外学者在这一时期开展了系统深入的研究，提出了“更为完善、数值稳定性更好、适用范围更广”的新模型和边界处理格式等，对推动格子 Boltzmann 方法的应用发挥了重要的作用。下面以热模型和多相流模型为例，简要介绍这一时期格子 Boltzmann 方法在模型方面的发展。

热模型方面,尽管 He 等[39]所设计的热力学能分布函数模型能够包含黏性热耗散和可压缩功,但模型的源项中含有复杂梯度项。一方面,梯度项破坏了格子 Boltzmann 方法的“简单”特性;另一方面,对源项中的梯度项进行有限差分计算会影响模型的数值稳定性并引入额外误差。为了解决这一问题,2007 年,Guo 等[60]提出了一种总能分布函数模型。相应地,总能分布函数的动理学方程亦可从连续 Boltzmann-BGK 方程导出。与热力学能分布函数模型相比,总能分布函数模型演化方程中的源项不含任何梯度项,因此计算更简单,数值稳定性更好。同年,针对热可压缩流动,Li 等[61]通过将双分布函数模型和多速度模型结合起来,构造了满足理想气体状态方程、适用于高马赫数可压缩流动的耦合双分布函数格子 Boltzmann 模型。近年来,出现了多个基于多松弛碰撞算子的双分布函数格子 Boltzmann 模型[62-64],即密度分布函数和温度或能量分布函数的演化方程均采用多松弛碰撞算子,从而可以有效地提高模型对高雷诺数和高瑞利数问题的适应性和数值稳定性。

多相流模型方面,在这一时期四类多相模型均得到了进一步的完善。针对伪势模型在采用平方根形式的伪势时所存在的热力学不一致性的问题,Li 等[65]提出了通过调整力学稳定性条件来实现伪势模型热力学一致性的方法,根据这一方法,提出了新的作用力格式来调整伪势模型的力学稳定性条件,继而实现了伪势模型的热力学一致性,并据此构造了基于多松弛碰撞算子的大密度比伪势多相模型[66];随后,Li 和 Luo[67]提出在多松弛碰撞算子的碰撞步添加一个源项,以实现伪势模型表面张力的自由可调。针对早期自由能模型不满足伽利略不变性的问题,Wagner 和 Li[68]提出通过作用力项将热力学压力张量的散度引入自由能模型,从而克服了原有模型的缺陷;但 Guo 等[69]发现,采用作用力项的自由能模型在离散层面存在力的不平衡,从而导致化学势梯度不为零,继而引发热力学不一致性的问题。为了解决这一问题,Guo[70]和 Li 等[71]分别从不同的角度对自由能模型进行了改进,实现了自由能模型的热力学一致性。此外,Mazloomi 等[72]和 Wen 等[73]构造了大密度比自由能模型。颜色梯度模型方面,Reis 和 Phillips[74]发现早期颜色梯度模型中的扰动算子并不能恢复正确的表面张力项,继而提出了一个新的扰动算子;Liu 等[75]将该扰动算子拓展到了三维颜色梯度模型;此外,对于密度比可调的颜色梯度模型,由于状态方程是通过平衡态密度分布函数来进行调整,从而破坏了模型的伽利略不变性。为了解决这一问题,Ba 等[76]和 Wen 等[77]基于 Li 等[78]的研究工作分别提出了改进的二维和三维颜色梯度模型:针对标准格子六阶张量不满足各向同性的问题,添加了相应的修正项,从而恢复了颜色梯度模型的伽利略不变性。相场模型方面,2006 年 Zheng 等[79]指出,此前的相场格子 Boltzmann 模型并不能恢复正确的 Cahn-Hilliard 方程,为此他们设计了一个新的模型,可以准确地恢复 Cahn-Hilliard 方程,但该模型随后被证明只能用于密度匹配的二元流体[80],并不适用于具有明显密度差的两相流动;2012 年,Li 等[81]发现,已有的相场模型所恢复的动量方程中存在一个附加界面力,当雷诺数较大时,该附加界面力会引入较大的数值误差;2015 年,Wang 等[82-83]基于通量求解器提出了一个大密度比相场模型;近年来,越来越多的相场模型采用 Allen-Cahn 方程[84-86]作为界面捕捉方程,相关研究表明该方程在相场格子 Boltzmann 模型中的表现优于 Cahn-Hilliard 方程[87]。此外,Chai 等[88]研究了局部及非局部 Allen-Cahn 方程的质量守恒特性。

经过了上述两个时期的发展,格子 Boltzmann 方法的理论体系日趋完善,它的应用也越来越广泛,在微流动[89-91]、多孔介质流动[92-97]、多相/多组分流动[53-59,98-110]、非牛顿流体[111-114]、纳米流体[115-117]、湍流[118-121]、化学反应流[122-123]、晶体生长[124-127]、燃料电池[128-133]、沸腾[134-144]、冷凝[145-152]等诸多领域都出现了格子 Boltzmann 方法的应用。以多孔介质流动为例,它通常涉及三个尺度,即孔隙尺度、表征体元尺度和宏观区域尺度。由于格子 Boltzmann 方法继承了格子气自动机的反弹格式,在模拟具有复杂几何构型的问题时具有显著的优势,因而在这个方向的发展非常迅速。目前,采用格子 Boltzmann 方法对多孔介质流动进行模拟主要在孔隙尺度和表征体元尺度上进行。在孔隙尺度上,可以直接使用格子 Boltzmann 方法描述孔隙内的流体流动,固体骨架通常被视作流场的边界,流固之间的相互作用通过适当的边界条件予以反映[92]。在表征体元尺度上,格子 Boltzmann 方法不需要详细的多孔介质结构信息,而是直接对表征体元尺度上的平均流动进行模拟,多孔介质对流动的影响通过相关的模型进行描述[93]。

池沸腾的 LBM 模拟

在格子 Boltzmann 方法的发展历史上,陆续出现了一些综述性文献[153-161]和相关书籍[162-171]。其中,1992 年 Benzi 等[153]在《Physics Reports》杂志上首次对格子 Boltzmann 方法的基本理论与应用做了简要回顾,而 Wolf-Gladrow[162]于 2000 年所著的《Lattice-Gas Cellular Automata and Lattice Boltzmann Models:An Introduction》一书则是国际上第一本格子 Boltzmann 方法的著作。其他综述及著作可参阅相关文献。此外,格子 Boltzmann 方法领域已形成两个专门的国际会议:Discrete Simulation of Fluid Dynamics(DSFD)以及 International Conference for Mesoscopic Methods in Engineering and Science(ICMMES),并发展出了专门的商业软件 PowerFlow,其应用结果在国际顶级杂志《Science》上发表[118]。

随着格子 Boltzmann 方法的飞速发展,它在国内也日益受到重视。从 20 世纪 80 年代末,朱照宣[172]、胡守信[173]以及钱跃竑等[174]在《力学与实践》等杂志上介绍了格子气方法以来,国内许多高校及研究机构在格子气和格子 Boltzmann 方法的理论与应用方面都开展了卓有成效的研究工作。对此,国家也给予了很大的关注和支持,如国家自然科学基金委员会在力学[175]、工程热物理与能源利用[176-177]等学科的发展战略调研报告中多次提到,研究诸如跨尺度问题、多相流动与传热等复杂物理问题,采用格子 Boltzmann 方法可以对某些流动图案或机制做出定性的说明,且随着计算机技术和并行算法的飞速发展,定量计算也已逐步得以实现。此外,近年来,国家批准的关于格子 Boltzmann 方法的基金项目也越来越多。在格子 Boltzmann 方法领域,华人学者做出了重要的贡献,相信在未来的研究工作中会发挥出更大的力量。

1.3 本书的目的和内容

1.3.1 本书目的

自 21 世纪初以来,我们在格子 Boltzmann 方法的理论及应用方面开展了一系列相关研究。本书除介绍格子 Boltzmann 方法的基本原理以外,也包含了我们在格子 Boltzmann 方法领域所取得的部分研究成果,以及在相关科研实践中所积累的经验。现将其整理于此,愿与国内外同行以及感兴趣的读者分享交流。

近年来,国内越来越多的高校以及科研机构开始对格子 Boltzmann 方法产生兴趣。在与同行交流时,我们深切感受到,作为格子 Boltzmann 方法领域的一本入门著作,应当细致明了地介绍格子 Boltzmann 方法的基本概念、基础理论以及发展现状，并配以适当的算例帮助读者快速入门。因此，本书的定位不仅仅是我们科研工作的总结，还尽可能全面地介绍格子 Boltzmann 方法的发展现状，并阐明该方法的统计物理与流体力学背景知识、模型推导、边界处理、数值实施等问题。我们认为，只有牢固掌握相关基础知识，并有着全局的视角，才能做出更多更出色的研究成果。

1.3.2 本书内容

采用格子 Boltzmann 方法求解物理问题时的基本过程如图 1.3 所示，具体步骤如下：

（1）基于各种简化假设，建立物理模型，确定计算区域、初始条件以及边界条件等，并根据物理问题的不同，选择相应的格子 Boltzmann 模型。

（2）进行网格划分。

（3）根据物理问题，进行计算区域的初始化（密度、速度、温度等），并由此计算所有节点上各个离散速度方向的平衡态分布函数，以此作为计算的初场。

（4）执行格子 Boltzmann 方程的碰撞步。

（5）根据边界条件，在相应边界节点上实施边界处理格式。

（6）执行格子 Boltzmann 方程的迁移步，并计算各节点上的宏观量。需要注意的是，根据所采用的边界处理格式的不同，第（5）步有可能需要在执行格子 Boltzmann 方程的迁移步之后再进行，例如非平衡态反弹格式。

（7）判断计算是否收敛。

（8）若计算收敛，则输出计算结果；否则返回第（4）步，继续求解，直到收敛为止。

本书正是基于上述流程来组织相关内容的。

全书共 10 章，除绪论外分四个部分。第 2 和第 3 章为第一部分，介绍格子 Boltzmann 方法的背景理论知识。第 4~7 章为第二部分，介绍格子 Boltzmann 方法在单相流动与传热、多相流以及多孔介质流动与传热方面的应用基础模型。第 8 和第 9 章为第三部分，介绍格子 Boltzmann 方法的边界条件和网格技术。第 10 章为第四部分，即格子 Boltzmann 方法的应用举例。具体地，第 2 章总结了从连续介质层面出发的流体力学基本方程，包括连续方程、动量方程以及各种形式的能量或温度方程，这些方程是计算流体力学与计算传热学的基础，也是格子 Boltzmann 方法在宏观层面所应恢复的方程（任何介观方法在宏观层面应当与相应的宏观方程保持一致）。第 3 章从介观的气体动理论出发，系统地阐

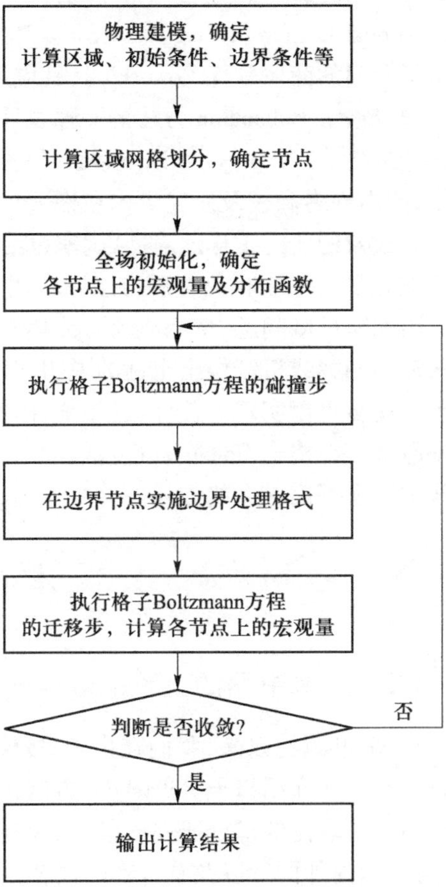

图 1.3 采用格子 Boltzmann 方法模拟求解的基本流程

述了格子 Boltzmann 方法的基础理论和基本模型。第 4 章介绍用于单相流动与传热的格子 Boltzmann 模型，包括等温不可压缩模型、热模型以及用于高马赫数可压缩流动的耦合双分布函数模型。第 5 章对气液两相流的四类格子 Boltzmann 模型进行了较为详尽的介绍，包括自由能模型、伪势模型、颜色梯度模型和相场模型。第 6 章介绍流固两相流格子 Boltzmann 模型以及固液相变格子 Boltzmann 模型。第 7 章介绍多孔介质流动与传热的格子 Boltzmann 模型。第 8 章介绍边界处理格式，从简单的反弹处理和镜面反射处理，到适用于复杂曲线边界的处理格式，均进行了介绍。第 9 章介绍格子 Boltzmann 方法的网格处理技术，涉及均分网格、多块网格、多重网格、结构化网格、非结构化网格等技术。最后，第 10 章给出了格子 Boltzmann 方法的一些算例以及在工程热物理领域的一些应用。

书末，附录中列出了读者在学习过程中可能用到的数学基础知识，以及格子单位与物理单位的转换，并公布了作者编制的两个算例的程序代码；最后列出了本书的主题索引。希望通过这些内容能够给读者提供方便，促使初学者更快地掌握格子 Boltzmann 方法。

参考文献

[1] Batchelor G K. An Introduction to Fluid Dynamics[M]. Cambridge: Cambridge University Press, 1970.

[2] 潘永祥，李慎. 自然科学发展史纲[M]. 北京：首都师范大学出版社，1996.

[3] Fourier J B J. Théorie Analytique de la Chaleur [M]. Paris: Firmin Didot Père et Fils, 1822.

[4] Anderson J D. Computational Fluid Dynamics: The Basics with Application[M]. New York: McGraw-Hill, 1995.

[5] 傅德薰，马延文. 计算流体力学[M]. 北京：高等教育出版社，2002.

[6] 陶文铨. 数值传热学[M]. 2 版. 西安：西安交通大学出版社，2001.

[7] 陶文铨. 计算传热学的近代进展[M]. 北京：科学出版社，2000.

[8] 宇波，李敬法，孙东亮. 数值传热学实训[M]. 北京：科学出版社，2018.

[9] Moin P, Kim J. The structure of the vorticity field in turbulent channel flow. Part 1. Analysis of instantaneous fields and statistical correlations[J]. Journal of Fluid Mechanics, 1985, 155: 441-464.

[10] Kim J, Moin P. The structure of the vorticity field in turbulent channel flow. Part 2. Study of ensemble-averaged fields[J]. Journal of Fluid Mechanics, 1986, 162: 339-363.

[11] Alder B J, Wainwright T E. Phase transition for a hard sphere system[J]. Journal of Chemical Physics, 1957, 27(5): 1208-1209.

[12] Rapaport D C. The Art of Molecular Dynamics Simulation [M]. Cambridge: Cambridge University Press, 1995.

[13] Meier K, Laesecke A, Kabelac S. A molecular dynamics simulation study of the self-diffusion coefficient and viscosity of the Lennard-Jones fluid[J]. International Journal of Thermophysics, 2001, 22: 161-173.

[14] Meier K, Laesecke A, Kabelac S. Transport coefficients of the Lennard-Jones model fluid. Ⅱ Self-diffusion [J]. Journal of Chemical Physics, 2004, 121(19): 9526-9535.

[15] Sun J, He Y L, Tao W Q. Molecular dynamics-continuum hybrid simulation for condensation of gas flow in a microchannel[J]. Microfluidics and Nanofluidics, 2009, 7(3): 407-422.

[16] Wang B B, Wang X D, Chen M, et al. Molecular dynamics simulations on evaporation of droplets with dissolved salts[J]. Entropy, 2013, 15(4): 1232-1246.

[17] Tang G H, Niu D, Guo L, et al. Failure and recovery of droplet nucleation and growth on damaged nanostructures: A molecular dynamics study[J]. Langmuir, 2020, 36(45): 13716-13724.

[18] Chen Y J, Yu B, Zou Y, et al. Molecular dynamics studies of bubble nucleation on a grooved substrate [J]. International Journal of Heat and Mass Transfer, 2020, 158: 119850.

[19] Xu J L, Zhou Z Q, Xu X D. Molecular dynamics simulation of micro-Poiseuille flow for liquid argon in nanoscale[J]. International Journal of Numerical Methods for Heat and Fluid Flow, 2004, 14(5): 664-688.

[20] Barisik M, Beskok A. Molecular dynamics simulations of shear-driven gas flows in nano-channels [J]. Microfluidics and Nanofluidics, 2011, 11(5): 611-622.

[21] Chapman S, Cowling T G. The Mathematical Theory of Non-uniform Gases[M]. 3rd edition. Cambridge: Cambridge University Press, 1970.

[22] 李元香,康立山,陈毓屏. 格子气自动机[M]. 北京:清华大学出版社,1994.

[23] Broadwell J E. Shock structure in a simple discrete velocity gas[J]. The Physics of Fluids, 1964, 7: 1243-1247.

[24] Hardy J, Pomeau Y, de Pazzis O. Time evolution of a two-dimensional model system. I. Invariant states and time correlation functions[J]. Journal of Mathematical Physics, 1973, 14(12): 1746-1759.

[25] Hardy J, de Pazzis O, Pomeau Y. Molecular dynamics of a classical lattice gas: Transport properties and time correlation functions[J]. Physical Review A, 1976, 13(5): 1949-1961.

[26] Frisch U, Hasslacher B, Pomeau Y. Lattice-gas automata for the Navier-Stokes equation[J]. Physical Review Letters, 1986, 56(14): 1505-1508.

[27] d'Humières D, Lallemand P, Frisch U. Lattice gas models for 3D hydrodynamics[J]. Europhysics Letters, 1986, 2(4): 291-297.

[28] McNamara G R, Zanetti G. Use of the Boltzmann equation to simulate lattice-gas automata[J]. Physical Review Letters, 1988, 61: 2332-2335.

[29] Higuera F J, Jimenez J. Boltzmann approach to lattice gas simulations[J]. Europhysics Letters, 1989, 9(7): 663-668.

[30] Higuera F J, Succi S, Benzi R. Lattice gas dynamics with enhanced collisions[J]. Europhysics Letters, 1989, 9(4): 345-349.

[31] Qian Y H, d'Humières D, Lallemand P. Lattice BGK models for Navier-Stokes equation[J]. Europhysics Letters, 1992, 17(6): 479-484.

[32] Chen S Y, Chen H D, Martnez D, et al. Lattice Boltzmann model for simulation of magnetohydrodynamics [J]. Physical Review Letters, 1991, 67: 3776-3779.

[33] Koelman J M V A. A simple lattice Boltzmann scheme for Navier-Stokes fluid flow[J]. Europhysics Letters, 1991, 15(6): 603-607.

[34] Bhatnagar P L, Gross E P, Krook M. A model for collision processes in gases. I. Small amplitude processes in charged and neutral one-component systems[J]. Physical Review, 1954, 94(3): 511-525.

[35] He X, Luo L S. A priori derivation of the lattice Boltzmann equation[J]. Physical Review E, 1997, 55(6): R6333-R6336.

[36] He X Y, Luo L S. Theory of the lattice Boltzmann method: From the Boltzmann equation to the lattice Boltzmann equation[J]. Physical Review E, 1997, 56(6): 6811-6817.

[37] Abe T. Derivation of the Lattice Boltzmann method by means of the discrete ordinate method for the Boltzmann equation[J]. Journal of Computational Physics, 1997, 131(1): 241-246.

[38] Shan X W, He X Y. Discretization of the velocity space in the solution of the Boltzmann equation [J]. Physical Review Letters, 1997, 80: 65-68.

[39] He X Y, Chen S Y, Doolen G D. A novel thermal model for the lattice Boltzmann method in incompressible limit[J]. Journal of Computational Physics, 1998, 146(1): 282-300.

[40] Lallemand P, Luo L S. Theory of the lattice Boltzmann method: Dispersion, dissipation, isotropy, Galilean invariance, and stability[J]. Physical Review E, 2000, 61(6): 6546–6562.

[41] Krafczyk M, Tolke J, Luo L S. Large-eddy simulations with a multiple-relaxation-time LBE model [J]. International Journal of Modern Physics B, 2003, 17(1–2): 33–39.

[42] McCracken M E, Abraham J. Simulations of liquid break up with an axisymmetric, multiple relaxation time, index-function lattice Boltzmann model[J]. International Journal of Modern Physics C, 2005, 16(11): 1671–1692.

[43] McCracken M E, Abraham J. Multiple-relaxation-time lattice-Boltzmann model for multiphase flow [J]. Physical Review E, 2005, 71(3): 036701.

[44] Premnath K N, Abraham J. Simulations of binary drop collisions with a multiple-relaxation-time lattice-Boltzmann model[J]. Physics of Fluids, 2005, 17(12): 122105.

[45] Asinari P. Semi-implicit-linearized multiple-relaxation-time formulation of lattice Boltzmann schemes for mixture modeling[J]. Physical Review E, 2006, 73(5): 056705.

[46] Shan X W, Chen H D. A general multiple-relaxation-time Boltzmann collision model[J]. International Journal of Modern Physics C, 2007, 18(4): 635–643.

[47] Asinari P. Multiple-relaxation-time lattice Boltzmann scheme for homogeneous mixture flows with external force[J]. Physical Review E, 2008, 77(5): 056706.

[48] Zheng L, Shi B C, Guo Z L. Multiple-relaxation-time model for the correct thermohydrodynamic equations [J]. Physical Review E, 2008, 78(2): 026705.

[49] Shan X W, Yuan X F, Chen H D. Kinetic theory representation of hydrodynamics: A way beyond the Navier-Stokes equation[J]. Journal of Fluid Mechanics, 2006, 550: 413–441.

[50] Philippi P C, Hegele L A, dos Santos L O E, et al. From the continuous to the lattice Boltzmann equation: The discretization problem and thermal models[J]. Physical Review E, 2006, 73(5): 056702.

[51] Chikatamarla S S, Karlin I V. Complete Galilean invariant lattice Boltzmann models[J]. Computer Physics Communications, 2008, 179(1–3): 140–143.

[52] Chikatamarla S S, Karlin I V. Entropy and Galilean invariance of lattice Boltzmann theories[J]. Physical Review Letters, 2006, 97(19): 190601.

[53] Gunstensen A K, Rothman D H, Zaleski S, et al. Lattice Boltzmann model of immiscible fluids[J]. Physical Review A, 1991, 43(8): 4320–4327.

[54] Shan X W, Chen H D. Lattice Boltzmann model for simulating flows with multiple phases and components [J]. Physical Review E, 1993, 47(3): 1815–1820.

[55] Shan X W, Chen H D. Simulation of nonideal gases and liquid-gas phase transitions by the lattice Boltzmann equation[J]. Physical Review E, 1994, 49(4): 2941–2948.

[56] Shan X W, Doolen G. Multicomponent lattice-Boltzmann model with interparticle interaction[J]. Journal of Statistical Physics, 1995, 81(1–2): 379–393.

[57] Swift M R, Osborn W R, Yeomans J M. Lattice Boltzmann simulation of nonideal fluids[J]. Physical Review Letters, 1995, 75(5): 830–833.

[58] He X Y, Shan X W, Doolen G D. Discrete Boltzmann equation model for nonideal gases[J]. Physical Review E, 1998, 57(1): R13–R16.

[59] He X Y, Chen S Y, Zhang R Y. A lattice Boltzmann scheme for incompressible multiphase flow and its application in simulation of Rayleigh-Taylor instability[J]. Journal of Computational Physics, 1999, 152(2): 642–663.

[60] Guo Z L, Zheng C G, Shi B C, et al. Thermal lattice Boltzmann equation for low Mach number flows: Decoupling model[J]. Physical Review E, 2007, 75(3): 036704.

[61] Li Q, He Y L, Wang Y, et al. Coupled double-distribution-function lattice Boltzmann method for the compressible Navier-Stokes equations[J]. Physical Review E, 2007, 76(5): 056705.

[62] Mezrhab A, Moussaoui M A, Jami M, et al. Double MRT thermal lattice Boltzmann method for simulating convective flows[J]. Physics Letters A, 2010, 374(34): 3499-3507.

[63] Liu Q, Feng X B, He Y L, et al. Three-dimensional multiple-relaxation-time lattice Boltzmann models for single-phase and solid-liquid phase-change heat transfer in porous media at the REV scale[J]. Applied Thermal Engineering, 2019, 152: 319-337.

[64] Huang R Z, Lan L J, Li Q. Lattice Boltzmann simulations of thermal flows beyond the Boussinesq and ideal-gas approximations[J]. Physical Review E, 2020, 102(4): 043304.

[65] Li Q, Luo K H, Li X J. Forcing scheme in pseudopotential lattice Boltzmann model for multiphase flows [J]. Physical Review E, 2012, 86(1): 016709.

[66] Li Q, Luo K H, Li X J. Lattice Boltzmann modeling of multiphase flows at large density ratio with an improved pseudopotential model[J]. Physical Review E, 2013, 87(5): 053301.

[67] Li Q, Luo K H. Achieving tunable surface tension in the pseudopotential lattice Boltzmann modeling of multiphase flows[J]. Physical Review E, 2013, 88(5): 053307.

[68] Wagner A J, Li Q. Investigation of Galilean invariance of multi-phase lattice Boltzmann methods[J]. Physica A, 2006, 362(1): 105-110.

[69] Guo Z L, Zheng C G, Shi B C. Force imbalance in lattice Boltzmann equation for two-phase flows [J]. Physical Review E, 2011, 83(3): 036707.

[70] Guo Z L. Well-balanced lattice Boltzmann model for two-phase systems[J]. Physics of Fluids, 2021, 33 (3): 031709.

[71] Li Q, Yu Y, Huang R Z. Achieving thermodynamic consistency in a class of free-energy multiphase lattice Boltzmann models[J]. Physical Review E, 2021, 103(1): 013304.

[72] Mazloomi M A, Chikatamarla S S, Karlin I V. Entropic lattice Boltzmann method for multiphase flows [J]. Physical Review Letters, 2015, 114(17): 174502.

[73] Wen B H, Zhao L, Qiu W, et al. Chemical-potential multiphase lattice Boltzmann method with superlarge density ratios[J]. Physical Review E, 2020, 102(1): 013303.

[74] Reis T, Phillips T N. Lattice Boltzmann model for simulating immiscible two-phase flows[J]. Journal of Physics A: Mathematical and Theoretical, 2007, 40: 4033-4053.

[75] Liu H H, Valocchi A J, Kang Q. Three-dimensional lattice Boltzmann model for immiscible two-phase flow simulations[J]. Physical Review E, 2012, 85(4): 046309.

[76] Ba Y, Liu H H, Li Q, et al. Multiple-relaxation-time color-gradient lattice Boltzmann model for simulating two-phase flows with high density ratio[J]. Physical Review E, 2016, 94(2): 023310.

[77] Wen Z X, Li Q, Yu Y, et al. Improved three-dimensional color-gradient lattice Boltzmann model for immiscible multiphase flows[J]. Physical Review E, 2019, 100(2): 023301.

[78] Li Q, Luo K H, He Y L, et al. Coupling lattice Boltzmann model for simulation of thermal flows on standard lattices[J]. Physical Review E, 2012, 85(1): 016710.

[79] Zheng H W, Shu C, Chew Y T. A lattice Boltzmann model for multiphase flows with large density ratio [J]. Journal of Computational Physics, 2006, 218(1): 353-371.

[80] Fakhari A, Rahimian M H. Phase-field modeling by the method of lattice Boltzmann equations[J]. Physical Review E, 2010, 81(3): 036707.

[81] Li Q, Luo K H, Gao Y J, et al. Additional interfacial force in lattice Boltzmann models for incompressible multiphase flows[J]. Physical Review E, 2012, 85(2): 026704.

[82] Wang Y, Shu C, Huang H B, et al. Multiphase lattice Boltzmann flux solver for incompressible multiphase

flows with large density ratio[J]. Journal of Computational Physics,2015,280:404-423.

[83] Wang Y,Shu C,Yang L M. An improved multiphase lattice Boltzmann flux solver for three-dimensional flows with large density ratio and high Reynolds number[J]. Journal of Computational Physics,2015,302:41-58.

[84] Ren F,Song B W,Sukop M C,et al. Improved lattice Boltzmann modeling of binary flow based on the conservative Allen-Cahn equation[J]. Physical Review E,2016,94(2):023311.

[85] Fakhari A,Bolster D. Diffuse interface modeling of three-phase contact line dynamics on curved boundaries:A lattice Boltzmann model for large density and viscosity ratios[J]. Journal of Computational Physics,2017,334:620-638.

[86] Liang H,Xu J R,Chen J X,et al. Phase-field-based lattice Boltzmann modeling of large-density-ratio two-phase flows[J]. Physical Review E,2018,97(3):033309.

[87] Wang H L,Chai Z H,Shi B C,et al. Comparative study of the lattice Boltzmann models for Allen-Cahn and Cahn-Hilliard equations[J]. Physical Review E,2016,94(3):033304.

[88] Chai Z H,Sun D K,Wang H L,et al. A comparative study of local and nonlocal Allen-Cahn equations with mass conservation[J]. International Journal of Heat and Mass Transfer,2018,122:631-642.

[89] Raabe D. Overview of the lattice Boltzmann method for nano- and microscale fluid dynamics in materials science and engineering[J]. Modelling and Simulation in Materials Science and Engineering,2004,12:13-46.

[90] Zhang J F. Lattice Boltzmann method for microfluidics:Models and applications[J]. Microfluidics and Nanofluidics,2011,10(1):1-28.

[91] Ansumali S,Karlin I V,Frouzakis C E,et al. Entropic lattice Boltzmann method for microflows[J]. Physica A:Statistical Mechanics and its Applications,2006,359:289-305.

[92] Tang G H,Tao W Q,He Y L. Gas slippage effect on microscale porous flow using the lattice Boltzmann method[J]. Physical Review E,2005,72(5):056301.

[93] Guo Z L,Zhao T S. Lattice Boltzmann model for incompressible flows through porous media[J]. Physical Review E,2002,66(3):036304.

[94] Pan C X,Luo L S,Miller C T. An evaluation of lattice Boltzmann schemes for porous medium flow simulation[J]. Computers & Fluids,2006,35(8):898-909.

[95] Liu Q,He Y L,Li Q,et al. A multiple-relaxation-time lattice Boltzmann model for convection heat transfer in porous media[J]. International Journal of Heat and Mass Transfer,2014,73:761-775.

[96] Liao Q,Yang Y X,Zhu X,et al. Lattice Boltzmann simulation of substrate solution through a porous granule immobilized PSB-cell for biohydrogen production[J]. International Journal of Hydrogen Energy,2013,38(35):15700-15709.

[97] Gao C,Xu R N,Jiang P X. Pore-scale numerical investigations of fluid flow in porous media using lattice Boltzmann method[J]. International Journal of Numerical Methods for Heat and Fluid Flow,2015,25:1957-1977.

[98] Grunau D,Chen S Y,Eggert K. A lattice Boltzmann model for multiphase fluid flows[J]. Physics of Fluids A,1993,5(10):2557-2562.

[99] Swift M R,Orlandini E,Osborn W R,Yeomans J M. Lattice Boltzmann simulations of liquid-gas and binary fluid systems[J]. Physical Review E,1996,54(5):5041-5052.

[100] Inamuro T,Ogata T,Tajima S,et al. A lattice Boltzmann method for incompressible two-phase flows with large density differences[J]. Journal of Computational Physics,2004,198(2):628-644.

[101] Luo L S. Theory of the lattice Boltzmann method:Lattice Boltzmann models for nonideal gases[J]. Physical Review E,2000,62(4):4982-4996.

[102] Ridl K S, Wagner A J. Lattice Boltzmann simulation of mixtures with multicomponent van der Waals equation of state[J]. Physical Review E,2018,98(4):043305.

[103] Zhang L Z, Yuan W Z. A lattice Boltzmann simulation of coalescence-induced droplet jumping on superhydrophobic surfaces with randomly distributed structures[J]. Applied Surface Science,2018,436:172-182.

[104] Zhang X Y,Huang M Y,Ji Q,et al. Lattice Boltzmann modeling and experimental study of water droplet spreading on wedge-shaped pattern surface[J]. International Journal of Heat and Mass Transfer,2019,130:857-861.

[105] Wang H M,Zhao H B,Guo Z L,et al. Lattice Boltzmann method for simulations of gas-particle flows over a backward-facing step[J]. Journal of Computational Physics,2013,239:57-71.

[106] Xu A,Zhao T S,An L,et al. A three-dimensional pseudo-potential-based lattice Boltzmann model for multiphase flows with large density ratio and variable surface tension[J]. International Journal of Heat and Fluid Flow,2015,56:261-271.

[107] Lycett-Brown D,Luo K H. Improved forcing scheme in pseudopotential lattice Boltzmann methods for multiphase flow at arbitrarily high density ratios[J]. Physical Review E,2015,91(2):023305.

[108] Fei L L,Du J Y,Luo K H,et al. Modeling realistic multiphase flows using a non-orthogonal multiple-relaxation-time lattice Boltzmann method[J]. Physics of Fluids,2019,31(4):042105.

[109] Yuan H Z,Wang Y,Shu C. An adaptive mesh refinement-multiphase lattice Boltzmann flux solver for simulation of complex binary fluid flows[J]. Physics of Fluids,2017,29(12):123604.

[110] Shi Y, Tang G H, Wang Y. Simulation of three-component fluid flows using the multiphase lattice Boltzmann flux solver[J]. Journal of Computational Physics,2016,314:228-243.

[111] Rakotomalala N,Salin D,Watzky P. Simulations of viscous flows of complex fluids with a Bhatnagar, Gross,and Krook lattice gas[J]. Physics of Fluids,1996,8(11):3200-3202.

[112] Boek E S. Lattice Boltzmann simulation of the flow of non-Newtonian fluids in porous media[J]. International Journal of Modern Physics B,2003,17:99-102.

[113] Gabbanelli S, Drazer G, Koplik J. Lattice Boltzmann method for non-Newtonian (power-law) fluids [J]. Physics Review E,2005,72(4):046312.

[114] Tang G H. Non-Newtonian flow in microporous structures under the electroviscous effect[J]. Journal of Non-Newtonian Fluid Mechanics,2011,166(14):875-881.

[115] Abbassi M A, Safaei M R, Djebali R, et al. LBM simulation of free convection in a nanofluid filled incinerator containing a hot block [J]. International Journal of Mechanical Sciences, 2018, 144: 172-185.

[116] Xuan Y M,Yao Z P. Lattice Boltzmann model for nanofluids[J]. Heat and Mass Transfer,2005,41(3):199-205.

[117] Zhou L J,Xuan Y M,Li Q. Multiscale simulation of flow and heat transfer of nanofluid with lattice Boltzmann method[J]. International Journal of Multiphase Flow,2010,36(5):364-374.

[118] Chen H D,Kandasamy S,Orszag S,et al. Extended Boltzmann kinetic equation for turbulent flows [J]. Science,2003,301:633-636.

[119] Strumolo G,Viswanathan B. New directions in computational aerodynamics[J]. Physics World,1997,10:45-49.

[120] Yu H D,Luo L S,Girimaji S S. LES of turbulent square jet flow using an MRT lattice Boltzmann model [J]. Computers & Fluids,2006,35(8):957-965.

[121] Geller S, Uphoff S, Krafczyk M. Turbulent jet computations based on MRT and cascaded lattice Boltzmann models[J]. Computers & Mathematics with Applications,2013,65(12):1956-1966.

[122] Chen S, Dawson S P, Doolen G D, et al. Lattice methods and their applications to reacting systems [J]. Computers and Chemical Engineering, 1995, 19: 617-646.

[123] Lei T, Luo K H. Lattice Boltzmann smulation of multicomponent porous media flows with chemical reaction[J]. Frontiers in Physics, 2021, 9: 715791.

[124] Miller W, Succi S, Mansutti D. A lattice Boltzmann model for anisotropic liquid/solid phase transition [J]. Physics Review Letters, 2001, 86: 8357-8381.

[125] Miller W, Succi S. A lattice Boltzmann model for anisotropic crystal growth from melt[J]. Journal of Statistical Physics, 2002, 107: 173-186.

[126] Sun D K, Zhu M F, Wang J, et al. Lattice Boltzmann modeling of bubble formation and dendritic growth in solidification of binary alloys[J]. International Journal of Heat and Mass Transfer, 2016, 94: 474-487.

[127] Sun D K, Xing H, Dong X L, et al. An anisotropic lattice Boltzmann-phase field scheme for numerical simulations of dendritic growth with melt convection [J]. International Journal of Heat and Mass Transfer, 2019, 133: 1240-1250.

[128] Park J, Li X. Multi-phase micro-scale flow simulation in the electrodes of a PEM fuel cell by lattice Boltzmann method[J]. Journal of Power Sources, 2008, 178(1): 248-257.

[129] Hao L, Cheng P. Lattice Boltzmann simulations of water transport in gas diffusion layer of a polymer electrolyte membrane fuel cell[J]. Journal of Power Sources, 2010, 195(12): 3870-3881.

[130] Han B, Meng H. Lattice Boltzmann simulation of liquid water transport in turning regions of serpentine gas channels in proton exchange membrane fuel cells [J]. Journal of Power Sources, 2012, 217: 268-279.

[131] Xu A, Shyy W, Zhao T S. Lattice Boltzmann modeling of transport phenomena in fuel cells and flow batteries[J]. Acta Mechanica Sinica, 2017, 33(3): 555-574.

[132] Hou Y Z, Deng H, Du Q, et al. Multi-component multi-phase lattice Boltzmann modeling of droplet coalescence in flow channel of fuel cell[J]. Journal of Power Sources, 2018, 393: 83-91.

[133] Chen L, Luan H B, He Y L, et al. Pore-scale flow and mass transport in gas diffusion layer of proton exchange membrane fuel cell with interdigitated flow fields [J]. International Journal of Thermal Sciences, 2012, 51: 132-144.

[134] Zhang R Y, Chen H D. Lattice Boltzmann method for simulations of liquid-vapor thermal flows[J]. Physical Review E, 2003, 67(6): 066711.

[135] Gong S, Cheng P. A lattice Boltzmann method for simulation of liquid-vapor phase-change heat transfer [J]. International Journal of Heat and Mass Transfer, 2012, 55(17-18): 4923-4927.

[136] Li Q, Kang Q J, Francois M M, et al. Lattice Boltzmann modeling of boiling heat transfer: The boiling curve and the effects of wettability [J]. International Journal of Heat and Mass Transfer, 2015, 85: 787-796.

[137] Gong S, Cheng P. Lattice Boltzmann simulations for surface wettability effects in saturated pool boiling heat transfer[J]. International Journal of Heat and Mass Transfer, 2015, 85: 635-646.

[138] Li Q, Yu Y, Zhou P, et al. Enhancement of boiling heat transfer using hydrophilic-hydrophobic mixed surfaces: A lattice Boltzmann study[J]. Applied Thermal Engineering, 2018, 132: 490-499.

[139] Ma X J, Cheng P, Quan X J. Simulations of saturated boiling heat transfer on bio-inspired two-phase heat sinks by a phase-change lattice Boltzmann method[J]. International Journal of Heat and Mass Transfer, 2018, 127: 1013-1024.

[140] Ma X J, Cheng P. Dry spot dynamics and wet area fractions in pool boiling on micro-pillar and micro-cavity hydrophilic heaters: A 3D lattice Boltzmann phase-change study[J]. International Journal of Heat and Mass Transfer, 2019, 141: 407-418.

[141] Li W X, Li Q, Yu Y, et al. Enhancement of nucleate boiling by combining the effects of surface structure and mixed wettability: A lattice Boltzmann study[J]. Applied Thermal Engineering, 2020, 180: 115849.

[142] Li Q, Yu Y, Wen Z X. How does boiling occur in lattice Boltzmann simulations? [J]. Physics of Fluids, 2020, 32(9): 093306.

[143] Li W X, Li Q, Yu Y, et al. Nucleate boiling enhancement by structured surfaces with distributed wettability-modified regions: A lattice Boltzmann study[J]. Applied Thermal Engineering, 2021, 194: 117130.

[144] Deng Z L, Liu X D, Wu S C, et al. Pool boiling heat transfer enhancement by bi-conductive surfaces [J]. International Journal of Thermal Sciences, 2021, 167: 107041.

[145] Liu X L, Cheng P. Lattice Boltzmann simulation of steady laminar film condensation on a vertical hydrophilic subcooled flat plate[J]. International Journal of Heat and Mass Transfer, 2013, 62: 507–514.

[146] Zhang Q Y, Sun D K, Zhang Y F, et al. Lattice Boltzmann modeling of droplet condensation on superhydrophobic nanoarrays[J]. Langmuir, 2014, 30(42): 12559–12569.

[147] Li X P, Cheng P. Lattice Boltzmann simulations for transition from dropwise to filmwise condensation on hydrophobic surfaces with hydrophilic spots[J]. International Journal of Heat and Mass Transfer, 2017, 110: 710–722.

[148] Li M J, Huber C, Mu Y T, et al. Lattice Boltzmann simulation of condensation in the presence of noncondensable gas[J]. International Journal of Heat and Mass Transfer, 2017, 109: 1004–1013.

[149] Wu S C, Yu C, Yu F W, et al. Lattice Boltzmann simulation of co-existing boiling and condensation phase changes in a confined micro-space[J]. International Journal of Heat and Mass Transfer, 2018, 126: 773–782.

[150] Zheng S F, Eimann F, Fieback T, et al. Numerical investigation of convective dropwise condensation flow by a hybrid thermal lattice Boltzmann method[J]. Applied Thermal Engineering, 2018, 145: 590–602.

[151] Shen L Y, Tang G H, Li Q, et al. Hybrid wettability-induced heat transfer enhancement for condensation with noncondensable gas[J]. Langmuir, 2019, 35(29): 9430–9440.

[152] Tang S, Li Q, Yu Y, et al. Enhancing dropwise condensation on downward-facing surfaces through the synergistic effects of surface structure and mixed wettability[J]. Physics of Fluids, 2021, 33(8): 083301.

[153] Benzi R, Succi S, Vergassola M. The lattice Boltzmann equation: Theory and applications[J]. Physics Reports, 1992, 222: 145–197.

[154] Chen S Y, Doolen G D. Lattice Boltzmann method for fluid flows[J]. Annual Review of Fluid Mechanics, 1998, 30: 329–364.

[155] Aidun C K, Clausen J R. Lattice-Boltzmann method for complex flows[J]. Annual Review of Fluid Mechanics, 2010, 42: 439–472.

[156] Li Q, Luo K H, Kang Q J, et al. Lattice Boltzmann methods for multiphase flow and phase-change heat transfer[J]. Progress in Energy and Combustion Science, 2016, 52: 62–105.

[157] Nourgaliev R R, Dinh T-N, Theofanous T, et al. The lattice Boltzmann equation method: theoretical interpretation, numerics and implications[J]. International Journal of Multiphase Flow, 2003, 29(1): 117–169.

[158] Chen L, Kang Q J, Mu Y T, et al. A critical review of the pseudopotential multiphase lattice Boltzmann model: Methods and applications[J]. International Journal of Heat and Mass Transfer, 2014, 76: 210–236.

[159] Liu H H, Kang Q J, Leonardi C R, et al. Multiphase lattice Boltzmann simulations for porous media applications[J]. Computational Geosciences, 2016, 20(4): 777–805.

[160] He Y L, Liu Q, Li Q, et al. Lattice Boltzmann methods for single-phase and solid-liquid phase-change heat transfer in porous media: A review[J]. International Journal of Heat and Mass Transfer, 2019, 129: 160-197.

[161] 李庆,余悦,唐诗. 多相格子 Boltzmann 方法及其在相变传热中的应用[J]. 科学通报, 2020, 65(17): 1677-1693.

[162] Wolf-Gladrow D A. Lattice-Gas Cellular Automata and Lattice Boltzmann Models: An Introduction[M]. Berlin: Springer, 2000.

[163] Succi S. Lattice Boltzmann Equation for Fluid Dynamics and Beyond[M]. Oxford: Clarendon Press, 2001.

[164] 郭照立,郑楚光,李青,等. 流体动力学的格子 Boltzmann 方法[M]. 武汉:湖北科学技术出版社, 2002.

[165] 郭照立,郑楚光. 格子 Boltzmann 方法的原理及应用[M]. 北京:科学出版社, 2009.

[166] 何雅玲,王勇,李庆. 格子 Boltzmann 方法的理论及应用[M]. 北京:科学出版社, 2009.

[167] Guo Z L, Shu C. Lattice Boltzmann Method and Its Applications in Engineering[M]. World Scientific, 2013.

[168] Huang H B, Sukop M C, Lu X Y. Multiphase Lattice Boltzmann Methods: Theory and Application[M]. John Wiley & Sons, 2015.

[169] Mohamad A A. Lattice Boltzmann Method: Fundamentals and Engineering Applications with Computer Codes[M]. Springer Science & Business Media, 2011.

[170] Sukop M C, Thorne D T. Lattice Boltzmann Modeling: An Introduction for Geoscientists and Engineers[M]. New York: Springer, 2006.

[171] Krüger T, Kusumaatmaja H, Kuzmin A, et al. The Lattice Boltzmann Method-Principles and Practice[M]. Springer Nature, 2017.

[172] 朱照宣. 点格自动机[J]. 力学与实践, 1987, 9(2): 1-6.

[173] 胡守信. 线性点格自动机[J]. 力学与实践, 1988, 10(4): 34-37.

[174] 钱跃竑, d'Humières D, Pomeau Y, 等. 格子气流体动力学及其最新进展[J]. 力学与实践, 1990, 12(1): 7-16.

[175] 国家自然科学基金委员会. 自然科学学科发展战略调研报告: 力学[M]. 北京:科学出版社, 1997.

[176] 国家自然科学基金委员会工程与材料科学部. 学科发展战略研究报告(2006 年~2010 年):工程热物理与能源利用[M]. 北京:科学出版社, 2006.

[177] 国家自然科学基金委员会工程与材料科学部. 工程热物理与能源利用学科发展战略研究报告(2011~2020)[M]. 北京:科学出版社, 2011.

第 2 章　流体力学基本方程

在格子 Boltzmann 方法中，经常会遇到根据格子 Boltzmann 模型推导其所对应的宏观方程的问题。通常情况下，从格子 Boltzmann 模型所导出的宏观方程是以守恒形式出现的，且能量方程中的黏性热耗散一般表示为黏性应力张量和速度梯度张量的双点积。这与一般流体力学文献所给出的方程形式有所不同，因此给学习者带来了一定的困难。当然，经过一些基本的笛卡儿张量运算（参见附录 A），可以发现实际上是一致的。考虑到读者在学习格子 Boltzmann 方法的过程中需要用到相关的流体力学知识，本章将介绍流体力学的基本方程，包括连续性方程、动量方程以及各种形式的能量方程。

2.1　连续介质假设

我们知道，流体力学是一门研究流体宏观运动的学科。虽然流体的微观运动在时间和空间上都非常复杂，具有不均匀性、离散性和随机性，但流体的宏观运动一般总是呈现出均匀性、连续性和确定性。流体力学通常不考虑单个粒子的运动，而是考虑大量粒子的平均运动及其统计特性，这就涉及平均尺度的问题[1]。

流体平均尺度的选取，以一个立方体为例，当立方体的边长向某一点趋于足够小的尺度 a 时，就可以得到该点单位体积的平均质量即平均密度。取过大或过小的 a，其相应的平均密度都不能反映流体在该处的密度值。因为过大的 a，会把流体空间分布的不均匀性因素包括进来，从而不能反映该点的密度值；而过小的 a，则会使所取体积中的分子数太少，由于分子随机地进入或离开，该体积内的分子数会随机变化，从而导致平均密度波动较大。由此可见，只有当 a 选取适当时，例如对标准状况下的气体取 $a=10^{-3}$ cm，其平均值取值才会稳定，才能代表该点的平均密度。这时，一方面，a 值已远小于密度有显著变化的空间尺度，因而以这样的 a 所得的平均密度可以反映质量在空间各点处的不同分布状况。另一方面，在相应的立方体体积10^{-9} cm^3 中，标准状况下的气体已包含了约 2.7×10^{10}个分子，完全可以获得一个确定的统计平均密度。因此，如果以分子间的平均距离或分子平均自由程作为粒子结构的一个长度尺度，并记作 l，将宏观量有显著变化的尺度记为 L，a 值的选取应当满足

$$l \ll a \ll L \tag{2.1}$$

这样的 a 作为流体宏观物理量的统计平均尺度是合理的。

在大多数流体力学问题中，L 的尺度一般远大于 1 mm，比粒子结构的尺度 l 大得多。采用 $l \ll a \ll L$ 这样的 a 值，意味着采用比粒子结构尺度大得多的尺度，因此粒子结构的尺度微乎其微，可以忽略，从而可以不把物质看作是由离散粒子构成，而是假设物质连续地

占满整个空间，它们在所取尺度上具有物理量的统计平均值，并在空间上连续分布。这样的假想介质就是**连续介质**(continuum)，其对应的假设就是**连续介质假设**(continuum hypothesis)。一般地，对于气体是否可以作为连续介质处理需要根据气体分子间的平均自由程(记为 λ_m)与气体所处空间的代表性尺度(记为 L)而定，它们的比值称为克努森数，表示为 $Kn=\lambda_m/L$。据钱学森的研究[2]，当 Kn 数小于等于 0.001 时，气体可以作为连续介质处理。对于液体，相关研究表明，一直到微米尺度的液体层厚度，液体仍然可以作为连续介质处理。

气体流动的分区

在连续介质中，比微观粒子结构尺度大得多而比宏观特征尺度小得多的流体微团，称为**流体质点**(fluid particle)。流体质点是流体力学所研究的最小物质实体，在流体力学中所讨论的流体密度、速度和温度，实际上是指流体质点的密度、速度和温度。流体质点包含着大量分子，但与宏观特征尺度相比，它又非常小。概括起来，流体质点在微观上充分大，在宏观上充分小。流体质点具有的物理量是均匀的，它是质点中大量流体分子的微观物理量的统计平均值。

既然流体质点在宏观上充分小，且连续地占满整个空间，那么流体质点所在的空间也就相当于一个点空间，于是流体质点的物理量也就是流体所在空间的连续函数。同样地，这些物理量也是时间的连续函数。事实上，引入连续介质假设的目的就在于此，它为在流体力学中充分使用连续函数的数学分析方法奠定了基础。实验和实践表明，基于连续介质假设而建立起来的流体力学理论对绝大多数工程问题都是正确的。流体运动与能量传递的基本方程是根据质量守恒、动量守恒和能量守恒定律所建立起来的连续性方程、动量方程和能量方程。

2.2 雷诺输运方程

为了从守恒定律导出流体力学的基本方程，需要解决系统的有关物理量在欧拉空间运动中对时间的全导数问题，这涉及系统和控制体以及描述流体运动的拉格朗日法和欧拉法的概念。**系统**(system)是指某一确定流体质点集合的总体，它随流体的运动而运动，且体积和形状可能变化但包含的流体质点不变。**控制体**(control volume)是指流场中某一确定的空间区域，流体质点随时间流入和流出这个空间体积。因此占据控制体的流体质点是随时间改变的。跟随流体质点去研究流体运动的方法称为**拉格朗日法**(Lagrangian method)，而着眼于空间坐标去研究流体运动的方法则称为**欧拉法**(Eulerian method)。这些概念的详细介绍，可参考相关流体力学书籍[3-4]。这里着重介绍推导流体力学基本方程所要用到的雷诺输运方程，它的作用是将一个流体系统的拉格朗日变化率表示为欧拉导数，从而将针对系统的物理量转换到易于研究的欧拉参考系中。

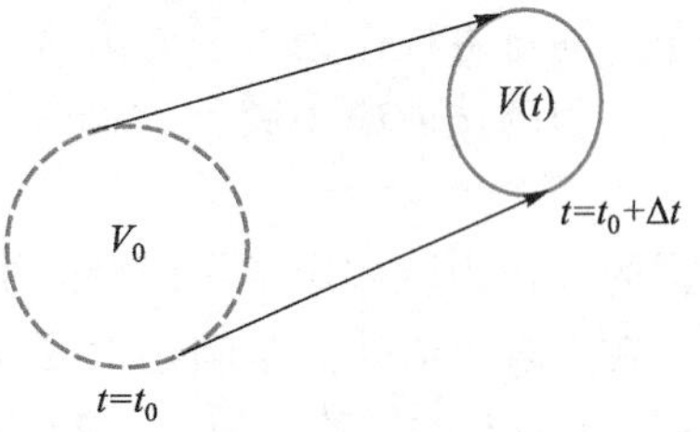

图 2.1 流动中取定的系统图

在流动中取定一个系统，如图 2.1 所示，系统在流动过程中 $t=t_0+\Delta t$ 时所占据的空间为控制体 $V(t)$，系统在 $t=t_0$ 时所占据的控制体 $V_0=V(t_0)$ 作为识别这一系统的标志。取微元体 $\delta V_0=\delta x_0\delta y_0\delta z_0$，$\delta V(t)=\delta x\delta y\delta z$，则有

$$\delta V(t)=\delta x\delta y\delta z=J\delta x_0\delta y_0\delta z_0=J\delta V_0 \qquad (2.2)$$

式中，

$$J=\frac{\partial(x,y,z)}{\partial(x_0,y_0,z_0)}=\begin{vmatrix}\frac{\partial x}{\partial x_0} & \frac{\partial y}{\partial x_0} & \frac{\partial z}{\partial x_0}\\ \frac{\partial x}{\partial y_0} & \frac{\partial y}{\partial y_0} & \frac{\partial z}{\partial y_0}\\ \frac{\partial x}{\partial z_0} & \frac{\partial y}{\partial z_0} & \frac{\partial z}{\partial z_0}\end{vmatrix} \tag{2.3}$$

J 称为 x、y、z 相对于 x_0、y_0、z_0 的**雅可比行列式**。由 $J=\delta V/\delta V_0$ 可知，它表示流体微团在 t 时刻的体积与初始时刻 t_0 的体积之比。在流体力学中，流体微团的相对体积膨胀率等于速度的散度，即

$$\frac{1}{\delta V}\frac{\mathrm{D}(\delta V)}{\mathrm{D}t}=\nabla\cdot\boldsymbol{u} \tag{2.4}$$

式中，$\boldsymbol{u}$ 为流体的速度矢量，m/s。

结合 J 的定义，式(2.4)可转化为

$$\frac{\delta V_0}{\delta V}\frac{\mathrm{D}}{\mathrm{D}t}\left(\frac{\delta V}{\delta V_0}\right)=\frac{1}{J}\frac{\mathrm{D}J}{\mathrm{D}t}=\nabla\cdot\boldsymbol{u} \tag{2.5}$$

系统的质量、动量或能量，对时间的全导数可推导如下：

$$\begin{aligned}\frac{\mathrm{D}}{\mathrm{D}t}\int_{V(t)}\phi\mathrm{d}V &= \frac{\mathrm{D}}{\mathrm{D}t}\int_{V_0}\phi J\mathrm{d}V_0 = \int_{V_0}\frac{\mathrm{D}}{\mathrm{D}t}(\phi J)\,\mathrm{d}V_0\\ &= \int_{V_0}\left(J\frac{\mathrm{D}\phi}{\mathrm{D}t}+\phi\frac{\mathrm{D}J}{\mathrm{D}t}\right)\mathrm{d}V_0\\ &= \int_{V_0}\left[\frac{\mathrm{D}\phi}{\mathrm{D}t}+\phi(\nabla\cdot\boldsymbol{u})\right]J\mathrm{d}V_0\\ &= \int_{V(t)}\left[\frac{\mathrm{D}\phi}{\mathrm{D}t}+\phi(\nabla\cdot\boldsymbol{u})\right]\mathrm{d}V\end{aligned} \tag{2.6}$$

根据物质导数的定义：$\frac{\mathrm{D}}{\mathrm{D}t}=\frac{\partial}{\partial t}+\boldsymbol{u}\cdot\nabla$，式(2.6)可写为

$$\begin{aligned}\frac{\mathrm{D}}{\mathrm{D}t}\int_{V(t)}\phi\mathrm{d}V &= \int_{V(t)}\left[\frac{\partial\phi}{\partial t}+\boldsymbol{u}\cdot\nabla\phi+\phi(\nabla\cdot\boldsymbol{u})\right]\mathrm{d}V\\ &= \int_{V(t)}\left[\frac{\partial\phi}{\partial t}+\nabla\cdot(\phi\boldsymbol{u})\right]\mathrm{d}V\\ &= \int_{V(t)}\frac{\partial\phi}{\partial t}\mathrm{d}V+\int_{V(t)}\nabla\cdot(\phi\boldsymbol{u})\,\mathrm{d}V\end{aligned} \tag{2.7}$$

进一步，根据高斯散度公式，式(2.7)可转化为

$$\frac{\mathrm{D}}{\mathrm{D}t}\int_{V(t)}\phi\mathrm{d}V=\int_{V(t)}\frac{\partial\phi}{\partial t}\mathrm{d}V+\int_{A(t)}\phi\boldsymbol{u}\cdot\boldsymbol{n}\mathrm{d}A \tag{2.8}$$

式中：$A(t)$ 为控制体 $V(t)$ 的表面；$\boldsymbol{n}$ 为控制体表面的法向单位矢量。

式(2.8)即为**雷诺输运方程**[3-7]，它表明系统的物质导数由两部分组成：右端第一项表示 t 时刻与系统重合的固定控制体内所含物理量 $\int_{V(t)}\phi\mathrm{d}V$ 在单位时间内的变化量，这一

变化是由流场中 ϕ 的不恒定性所引起的；右端第二项则表示，在单位时间内，流体通过控制体表面 $A(t)$ 所引起的控制体内物理量 $\int_{V(t)} \phi \mathrm{d}V$ 的变化，也就是系统由一个位置变化到另一个位置时，由于流场的不均匀性而引起的 $\int_{V(t)} \phi \mathrm{d}V$ 的迁移变化率。

2.3 连续性方程

根据质量守恒，一个系统无论经历何种变化与过程，其总质量保持不变。流体在流动过程中也遵循这一物质守恒定律。系统的总质量可表示为

$$m = \int_{V(t)} \rho \mathrm{d}V \tag{2.9}$$

质量守恒要求

$$\frac{\mathrm{D}m}{\mathrm{D}t} = \frac{\mathrm{D}}{\mathrm{D}t}\int_{V(t)} \rho \mathrm{d}V = 0 \tag{2.10}$$

根据雷诺输运方程有

$$\frac{\mathrm{D}}{\mathrm{D}t}\int_{V(t)} \rho \mathrm{d}V = \int_{V(t)} \left[\frac{\partial \rho}{\partial t} + \nabla \cdot (\rho \boldsymbol{u})\right] \mathrm{d}V = 0 \tag{2.11}$$

由于积分区域 $V(t)$ 是任意选取的，为了使积分式恒等于零，可得**连续性方程**（continuity equation）：

$$\frac{\partial \rho}{\partial t} + \nabla \cdot (\rho \boldsymbol{u}) = 0 \tag{2.12}$$

或

$$\frac{\mathrm{D}\rho}{\mathrm{D}t} + \rho \nabla \cdot \boldsymbol{u} = 0 \tag{2.13}$$

对不可压缩流体，流体质点的密度在运动过程中保持不变，即密度的物质导数为零，即 $\mathrm{D}\rho/\mathrm{D}t = 0$，于是不可压缩流体的连续性方程为

$$\nabla \cdot \boldsymbol{u} = 0 \tag{2.14}$$

结合雷诺输运方程和连续性方程可导出雷诺第二输运方程。如把 $\rho\phi$ 看作某一物理量，则

$$\begin{aligned}\frac{\mathrm{D}}{\mathrm{D}t}\int_{V(t)} (\rho\phi) \mathrm{d}V &= \int_{V(t)} \left[\frac{\mathrm{D}(\rho\phi)}{\mathrm{D}t} + (\rho\phi)(\nabla \cdot \boldsymbol{u})\right] \mathrm{d}V \\ &= \int_{V(t)} \left[\rho \frac{\mathrm{D}\phi}{\mathrm{D}t} + \phi \frac{\mathrm{D}\rho}{\mathrm{D}t} + \rho\phi(\nabla \cdot \boldsymbol{u})\right] \mathrm{d}V\end{aligned} \tag{2.15}$$

式(2.15)右端第二、三项可归纳为 $\phi\left[\frac{\mathrm{D}\rho}{\mathrm{D}t} + \rho(\nabla \cdot \boldsymbol{u})\right]$，由连续性方程(2.13)可知此项为零，于是有

$$\frac{\mathrm{D}}{\mathrm{D}t}\int_{V(t)} (\rho\phi) \mathrm{d}V = \int_{V(t)} \rho \frac{\mathrm{D}\phi}{\mathrm{D}t} \mathrm{d}V \tag{2.16}$$

式(2.16)即为**雷诺第二输运方程**。

2.4 动量方程

根据牛顿第二定律,系统总动量的变化率等于作用在系统上的总质量力和作用在表面上的总表面力之和,即

$$\frac{\mathrm{D}}{\mathrm{D}t}\int_{V(t)}(\rho\boldsymbol{u})\,\mathrm{d}V=\int_{V(t)}\rho\boldsymbol{F}\mathrm{d}V+\int_{A(t)}\boldsymbol{\sigma}\cdot\boldsymbol{n}\mathrm{d}A \tag{2.17}$$

式中:$\boldsymbol{F}$ 为质量力,$\mathrm{m/s^2}$;$\boldsymbol{\sigma}$ 为应力张量,Pa;$\boldsymbol{n}$ 为控制体表面的法向单位矢量。

根据雷诺第二输运方程和高斯散度公式有

$$\int_{V(t)}\rho\frac{\mathrm{D}\boldsymbol{u}}{\mathrm{D}t}\mathrm{d}V=\int_{V(t)}\rho\boldsymbol{F}\mathrm{d}V+\int_{V(t)}\nabla\cdot\boldsymbol{\sigma}\mathrm{d}V \tag{2.18}$$

即

$$\int_{V(t)}\left(\rho\frac{\mathrm{D}\boldsymbol{u}}{\mathrm{D}t}-\rho\boldsymbol{F}-\nabla\cdot\boldsymbol{\sigma}\right)\mathrm{d}V=0 \tag{2.19}$$

由于积分区域 $V(t)$ 是任意选取的,于是可得**微分形式的动量方程**:

$$\rho\frac{\mathrm{D}\boldsymbol{u}}{\mathrm{D}t}=\rho\boldsymbol{F}+\nabla\cdot\boldsymbol{\sigma} \tag{2.20}$$

利用连续性方程,可将式(2.20)改写成守恒形式(conservation form)的动量方程

$$\frac{\partial(\rho\boldsymbol{u})}{\partial t}+\nabla\cdot(\rho\boldsymbol{u}\boldsymbol{u})=\rho\boldsymbol{F}+\nabla\cdot\boldsymbol{\sigma} \tag{2.21}$$

上述方程是从牛顿第二定律导出的,它适用于任何一种流体。

对牛顿流体(应力与应变率之间符合线性关系的流体称为牛顿流体)而言,要得到更具体的动量方程,需引入牛顿流体的本构方程。

2.4.1 牛顿流体的本构方程

广义地讲,反映物质物理性质之间的关系式,统称**本构方程**(constitutive equation)。在流体力学中,本构方程一般专指应力张量与应变率张量之间的关系。牛顿根据最简单的平面剪切运动得出实验定律:两层流体间的切向应力与其速度梯度成正比,即

牛顿流体与非牛顿流体

$$\tau=\mu\frac{\partial u_x}{\partial y} \tag{2.22}$$

式中:τ 为切应力,Pa;$\partial u_x/\partial y$ 为切应变率,1/s;比例系数 μ 称为动力黏度,Pa · s;u_x 为速度矢量 $\boldsymbol{u}$ 在 x 方向的分量,m/s。在平面剪切运动中,y 方向和 z 方向的速度分量 u_y 和 u_z 均为零。

对于一般形式的应力张量与应变率张量之间的关系,应采用理论推演方法。具体地,牛顿流体应力张量与应变率张量之间的关系满足斯托克斯假设[1,8-9],即

(1) 当流体静止时,应变率为零,流体的应力就是流体的静压力;

(2) 应力张量是应变率张量的线性函数;

(3) 流体是各向同性的,即流体的物理性质与方向无关。

根据假设(1),应力张量可写为

$$\boldsymbol{\sigma}=-p\boldsymbol{I}+\boldsymbol{\Pi} \tag{2.23}$$

式中:p 为流体静压力,Pa;$\boldsymbol{\Pi}$ 为黏性应力张量,Pa;$\boldsymbol{I}$ 为单位二阶张量。黏性应力是由流体运动引起的,当运动停止后,黏性应力张量为零。式(2.23)亦可写成张量下标的形式:

$$\sigma_{ij}=-p\delta_{ij}+\Pi_{ij} \tag{2.24}$$

式中,δ_{ij}为**克罗内克符号**(Kronecker symbol)。

可以证明,应力张量和黏性应力张量均为对称张量。根据假设(2),可将黏性应力张量写成

$$\boldsymbol{\Pi}=a\boldsymbol{S}+b\boldsymbol{I} \tag{2.25}$$

式中:a 及 b 为标量常数;应变率张量 $\boldsymbol{S}$ 为

$$\boldsymbol{S}=\frac{1}{2}[\nabla\boldsymbol{u}+(\nabla\boldsymbol{u})^{\mathrm{T}}]=\frac{1}{2}\left(\frac{\partial u_i}{\partial x_j}+\frac{\partial u_j}{\partial x_i}\right) \tag{2.26}$$

根据式(2.22)有

$$\sigma_{xy}=2\mu\left(\frac{1}{2}\frac{\partial u_x}{\partial y}\right)=2\mu S_{xy} \tag{2.27}$$

对比式(2.25)和式(2.27)可得

$$a=2\mu \tag{2.28}$$

由式(2.24)和式(2.25),有

$$\sigma_{xx}=2\mu\frac{\partial u_x}{\partial x}+b-p \tag{2.29}$$

$$\sigma_{yy}=2\mu\frac{\partial u_y}{\partial y}+b-p \tag{2.30}$$

$$\sigma_{zz}=2\mu\frac{\partial u_z}{\partial z}+b-p \tag{2.31}$$

将式(2.29)~式(2.31)相加可得

$$\sigma_{xx}+\sigma_{yy}+\sigma_{zz}=2\mu(\nabla\cdot\boldsymbol{u})+3(b-p) \tag{2.32}$$

于是

$$b=\frac{1}{3}(\sigma_{xx}+\sigma_{yy}+\sigma_{zz})+p-\frac{2}{3}\mu(\nabla\cdot\boldsymbol{u}) \tag{2.33}$$

当流体静止时,流体的应力应趋于流体静压力,从而$(\sigma_{xx}+\sigma_{yy}+\sigma_{zz})/3$ 应包含$-p$ 这一项。此外,由于 $\sigma_{xx}+\sigma_{yy}+\sigma_{zz}$是应力张量的一个不变量,根据各向同性假设,它应与应变率张量的不变量$\nabla\cdot\boldsymbol{u}$ 有关,故有

$$\frac{1}{3}(\sigma_{xx}+\sigma_{yy}+\sigma_{zz})=-p+\mu_{\mathrm{B}}(\nabla\cdot\boldsymbol{u}) \tag{2.34}$$

式中,μ_{B} 为体积黏度(bulk viscosity),Pa·s。

根据式(2.33)和式(2.34)可得应力张量为

$$\boldsymbol{\sigma}=-p\boldsymbol{I}+2\mu\boldsymbol{S}+\mu'(\nabla\cdot\boldsymbol{u})\boldsymbol{I} \tag{2.35}$$

式中,μ'为第二黏度系数。

$$\mu'=-\frac{2}{3}\mu+\mu_B \tag{2.36}$$

为了解体积黏度的物理意义,定义一点的力学压力为

$$p_m=-\frac{1}{3}(\sigma_{xx}+\sigma_{yy}+\sigma_{zz}) \tag{2.37}$$

由式(2.34)有

$$p-p_m=\mu_B(\nabla\cdot\boldsymbol{u}) \tag{2.38}$$

可见热力学压力与力学压力之差正比于速度散度,其比例系数为体积黏度。对不可压缩流体,$\nabla\cdot\boldsymbol{u}=0$,从而力学压力等于热力学压力。对可压缩流体,$\nabla\cdot\boldsymbol{u}\neq0$,此时力学压力不等于热力学压力。体积黏度的微观机理与体积变化时的能量耗散机制有关,除高温和高频声波等极端情况外,在一般的气体运动中均可采用斯托克斯假设,即取 $\mu_B=0$[9]。此时黏性应力张量 $\boldsymbol{\Pi}$ 可写成

$$\boldsymbol{\Pi}=\mu\left\{[\nabla\boldsymbol{u}+(\nabla\boldsymbol{u})^{\mathrm{T}}]-\frac{2}{3}(\nabla\cdot\boldsymbol{u})\boldsymbol{I}\right\} \tag{2.39}$$

2.4.2 Navier-Stokes 方程

将应力张量的表达式(2.35)代入微分形式的动量方程(2.20),可得

$$\rho\frac{\mathrm{D}\boldsymbol{u}}{\mathrm{D}t}=\rho\boldsymbol{F}-\nabla p+\nabla\cdot\{\mu[\nabla\boldsymbol{u}+(\nabla\boldsymbol{u})^{\mathrm{T}}]\}+\nabla(\mu'\nabla\cdot\boldsymbol{u}) \tag{2.40}$$

式(2.40)即为 **Navier-Stokes(纳维-斯托克斯)方程**。这一方程是由纳维于 1821 年提出的,并于 1845 年由斯托克斯完成最终形式。写成直角坐标系的约定求和形式即为

纳维和斯托克斯的人物小传

$$\rho\left(\frac{\partial u_i}{\partial t}+u_j\frac{\partial u_i}{\partial x_j}\right)=\rho F_i-\frac{\partial p}{\partial x_i}+\frac{\partial}{\partial x_j}\left[\mu\left(\frac{\partial u_i}{\partial x_j}+\frac{\partial u_j}{\partial x_i}\right)\right]+\frac{\partial}{\partial x_i}\left(\mu'\frac{\partial u_j}{\partial x_j}\right) \tag{2.41}$$

当温度变化较小时,可认为 μ 和 μ' 为常数,此时式(2.40)可写为

$$\rho\frac{\mathrm{D}\boldsymbol{u}}{\mathrm{D}t}=\rho\boldsymbol{F}-\nabla p+\mu\nabla\cdot[\nabla\boldsymbol{u}+(\nabla\boldsymbol{u})^{\mathrm{T}}]+\mu'\nabla(\nabla\cdot\boldsymbol{u}) \tag{2.42}$$

而

$$\begin{aligned}\nabla\cdot(\nabla\boldsymbol{u})^{\mathrm{T}}&=\boldsymbol{e}_k\frac{\partial}{\partial x_k}\cdot\left(\frac{\partial u_i}{\partial x_j}\boldsymbol{e}_i\boldsymbol{e}_j\right)=\frac{\partial}{\partial x_i}\left(\frac{\partial u_i}{\partial x_j}\right)\boldsymbol{e}_j\\&=\boldsymbol{e}_j\frac{\partial}{\partial x_j}\left(\frac{\partial u_i}{\partial x_i}\right)=\nabla(\nabla\cdot\boldsymbol{u})\end{aligned} \tag{2.43}$$

式中,$\boldsymbol{e}_i$、$\boldsymbol{e}_j$ 和 $\boldsymbol{e}_k$ 分别为 i、j 和 k 方向的单位矢量。于是式(2.42)可写为

$$\rho\frac{\mathrm{D}\boldsymbol{u}}{\mathrm{D}t}=\rho\boldsymbol{F}-\nabla p+\mu\nabla^2\boldsymbol{u}+(\mu+\mu')\nabla(\nabla\cdot\boldsymbol{u}) \tag{2.44}$$

对不可压缩流体,式(2.44)化为

$$\frac{\mathrm{D}\boldsymbol{u}}{\mathrm{D}t}=\boldsymbol{F}-\frac{1}{\rho}\nabla p+\nu\nabla^2\boldsymbol{u} \tag{2.45}$$

式中,$\nu=\mu/\rho$ 为运动黏度,$\mathrm{m^2/s}$。利用连续性方程,可得式(2.44)和式(2.45)的守恒形式分别为

$$\frac{\partial(\rho\boldsymbol{u})}{\partial t}+\nabla\cdot(\rho\boldsymbol{u}\boldsymbol{u})=\rho\boldsymbol{F}-\nabla p+\mu\nabla^2\boldsymbol{u}+(\mu+\mu')\nabla(\nabla\cdot\boldsymbol{u}) \tag{2.46}$$

$$\frac{\partial\boldsymbol{u}}{\partial t}+\nabla\cdot(\boldsymbol{u}\boldsymbol{u})=\boldsymbol{F}-\frac{1}{\rho}\nabla p+\nu\nabla^2\boldsymbol{u} \tag{2.47}$$

2.5 能量方程

流体流动过程中,机械能可转换为流体的热力学能(即内能),流体的热力学能在一定条件下也可转换为机械能,同时流体在流动过程中还可能与外界发生热量交换。因此,仅从流体力学的角度描述流体运动是不够的,还需引入热力学第一定律。根据热力学第一定律,一个系统的热力学能变化等于外力对该系统所做的功与外界传递给系统的热量之和。但由于流体处于连续的运动中,因此在研究流体系统的能量守恒时需考虑流体总能(热力学能和动能之和)的变化。

2.5.1 总能方程

设单位质量的流体具有的比热力学能为 ε,比动能为 $u^2/2$,则 $\rho\varepsilon$ 表示单位体积的热力学能,$\rho u^2/2$ 表示单位体积的动能。略去位能不计,流体单位体积的总能为

$$\rho E=\rho\varepsilon+\frac{1}{2}\rho u^2 \tag{2.48}$$

根据**热力学第一定律**:系统总能的变化 = 单位时间内外力对系统所做的功 + 单位时间内系统增加的热量(包括外界传递给系统的热量以及系统内热源所产生热量),则有[3-4]

$$\frac{\mathrm{D}}{\mathrm{D}t}\int_{V(t)}\rho\left(\varepsilon+\frac{1}{2}u^2\right)\mathrm{d}V=\int_{V(t)}\rho\boldsymbol{F}\cdot\boldsymbol{u}\mathrm{d}V+\int_{A(t)}(\boldsymbol{n}\cdot\boldsymbol{\sigma})\cdot\boldsymbol{u}\mathrm{d}A+\int_{V(t)}\rho Q\mathrm{d}V-\int_{A(t)}\boldsymbol{q}\cdot\boldsymbol{n}\mathrm{d}A \tag{2.49}$$

式中:Q 表示由辐射或化学能释放等因素而引起的单位质量流体的热量增量;$-\int_{A(t)}\boldsymbol{q}\cdot\boldsymbol{n}\mathrm{d}A$ 表示热传导所引起的热量增量,负号表示热流的方向与控制体表面外法线的方向 $\boldsymbol{n}$ 相反。根据**傅里叶定律**(Fourier's law)有:$\boldsymbol{q}=-\lambda\nabla T$,其中 λ 为导热系数,W/(m·K)。

根据雷诺第二输运方程有

$$\frac{\mathrm{D}}{\mathrm{D}t}\int_{V(t)}\rho\left(\varepsilon+\frac{1}{2}u^2\right)\mathrm{d}V=\int_{V(t)}\rho\frac{\mathrm{D}}{\mathrm{D}t}\left(\varepsilon+\frac{1}{2}u^2\right)\mathrm{d}V \tag{2.50}$$

根据高斯散度公式有

$$\int_{A(t)}(\boldsymbol{n}\cdot\boldsymbol{\sigma})\cdot\boldsymbol{u}\mathrm{d}A=\int_{V(t)}\nabla\cdot(\boldsymbol{\sigma}\cdot\boldsymbol{u})\mathrm{d}V \tag{2.51}$$

$$\int_{A(t)}\boldsymbol{q}\cdot\boldsymbol{n}\mathrm{d}A=\int_{V(t)}\nabla\cdot\boldsymbol{q}\mathrm{d}V \tag{2.52}$$

于是,能量方程可写成

$$\int_{V(t)}\rho\frac{\mathrm{D}}{\mathrm{D}t}\left(\varepsilon+\frac{1}{2}u^2\right)\mathrm{d}V=\int_{V(t)}\rho\boldsymbol{F}\cdot\boldsymbol{u}\mathrm{d}V+\int_{V(t)}\nabla\cdot(\boldsymbol{\sigma}\cdot\boldsymbol{u})\mathrm{d}V+\int_{V(t)}\rho Q\mathrm{d}V-\int_{V(t)}\nabla\cdot\boldsymbol{q}\mathrm{d}V \tag{2.53}$$

由于 $V(t)$ 的任意性,可得**微分形式的总能方程**(total energy equation)为

$$\rho\frac{\mathrm{D}}{\mathrm{D}t}\left(\varepsilon+\frac{u^2}{2}\right)=\rho\boldsymbol{F}\cdot\boldsymbol{u}-\nabla\cdot\boldsymbol{q}+\nabla\cdot(\boldsymbol{\sigma}\cdot\boldsymbol{u})+\rho Q \tag{2.54}$$

式中,

$$\nabla\cdot(\boldsymbol{\sigma}\cdot\boldsymbol{u})=\boldsymbol{e}_l\frac{\partial}{\partial x_l}(\sigma_{ij}\boldsymbol{e}_i\boldsymbol{e}_j u_k\boldsymbol{e}_k)$$

$$=\boldsymbol{e}_l\frac{\partial}{\partial x_l}(\sigma_{ij}u_j\boldsymbol{e}_i)=\frac{\partial}{\partial x_i}(\sigma_{ij}u_j)=u_j\frac{\partial\sigma_{ij}}{\partial x_i}+\sigma_{ij}\frac{\partial u_j}{\partial x_i} \tag{2.55}$$

于是式(2.54)可化为

$$\rho\frac{\mathrm{D}}{\mathrm{D}t}\left(\varepsilon+\frac{u^2}{2}\right)=\rho\boldsymbol{F}\cdot\boldsymbol{u}-\nabla\cdot\boldsymbol{q}+\boldsymbol{u}\cdot(\nabla\cdot\boldsymbol{\sigma})+\boldsymbol{\sigma}:\nabla\boldsymbol{u}+\rho Q \tag{2.56}$$

式中，$\boldsymbol{\sigma}:\nabla\boldsymbol{u}$ 是应力张量和速度梯度张量的数量积。

$$\boldsymbol{\sigma}:\nabla\boldsymbol{u}=\sigma_{mn}\boldsymbol{e}_m\boldsymbol{e}_n:\frac{\partial u_j}{\partial x_i}\boldsymbol{e}_i\boldsymbol{e}_j=\sigma_{ij}\frac{\partial u_j}{\partial x_i} \tag{2.57}$$

由于应力张量为对称张量，从而有

$$\begin{aligned}\boldsymbol{\sigma}:\nabla\boldsymbol{u}&=\sigma_{ij}\frac{\partial u_j}{\partial x_i}=\frac{1}{2}\sigma_{ij}\frac{\partial u_j}{\partial x_i}+\frac{1}{2}\sigma_{ij}\frac{\partial u_j}{\partial x_i}\\&=\frac{1}{2}\sigma_{ij}\frac{\partial u_j}{\partial x_i}+\frac{1}{2}\sigma_{ji}\frac{\partial u_j}{\partial x_i}\\&=\frac{1}{2}\sigma_{ij}\frac{\partial u_j}{\partial x_i}+\frac{1}{2}\sigma_{ij}\frac{\partial u_i}{\partial x_j}\\&=\frac{1}{2}\sigma_{ij}\left(\frac{\partial u_j}{\partial x_i}+\frac{\partial u_i}{\partial x_j}\right)=\boldsymbol{\sigma}:\boldsymbol{S}\end{aligned} \tag{2.58}$$

即应力张量和应变率张量的数量积与应力张量和速度梯度张量的数量积相等。

2.5.2 热力学能方程

为了导出系统的热力学能方程，先对动量方程两边点乘 $\boldsymbol{u}$，有

$$\rho\boldsymbol{u}\cdot\frac{\mathrm{D}\boldsymbol{u}}{\mathrm{D}t}=\rho\frac{\mathrm{D}}{\mathrm{D}t}\left(\frac{u^2}{2}\right)=\boldsymbol{u}\cdot\rho\boldsymbol{F}+\boldsymbol{u}\cdot(\nabla\cdot\boldsymbol{\sigma}) \tag{2.59}$$

从式(2.56)减去式(2.59)，可得

$$\rho\frac{\mathrm{D}\varepsilon}{\mathrm{D}t}=-\nabla\cdot\boldsymbol{q}+\boldsymbol{\sigma}:\nabla\boldsymbol{u}+\rho Q \tag{2.60}$$

写成守恒形式为

$$\frac{\partial(\rho\varepsilon)}{\partial t}+\nabla\cdot(\rho\varepsilon\boldsymbol{u})=-\nabla\cdot\boldsymbol{q}+\boldsymbol{\sigma}:\nabla\boldsymbol{u}+\rho Q \tag{2.61}$$

式(2.61)即**热力学能方程**(internal energy equation)。引入傅里叶定律，$-\nabla\cdot\boldsymbol{q}$ 可写成 $\nabla\cdot(\lambda\nabla T)$。利用牛顿流体的本构方程，可对表面力做功项做如下处理[7]：

$$\begin{aligned}\boldsymbol{\sigma}:\nabla\boldsymbol{u}&=(-p\boldsymbol{I}+\boldsymbol{\Pi}):\nabla\boldsymbol{u}=(-p\boldsymbol{I}+\boldsymbol{\Pi}):\boldsymbol{S}\\&=(-p+\mu'\nabla\cdot\boldsymbol{u})\boldsymbol{I}:\boldsymbol{S}+2\mu\boldsymbol{S}:\boldsymbol{S}\\&=(-p+\mu'\nabla\cdot\boldsymbol{u})\nabla\cdot\boldsymbol{u}+2\mu\boldsymbol{S}:\boldsymbol{S}\\&=-p\nabla\cdot\boldsymbol{u}+\mu'(\nabla\cdot\boldsymbol{u})^2+2\mu\boldsymbol{S}:\boldsymbol{S}\end{aligned} \tag{2.62}$$

式(2.62)表明，$\boldsymbol{\sigma}:\nabla\boldsymbol{u}$ 由两项组成，其中 $-p\nabla\cdot\boldsymbol{u}$ 代表流体膨胀时克服法向压力 p 所做的功；另一项 $\Phi=\boldsymbol{\Pi}:\nabla\boldsymbol{u}=\mu'(\nabla\cdot\boldsymbol{u})^2+2\mu\boldsymbol{S}:\boldsymbol{S}$ 是克服黏性力所做的功，此功转化为热能耗散掉了，故 Φ 称为耗散函数。引入傅里叶定律和耗散函数，式(2.60)可写成

$$\rho\frac{\mathrm{D}\varepsilon}{\mathrm{D}t}=-p\ \nabla\cdot\boldsymbol{u}+\nabla\cdot(\lambda\nabla T)+\Phi+\rho Q \tag{2.63}$$

式中,$\Phi=\mu(2S_{xx}^2+2S_{yy}^2+2S_{zz}^2+4S_{xy}^2+4S_{xz}^2+4S_{yz}^2)+\mu'(S_{xx}+S_{yy}+S_{zz})^2$。在直角坐标中,$\Phi$ 为

$$\Phi=\mu\left[2\left(\frac{\partial u_x}{\partial x}\right)^2+2\left(\frac{\partial u_y}{\partial y}\right)^2+2\left(\frac{\partial u_z}{\partial z}\right)^2+\left(\frac{\partial u_x}{\partial y}+\frac{\partial u_y}{\partial x}\right)^2+\right.$$
$$\left.\left(\frac{\partial u_x}{\partial z}+\frac{\partial u_z}{\partial x}\right)^2+\left(\frac{\partial u_y}{\partial z}+\frac{\partial u_z}{\partial y}\right)^2\right]+\mu'\left(\frac{\partial u_x}{\partial x}+\frac{\partial u_y}{\partial y}+\frac{\partial u_z}{\partial z}\right)^2 \tag{2.64}$$

2.5.3 其他形式的能量方程

热力学中比焓的定义为 $h=\varepsilon+p/\rho$,ρh 表示单位体积的焓。相应地,焓方程可从热力学能方程导出。将 $\varepsilon=h-p/\rho$ 代入式(2.63)可得

$$\rho\frac{\mathrm{D}h}{\mathrm{D}t}-\rho\left[\frac{1}{\rho}\frac{\mathrm{D}p}{\mathrm{D}t}+p\frac{\mathrm{D}}{\mathrm{D}t}\left(\frac{1}{\rho}\right)\right]=-p\ \nabla\cdot\boldsymbol{u}+\nabla\cdot(\lambda\nabla T)+\Phi+\rho Q \tag{2.65}$$

结合连续性方程(2.13)可得

$$\rho\frac{\mathrm{D}h}{\mathrm{D}t}=\nabla\cdot(\lambda\nabla T)+\frac{\mathrm{D}p}{\mathrm{D}t}+\Phi+\rho Q \tag{2.66}$$

写成守恒形式为

$$\frac{\partial(\rho h)}{\partial t}+\nabla\cdot(\rho h\boldsymbol{u})=\nabla\cdot(\lambda\nabla T)+\frac{\partial p}{\partial t}+\boldsymbol{u}\cdot\nabla p+\Phi+\rho Q \tag{2.67}$$

式(2.67)即为**焓方程**。结合式(2.67)和总焓的定义可导出相应的总焓方程。

此外,根据热力学能方程亦可导出用熵表示的能量方程。在热力学中,比热力学能 ε 和比熵 s 存在如下关系式:

$$\mathrm{d}\varepsilon=T\mathrm{d}s-p\mathrm{d}\left(\frac{1}{\rho}\right) \tag{2.68}$$

于是有

$$\frac{\mathrm{D}\varepsilon}{\mathrm{D}t}=T\frac{Ds}{\mathrm{D}t}-p\frac{\mathrm{D}}{\mathrm{D}t}\left(\frac{1}{\rho}\right) \tag{2.69}$$

结合式(2.69)和连续性方程(2.13),可从式(2.63)导出如下方程:

$$\rho T\frac{Ds}{\mathrm{D}t}=-\nabla\cdot\boldsymbol{q}+\Phi+\rho Q \tag{2.70}$$

结合连续性方程,式(2.70)可写成[10]

$$\frac{\partial(\rho s)}{\partial t}+\nabla\cdot(\rho s\boldsymbol{u})=-\nabla\cdot\left(\frac{\boldsymbol{q}}{T}\right)-\frac{1}{T^2}(\boldsymbol{q}\cdot\nabla T)+\frac{1}{T}(\Phi+\rho Q) \tag{2.71}$$

式(2.71)即为**用熵表示的能量方程**。

更进一步,根据热力学有

$$\mathrm{d}\varepsilon=c_V\mathrm{d}T+\left[T\left(\frac{\partial p}{\partial T}\right)_\rho-p\right]\mathrm{d}\left(\frac{1}{\rho}\right) \tag{2.72}$$

式中,c_V 为比定容热容,J/(kg · K)。式(2.72)通常被称为第一 dε 方程。依据式(2.72)和连续性方程(2.13),可从式(2.63)导出如下形式的温度方程:

$$\rho c_V\frac{\mathrm{D}T}{\mathrm{D}t}=\nabla\cdot(\lambda\nabla T)-T\left(\frac{\partial p}{\partial T}\right)_\rho\nabla\cdot\boldsymbol{u}+\Phi+\rho Q \tag{2.73}$$

式(2.73)即为基于比定容热容的**温度方程**。对理想气体，$T\left(\partial p/\partial T\right)_\rho=p$；对实际气体和超临界流体，可按照相应的状态方程计算$(\partial p/\partial T)_\rho$。

类似地，根据热力学中的 dh 方程可从式(2.66)导出基于比定压热容 c_p 的温度方程，即

$$\rho c_p \frac{\mathrm{D}T}{\mathrm{D}t}=\nabla\cdot(\lambda\nabla T)+T\beta\frac{\mathrm{D}p}{\mathrm{D}t}+\Phi+\rho Q \tag{2.74}$$

式中，β 为体胀系数。

$$\beta=-\frac{1}{\rho}\left(\frac{\partial\rho}{\partial T}\right)_p \tag{2.75}$$

式(2.74)写成守恒形式为

$$\frac{\partial(\rho c_p T)}{\partial t}+\nabla\cdot(\rho c_p T\boldsymbol{u})=\nabla\cdot(\lambda\nabla T)+\rho T\frac{\mathrm{D}c_p}{\mathrm{D}t}+T\beta\frac{\mathrm{D}p}{\mathrm{D}t}+\Phi+\rho Q \tag{2.76}$$

当比定压热容 c_p 随空间变化比较显著时，$\mathrm{D}c_p/\mathrm{D}t$ 的影响不可忽略。对理想气体，$T\beta=1$。

各种形式的能量方程

超临界流体

2.6 小 结

本章介绍了流体力学的基本方程：连续性方程、动量方程和能量方程。以**守恒形式**的流体力学基本方程组为例，可总结如下：

$$\frac{\partial\rho}{\partial t}+\nabla\cdot(\rho\boldsymbol{u})=0 \tag{2.77}$$

$$\frac{\partial(\rho\boldsymbol{u})}{\partial t}+\nabla\cdot(\rho\boldsymbol{u}\boldsymbol{u})=\rho\boldsymbol{F}-\nabla p+\nabla\cdot\{\mu[\nabla\boldsymbol{u}+(\nabla\boldsymbol{u})^{\mathrm{T}}]\}+\nabla(\mu'\nabla\cdot\boldsymbol{u}) \tag{2.78}$$

$$\frac{\partial(\rho\varepsilon)}{\partial t}+\nabla\cdot(\rho\varepsilon\boldsymbol{u})=-p\,\nabla\cdot\boldsymbol{u}+\nabla\cdot(\lambda\nabla T)+\boldsymbol{\Pi}:\nabla\boldsymbol{u}+\rho Q \tag{2.79}$$

在计算流体力学中，常把上述方程组称为 **Navier-Stokes 方程组**。方程组中的压力和热力学能是密度和温度的函数，即

$$p=p(\rho,T),\quad \varepsilon=\varepsilon(\rho,T) \tag{2.80}$$

对理想气体，则有

$$p=\rho R_{\mathrm{g}}T,\quad \mathrm{d}\varepsilon=c_V\mathrm{d}T \tag{2.81}$$

式中：$R_{\mathrm{g}}=R/M$ 为气体常数，J/(kg·K)；R 为摩尔气体常数，其值为 $R\approx8.31$ J/(mol·K)；M 为气体摩尔质量，g/mol。

在方程(2.77)~(2.80)中，动量方程是矢量方程，其余的连续性方程、能量方程以及压力和热力学能的表达式都是标量方程。共计 7 个标量方程，未知量也是 7 个，包括密度、三个速度分量、压力、热力学能和温度，因此上述方程组是封闭的。

为了确定一个流场，通常并不需要同时求解上述 7 个方程。对不可压缩流动，由于密度可以近似看作常数，连续性方程和动量方程包含 4 个未知量($\boldsymbol{u}$ 和 p)，所以方程组是封闭的，求出 $\boldsymbol{u}$ 和 p 后，根据能量方程可求得温度。由于忽略了流体的可压缩性，流体的动力学问题和热力学问题可以分开求解，简化了求解过程。但在可压缩流动中，流体的密度不再为常数，此时为了求得流场的速度分布和压力分布，连续性方程和动量方程必须与能量方程耦合求解。

需要指出的是,为了表述的简洁,也为下文分析格子 Boltzmann 方法与连续介质流体力学基本方程关系的需要,本章用到了笛卡儿张量运算。对笛卡儿张量生疏的读者可先阅读本书的附录 A 或其他的参考文献[11]。掌握张量的基本知识对学习格子 Boltzmann 方法和阅读相关科技文献会有很大帮助。

习　题

2-1　试证明流体微团的相对体积膨胀率等于速度的散度。

2-2　标准状况下,某一气体分子的平均自由程约为 6.8×10^{-8} m,在该状况下当气体所处空间的特征尺度大于多少时该气体可以作为连续介质处理?

2-3　结合式(2.67)和总焓的定义导出相应的总焓方程,并与总能方程即式(2.54)进行对比。

2-4　有人指出,理想气体的热力学能可以表示为 $\varepsilon=c_V T$,当比定容热容 c_V 为常数时,式(2.63)和式(2.73)是一致的,但当 c_V 为变量时,这两个方程不一致,因此必有一个是错误的。这一说法对吗?为什么?

参 考 文 献

[1] 周光坰,严宗毅,许世雄,等. 流体力学:上册[M].2 版. 北京:高等教育出版社,2003.

[2] Tsien H S. Superaerodynamics, mechanics of rarefied gases[J]. Journal of the Aeronautical Sciences, 1946, 13:653-664.

[3] 章梓雄,董曾南. 粘性流体力学[M].2 版. 北京:清华大学出版社,2011.

[4] 董曾南,章梓雄. 非粘性流体力学[M]. 北京:清华大学出版社,2003.

[5] 丁祖荣. 流体力学:上册 [M]. 北京:高等教育出版社,2003.

[6] White F M. Fluid Mechanics[M]. 5th edition. Boston: McGrawHill, 2003.

[7] 刘树红,吴玉林,左志钢. 应用流体力学[M].2 版. 北京:清华大学出版社,2012.

[8] 张鸣远,景思睿,李国君. 高等工程流体力学[M]. 西安:西安交通大学出版社,2006.

[9] Currie I G. Fundamental Mechanics of Fluids[M]. 3rd edition. New York: Marcel Dekker Inc., 2003.

[10] Bird R B, Stewart W E, Lightfoot E N. Transport Phenomena[M]. 2nd edition. John Wiley & Sons, Inc., 2001.

[11] 李开泰,黄艾香. 张量理论与应用[M]. 北京:科学出版社,2004.

第3章 格子Boltzmann方法的基础理论和基本模型

尽管连续介质只是一种假设，但当研究对象的尺度比粒子的结构尺度大得多时，连续介质假设是成立的，所以它在流体力学中被广泛采用，并获得了很大的成功。例如，飞机在空气中的飞行、船舶在水中的航行，相应的特征尺度（飞机和船舶的长度）远大于粒子的结构尺度，此时空气和水可以被认为是连续介质。类似地，血液在动脉中的流动，血液可以看作是连续介质。甚至当研究星系结构时，恒星间的距离约为 4×10^{16} m，它们在半径约为 4×10^{20} m 的银河系中运动，星系也可以看作是一种连续介质。然而，在研究高空稀薄气体中的物体运动时，分子的平均自由程很大，它与物体的特征尺度相比为同阶量，这时便不能将高空稀薄气体视为连续介质。同样，血液在微血管中的流动，也不能当作连续介质处理。这些都是连续介质假设应用范围的限制。此外，尽管使用 Navier-Stokes 方程组描述流体的运动是比较方便的，但在某些情况下，求解这组偏微分方程是非常困难或根本不可能的。还有一些流体系统，我们甚至不知道其数学模型和宏观控制方程，更不用说进行数值模拟研究了。

对连续介质假设不适用或宏观方程难以描述的系统，采用微观的分子动力学或者**气体动理论**（gas kinetic theory）进行描述更为恰当。但分子动力学模拟方法需要跟踪许多粒子的运动，计算量非常之大，对计算机的存储量和计算速度都有着很高的要求。而气体动理论的基本方程——连续 **Boltzmann 方程**是一个比 Navier-Stokes 方程更加复杂的积分微分方程，它的精确求解通常是比较困难的。在这种情况下，格子方法成为流体计算和建模的一种有效方法。格子气自动机实际上是一种简化的分子动力学模型，而格子 Boltzmann 方法则是从格子气自动机演变而来的。正如本书绪论中所述，1988 年，McNamara 和 Zanetti[1] 提出把格子气自动机中的整数运算变成实数运算，格点上的粒子数不再用 0 和 1 来表征，而是用实数来表示的**局部粒子分布函数**，这标志着格子 Boltzmann 方法的诞生。

实际上，格子方法是基于这样的事实，即流体的宏观运动是分子微观热运动统计平均的结果，宏观行为对每个具体分子的运动细节是不敏感的。Navier-Stokes 方程组所描述的守恒定律与微观分子的运动规律是一致的，流体分子内部相互作用的差别反映在 Navier-Stokes 方程组的输运系数上。因此，可以构造微观或介观的模型，使之在遵循基本守恒定律的前提下尽可能地简单。各种格子模型正是依据这一原则构造出来的。虽然格子 Boltzmann 方法诞生时并没有与气体动理论建立直接的联系，不过 He 和 Luo[2-3] 于 1997 年连续撰文，分析了格子 Boltzmann 方程与 Boltzmann 方程的内在联系，他们发现格子 Boltzmann 方程可以看作 Boltzmann 方程的一种特殊离散格式。如今，格子 Boltzmann 方法已经独立于格子气自动机[4] 发展成为一种自成体系的介观方法。

因此,本章选择从气体动理论出发并沿着 Boltzmann 方程—Maxwell 分布—Boltzmann-BGK 方程—格子 Boltzmann-BGK 方程这条主线对格子 Boltzmann 方法的基础理论进行介绍。Boltzmann 方程是气体动理论的基本方程,而 Maxwell 分布则是 Boltzmann 方程的一个特殊解,它是单组分单原子气体的平衡态分布。由于 Boltzmann 方程是复杂的微分积分方程,它的直接求解通常是非常困难的,于是出现了许多近似解法,例如用 Boltzmann-BGK 方程代替 Boltzmann 方程,其中 BGK 指的是 1954 年 Bhatnagar、Gross 和 Krook 针对 Boltzmann 方程的碰撞项所提出的一种近似[5],这种近似通常也被称为**单松弛碰撞算子**,它使得 Boltzmann 方程线性化,从而大大简化了方程的求解。格子 Boltzmann 方法中的格子 Boltzmann-BGK 方程正是 Boltzmann-BGK 方程的一种特殊离散形式。

在本章,我们还将介绍 Maxwell 输运方程以及如何通过 Chapman-Enskog 多尺度展开从格子 Boltzmann-BGK 方程导出宏观方程。此外,还将介绍另外两种碰撞算子,即**多松弛碰撞算子和中心矩碰撞算子**,以及格子 Boltzmann 方法中常用的作用力格式。

3.1 Boltzmann 方程

克劳修斯的人物小传

麦克斯韦的人物小传

吉布斯的人物小传

气体动理论的飞跃发展发生在 19 世纪中后期,它的主要奠基人是 Clausius(克劳修斯)、Maxwell(麦克斯韦)和 Boltzmann(玻尔兹曼)。Clausius 第一次正确地证明了玻意耳定律,并且第一个引入了自由程的概念;Maxwell 首先认识到分子的速度各不相同从而得到了速度分布律,之后又建立了气体输运过程理论;Boltzmann 在速度分布律中引入了重力场,并于 1872 年提出了 ***H* 定理**(H theorem)用以证明了速度分布律并给出了熵的统计意义,随后他又导出了著名的输运方程,并且完成了输运过程的数学理论。这三位奠基人所做工作的重要意义在于引入了统计概念,把宏观现象和微观基础联系了起来。1902 年,Gibbs(吉布斯)把 Boltzmann 和 Maxwell 所创立的统计方法发展成为系统的**系综理论**(ensemble theory),将原来仅适用于气体的理论,推广为对气体、液体和固体都普遍适用的统计力学。但是,当统计力学应用于热辐射问题时,却得不到与实验相符的结果。为了解决这个问题,经过多方面的努力,终于在 1926 年建立了量子力学,在量子力学基础上建立起来的统计理论成为量子统计力学。至此,统计物理学形成了比较完善的理论体系。

在任何一个宏观体系中,每个分子的微观运动都遵守力学规律,因此只要算出大量粒子的运动,就可以确定系统的宏观参数,这是分子动力学模拟的基本出发点。从另一方面考虑,我们可以不去确定每个分子的运动状态,而是求出分子处在某一状态的概率,继而通过统计方法得到系统的宏观参数,这正是 Boltzmann 方程的基本思想。Boltzmann 方程是统计力学中描述分布函数时空演化的方程,它是基于以下三个重要假设建立的[6]:

(1) 分子碰撞时只考虑二体碰撞,即认为三个分子或三个以上分子同时碰撞在一起的概率很小;

(2) 分子混沌假设,这一假设认为各个分子的速度分布是不依赖于另外的分子而独立的,即分子在碰撞之前速度不相关;

(3) 外力不影响局部碰撞的动力学行为。

假设(1)要求所研究的问题只能限于不太高的密度,而假设(2)则要求碰撞时间非常短。为了简单起见,这里仅讨论单组分气体。设粒子分布函数为 f,它是空间位置矢量 $\boldsymbol{r}(x,y,z)$、分子速度矢量 $\boldsymbol{\xi}(\xi_x,\xi_y,\xi_z)$ 及时间 t 的函数。$f(\boldsymbol{r},\boldsymbol{\xi},t)\mathrm{d}\boldsymbol{r}\mathrm{d}\boldsymbol{\xi}$ 表示 t 时刻、位于 $\boldsymbol{r}$

与 $\boldsymbol{r}+\mathrm{d}\boldsymbol{r}$ 间的体积元中、速度在 $\boldsymbol{\xi}$ 与 $\boldsymbol{\xi}+\mathrm{d}\boldsymbol{\xi}$ 之间的分子数。考虑体积力时,可设 $m\boldsymbol{a}$ 为作用在每个分子上的外力,其中 m 是分子质量,$\boldsymbol{a}$ 是外力加速度。根据 f 的定义有

$$n=\int_{\Re} f(\boldsymbol{r},\boldsymbol{\xi},t)\,\mathrm{d}\boldsymbol{\xi} \tag{3.1}$$

式中:$\Re$ 为分子速度空间;n 为 t 时刻、空间位置 $\boldsymbol{r}$ 处单位体积内的分子数,亦称数密度。

对任一分子,如果在时间间隔 $\mathrm{d}t$ 内无碰撞,则分子的位置矢量将由 $\boldsymbol{r}$ 变为 $\boldsymbol{r}+\mathrm{d}\boldsymbol{r}$,而它的速度将由 $\boldsymbol{\xi}$ 变为 $\boldsymbol{\xi}+\boldsymbol{a}\mathrm{d}t$。因此,如果 $\mathrm{d}t$ 时间间隔内无碰撞,原来 t 时刻在 $\mathrm{d}\boldsymbol{r}\mathrm{d}\boldsymbol{\xi}$ 中的气体分子 $f(\boldsymbol{r},\boldsymbol{\xi},t)\,\mathrm{d}\boldsymbol{r}\mathrm{d}\boldsymbol{\xi}$ 将全部既不增加也不减少地转移到 $t+\mathrm{d}t$ 时刻 $\boldsymbol{r}+\mathrm{d}\boldsymbol{r}$、$\boldsymbol{\xi}+\boldsymbol{a}\mathrm{d}t$ 的 $\mathrm{d}\boldsymbol{r}\mathrm{d}\boldsymbol{\xi}$ 中,于是有

$$f(\boldsymbol{r}+\mathrm{d}\boldsymbol{r},\boldsymbol{\xi}+\boldsymbol{a}\mathrm{d}t,t+\mathrm{d}t)\,\mathrm{d}\boldsymbol{r}\mathrm{d}\boldsymbol{\xi}=f(\boldsymbol{r},\boldsymbol{\xi},t)\,\mathrm{d}\boldsymbol{r}\mathrm{d}\boldsymbol{\xi} \tag{3.2}$$

对式(3.2)左端做泰勒级数展开,然后两边同除以 $\mathrm{d}t$,并令 $\mathrm{d}t\to 0$ 可得

$$\frac{\partial f}{\partial t}+\boldsymbol{\xi}\cdot\frac{\partial f}{\partial \boldsymbol{r}}+\boldsymbol{a}\cdot\frac{\partial f}{\partial \boldsymbol{\xi}}=0 \tag{3.3}$$

即

$$\left(\frac{\partial f}{\partial t}\right)_{\text{move}}=-\boldsymbol{\xi}\cdot\frac{\partial f}{\partial \boldsymbol{r}}-\boldsymbol{a}\cdot\frac{\partial f}{\partial \boldsymbol{\xi}} \tag{3.4}$$

式中,$(\partial f/\partial t)_{\text{move}}$ 表示分子运动所引起的 $f(\boldsymbol{r},\boldsymbol{\xi},t)$ 的增加。除此之外,式(3.2)右端还应考虑由分子碰撞所引起的分子数的变化量:$(\partial f/\partial t)_{\text{collision}}\mathrm{d}\boldsymbol{r}\mathrm{d}\boldsymbol{\xi}\mathrm{d}t$,所以

$$\left(\frac{\partial f}{\partial t}\right)=\left(\frac{\partial f}{\partial t}\right)_{\text{move}}+\left(\frac{\partial f}{\partial t}\right)_{\text{collision}} \tag{3.5}$$

或

$$\frac{\partial f}{\partial t}+\boldsymbol{\xi}\cdot\frac{\partial f}{\partial \boldsymbol{r}}+\boldsymbol{a}\cdot\frac{\partial f}{\partial \boldsymbol{\xi}}=\left(\frac{\partial f}{\partial t}\right)_{\text{collision}} \tag{3.6}$$

Maxwell 和 Boltzmann 所采用的碰撞模型主要有两个,一个是刚球模型,另一个是力心点模型。在刚球模型中,分子被认为是弹性刚球,碰撞时分子大小和形状不变,并且球面光滑,碰撞时相互作用力在两个球心的连线上,不影响切向的速度。在力心点模型中,分子被假定为质点,分子间的作用力是有心力,并且是分子距离的函数。这里我们讨论刚球模型,并且忽略三个或三个以上分子同时碰撞的机会,只考虑二体碰撞。

首先讨论碰撞前后分子速度的改变。设分子质量为 m,分子直径为 d_{D},碰撞前后的速度分别为 $\boldsymbol{\xi}_1$、$\boldsymbol{\xi}_2$ 和 $\boldsymbol{\xi}_1'$、$\boldsymbol{\xi}_2'$。由于碰撞是弹性的,碰撞前后的动量和能量都应守恒,故有

$$m\boldsymbol{\xi}_1+m\boldsymbol{\xi}_2=m\boldsymbol{\xi}_1'+m\boldsymbol{\xi}_2' \tag{3.7}$$

$$\frac{1}{2}m\xi_1^2+\frac{1}{2}m\xi_2^2=\frac{1}{2}m\xi_1'^2+\frac{1}{2}m\xi_2'^2 \tag{3.8}$$

从上述关系不难推出,两个光滑球在弹性碰撞时,沿球心连线方向的分速度彼此交换,而与球心连线垂直的分速度则保持不变。若取球心连线方向为 x 轴,则有

$$\left.\begin{aligned}&\xi_{1x}'=\xi_{2x},\quad \xi_{1y}'=\xi_{1y},\quad \xi_{1z}'=\xi_{1z}\\&\xi_{2x}'=\xi_{1x},\quad \xi_{2y}'=\xi_{2y},\quad \xi_{2z}'=\xi_{2z}\end{aligned}\right\} \tag{3.9}$$

由式(3.9)可得

$$\xi_{2x}'-\xi_{1x}'=-(\xi_{2x}-\xi_{1x}) \tag{3.10}$$

若取 $\boldsymbol{n}$ 为球心连线方向的单位矢量,则有

$$\left.\begin{aligned}(\boldsymbol{\xi}_2'-\boldsymbol{\xi}_1')\cdot\boldsymbol{n}&=-(\boldsymbol{\xi}_2-\boldsymbol{\xi}_1)\cdot\boldsymbol{n}\\ |\boldsymbol{\xi}_2'-\boldsymbol{\xi}_1'|&=|\boldsymbol{\xi}_2-\boldsymbol{\xi}_1|=|\boldsymbol{g}|\end{aligned}\right\}\tag{3.11}$$

即在碰撞过程中两个分子的相对速率不变，但在球心连线方向上相对速度的投影改变符号。此外，根据式(3.9)还可得到

$$\begin{aligned}\mathrm{d}\boldsymbol{\xi}_1'\mathrm{d}\boldsymbol{\xi}_2'&=\mathrm{d}\xi_{1x}'\mathrm{d}\xi_{1y}'\mathrm{d}\xi_{1z}'\mathrm{d}\xi_{2x}'\mathrm{d}\xi_{2y}'\mathrm{d}\xi_{2z}'\\&=\mathrm{d}\xi_{2x}\mathrm{d}\xi_{1y}\mathrm{d}\xi_{1z}\mathrm{d}\xi_{1x}\mathrm{d}\xi_{2y}\mathrm{d}\xi_{2z}\\&=\mathrm{d}\boldsymbol{\xi}_1\mathrm{d}\boldsymbol{\xi}_2\end{aligned}\tag{3.12}$$

式(3.11)和式(3.12)将在后面的推导中用到。

刚球模型分子碰撞的示意图如图3.1所示，两个分子中心的连线为 x 轴，$\boldsymbol{g}=\boldsymbol{\xi}_2-\boldsymbol{\xi}_1$ 是相对速度，θ 是 $\boldsymbol{g}$ 与 x 轴之间的夹角。为叙述方便，将第二个分子称为“来打分子”，第一个分子称为“挨打分子”。当分子碰撞时，来打分子的中心所在的球面微元在挨打分子的**立体角**(solid angle)为 $\mathrm{d}\Omega$，相应的球面微元面积为 $d_{\mathrm{D}}^2\mathrm{d}\Omega$。当来打分子以相对速度 $\boldsymbol{\xi}_2-\boldsymbol{\xi}_1$ 飞向挨打分子时，在 $\mathrm{d}t$ 时间间隔内，要与挨打分子相撞，则它必须位于以 $\boldsymbol{\xi}_2-\boldsymbol{\xi}_1$ 为轴线，高为 $|\boldsymbol{g}|\cos\theta\mathrm{d}t$，底面积为 $d_{\mathrm{D}}^2\mathrm{d}\Omega$ 的柱体内，这个柱体的体积为 $d_{\mathrm{D}}^2|\boldsymbol{g}|\cos\theta\mathrm{d}\Omega\mathrm{d}t$，此时存在的来打分子数为

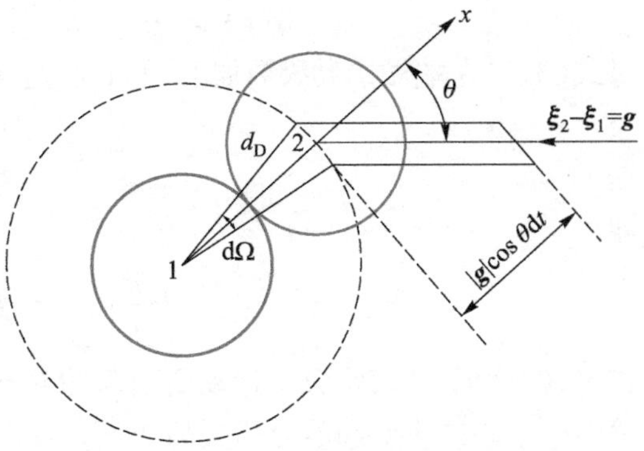

图3.1 刚球模型分子碰撞示意图

$$f(\boldsymbol{r},\boldsymbol{\xi}_2,t)d_{\mathrm{D}}^2|\boldsymbol{g}|\cos\theta\mathrm{d}\Omega\mathrm{d}t\mathrm{d}\boldsymbol{\xi}_2\tag{3.13}$$

式(3.13)乘以 $f(\boldsymbol{r},\boldsymbol{\xi}_1,t)\mathrm{d}\boldsymbol{r}\mathrm{d}\boldsymbol{\xi}_1$ 即为 $\mathrm{d}t$ 时间间隔内、处在体积元 $\mathrm{d}\boldsymbol{r}$ 内、速度间隔在 $\mathrm{d}\boldsymbol{\xi}_1$ 中的分子和速度间隔在 $\mathrm{d}\boldsymbol{\xi}_2$ 中的分子在 $\mathrm{d}\Omega$ 内的碰撞次数

$$f(\boldsymbol{r},\boldsymbol{\xi}_1,t)f(\boldsymbol{r},\boldsymbol{\xi}_2,t)d_{\mathrm{D}}^2|\boldsymbol{g}|\cos\theta\mathrm{d}\Omega\mathrm{d}t\mathrm{d}\boldsymbol{\xi}_2\mathrm{d}\boldsymbol{r}\mathrm{d}\boldsymbol{\xi}_1\tag{3.14}$$

碰撞之后，原来速度在 $\mathrm{d}\boldsymbol{\xi}_1$ 和 $\mathrm{d}\boldsymbol{\xi}_2$ 中的分子变为速度 $\mathrm{d}\boldsymbol{\xi}_1'$ 和 $\mathrm{d}\boldsymbol{\xi}_2'$ 中的分子。由此可见，碰撞次数即为分子数的减少。将式(3.14)对一切可能的 $\boldsymbol{\xi}_2$ 和 $\mathrm{d}\Omega$ 积分，就可得到 $\mathrm{d}t$ 时间内、$\mathrm{d}\boldsymbol{r}\mathrm{d}\boldsymbol{\xi}_1$ 中因为碰撞而减少的分子数

$$\mathrm{d}\boldsymbol{r}\mathrm{d}\boldsymbol{\xi}_1\mathrm{d}t\int_{\Re}\int_{\Omega}f_1f_2d_{\mathrm{D}}^2|\boldsymbol{g}|\cos\theta\mathrm{d}\Omega\mathrm{d}\boldsymbol{\xi}_2\tag{3.15}$$

式中，f_1 和 f_2 分别为 $f(\boldsymbol{r},\boldsymbol{\xi}_1,t)$ 和 $f(\boldsymbol{r},\boldsymbol{\xi}_2,t)$ 的简写。

对刚体碰撞，按照力学知识，每个如下碰撞：

$$\boldsymbol{\xi}_1+\boldsymbol{\xi}_2\rightarrow\boldsymbol{\xi}_1'+\boldsymbol{\xi}_2'\tag{3.16}$$

必有一回复碰撞

$$\boldsymbol{\xi}_1'+\boldsymbol{\xi}_2'\rightarrow\boldsymbol{\xi}_1+\boldsymbol{\xi}_2\tag{3.17}$$

对照式(3.15)可知，由该回复碰撞增加的分子数为

$$\mathrm{d}\boldsymbol{r}\mathrm{d}\boldsymbol{\xi}_1'\mathrm{d}t\int_{\Re}\int_{\Omega}f_1'f_2'd_{\mathrm{D}}^2|\boldsymbol{g}|\cos\theta\mathrm{d}\Omega\mathrm{d}\boldsymbol{\xi}_2'\tag{3.18}$$

根据式(3.12)，式(3.18)亦可写为

$$\mathrm{d}\boldsymbol{r}\mathrm{d}\boldsymbol{\xi}_1\mathrm{d}t\int_{\Re}\int_{\Omega}f_1'f_2'd_{\mathrm{D}}^2|\boldsymbol{g}|\cos\theta\mathrm{d}\Omega\mathrm{d}\boldsymbol{\xi}_2\tag{3.19}$$

因此，由碰撞净增加的分子数为

$$\left(\frac{\partial f_1}{\partial t}\right)_{\text{collision}} \mathrm{d}\boldsymbol{r}\mathrm{d}\boldsymbol{\xi}_1\mathrm{d}t = \mathrm{d}\boldsymbol{r}\mathrm{d}\boldsymbol{\xi}_1\mathrm{d}t\int_{\Re}\int_{\Omega}(f_1'f_2' - f_1f_2)d_{\mathrm{D}}^2|\boldsymbol{g}|\cos\theta\mathrm{d}\Omega\mathrm{d}\boldsymbol{\xi}_2 \tag{3.20}$$

消去 $\mathrm{d}\boldsymbol{r}\mathrm{d}\boldsymbol{\xi}_1\mathrm{d}t$，并写成通用形式可得

$$\left(\frac{\partial f}{\partial t}\right)_{\text{collision}} = \int_{\Re}\int_{\Omega}(f'f_1' - ff_1)d_{\mathrm{D}}^2|\boldsymbol{g}|\cos\theta\mathrm{d}\Omega\mathrm{d}\boldsymbol{\xi}_1 \tag{3.21}$$

结合式(3.6)和式(3.21)可得分布函数 f 的控制方程，即 **Boltzmann 方程**[6]：

$$\frac{\partial f}{\partial t} + \boldsymbol{\xi}\cdot\frac{\partial f}{\partial \boldsymbol{r}} + \boldsymbol{a}\cdot\frac{\partial f}{\partial \boldsymbol{\xi}} = \int_{\Re}\int_{\Omega}(f'f_1' - ff_1)d_{\mathrm{D}}^2|\boldsymbol{g}|\cos\theta\mathrm{d}\Omega\mathrm{d}\boldsymbol{\xi}_1 \tag{3.22}$$

该方程是稀薄气体动理论的基本方程，它是一个复杂的微分积分方程，其右端项称为**碰撞积分或碰撞项**(collision term)，通常亦用 $J(ff_1)$ 表示。这一积分的存在给 Boltzmann 方程的求解带来了很大的困难。因此，人们便提出了许多近似解法，如碰撞项线性化、Hilbert 扰动法、Chapman-Enskog 多尺度展开等。Boltzmann 方程在整个稀薄气体动力学中占据中心地位[7]：在自由分子流区中，通常采用无碰撞项的 Boltzmann 方程；在滑移区，可采用从解 Boltzmann 方程的 Chapman-Enskog 展开所得到的一阶近似 Navier-Stokes 方程组或二阶近似 Burnett 方程组；在过渡区，则利用 Boltzmann 方程或用与它等价的方法求解气体流动的问题[7]。

3.2 Maxwell 分布及 Boltzmann H 定理

一般情况下要获得 Boltzmann 方程的解析解是非常困难的，但对不受外力作用的单组分单原子气体，Boltzmann 方程存在一个解析解，即单组分单原子气体的平衡态分布——Maxwell 分布。下面给出 Maxwell 分布的数学推导。

3.2.1 Maxwell 分布

对不受外力作用并处于均匀状态的单组分单原子气体，Boltzmann 方程(3.22)可简化为

$$\frac{\partial f}{\partial t} = J(ff_1) = \int_{\Re}\int_{\Omega}(f'f_1' - ff_1)d_{\mathrm{D}}^2|\boldsymbol{g}|\cos\theta\mathrm{d}\Omega\mathrm{d}\boldsymbol{\xi}_1 \tag{3.23}$$

Maxwell 分布是一种平衡态分布，对平衡态来说分布函数不随时间变化，即 $\partial f/\partial t=0$。结合式(3.23)可知，等式 $\partial f/\partial t=0$ 成立需满足如下条件：

$$f'f_1' - ff_1 = 0 \tag{3.24}$$

分子碰撞时，两个分子的质量之和、动量之和及能量之和，在碰撞前后皆相等，所以有

$$\begin{gathered}1+1=1+1\\ m\boldsymbol{\xi}_1+m\boldsymbol{\xi}_2=m\boldsymbol{\xi}_1'+m\boldsymbol{\xi}_2'\\ \frac{1}{2}m\xi_1^2+\frac{1}{2}m\xi_2^2=\frac{1}{2}m\xi_1'^2+\frac{1}{2}m\xi_2'^2\end{gathered} \tag{3.25}$$

故通常称 1、$\boldsymbol{\xi}$、$\xi^2/2$ 为**碰撞不变量**(collision invariant)。由式(3.24)可知，$\ln f$ 亦有此特性：

$$\ln f+\ln f_1 = \ln f'+\ln f_1' \tag{3.26}$$

满足这一条件的量通常称为求和不变量。相关研究表明,碰撞不变量或它们的线性组合是仅有的求和不变量[7-8],因而有

$$\ln f=A\xi^2+\boldsymbol{B}\cdot\boldsymbol{\xi}+C \tag{3.27}$$

分子速度 $\boldsymbol{\xi}$ 存在如下关系:

$$\boldsymbol{\xi}=\boldsymbol{c}+\boldsymbol{u} \tag{3.28}$$

式中:$\boldsymbol{c}=\boldsymbol{\xi}-\boldsymbol{u}$ 为分子**热运动速度**,有时又称为分子**随机速度**;$\boldsymbol{u}$ 为气体的宏观流动速度,其定义为

$$\boldsymbol{u}=\frac{\int_{\mathfrak{R}}f\boldsymbol{\xi}\,\mathrm{d}\boldsymbol{\xi}}{\int_{\mathfrak{R}}f\,\mathrm{d}\boldsymbol{\xi}} \tag{3.29}$$

很明显,热运动速度的统计平均值为零。同时,由于 $\boldsymbol{c}$ 与分子热运动有关,对单原子气体有

$$\frac{3}{2}R_gT=\frac{1}{n}\int_{\mathfrak{R}}\frac{1}{2}c^2f\,\mathrm{d}\boldsymbol{\xi} \tag{3.30}$$

式中:R_g 为气体常数;T 为气体动理温度。

将式(3.28)代入式(3.27)可得

$$\ln f=Ac^2+(2A\boldsymbol{u}+\boldsymbol{B})\cdot\boldsymbol{c}+\boldsymbol{B}\cdot\boldsymbol{u}+Au^2+C \tag{3.31}$$

由于平衡态气体中不能区分出任何特定的方向,故分布函数应该是各向同性的,它不应该显式地依赖于 $\boldsymbol{c}$,所以应该有

$$\boldsymbol{B}=-2A\boldsymbol{u} \tag{3.32}$$

于是

$$\ln f=Ac^2-Au^2+C \tag{3.33}$$

即

$$f=\exp(Ac^2-Au^2+C) \tag{3.34}$$

令 $A=-\beta^2$(负号的引入是为了满足 f 应有界,故 c^2 的系数应为负的条件)。同时,为了简单令 $\exp(\beta^2u^2+C)=\alpha$,有

$$f=\alpha\exp(-\beta^2c^2) \tag{3.35}$$

式中,常数 α 和 β 可通过数密度和温度的表达式(3.1)和式(3.30)确定。

$$\begin{aligned}n&=\int_{\mathfrak{R}}f\,\mathrm{d}\boldsymbol{\xi}=\alpha\int_{\mathfrak{R}}\exp(-\beta^2c^2)\,\mathrm{d}\boldsymbol{c}\\&=\alpha\int_{-\infty}^{+\infty}\int_{-\infty}^{+\infty}\int_{-\infty}^{+\infty}\exp[-\beta^2(c_x^2+c_y^2+c_z^2)]\,\mathrm{d}c_x\mathrm{d}c_y\mathrm{d}c_z\\&=\alpha\left[\int_{-\infty}^{+\infty}\exp(-\beta^2c_x^2)\,\mathrm{d}c_x\right]^3\end{aligned} \tag{3.36}$$

$$\begin{aligned}\frac{3}{2}R_gT&=\frac{1}{n}\int_{\mathfrak{R}}\frac{1}{2}c^2f\,\mathrm{d}\boldsymbol{\xi}=\frac{\alpha}{2n}\int_{\mathfrak{R}}c^2\exp(-\beta^2c^2)\,\mathrm{d}\boldsymbol{c}\\&=\frac{\alpha}{2n}\int_{-\infty}^{+\infty}\int_{-\infty}^{+\infty}\int_{-\infty}^{+\infty}(c_x^2+c_y^2+c_z^2)\exp[-\beta^2(c_x^2+c_y^2+c_z^2)]\,\mathrm{d}c_x\mathrm{d}c_y\mathrm{d}c_z\\&=\frac{\alpha}{2n}\Big[\int_{-\infty}^{+\infty}c_x^2\exp(-\beta^2c_x^2)\,\mathrm{d}c_x\int_{-\infty}^{+\infty}\exp(-\beta^2c_y^2)\,\mathrm{d}c_y\int_{-\infty}^{+\infty}\exp(-\beta^2c_z^2)\,\mathrm{d}c_z+\end{aligned}$$

$$\int_{-\infty}^{+\infty}\exp(-\beta^2c_x^2)\,\mathrm{d}c_x\int_{-\infty}^{+\infty}c_y^2\exp(-\beta^2c_y^2)\,\mathrm{d}c_y\int_{-\infty}^{+\infty}\exp(-\beta^2c_z^2)\,\mathrm{d}c_z+$$

$$\left.\int_{-\infty}^{+\infty}\exp(-\beta^2c_x^2)\,\mathrm{d}c_x\int_{-\infty}^{+\infty}\exp(-\beta^2c_y^2)\,\mathrm{d}c_y\int_{-\infty}^{+\infty}c_z^2\exp(-\beta^2c_z^2)\,\mathrm{d}c_z\right]$$

$$=\frac{\alpha}{2n}3\int_{-\infty}^{+\infty}c_x^2\exp(-\beta^2c_x^2)\,\mathrm{d}c_x\int_{-\infty}^{+\infty}\exp(-\beta^2c_y^2)\,\mathrm{d}c_y\int_{-\infty}^{+\infty}\exp(-\beta^2c_z^2)\,\mathrm{d}c_z \tag{3.37}$$

根据数学积分，

$$\int_{-\infty}^{+\infty}\exp(-bx^2)\,\mathrm{d}x=\sqrt{\frac{\pi}{b}},\quad \int_{-\infty}^{+\infty}x^2\exp(-bx^2)\,\mathrm{d}x=\frac{1}{2}\sqrt{\frac{\pi}{b^3}} \tag{3.38}$$

可将式(3.36)和式(3.37)分别转化为

$$n=\alpha\left(\frac{\sqrt{\pi}}{\beta}\right)^3 \tag{3.39}$$

$$\frac{3}{2}R_gT=\frac{\alpha}{2n}\frac{3}{2}\frac{\pi^{3/2}}{\beta^5} \tag{3.40}$$

将式(3.39)代入式(3.40)可得

$$\beta^2=\frac{1}{2R_gT} \tag{3.41}$$

将式(3.41)代入式(3.39)可得

$$\alpha=n\left(\frac{1}{2\pi R_gT}\right)^{3/2} \tag{3.42}$$

于是可得 **Maxwell 分布**：

$$f=n\frac{1}{(2\pi R_gT)^{3/2}}\exp\left[-\frac{(\boldsymbol{\xi}-\boldsymbol{u})^2}{2R_gT}\right] \tag{3.43}$$

根据上述推导，对 D 维空间有

$$f=n\frac{1}{(2\pi R_gT)^{D/2}}\exp\left[-\frac{(\boldsymbol{\xi}-\boldsymbol{u})^2}{2R_gT}\right] \tag{3.44}$$

式(3.44)所表达的 Maxwell 分布是气体动理论中基本的物理定律之一，它规定了单组分单原子气体在平衡态时的最可几速度分布。分子束技术可以精确地测量出分子速度的 Maxwell 分布，并证明它的正确性。以上证明或推导是针对单组分单原子气体进行的，不过 Maxwell 分布对于双原子与多原子气体的平动速度也是正确的[7]。

3.2.2　*H* 定理

Boltzmann 定义的 ***H* 函数**为 $H(t)=\overline{\ln f}$，其中“—”表示统计平均，即

$$H(t)=\overline{\ln f}=\frac{\int_{\Re}f\ln f\mathrm{d}\boldsymbol{\xi}}{\int_{\Re}f\mathrm{d}\boldsymbol{\xi}}=\frac{1}{n}\int_{\Re}f\ln f\mathrm{d}\boldsymbol{\xi} \tag{3.45}$$

根据式(3.45)可得

$$\frac{\partial H}{\partial t}=\frac{1}{n}\int_{\Re}(1+\ln f)\left(\frac{\partial f}{\partial t}\right)\mathrm{d}\boldsymbol{\xi} \tag{3.46}$$

结合式(3.23)可得

$$\frac{\partial H}{\partial t}=\frac{1}{n}\int_{\Re}\int_{\Re}\int_{\Omega}(1+\ln f)(f'f_1'-ff_1)d_{\mathrm{D}}^2|\boldsymbol{g}|\cos\theta \mathrm{d}\Omega \mathrm{d}\boldsymbol{\xi}_1\mathrm{d}\boldsymbol{\xi} \tag{3.47}$$

交换 $\boldsymbol{\xi}$ 和 $\boldsymbol{\xi}_1$ 不会改变$\partial H/\partial t$ 的值,故有

$$\frac{\partial H}{\partial t}=\frac{1}{n}\int_{\Re}\int_{\Re}\int_{\Omega}(1+\ln f_1)(f_1'f'-f_1f)d_{\mathrm{D}}^2|\boldsymbol{g}|\cos\theta \mathrm{d}\Omega \mathrm{d}\boldsymbol{\xi}\mathrm{d}\boldsymbol{\xi}_1 \tag{3.48}$$

将式(3.47)与式(3.48)相加,取平均

$$\frac{\partial H}{\partial t}=\frac{1}{2n}\int_{\Re}\int_{\Re}\int_{\Omega}(2+\ln ff_1)(f'f_1'-ff_1)d_{\mathrm{D}}^2|\boldsymbol{g}|\cos\theta \mathrm{d}\Omega \mathrm{d}\boldsymbol{\xi}_1\mathrm{d}\boldsymbol{\xi} \tag{3.49}$$

更进一步,$\boldsymbol{\xi}$、$\boldsymbol{\xi}_1$ 分别与 $\boldsymbol{\xi}'$、$\boldsymbol{\xi}_1'$ 交换可得

$$\frac{\partial H}{\partial t}=-\frac{1}{2n}\int_{\Re}\int_{\Re}\int_{\Omega}(2+\ln f'f_1')(f'f_1'-ff_1)d_{\mathrm{D}}^2|\boldsymbol{g}|\cos\theta \mathrm{d}\Omega \mathrm{d}\boldsymbol{\xi}_1'\mathrm{d}\boldsymbol{\xi}' \tag{3.50}$$

将式(3.49)与式(3.50)相加,并依据式(3.12)可得

$$\frac{\partial H}{\partial t}=\frac{1}{4n}\int_{\Re}\int_{\Re}\int_{\Omega}(\ln ff_1-\ln f'f_1')(f'f_1'-ff_1)d_{\mathrm{D}}^2|\boldsymbol{g}|\cos\theta \mathrm{d}\Omega \mathrm{d}\boldsymbol{\xi}_1\mathrm{d}\boldsymbol{\xi} \tag{3.51}$$

根据数学关系式$(a-b)(\ln b-\ln a)\leqslant 0$ 有

$$\frac{\partial H}{\partial t}\leqslant 0 \tag{3.52}$$

式(3.52)即为 **Boltzmann *H* 定理**,它是由 Boltzmann 于 1872 年得到的。这一定理表明,**当时间变化而分布函数发生变化时,*H* 总是减少的,当 *H* 减少到它的极小值而不再改变时,系统就达到了平衡态,这体现了不可逆性**。按理说,一个系统的全体分子都反方向运动时,分子运动应该按照原来进程的反方向进行,这是微观运动的可逆性。然而 H 定理却表明往一个方向是不可逆的,因此可以认为 H 定理阐述的是一个统计结果,即往 H 减少方向的几率远远大于往 H 增加方向的几率,宏观过程的不可逆性是统计平均的结果。

3.3 Maxwell 输运方程及宏观量守恒方程组

前面给出了 Boltzmann 方程及 Maxwell 分布的推导过程。在这一节,将介绍 Maxwell 输运方程并通过它揭示 Boltzmann 方程与流体力学基本方程之间的联系。

将一个与分子相关联的物理量 ϕ(这一物理量可以是气体分子的质量、动量、能量等)与 Boltzmann 方程相乘并逐项对速度空间积分,即可得到 Boltzmann 方程的矩,称之为 Boltzmann 方程的**矩方程**(moment equation)。在历史上,这一方程是 Maxwell 在 Boltzmann 导出 Boltzmann 方程之前建立的,所以通常被称为 Maxwell 输运方程[6],但为了叙述方便,我们基于 Boltzmann 方程给出 Maxwell 输运方程的数学推导。

对 Boltzmann 方程(3.22)两边同乘 ϕ,并基于分子速度空间 $\Re$ 积分可得

$$\int_{\Re}\left(\phi\frac{\partial f}{\partial t}+\phi\boldsymbol{\xi}\cdot\frac{\partial f}{\partial \boldsymbol{r}}+\phi\boldsymbol{a}\cdot\frac{\partial f}{\partial \boldsymbol{\xi}}\right)\mathrm{d}\boldsymbol{\xi}=\int_{\Re}\phi J(ff_1)\mathrm{d}\boldsymbol{\xi} \tag{3.53}$$

根据平均值的定义,式(3.53)左端第一项可转化为

$$\int_{\Re}\phi\frac{\partial f}{\partial t}\mathrm{d}\boldsymbol{\xi}=\int_{\Re}\frac{\partial}{\partial t}(\phi f)\mathrm{d}\boldsymbol{\xi}-\int_{\Re}f\frac{\partial \phi}{\partial t}\mathrm{d}\boldsymbol{\xi}$$

$$= \frac{\partial}{\partial t}\int_{\Re}(\phi f)\,\mathrm{d}\boldsymbol{\xi} - \int_{\Re} f\frac{\partial \phi}{\partial t}\mathrm{d}\boldsymbol{\xi}$$

$$= \frac{\partial(n\overline{\phi})}{\partial t} - n\,\overline{\frac{\partial \phi}{\partial t}} \tag{3.54}$$

由于$\partial(\ )/\partial t$ 和积分变量 $\boldsymbol{\xi}$ 无关,因此$\partial(\ )/\partial t$ 和积分可以变换次序。式(3.53)左端第二项可写为

$$\int_{\Re}\phi\boldsymbol{\xi}\cdot\frac{\partial f}{\partial \boldsymbol{r}}\mathrm{d}\boldsymbol{\xi} = \int_{\Re}\frac{\partial}{\partial \boldsymbol{r}}\cdot(\phi\boldsymbol{\xi}f)\,\mathrm{d}\boldsymbol{\xi} - \int_{\Re} f\boldsymbol{\xi}\cdot\frac{\partial \phi}{\partial \boldsymbol{r}}\mathrm{d}\boldsymbol{\xi} - \int_{\Re}\phi f\,\nabla_r\cdot\boldsymbol{\xi}\,\mathrm{d}\boldsymbol{\xi}$$

$$= \frac{\partial}{\partial \boldsymbol{r}}\cdot(n\,\overline{\phi\boldsymbol{\xi}}) - n\overline{\boldsymbol{\xi}\cdot\frac{\partial \phi}{\partial \boldsymbol{r}}} \tag{3.55}$$

由于 $\boldsymbol{\xi}$ 与 $\boldsymbol{r}$ 无关,因此$\nabla_r\cdot\boldsymbol{\xi}=0$。对式(3.53)左端第三项有

$$\int_{\Re}\phi\boldsymbol{a}\cdot\frac{\partial f}{\partial \boldsymbol{\xi}}\mathrm{d}\boldsymbol{\xi} = \int_{\Re}\boldsymbol{a}\cdot\frac{\partial(f\phi)}{\partial \boldsymbol{\xi}}\mathrm{d}\boldsymbol{\xi} - \int_{\Re} f\boldsymbol{a}\cdot\frac{\partial \phi}{\partial \boldsymbol{\xi}}\mathrm{d}\boldsymbol{\xi}$$

$$= \boldsymbol{a}\cdot\int_{\Re}\frac{\partial(f\phi)}{\partial \boldsymbol{\xi}}\mathrm{d}\boldsymbol{\xi} - \int_{\Re} f\boldsymbol{a}\cdot\frac{\partial \phi}{\partial \boldsymbol{\xi}}\mathrm{d}\boldsymbol{\xi} \tag{3.56}$$

当 $|\boldsymbol{\xi}|\to\infty$时,$f\to 0$,因而有 $\int_{\Re}\frac{\partial(f\phi)}{\partial \boldsymbol{\xi}}\mathrm{d}\boldsymbol{\xi} = 0$, 故

$$\int_{\Re}\phi\boldsymbol{a}\cdot\frac{\partial f}{\partial \boldsymbol{\xi}}\mathrm{d}\boldsymbol{\xi} = -\int_{\Re} f\boldsymbol{a}\cdot\frac{\partial \phi}{\partial \boldsymbol{\xi}}\mathrm{d}\boldsymbol{\xi} = -n\boldsymbol{a}\cdot\overline{\frac{\partial \phi}{\partial \boldsymbol{\xi}}} \tag{3.57}$$

结合式(3.54)~式(3.57)可得

$$\frac{\partial(n\overline{\phi})}{\partial t} + \frac{\partial}{\partial \boldsymbol{r}}\cdot(n\overline{\phi\boldsymbol{\xi}}) - n\left(\overline{\frac{\partial \phi}{\partial t}} + \overline{\boldsymbol{\xi}\cdot\frac{\partial \phi}{\partial \boldsymbol{r}}} + \boldsymbol{a}\cdot\overline{\frac{\partial \phi}{\partial \boldsymbol{\xi}}}\right) = \int_{\Re}\phi J(ff_1)\,\mathrm{d}\boldsymbol{\xi} \tag{3.58}$$

式(3.58)即为 **Maxwell 输运方程**。该方程是宏观统计平均量 $\overline{\phi}$ 满足的方程,如果微观量 ϕ 只和分子速度 $\boldsymbol{\xi}$ 有关而与位置矢量 $\boldsymbol{r}$ 和时间 t 无关,则式(3.58)变为

$$\frac{\partial(n\overline{\phi})}{\partial t} + \frac{\partial}{\partial \boldsymbol{r}}\cdot(n\overline{\phi\boldsymbol{\xi}}) - n\boldsymbol{a}\cdot\overline{\frac{\partial \phi}{\partial \boldsymbol{\xi}}} = \int_{\Re}\phi J(ff_1)\,\mathrm{d}\boldsymbol{\xi} \tag{3.59}$$

对式(3.59)的右端项,根据式(3.47),用 ϕ 替换$(1+\ln f)$,类似地,可得

$$\int_{\Re}\phi J(ff_1)\,\mathrm{d}\boldsymbol{\xi} = \frac{1}{4}\int_{\Re}\int_{\Re}\int_{\Omega}(\phi + \phi_1 - \phi' - \phi_1')(f'f_1' - ff_1)\,d_{\mathrm{D}}^2\,|\boldsymbol{g}|\cos\theta\,\mathrm{d}\Omega\,\mathrm{d}\boldsymbol{\xi}_1\mathrm{d}\boldsymbol{\xi} \tag{3.60}$$

式(3.60)将 $\int_{\Re}\phi J(ff_1)\,\mathrm{d}\boldsymbol{\xi}$ 和碰撞前后物理量总和之差$(\phi+\phi_1-\phi'-\phi_1')$联系了起来。因此,Maxwell 输运方程(3.58)对碰撞前后守恒的物理量具有重要的意义。

如果某一物理量在碰撞前后是守恒的,可写成

$$\phi(\boldsymbol{\xi})+\phi(\boldsymbol{\xi}_1)=\phi(\boldsymbol{\xi}')+\phi(\boldsymbol{\xi}_1') \tag{3.61}$$

对这类物理量,根据式(3.60)有

$$\int_{\Re}\phi J(ff_1)\,\mathrm{d}\boldsymbol{\xi} = 0 \tag{3.62}$$

关于碰撞前后守恒的量,很自然地便会想到质量、动量及能量这些碰撞不变量。于是,它们的输运方程可写为

$$\frac{\partial(n\overline{\phi})}{\partial t}+\frac{\partial}{\partial \boldsymbol{r}}\cdot(n\overline{\phi\boldsymbol{\xi}})-n\boldsymbol{a}\cdot\overline{\frac{\partial\phi}{\partial\boldsymbol{\xi}}}=0 \tag{3.63}$$

或

$$\frac{\partial}{\partial t}\int_{\Re}\phi f \mathrm{d}\boldsymbol{\xi}+\frac{\partial}{\partial \boldsymbol{r}}\cdot\int_{\Re}(\phi\boldsymbol{\xi}f)\mathrm{d}\boldsymbol{\xi}-\boldsymbol{a}\cdot\int_{\Re}f\frac{\partial\phi}{\partial\boldsymbol{\xi}}\mathrm{d}\boldsymbol{\xi}=0 \tag{3.64}$$

以分子质量 $\phi=m$ 代入式(3.64),可得

$$\frac{\partial\rho}{\partial t}+\nabla\cdot(\rho\boldsymbol{u})=0 \tag{3.65}$$

式中:$\rho=nm$ 为气体密度;$\boldsymbol{u}$ 为气体的宏观平均速度,

$$\boldsymbol{u}=\frac{1}{n}\int_{\Re}f\boldsymbol{\xi}\mathrm{d}\boldsymbol{\xi} \tag{3.66}$$

式(3.65)即为流体力学中的连续方程。

以分子动量 $\phi=m\boldsymbol{\xi}$ 代入式(3.64),可得

$$\frac{\partial}{\partial t}(\rho\boldsymbol{u})+\frac{\partial}{\partial \boldsymbol{r}}\cdot\int_{\Re}m\boldsymbol{\xi}\boldsymbol{\xi}f\mathrm{d}\boldsymbol{\xi}-\rho\boldsymbol{a}=0 \tag{3.67}$$

根据式(3.28)有

$$\begin{aligned}\int_{\Re}m\boldsymbol{\xi}\boldsymbol{\xi}f\mathrm{d}\boldsymbol{\xi}&=\int_{\Re}mf(\boldsymbol{c}+\boldsymbol{u})(\boldsymbol{c}+\boldsymbol{u})\mathrm{d}\boldsymbol{\xi}\\&=m\int_{\Re}f\boldsymbol{c}\boldsymbol{c}\mathrm{d}\boldsymbol{\xi}+m\int_{\Re}f\boldsymbol{u}\boldsymbol{u}\mathrm{d}\boldsymbol{\xi}+2m\int_{\Re}f\boldsymbol{c}\boldsymbol{u}\mathrm{d}\boldsymbol{\xi}\\&=-\boldsymbol{\sigma}+\rho\boldsymbol{u}\boldsymbol{u}\end{aligned} \tag{3.68}$$

式中,$\boldsymbol{\sigma}=-m\int_{\Re}f\boldsymbol{c}\boldsymbol{c}\mathrm{d}\boldsymbol{\xi}$ 为应力张量。在推导式(3.68)时,用到了平均热运动速度为零的性质,即

$$\int_{\Re}f\boldsymbol{c}\boldsymbol{u}\mathrm{d}\boldsymbol{\xi}=\boldsymbol{u}\int_{\Re}f\boldsymbol{c}\mathrm{d}\boldsymbol{\xi}=0 \tag{3.69}$$

将式(3.68)代入式(3.67)可得

$$\frac{\partial}{\partial t}(\rho\boldsymbol{u})+\nabla\cdot(\rho\boldsymbol{u}\boldsymbol{u})=\nabla\cdot\boldsymbol{\sigma}+\rho\boldsymbol{a} \tag{3.70}$$

式(3.70)即为流体力学中的动量方程。

以分子能量 $\phi=\frac{1}{2}m\xi^2$ 代入式(3.64),可得

$$\frac{\partial}{\partial t}\int_{\Re}\frac{1}{2}m\xi^2f\mathrm{d}\boldsymbol{\xi}+\frac{\partial}{\partial \boldsymbol{r}}\cdot\int_{\Re}\frac{1}{2}m\xi^2\boldsymbol{\xi}f\mathrm{d}\boldsymbol{\xi}-\boldsymbol{a}\cdot\int_{\Re}fm\boldsymbol{\xi}\mathrm{d}\boldsymbol{\xi}=0 \tag{3.71}$$

式(3.71)左端第一项中的积分为

$$\begin{aligned}\int_{\Re}\frac{1}{2}m\xi^2f\mathrm{d}\boldsymbol{\xi}&=\int_{\Re}\frac{1}{2}m(\boldsymbol{c}+\boldsymbol{u})\cdot(\boldsymbol{c}+\boldsymbol{u})f\mathrm{d}\boldsymbol{\xi}\\&=\int_{\Re}\frac{1}{2}mc^2f\mathrm{d}\boldsymbol{\xi}+\int_{\Re}\frac{1}{2}mu^2f\mathrm{d}\boldsymbol{\xi}\\&=\rho\varepsilon+\frac{1}{2}\rho u^2\end{aligned} \tag{3.72}$$

式中,热力学能 ε 为

$$\varepsilon = \frac{1}{2n}\int_{\Re} c^2 f \mathrm{d}\boldsymbol{\xi} \tag{3.73}$$

根据式(3.30),对单原子气体有 $\varepsilon = 3R_g T/2$。式(3.71)左端第二项中的积分为

$$\begin{aligned}\int_{\Re}\frac{1}{2}m\xi^2\boldsymbol{\xi}f\mathrm{d}\boldsymbol{\xi} &= \int_{\Re}\frac{1}{2}m(c^2+2\boldsymbol{c}\cdot\boldsymbol{u}+u^2)(\boldsymbol{c}+\boldsymbol{u})f\mathrm{d}\boldsymbol{\xi}\\ &= \int_{\Re}\frac{1}{2}mc^2\boldsymbol{c}f\mathrm{d}\boldsymbol{\xi} + \int_{\Re}m(\boldsymbol{c}\cdot\boldsymbol{u})\boldsymbol{c}f\mathrm{d}\boldsymbol{\xi} + \int_{\Re}\frac{1}{2}mu^2\boldsymbol{c}f\mathrm{d}\boldsymbol{\xi} +\\ &\quad \int_{\Re}\frac{1}{2}mc^2\boldsymbol{u}f\mathrm{d}\boldsymbol{\xi} + \int_{\Re}m(\boldsymbol{c}\cdot\boldsymbol{u})\boldsymbol{u}f\mathrm{d}\boldsymbol{\xi} + \int_{\Re}\frac{1}{2}mu^2\boldsymbol{u}f\mathrm{d}\boldsymbol{\xi}\end{aligned} \tag{3.74}$$

根据平均热运动速度为零的性质有

$$\int_{\Re}\frac{1}{2}mu^2\boldsymbol{c}f\mathrm{d}\boldsymbol{\xi} = 0,\quad \int_{\Re}m(\boldsymbol{c}\cdot\boldsymbol{u})\boldsymbol{u}f\mathrm{d}\boldsymbol{\xi} = 0 \tag{3.75}$$

于是式(3.74)可转化为

$$\int_{\Re}\frac{1}{2}m\xi^2\boldsymbol{\xi}f\mathrm{d}\boldsymbol{\xi} = \boldsymbol{q} + \rho\left(\varepsilon + \frac{1}{2}u^2\right)\boldsymbol{u} - \boldsymbol{\sigma}\cdot\boldsymbol{u} \tag{3.76}$$

式中,$\boldsymbol{q} = \int_{\Re}\frac{1}{2}mc^2\boldsymbol{c}f\mathrm{d}\boldsymbol{\xi}$ 为热流矢量。将式(3.72)和式(3.76)代入式(3.71)可得

$$\frac{\partial}{\partial t}\left(\rho\varepsilon+\frac{1}{2}\rho u^2\right)+\nabla\cdot\left[\boldsymbol{q}+\rho\left(\varepsilon+\frac{1}{2}u^2\right)\boldsymbol{u}-\boldsymbol{\sigma}\cdot\boldsymbol{u}\right]-\rho\boldsymbol{a}\cdot\boldsymbol{u}=0 \tag{3.77}$$

即

$$\frac{\partial}{\partial t}\left(\rho\varepsilon+\frac{1}{2}\rho u^2\right)+\nabla\cdot\left[\left(\rho\varepsilon+\frac{1}{2}\rho u^2\right)\boldsymbol{u}\right]=-\nabla\cdot\boldsymbol{q}+\nabla\cdot(\boldsymbol{\sigma}\cdot\boldsymbol{u})+\rho\boldsymbol{a}\cdot\boldsymbol{u} \tag{3.78}$$

式(3.78)即为流体力学中的总能方程。

从上述推导可知,流体力学中宏观量的守恒方程组可依据气体动理论直接从微观的守恒量导出。不过在流体力学中,为了使方程组封闭,要附加一些定律。例如,广义牛顿内摩擦定律或本构方程给出了应力张量 $\boldsymbol{\sigma}$ 和应变率张量的关系,而傅里叶定律则给出热流矢量 $\boldsymbol{q}$ 和温度梯度的关系。这些定律中出现的一些称为输运系数的正比系数,如动力黏度 μ、导热系数 λ 等,通常由实验确定。从气体动理论来看,如果能直接求解 Boltzmann 方程,则可以从分子间作用力模型出发求得相应的输运系数。关于这一点,读者可参考文献[6,9]。

3.4 从 Boltzmann 方程到格子 Boltzmann 方程

由于 Boltzmann 方程与流体力学基本方程之间有着密切的联系,因而可以通过数值求解 Boltzmann 方程来模拟流体的宏观运动。求解 Boltzmann 方程最主要的难点在于碰撞项,它是分布函数的非线性项,此外它还与具体的分子间作用力有关。因此,一种自然的想法是能否用比较简单的碰撞算子代替复杂的碰撞项,BGK 近似就是在这一背景下产生的。

3.4.1 BGK 近似

BGK 近似是由 Bhatnagar、Gross 和 Krook 于 1954 年提出的[5],他们认为以一个简单的

碰撞算子 Ω_f 代替 Boltzmann 方程的碰撞项 $J(ff_1)$，它应该具有以下两个特性[6]：

（1）对碰撞不变量 $\phi=(m,m\boldsymbol{\xi},m\xi^2/2)$，$\Omega_f$ 应满足

$$\int_{\Re}\phi\Omega_f\mathrm{d}\boldsymbol{\xi}=0 \tag{3.79}$$

（2）根据 Boltzmann H 定理有

$$\int_{\Re}(1+\ln f)J(ff_1)\mathrm{d}\boldsymbol{\xi}\leqslant 0 \tag{3.80}$$

式(3.80)反映的是气体趋于平衡态的性质，式中的等号当且仅当 $f=f^{\mathrm{eq}}$ 时成立。根据式(3.80)，碰撞算子 Ω_f 应当满足的第二个特性为

$$\int_{\Re}(1+\ln f)\Omega_f\mathrm{d}\boldsymbol{\xi}\leqslant 0 \tag{3.81}$$

最简单的碰撞算子 Ω_f 可以这样得到，即假定碰撞效应是改变 f 使其趋于 Maxwell 平衡态分布 f^{eq}，并设改变率的大小与 f^{eq} 和 f 的差值成正比，于是可引入

$$\Omega_f=\upsilon[f^{\mathrm{eq}}(\boldsymbol{r},\boldsymbol{\xi},t)-f(\boldsymbol{r},\boldsymbol{\xi},t)] \tag{3.82}$$

υ 是一个与分子速度 $\boldsymbol{\xi}$ 无关的比例系数。从而 Boltzmann 方程(3.22)可转化为

$$\frac{\partial f}{\partial t}+\boldsymbol{\xi}\cdot\frac{\partial f}{\partial \boldsymbol{r}}+\boldsymbol{a}\cdot\frac{\partial f}{\partial \boldsymbol{\xi}}=\upsilon(f^{\mathrm{eq}}-f) \tag{3.83}$$

式(3.83)即为 **Boltzmann-BGK 方程**。

根据第一个特性，对 $\phi=(m,m\boldsymbol{\xi},m\xi^2/2)$ 有

$$\int_{\Re}\phi f^{\mathrm{eq}}(\boldsymbol{r},\boldsymbol{\xi},t)\mathrm{d}\boldsymbol{\xi}=\int_{\Re}\phi f(\boldsymbol{r},\boldsymbol{\xi},t)\mathrm{d}\boldsymbol{\xi} \tag{3.84}$$

于是，f^{eq} 中的 ρ、$\boldsymbol{u}$ 和 T 就是由速度分布函数 f 求得的宏观密度、速度和温度。不过这里的平衡态分布函数随时间和空间变化，因此通常被称为**局部平衡态分布函数**。

由式(3.83)可知比例系数 υ 的量纲为时间的倒数。对比式(3.22)和式(3.83)，可认为

$$\upsilon f=\int_{\Re}\int_{\Omega}ff_1d_{\mathrm{D}}^2|\boldsymbol{g}|\cos\theta\mathrm{d}\Omega\mathrm{d}\boldsymbol{\xi}_1 \tag{3.85}$$

由于 $f(\boldsymbol{\xi})$ 不依赖于 $\boldsymbol{\xi}_1$，因此式(3.85)右端的 f 可以移到积分符号外面，故有

$$\upsilon=\int_{\Re}\int_{\Omega}f_1d_{\mathrm{D}}^2|\boldsymbol{g}|\cos\theta\mathrm{d}\Omega\mathrm{d}\boldsymbol{\xi}_1 \tag{3.86}$$

结合式(3.13)和式(3.86)可知，υ 是单位时间内空间位置 $\boldsymbol{r}$ 处一个速度为 $\boldsymbol{\xi}$ 的分子与别的分子的碰撞总数，亦称**碰撞频率**。

下面考察 $\upsilon(f^{\mathrm{eq}}-f)$ 是否满足 Ω_f 的第二个特性。对 $\int_{\Re}(1+\ln f)\Omega_f\mathrm{d}\boldsymbol{\xi}$ 有

$$\int_{\Re}(1+\ln f)\Omega_f\mathrm{d}\boldsymbol{\xi}=\int_{\Re}\Omega_f\mathrm{d}\boldsymbol{\xi}+\int_{\Re}\left(\ln\frac{f}{f^{\mathrm{eq}}}\right)\Omega_f\mathrm{d}\boldsymbol{\xi}+\int_{\Re}(\ln f^{\mathrm{eq}})\Omega_f\mathrm{d}\boldsymbol{\xi} \tag{3.87}$$

由于 $\ln f^{\mathrm{eq}}$ 是由碰撞不变量组成的，因此由式(3.84)可得

$$\int_{\Re}(\ln f^{\mathrm{eq}})\Omega_f\mathrm{d}\boldsymbol{\xi}=0 \tag{3.88}$$

取 $\phi=m$ 时，根据式(3.79)有 $\int_{\Re}m\Omega_f\mathrm{d}\boldsymbol{\xi}=0$，故

$$\int_{\Re}\Omega_f\mathrm{d}\boldsymbol{\xi}=0 \tag{3.89}$$

于是,式(3.87)可写成

$$\int_{\Re}(1+\ln f)\,\Omega_f \mathrm{d}\boldsymbol{\xi}=\int_{\Re}\left(\ln \frac{f}{f^{\mathrm{eq}}}\right)\Omega_f \mathrm{d}\boldsymbol{\xi}=\upsilon\int_{\Re}\left(\ln \frac{f}{f^{\mathrm{eq}}}\right)(f^{\mathrm{eq}}-f)\,\mathrm{d}\boldsymbol{\xi} \tag{3.90}$$

根据关系式$(a-b)(\ln b-\ln a)\leqslant 0$可知式(3.90)右端小于等于零,于是有

$$\int_{\Re}(1+\ln f)\left[\upsilon(f^{\mathrm{eq}}-f)\right]\mathrm{d}\boldsymbol{\xi}\leqslant 0 \tag{3.91}$$

即第二个特性也是满足的。式(3.91)中的等号当且仅当$f=f^{\mathrm{eq}}$时成立。

BGK 近似使得 Boltzmann 方程线性化,大大简化了方程的求解。根据υ的量纲,可引入碰撞时间τ_0:

$$\tau_0=\frac{1}{\upsilon} \tag{3.92}$$

τ_0是两次碰撞的平均时间间隔,也称**松弛时间**(relaxation time)或**弛豫时间**。在物理学中,从非平衡态分布向平衡态分布趋近的过程称为"弛豫"。根据式(3.83)和式(3.92),Boltzmann-BGK 方程可写成

$$\frac{\partial f}{\partial t}+\boldsymbol{\xi}\cdot\nabla f+\boldsymbol{a}\cdot\nabla_{\boldsymbol{\xi}} f=-\frac{1}{\tau_0}(f-f^{\mathrm{eq}}) \tag{3.93}$$

对分布函数f乘以粒子的质量,则式(3.93)中的分布函数表示粒子的密度分布而非数量分布。如不特别指出,下文的f均表示粒子的密度分布,相应的 Maxwell 平衡态分布f^{eq}变为

$$f^{\mathrm{eq}}=\frac{\rho}{(2\pi R_{\mathrm{g}}T)^{D/2}}\exp\left[-\frac{(\boldsymbol{\xi}-\boldsymbol{u})^2}{2R_{\mathrm{g}}T}\right] \tag{3.94}$$

为了了解松弛时间τ_0的物理意义,考虑不存在外力且空间均匀但随时间变化的情况,此时式(3.93)可简化为

$$\frac{\partial f}{\partial t}=-\frac{1}{\tau_0}(f-f^{\mathrm{eq}}) \tag{3.95}$$

求解式(3.95)可得

$$f-f^{\mathrm{eq}}=(f-f^{\mathrm{eq}})\big|_{t=0}\exp\left(-\frac{t}{\tau_0}\right) \tag{3.96}$$

由式(3.96)可知,τ_0是$(f-f^{\mathrm{eq}})$变为初始值的 exp(-1)倍所需要的时间[6]。

BGK 近似地把气体从当前状态向平衡态的过渡看作是一个简单的弛豫过程,因其简单而被广泛采用,但它终究是用近似的方式替代了以坚实物理为基础的碰撞项,因而必然存在缺陷[7]。不过对于许多偏离平衡态不远的问题来说,BGK 近似的精度是可以满足要求的。

3.4.2 格子 Boltzmann 方程

正如 He 和 Luo 所指出的,格子 Boltzmann 方程是连续 Boltzmann-BGK 方程的一种特殊的离散形式[2-3]。这一离散包括速度离散、时间离散和空间离散。首先,要在速度空间对 Boltzmann-BGK 方程进行离散。我们知道,微观粒子无时无刻不在做着不规则的热运动,因此微观粒子的速度是连续的,其速度在相空间是无穷维的。然而,微观粒子的运动细节并不显著地决定流体的宏观运动。因此,可以在相空间将粒子的速度$\boldsymbol{\xi}$简化为有限

维的离散速度,即$\{\boldsymbol{e}_0,\boldsymbol{e}_1,\cdots,\boldsymbol{e}_{m-1}\}$,其中 m 表示离散速度的个数。同时,分布函数f也相应地被离散为$\{f_0,f_1,\cdots,f_{m-1}\}$,它们可以表示为$f_\alpha=f_\alpha(\boldsymbol{r},\boldsymbol{e}_\alpha,t)$,其中 $\alpha=0,1,\cdots,m-1$。在此基础上可得速度离散的 Boltzmann-BGK 方程

$$\frac{\partial f_\alpha}{\partial t}+\boldsymbol{e}_\alpha\cdot\nabla f_\alpha=-\frac{1}{\tau_0}(f_\alpha-f_\alpha^{\mathrm{eq}})+F_\alpha \tag{3.97}$$

式中:f_α^{eq}为离散速度空间的平衡态分布函数;F_α 为离散速度空间的作用力项,它是 $\boldsymbol{a}\cdot\nabla_{\boldsymbol{\xi}}f$ 在离散速度空间的投影,即 $F_\alpha=\boldsymbol{a}\cdot\nabla_{\boldsymbol{\xi}}f_\alpha$。当流速较低时,平衡态分布函数$f_\alpha^{\mathrm{eq}}$可表示为

$$\begin{aligned}
f_\alpha^{\mathrm{eq}}&=\rho\,(2\pi R_{\mathrm{g}}T)^{-D/2}\exp\left[-\frac{(\boldsymbol{e}_\alpha-\boldsymbol{u})^2}{2R_{\mathrm{g}}T}\right]\\
&=\rho\,(2\pi R_{\mathrm{g}}T)^{-D/2}\exp\left(-\frac{e_\alpha^2}{2R_{\mathrm{g}}T}\right)\exp\left(\frac{\boldsymbol{e}_\alpha\cdot\boldsymbol{u}}{R_{\mathrm{g}}T}-\frac{u^2}{2R_{\mathrm{g}}T}\right)\\
&=\rho\,(2\pi R_{\mathrm{g}}T)^{-D/2}\exp\left(-\frac{e_\alpha^2}{2R_{\mathrm{g}}T}\right)\left[1+\frac{\boldsymbol{e}_\alpha\cdot\boldsymbol{u}}{R_{\mathrm{g}}T}+\frac{(\boldsymbol{e}_\alpha\cdot\boldsymbol{u})^2}{2R_{\mathrm{g}}^2T^2}-\frac{u^2}{2R_{\mathrm{g}}T}\right]\\
&=\rho\omega_\alpha\left[1+\frac{\boldsymbol{e}_\alpha\cdot\boldsymbol{u}}{R_{\mathrm{g}}T}+\frac{(\boldsymbol{e}_\alpha\cdot\boldsymbol{u})^2}{2R_{\mathrm{g}}^2T^2}-\frac{u^2}{2R_{\mathrm{g}}T}\right]+O(u^3)
\end{aligned} \tag{3.98}$$

式中,$\omega_\alpha=(2\pi R_{\mathrm{g}}T)^{-D/2}\exp(-e_\alpha^2/2R_{\mathrm{g}}T)$。

要数值求解式(3.97),还须对其在时间和空间上进行离散,从而得到完全离散化的 Boltzmann-BGK 方程。对式(3.97)沿特征线积分有

$$f_\alpha(\boldsymbol{r}+\boldsymbol{e}_\alpha\delta_t,t+\delta_t)-f_\alpha(\boldsymbol{r},t)=\int_0^{\delta_t}[\Omega_{f_\alpha}(\boldsymbol{r}+\boldsymbol{e}_\alpha t',t+t')+F_\alpha(\boldsymbol{r}+\boldsymbol{e}_\alpha t',t+t')]\,\mathrm{d}t' \tag{3.99}$$

式中:δ_t 为时间步长;$\Omega_{f_\alpha}=(f_\alpha^{\mathrm{eq}}-f_\alpha)/\tau_0$。对式(3.99)右端积分采用矩形法进行逼近可得

$$f_\alpha(\boldsymbol{r}+\boldsymbol{e}_\alpha\delta_t,t+\delta_t)-f_\alpha(\boldsymbol{r},t)=-\frac{1}{\tau}[f_\alpha(\boldsymbol{r},t)-f_\alpha^{\mathrm{eq}}(\boldsymbol{r},t)]+\delta_tF_\alpha(\boldsymbol{r},t) \tag{3.100}$$

式(3.100)即为**格子 Boltzmann-BGK 方程**,亦称**单松弛格子 Boltzmann 方程**,式中 $\tau=\tau_0/\delta_t$ 为量纲为 1 的松弛时间。在标准格子 Boltzmann 方法中,时间和空间的离散并不是独立的,它们通过粒子的离散速度联系在一起,即 $\delta_x=e_{\alpha x}\delta_t,\delta_y=e_{\alpha y}\delta_t,\delta_z=e_{\alpha z}\delta_t$。根据这一特性,可在物理空间将粒子运动分为碰撞(collision)和迁移(streaming)两个过程,即

名词术语总结

$$碰撞:\quad f_\alpha^*(\boldsymbol{r},t)=f_\alpha(\boldsymbol{r},t)-\frac{1}{\tau}[f_\alpha(\boldsymbol{r},t)-f_\alpha^{\mathrm{eq}}(\boldsymbol{r},t)]+\delta_tF_\alpha(\boldsymbol{r},t) \tag{3.101}$$

$$迁移:\quad f_\alpha(\boldsymbol{r}+\boldsymbol{e}_\alpha\delta_t,t+\delta_t)=f_\alpha^*(\boldsymbol{r},t) \tag{3.102}$$

即粒子在某一节点与其他粒子碰撞之后,经过一个时间步长,刚好由该节点运动到其相邻节点。从式(3.101)和式(3.102)可以看到,碰撞过程是完全局部的,而迁移过程则是线性的。"碰撞-迁移"特性是格子 Boltzmann 方法非常重要的特性,这一特性继承自格子气自动机,它使得格子 Boltzmann 方法具有很好的并行特性和较强的复杂边界处理能力。由于格子 Boltzmann 方法具有清晰的粒子运动图像,使得它能够非常直观地处理流体与固体的相互作用,从而对多孔介质流动、气固两相流等复杂流动的模拟具有一定的优势[10]。

碰撞-迁移过程演示

3.5 格子 Boltzmann 方法的基本模型

一个完整的格子 Boltzmann 模型通常由三部分组成[11]:**格子**,即**离散速度模型**(discrete velocity model,DVM),**平衡态分布函数**,以及分布函数的**演化方程**(evolution equation)。构造格子 Boltzmann 模型的关键是选择合适的平衡态分布函数,而平衡态分布函数的具体形式又与离散速度模型有关。离散速度的个数和对称性决定了格子 Boltzmann 模型在宏观层面能否与相应的宏观方程一致。如果离散速度的个数太少可能导致某些应当守恒的物理量不满足守恒律,个数太多则会造成计算量的增加。格子 Boltzmann 模型的建立可理解为如下一个过程:首先,平衡态分布函数应当满足一定的约束条件,导出这些约束条件是建模的第一步;接下来要选择合适的离散速度模型,并依据离散速度模型和约束条件确定平衡态分布函数的具体形式。

1992 年,Qian 等[12]提出的 **D*d*Q*m***(d 维空间,m 个离散速度)系列模型是格子 Boltzmann 方法的基本模型。下面以 D2Q9 模型(其离散速度如图 3.2 所示)为例,介绍如何确定模型的平衡态分布函数以及如何通过 Chapman-Enskog 展开导出模型所对应的宏观方程。

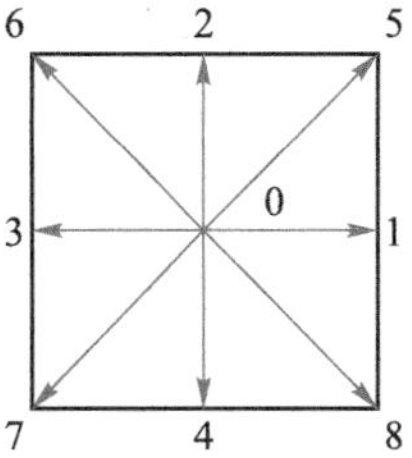

图 3.2 D2Q9 模型

3.5.1 D2Q9 模型平衡态分布函数的确定

D2Q9 模型的离散速度配置如下:

$$\boldsymbol{e}_\alpha=(e_{\alpha x},e_{\alpha y})^{\mathrm{T}}=c\begin{bmatrix}0 & 1 & 0 & -1 & 0 & 1 & -1 & -1 & 1\\ 0 & 0 & 1 & 0 & -1 & 1 & 1 & -1 & -1\end{bmatrix} \tag{3.103}$$

式中,$c=\delta_x/\delta_t$,δ_x 和 δ_t 分别为网格步长和时间步长,且通常 x 方向和 y 方向的网格步长相同,即 $\delta_x=\delta_y$。在格子 Boltzmann 方法中,c 通常取为 1。

He 和 Luo[2-3]把平衡态分布函数的各阶矩看作 Gauss 型积分,在连续时间和空间意义下,通过积分近似可以确定平衡态分布函数的权系数,从而在理论上给出了严格的证明和推导。但在实际应用中,采用待定系数法更为灵活有效。Qian 等[12]的 DdQm 系列模型采用如下形式的平衡态分布函数:

$$f_\alpha^{\mathrm{eq}}=\rho\omega_\alpha\left[1+\frac{\boldsymbol{e}_\alpha\cdot\boldsymbol{u}}{c_s^2}+\frac{(\boldsymbol{e}_\alpha\cdot\boldsymbol{u})^2}{2c_s^4}-\frac{u^2}{2c_s^2}\right] \tag{3.104}$$

式中:$u^2=|\boldsymbol{u}|^2$;c_s 为格子声速。一般来说,$|\boldsymbol{e}_\alpha|$相等的方向,权系数 ω_α 相同。例如,对 D2Q9 模型,$\omega_1=\omega_2=\omega_3=\omega_4$,$\omega_5=\omega_6=\omega_7=\omega_8$。

由附录 B 可知,对 D2Q9 模型,离散速度 $\boldsymbol{e}_\alpha$ 满足如下关系式:

$$\sum_{\alpha=1}^{4}e_{\alpha i}=0,\quad \sum_{\alpha=5}^{8}e_{\alpha i}=0 \tag{3.105}$$

$$\sum_{\alpha=1}^{4}e_{\alpha i}e_{\alpha j}=2c^2\delta_{ij},\quad \sum_{\alpha=5}^{8}e_{\alpha i}e_{\alpha j}=4c^2\delta_{ij} \tag{3.106}$$

$$\sum_{\alpha=1}^{4}e_{\alpha i}e_{\alpha j}e_{\alpha k}=0,\quad \sum_{\alpha=5}^{8}e_{\alpha i}e_{\alpha j}e_{\alpha k}=0 \tag{3.107}$$

$$\sum_{\alpha=1}^{4}e_{\alpha i}e_{\alpha j}e_{\alpha k}e_{\alpha l}=2c^4\delta_{ijkl},\quad \sum_{\alpha=5}^{8}e_{\alpha i}e_{\alpha j}e_{\alpha k}e_{\alpha l}=4c^4\Delta_{ijkl}-8c^4\delta_{ijkl} \tag{3.108}$$

式中:δ_{ij}为克罗内克符号;$\Delta_{ijkl}=(\delta_{ij}\delta_{kl}+\delta_{ik}\delta_{jl}+\delta_{il}\delta_{jk})$。平衡态分布函数需满足下列约束条件:

$$\sum_{\alpha=0}^{8}f_\alpha^{\mathrm{eq}}=\rho \tag{3.109}$$

$$\sum_{\alpha=0}^{8} f_\alpha^{eq}\boldsymbol{e}_\alpha=\rho\boldsymbol{u} \tag{3.110}$$

$$\sum_{\alpha=0}^{8} f_\alpha^{eq}e_{\alpha i}e_{\alpha j}=\rho u_i u_j+p\delta_{ij} \tag{3.111}$$

将式(3.104)代入式(3.109)~式(3.111),分别可得

$$\rho(\omega_0+4\omega_1+4\omega_5)+\rho u^2\left[\omega_1\left(\frac{c^2}{c_s^4}-\frac{2}{c_s^2}\right)+\omega_5\left(\frac{2c^2}{c_s^4}-\frac{2}{c_s^2}\right)-\omega_0\frac{1}{2c_s^2}\right]=\rho \tag{3.112}$$

$$\rho\boldsymbol{u}\left(\omega_1\frac{2c^2}{c_s^2}+\omega_5\frac{4c^2}{c_s^2}\right)=\rho\boldsymbol{u} \tag{3.113}$$

$$\rho\omega_1\left[\sum_{\alpha=1}^{4} e_{\alpha i}e_{\alpha j}\left(1-\frac{u^2}{2c_s^2}\right)+\sum_{\alpha=1}^{4}\frac{e_{\alpha i}e_{\alpha j}e_{\alpha k}e_{\alpha l}u_k u_l}{2c_s^4}\right]+$$
$$\rho\omega_5\left[\sum_{\alpha=5}^{8} e_{\alpha i}e_{\alpha j}\left(1-\frac{u^2}{2c_s^2}\right)+\sum_{\alpha=5}^{8}\frac{e_{\alpha i}e_{\alpha j}e_{\alpha k}e_{\alpha l}u_k u_l}{2c_s^4}\right]=\rho u_i u_j+p\delta_{ij} \tag{3.114}$$

整理可得

$$\omega_0+4\omega_1+4\omega_5=1 \tag{3.115}$$

$$\omega_1\left(\frac{c^2}{c_s^4}-\frac{2}{c_s^2}\right)+\omega_5\left(\frac{2c^2}{c_s^4}-\frac{2}{c_s^2}\right)-\omega_0\frac{1}{2c_s^2}=0 \tag{3.116}$$

$$(\omega_1+2\omega_5)\frac{2c^2}{c_s^2}=1 \tag{3.117}$$

$$\omega_5\frac{4c^4}{c_s^4}=1 \tag{3.118}$$

$$\omega_1\frac{c^4}{c_s^4}-\omega_5\frac{4c^4}{c_s^4}=0 \tag{3.119}$$

$$2c^2\omega_1\left(1-\frac{u^2}{2c_s^2}\right)+4c^2\omega_5\left(1-\frac{u^2}{2c_s^2}\right)+\omega_5\frac{2c^4u^2}{c_s^4}=\frac{p}{\rho} \tag{3.120}$$

推导中用到

$$\Delta_{ijkl}u_k u_l=(\delta_{ij}\delta_{kl}+\delta_{ik}\delta_{jl}+\delta_{il}\delta_{jk})u_k u_l=u^2\delta_{ij}+2u_i u_j \tag{3.121}$$

联立式(3.117)~式(3.119),并利用式(3.115)可得

$$\omega_0=4/9,\quad \omega_1=1/9,\quad \omega_5=1/36,\quad c_s^2=c^2/3 \tag{3.122}$$

将式(3.122)代入式(3.116)和式(3.120),可知式(3.116)成立,并有 $p=\rho c_s^2$。

D2Q9 模型的宏观密度和速度定义如下:

$$\rho=\sum_{\alpha=0}^{8} f_\alpha \tag{3.123}$$

$$\rho\boldsymbol{u}=\sum_{\alpha=0}^{8}\boldsymbol{e}_\alpha f_\alpha \tag{3.124}$$

宏观压力由状态方程 $p=\rho c_s^2$ 直接给出。

除了二维的 D2Q9 模型之外,还有两个常用的三维 D*d*Q*m* 模型,它们的离散速度、权系数以及格子声速(假定 $c=1$)分别为

D3Q15：

$$
\boldsymbol{e}_\alpha=\begin{bmatrix}0 & 1 & -1 & 0 & 0 & 0 & 0 & 1 & -1 & 1 & -1 & 1 & -1 & 1 & -1\\0 & 0 & 0 & 1 & -1 & 0 & 0 & 1 & 1 & -1 & -1 & 1 & 1 & -1 & -1\\0 & 0 & 0 & 0 & 0 & 1 & -1 & 1 & 1 & 1 & 1 & -1 & -1 & -1 & -1\end{bmatrix}
$$

$\omega_0=2/9,\omega_{1\text{-}6}=1/9,\omega_{7-14}=1/72,c_s=1/\sqrt{3}$。

D3Q19（其离散速度如图 3.3 所示）：

$$
\boldsymbol{e}_\alpha=\begin{bmatrix}0 & 1 & -1 & 0 & 0 & 0 & 0 & 1 & -1 & 1 & -1 & 1 & -1 & 1 & -1 & 0 & 0 & 0 & 0\\0 & 0 & 0 & 1 & -1 & 0 & 0 & 1 & -1 & -1 & 1 & 0 & 0 & 0 & 0 & 1 & -1 & 1 & -1\\0 & 0 & 0 & 0 & 0 & 1 & -1 & 0 & 0 & 0 & 0 & 1 & -1 & -1 & 1 & 1 & -1 & -1 & 1\end{bmatrix}.
$$

$\omega_0=1/3,\quad \omega_{1\text{-}6}=1/18,\quad \omega_{7-18}=1/36,\quad c_s=1/\sqrt{3}$。

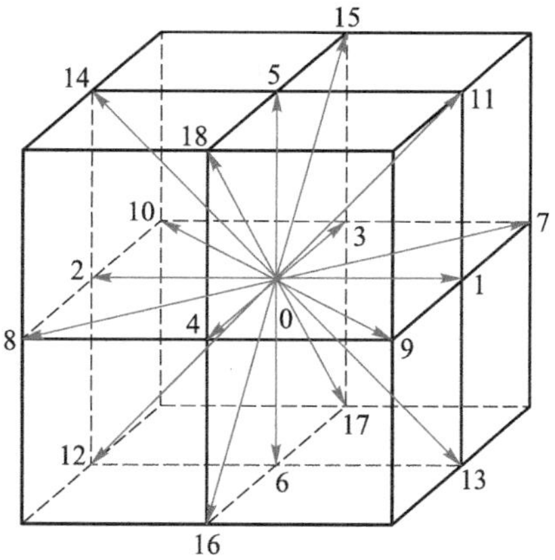

图 3.3 D3Q19 模型

3.5.2 Chapman-Enskog 展开

在本小节，将介绍如何通过 **Chapman-Enskog 展开**（Chapman-Enskog expansion）[9] 导出基本模型所对应的宏观方程。Chapman-Enskog 展开是由 Chapman 和 Enskog 提出的，它是一种**多尺度技术**[11]。在多尺度技术中，通常采用如下三种时间尺度[11]：

（1）碰撞时间尺度 K^0（K 是一个任意小的正数），即粒子碰撞过程非常迅速；

（2）对流时间尺度 K^{-1}，对流过程慢于粒子碰撞过程；

（3）扩散时间尺度 K^{-2}，扩散过程慢于对流过程。

以 $K=0.01$ 为例，假定粒子碰撞过程的时间为 t_c，则对流过程所需时间为 $100t_c$，而扩散过程所需时间则为 $10\ 000t_c$。在空间上，碰撞过程的空间尺度为分子平均自由程，而对流和扩散过程的空间尺度则为物理问题的宏观特征尺度。于是在多尺度技术中，通常使用以下三种形式的时间尺度：t^*、$t_1=Kt^*$ 和 $t_2=K^2t^*$，及两种空间尺度：$\boldsymbol{r}^*$ 和 $\boldsymbol{r}_1=K\boldsymbol{r}^*$。从而，时间导数和空间导数可以表示为

$$
\frac{\partial}{\partial \boldsymbol{r}}=K\frac{\partial}{\partial \boldsymbol{r}_1} \tag{3.125}
$$

$$
\frac{\partial}{\partial t}=K\frac{\partial}{\partial t_1}+K^2\frac{\partial}{\partial t_2} \tag{3.126}
$$

Chapman-Enskog 展开在采用上述多尺度技术的同时,还需对分布函数进行展开,即假定分布函数已离平衡态不远,并趋于平衡态。因此,f_α 可展开成 K 的幂级数形式

$$f_\alpha=f_\alpha^{\mathrm{eq}}+Kf_\alpha^{(1)}+K^2f_\alpha^{(2)}+\cdots \tag{3.127}$$

式中,f_α^{eq} 也可表示为 $f_\alpha^{(0)}$。

不考虑外力作用时,单松弛格子 Boltzmann 方程可写成如下形式:

$$f_\alpha(\boldsymbol{r}+\boldsymbol{e}_\alpha\delta_t,t+\delta_t)-f_\alpha(\boldsymbol{r},t)=-\frac{1}{\tau}[f_\alpha(\boldsymbol{r},t)-f_\alpha^{\mathrm{eq}}(\boldsymbol{r},t)] \tag{3.128}$$

对式(3.128)左端做泰勒级数展开可得

$$\delta_t\left(\frac{\partial}{\partial t}+\boldsymbol{e}_\alpha\cdot\nabla\right)f_\alpha+\frac{\delta_t^2}{2}\left(\frac{\partial}{\partial t}+\boldsymbol{e}_\alpha\cdot\nabla\right)^2f_\alpha+O(\delta_t^3)=-\frac{1}{\tau}(f_\alpha-f_\alpha^{\mathrm{eq}}) \tag{3.129}$$

将式(3.125)~式(3.127)代入式(3.129),并对比各阶系数可得

$$K^1:\quad \left(\frac{\partial}{\partial t_1}+\boldsymbol{e}_\alpha\cdot\nabla_1\right)f_\alpha^{\mathrm{eq}}+\frac{1}{\tau\delta_t}f_\alpha^{(1)}=0 \tag{3.130}$$

$$K^2:\quad \frac{\partial f_\alpha^{\mathrm{eq}}}{\partial t_2}+\left(\frac{\partial}{\partial t_1}+\boldsymbol{e}_\alpha\cdot\nabla_1\right)f_\alpha^{(1)}+\frac{\delta_t}{2}\left(\frac{\partial}{\partial t_1}+\boldsymbol{e}_\alpha\cdot\nabla_1\right)^2f_\alpha^{\mathrm{eq}}+\frac{1}{\tau\delta_t}f_\alpha^{(2)}=0 \tag{3.131}$$

利用式(3.130),可将式(3.131)简化为

$$K^2:\quad \frac{\partial f_\alpha^{\mathrm{eq}}}{\partial t_2}+\left(\frac{\partial}{\partial t_1}+\boldsymbol{e}_\alpha\cdot\nabla_1\right)\left(1-\frac{1}{2\tau}\right)f_\alpha^{(1)}+\frac{1}{\tau\delta_t}f_\alpha^{(2)}=0 \tag{3.132}$$

联立式(3.109)、式(3.110)、式(3.123)及式(3.124)可得

$$\sum_{\alpha=0}^{m-1}f_\alpha^{(n)}=0,\quad \sum_{\alpha=0}^{m-1}\boldsymbol{e}_\alpha f_\alpha^{(n)}=0,\quad n=1,2,\cdots \tag{3.133}$$

式中,m 为离散速度的个数。对式(3.130)求离散速度的零阶和一阶矩可得 t_1 尺度上的宏观方程为

$$\frac{\partial\rho}{\partial t_1}+\nabla_1\cdot(\rho\boldsymbol{u})=0 \tag{3.134}$$

$$\frac{\partial}{\partial t_1}(\rho u_j)+\frac{\partial}{\partial r_{1i}}(\rho u_iu_j)=-\frac{\partial p}{\partial r_{1j}} \tag{3.135}$$

类似地,对式(3.131)求离散速度的零阶矩和一阶矩可得 t_2 尺度上的宏观方程

$$\frac{\partial\rho}{\partial t_2}=0 \tag{3.136}$$

$$\frac{\partial}{\partial t_2}(\rho u_j)+\frac{\partial}{\partial r_{1i}}\left[\left(1-\frac{1}{2\tau}\right)\sum_{\alpha=0}^{m-1}e_{\alpha i}e_{\alpha j}f_\alpha^{(1)}\right]=0 \tag{3.137}$$

根据式(3.130)有

$$\begin{aligned}\sum_{\alpha=0}^{m-1}e_{\alpha i}e_{\alpha j}f_\alpha^{(1)}&=-\tau\delta_t\sum_{\alpha=0}^{m-1}e_{\alpha i}e_{\alpha j}\left(\frac{\partial}{\partial t_1}+\boldsymbol{e}_\alpha\cdot\nabla_1\right)f_\alpha^{\mathrm{eq}}\\&=-\tau\delta_t\left[\frac{\partial}{\partial t_1}(\rho u_iu_j+\rho c_s^2\delta_{ij})+\frac{\partial}{\partial r_{1k}}\left(\sum_{\alpha=0}^{m-1}e_{\alpha i}e_{\alpha j}e_{\alpha k}f_\alpha^{\mathrm{eq}}\right)\right]\end{aligned} \tag{3.138}$$

由式(3.134)可得

$$\frac{\partial}{\partial t_1}(\rho u_iu_j+\rho c_s^2\delta_{ij})=\frac{\partial(\rho u_iu_j)}{\partial t_1}-c_s^2\nabla_1\cdot(\rho\boldsymbol{u})\delta_{ij} \tag{3.139}$$

而$\partial(\rho u_i u_j)/\partial t_1$ 可写成

$$\frac{\partial(\rho u_i u_j)}{\partial t_1}=u_j\frac{\partial(\rho u_i)}{\partial t_1}+u_i\frac{\partial(\rho u_j)}{\partial t_1}-u_i u_j\frac{\partial\rho}{\partial t_1} \tag{3.140}$$

根据式(3.134)和式(3.135)可得

$$\begin{aligned}\frac{\partial(\rho u_i u_j)}{\partial t_1}&=u_j\left[-\frac{\partial}{\partial r_{1k}}(\rho u_i u_k+\rho c_s^2\delta_{ik})\right]+u_i\left[-\frac{\partial}{\partial r_{1k}}(\rho u_j u_k+\rho c_s^2\delta_{jk})\right]-u_i u_j\frac{\partial\rho}{\partial t_1}\\&=-u_j\frac{\partial}{\partial r_{1k}}(\rho u_i u_k)-c_s^2 u_j\frac{\partial\rho}{\partial r_{1i}}-u_i\frac{\partial}{\partial r_{1k}}(\rho u_j u_k)-c_s^2 u_i\frac{\partial\rho}{\partial r_{1j}}+u_i u_j\frac{\partial(\rho u_k)}{\partial r_{1k}}\\&=-c_s^2 u_j\frac{\partial\rho}{\partial r_{1i}}-c_s^2 u_i\frac{\partial\rho}{\partial r_{1j}}-\rho u_j u_k\frac{\partial u_i}{\partial r_{1k}}-u_i\frac{\partial}{\partial r_{1k}}(\rho u_j u_k)\\&=-c_s^2 u_j\frac{\partial\rho}{\partial r_{1i}}-c_s^2 u_i\frac{\partial\rho}{\partial r_{1j}}-\frac{\partial}{\partial r_{1k}}(\rho u_i u_j u_k)\end{aligned} \tag{3.141}$$

根据 D2Q9 模型的平衡态分布函数，经过数学运算有

$$\begin{aligned}\frac{\partial}{\partial r_{1k}}\left(\sum_{\alpha=0}^{m-1}e_{\alpha i}e_{\alpha j}e_{\alpha k}f_\alpha^{\mathrm{eq}}\right)&=\frac{\partial}{\partial r_{1k}}[\rho c_s^2(\delta_{ij}u_k+\delta_{ik}u_j+\delta_{jk}u_i)]\\&=c_s^2\nabla_1\cdot(\rho\boldsymbol{u})\delta_{ij}+\rho c_s^2\frac{\partial u_j}{\partial r_{1i}}+c_s^2 u_j\frac{\partial\rho}{\partial r_{1i}}+\rho c_s^2\frac{\partial u_i}{\partial r_{1j}}+c_s^2 u_i\frac{\partial\rho}{\partial r_{1j}}\end{aligned} \tag{3.142}$$

结合式(3.139)~式(3.142)可得

$$\sum_{\alpha=0}^{m-1}e_{\alpha i}e_{\alpha j}f_\alpha^{(1)}=-\tau\delta_t\left[\rho c_s^2\left(\frac{\partial u_j}{\partial r_{1i}}+\frac{\partial u_i}{\partial r_{1j}}\right)-\frac{\partial}{\partial r_{1k}}(\rho u_i u_j u_k)\right] \tag{3.143}$$

联立 t_1 和 t_2 时间尺度上的方程可得基本模型所对应的宏观方程为

$$\frac{\partial\rho}{\partial t}+\nabla\cdot(\rho\boldsymbol{u})=0 \tag{3.144}$$

$$\frac{\partial(\rho\boldsymbol{u})}{\partial t}+\nabla\cdot(\rho\boldsymbol{uu})=-\nabla p+\nabla\cdot\left\{\mu[\nabla\boldsymbol{u}+(\nabla\boldsymbol{u})^{\mathrm{T}}]-\frac{\mu}{\rho c_s^2}\nabla\cdot(\rho\boldsymbol{uuu})\right\} \tag{3.145}$$

式中，动力黏度 μ 为

$$\mu=\rho c_s^2(\tau-0.5)\delta_t \tag{3.146}$$

相应的运动黏度为 $\nu=\mu/\rho$。在导出格子 Boltzmann 方程(3.100)时，所用的矩形法只有一阶精度，而式(3.146)中的系数 0.5 所对应的正是这一离散格式带来的**数值黏性**。通过将数值黏性吸收到物理黏性中，格子 Boltzmann 方程实际上具有二阶精度。

式(3.144)和式(3.145)与标准的可压缩 Navier-Stokes 方程组相比，虽然连续性方程是一致的，但动量方程(3.145)存在着偏差，一是存在着与$\nabla\cdot(\rho\boldsymbol{uuu})$有关的偏差项，二是体积黏度非零。不过，如果流体的密度为常数，即 $\rho=\rho_0=\mathrm{const}$，且对低马赫数流动来说，$\nabla\cdot(\rho\boldsymbol{uuu})$可以忽略，此时模型的宏观方程转化为

数值黏性

$$\nabla\cdot\boldsymbol{u}=0 \tag{3.147}$$

$$\frac{\partial\boldsymbol{u}}{\partial t}+\nabla\cdot(\boldsymbol{uu})=-\frac{1}{\rho_0}\nabla p+\nu\nabla\cdot[\nabla\boldsymbol{u}+(\nabla\boldsymbol{u})^{\mathrm{T}}] \tag{3.148}$$

式中，$\nabla\cdot(\nabla\boldsymbol{u})^{\mathrm{T}}=\nabla(\nabla\cdot\boldsymbol{u})=0$。于是上述方程组即为标准的不可压缩 Navier-Stokes 方程组。需要指出的是，对格子 Boltzmann 方法来说，构造不可压缩格子 Boltzmann 模型并不

是直接将平衡态分布函数中的密度 ρ 改为 ρ_0。在标准格子 Boltzmann 模型中，压力与密度成正比，直接取 $\rho=\rho_0=\text{const}$ 将导致压力为常数，这与实际流动中往往是通过压力梯度来驱动流体运动的事实不符。为了解决这一问题，前人做了许多工作，我们将在下一章进行介绍。

3.6 多松弛格子 Boltzmann 模型

2000 年，Lallemand 和 Luo[13] 对 1992 年 d'Humiéres 在一次会议上所提出的广义格子 Boltzmann 模型进行了系统的理论分析，他们发现，与单松弛格子 Boltzmann 模型相比，广义格子 Boltzmann 模型在数值稳定性、参数选取等方面具有优势。广义格子 Boltzmann 模型与单松弛格子 Boltzmann 模型的主要区别在于它的碰撞算子采用多个松弛时间，因而也被称为**多松弛模型**（multiple-relaxation-time）或**多松弛碰撞算子**。Lallemand 和 Luo 的研究使得长期被忽视的多松弛碰撞算子开始受到人们的重视。在这一理论的指导下，一些新的格子 Boltzmann 模型迅速出现，并在许多领域得到了应用。

多松弛格子 Boltzmann 模型具有良好的数值稳定性，对格子 Boltzmann 方法在高雷诺数、高瑞利数流动与传热问题中的应用具有重要意义。但需要指出的是，不能因为多松弛格子 Boltzmann 模型的出现而无视单松弛格子 Boltzmann 模型的一些优点，更不能因此否定单松弛格子 Boltzmann 模型的历史意义和研究价值。单松弛格子 Boltzmann 方程可以从气体动理论的 Boltzmann-BGK 方程导出，方程简单直观且易于实施，有利于初学者入门，亦有利于研究者发现一些建模的基本规律。

3.6.1 基本原理

在单松弛碰撞算子中，不同离散速度方向的分布函数是以相同的松弛时间向其对应的平衡态分布过渡的，而在多松弛碰撞算子中，分布函数的各阶矩是以不同的松弛时间向其对应的平衡态矩函数过渡。根据离散速度的个数，可以定义相应个数的矩函数

$$\boldsymbol{m}=\boldsymbol{M}\boldsymbol{f}=(m_0,m_1,\cdots,m_{N-2},m_{N-1})^{\mathrm{T}} \tag{3.149}$$

式中：$\boldsymbol{M}$ 为转换矩阵；$\boldsymbol{f}=(f_0,f_1,\cdots,f_{N-2},f_{N-1})^{\mathrm{T}}$。相应地，平衡态矩函数可定义为

$$\boldsymbol{m}^{\mathrm{eq}}=\boldsymbol{M}\boldsymbol{f}^{\mathrm{eq}}=(m_0^{\mathrm{eq}},m_1^{\mathrm{eq}},\cdots,m_{N-2}^{\mathrm{eq}},m_{N-1}^{\mathrm{eq}})^{\mathrm{T}} \tag{3.150}$$

式中，$\boldsymbol{f}^{\mathrm{eq}}=(f_0^{\mathrm{eq}},f_1^{\mathrm{eq}},\cdots,f_{N-2}^{\mathrm{eq}},f_{N-1}^{\mathrm{eq}})^{\mathrm{T}}$。

在离散速度空间，不含外力项的多松弛格子 Boltzmann 方程可写为

$$f_\alpha(\boldsymbol{r}+\boldsymbol{e}_\alpha\delta_t,t+\delta_t)-f_\alpha(\boldsymbol{r},t)=-\overline{\Lambda}_{\alpha\beta}[f_\beta(\boldsymbol{r},t)-f_\beta^{\mathrm{eq}}(\boldsymbol{r},t)] \tag{3.151}$$

式中：$\overline{\boldsymbol{\Lambda}}=(\boldsymbol{M}^{-1}\boldsymbol{\Lambda}\boldsymbol{M})$ 为碰撞矩阵；$\boldsymbol{M}^{-1}$ 表示转换矩阵 $\boldsymbol{M}$ 的逆矩阵；$\boldsymbol{\Lambda}$ 为对角松弛矩阵。

在多松弛模型中，转换矩阵 $\boldsymbol{M}$ 是关键要素，它决定了每个矩的物理意义和具体形式，并且影响着模型的计算效率[10]。引入基向量 $\boldsymbol{\varphi}_i$ 表示转换矩阵 $\boldsymbol{M}$ 第 $i+1$ 行的行向量，则 $\boldsymbol{M}$ 可表示为

$$\boldsymbol{M}=[\boldsymbol{\varphi}_0,\boldsymbol{\varphi}_1,\cdots,\boldsymbol{\varphi}_{N-1}]^{\mathrm{T}} \tag{3.152}$$

一般地，可根据离散速度多项式向量 $\boldsymbol{a}=e_{\alpha x}^h e_{\alpha y}^l e_{\alpha z}^k(h,l,k\geqslant 0)$，通过 Gram-Schmidt 正交化[13] 构造 N 个正交基向量 $\boldsymbol{b}_i$，继而在此基础上得到 $\boldsymbol{\varphi}_i$。这样得到的转换矩阵是正交矩阵，相应的模型为正交多松弛模型。以 D2Q9 模型为例，从低阶到高阶的 9 个多项式向量分

别为

$$\boldsymbol{a}_0=e_{\alpha x}^0 e_{\alpha y}^0,\quad \boldsymbol{a}_1=e_{\alpha x},\quad \boldsymbol{a}_2=e_{\alpha y},\quad \boldsymbol{a}_3=e_{\alpha x}^2$$
$$\boldsymbol{a}_4=e_{\alpha x}e_{\alpha y},\quad \boldsymbol{a}_5=e_{\alpha y}^2,\quad \boldsymbol{a}_6=e_{\alpha x}^2 e_{\alpha y},\quad \boldsymbol{a}_7=e_{\alpha x}e_{\alpha y}^2,\quad \boldsymbol{a}_8=e_{\alpha x}^2 e_{\alpha y}^2 \tag{3.153}$$

根据 Gram-Schmidt 正交化,正交基向量 $\boldsymbol{b}_i$ 可通过如下式子得到[14]:

$$\boldsymbol{b}_0=\boldsymbol{a}_0,\quad \boldsymbol{b}_1=\boldsymbol{a}_1-\boldsymbol{b}_0\frac{\boldsymbol{b}_0\cdot\boldsymbol{a}_1}{\boldsymbol{b}_0\cdot\boldsymbol{b}_0}$$
$$\boldsymbol{b}_2=\boldsymbol{a}_2-\boldsymbol{b}_0\frac{\boldsymbol{b}_0\cdot\boldsymbol{a}_2}{\boldsymbol{b}_0\cdot\boldsymbol{b}_0}-\boldsymbol{b}_1\frac{\boldsymbol{b}_1\cdot\boldsymbol{a}_2}{\boldsymbol{b}_1\cdot\boldsymbol{b}_1},\quad \boldsymbol{b}_{q-1}=\boldsymbol{a}_{q-1}-\sum_{i=0}^{q-2}\boldsymbol{b}_i\frac{\boldsymbol{b}_i\cdot\boldsymbol{a}_{q-1}}{\boldsymbol{b}_i\cdot\boldsymbol{b}_i} \tag{3.154}$$

以正交基向量 $\boldsymbol{b}_i$ 为基础,可以得到转换矩阵 $\boldsymbol{M}$ 的基向量 $\boldsymbol{\varphi}_i$。

与单松弛模型一样,多松弛模型的格子 Boltzmann 方程也可以分成碰撞步和迁移步,分别表示为

$$\text{碰撞:}\quad f_\alpha^*(\boldsymbol{r},t)=f_\alpha(\boldsymbol{r},t)-\overline{\Lambda}_{\alpha\beta}[f_\beta(\boldsymbol{r},t)-f_\beta^{\mathrm{eq}}(\boldsymbol{r},t)] \tag{3.155}$$

$$\text{迁移:}\quad f_\alpha(\boldsymbol{r}+\boldsymbol{e}_\alpha\delta_t,t+\delta_t)=f_\alpha^*(\boldsymbol{r},t) \tag{3.156}$$

式(3.155)两端同乘转换矩阵 $\boldsymbol{M}$ 可得

$$\boldsymbol{m}^*=\boldsymbol{m}-\boldsymbol{\Lambda}(\boldsymbol{m}-\boldsymbol{m}^{\mathrm{eq}}) \tag{3.157}$$

式中,$\boldsymbol{m}^*$ 为碰撞后的矩函数。根据式(3.157),多松弛模型的碰撞步可以直接在矩空间进行,而迁移步则仍在离散速度空间执行,如图 3.4 所示,且有

$$\boldsymbol{f}^*=\boldsymbol{M}^{-1}\boldsymbol{m}^* \tag{3.158}$$

式中,$\boldsymbol{f}^*=(f_0^*,f_1^*,\cdots,f_{N-2}^*,f_{N-1}^*)^{\mathrm{T}}$。根据式(3.158)得到 $f_\alpha^*(\boldsymbol{r},t)$ 之后,即可执行迁移步式(3.156)。

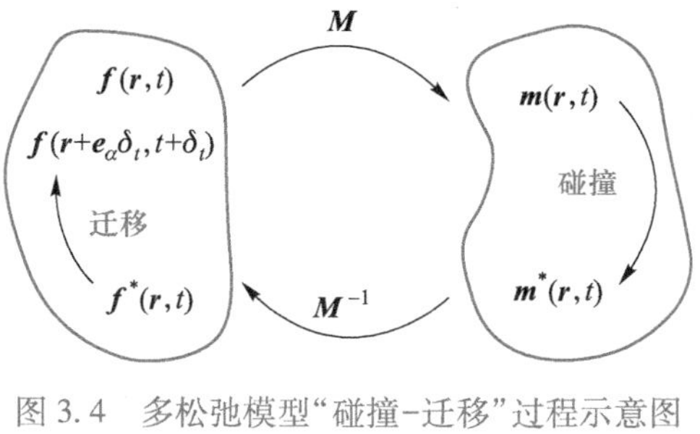

图 3.4 多松弛模型“碰撞-迁移”过程示意图

在矩空间执行碰撞步至少带来了两个好处。首先,不同的矩函数可以使用不同的松弛时间,这既符合物理规律,也是多松弛模型最突出的优点;其次,平衡态矩函数可直接根据目标方程选取,并不一定要按照式(3.150)选取。

3.6.2 正交多松弛模型

下面以 D2Q9 和 D3Q19 模型为例,介绍正交多松弛格子 Boltzmann 模型的基向量 $\boldsymbol{\varphi}_i$、转换矩阵 $\boldsymbol{M}$、平衡态矩函数 $\boldsymbol{m}^{\mathrm{eq}}$、对角松弛矩阵 $\boldsymbol{\Lambda}$ 以及相应的输运系数。

1）D2Q9 正交多松弛模型

在式(3.153)和式(3.154)的基础上，D2Q9 模型转换矩阵的基向量可选取为[13]

$$\varphi_{0,\alpha}=1,\quad \varphi_{1,\alpha}=3(e_{\alpha x}^2+e_{\alpha y}^2)-4,\quad \varphi_{2,\alpha}=4-\frac{21}{2}(e_{\alpha x}^2+e_{\alpha y}^2)+\frac{9}{2}(e_{\alpha x}^2+e_{\alpha y}^2)^2$$

$$\varphi_{3,\alpha}=e_{\alpha x},\quad \varphi_{4,\alpha}=[3(e_{\alpha x}^2+e_{\alpha y}^2)-5]e_{\alpha x},\quad \varphi_{5,\alpha}=e_{\alpha y}$$

$$\varphi_{6,\alpha}=[3(e_{\alpha x}^2+e_{\alpha y}^2)-5]e_{\alpha y},\quad \varphi_{7,\alpha}=e_{\alpha x}^2-e_{\alpha y}^2,\quad \varphi_{8,\alpha}=e_{\alpha x}e_{\alpha y} \tag{3.159}$$

相应的转换矩阵可表示为

$$\boldsymbol{M}=\begin{bmatrix} \varphi_{0,0} & \varphi_{0,1} & \cdots & \varphi_{0,8} \\ \varphi_{1,0} & \varphi_{1,1} & \cdots & \varphi_{1,8} \\ \vdots & \vdots & & \vdots \\ \varphi_{8,0} & \varphi_{8,1} & \cdots & \varphi_{8,8} \end{bmatrix} \tag{3.160}$$

结合式(3.159)和离散速度式(3.103)，D2Q9 模型对应的正交转换矩阵为

$$\boldsymbol{M}=\begin{bmatrix} 1 & 1 & 1 & 1 & 1 & 1 & 1 & 1 & 1 \\ -4 & -1 & -1 & -1 & -1 & 2 & 2 & 2 & 2 \\ 4 & -2 & -2 & -2 & -2 & 1 & 1 & 1 & 1 \\ 0 & 1 & 0 & -1 & 0 & 1 & -1 & -1 & 1 \\ 0 & -2 & 0 & 2 & 0 & 1 & -1 & -1 & 1 \\ 0 & 0 & 1 & 0 & -1 & 1 & 1 & -1 & -1 \\ 0 & 0 & -2 & 0 & 2 & 1 & 1 & -1 & -1 \\ 0 & 1 & -1 & 1 & -1 & 0 & 0 & 0 & 0 \\ 0 & 0 & 0 & 0 & 0 & 1 & -1 & 1 & -1 \end{bmatrix} \tag{3.161}$$

其逆矩阵 $\boldsymbol{M}^{-1}$也是正交矩阵[15]。可以看到，转换矩阵 $\boldsymbol{M}$ 中有许多项为零，在程序编写中可以不考虑这些项。例如，根据式(3.149)和式(3.161)，m_8 可写成

$$m_8(\boldsymbol{r},t)=f_5(\boldsymbol{r},t)-f_6(\boldsymbol{r},t)+f_7(\boldsymbol{r},t)-f_8(\boldsymbol{r},t) \tag{3.162}$$

对 D2Q9 正交多松弛模型，平衡态矩函数可选取为

$$\boldsymbol{m}^{\mathrm{eq}}=\boldsymbol{M}\boldsymbol{f}^{\mathrm{eq}}=\rho\,(1,-2+3u^2,1-3u^2,u_x,-u_x,u_y,-u_y,u_x^2-u_y^2,u_xu_y)^{\mathrm{T}} \tag{3.163}$$

式中，$u^2=u_x^2+u_y^2$。以 m_8^{eq} 为例，它是由 $m_8^{\mathrm{eq}}=\sum\limits_{\alpha=0}^{8}\varphi_{8,\alpha}f_\alpha^{\mathrm{eq}}$得到的，结合式(3.159)和式(3.111)可知，该项为 ρu_xu_y。该模型对应的对角松弛矩阵为

$$\Lambda=\mathrm{diag}\left(\frac{1}{\tau_{\mathrm{c}}},\frac{1}{\tau_e},\frac{1}{\tau_\zeta},\frac{1}{\tau_{\mathrm{c}}},\frac{1}{\tau_q},\frac{1}{\tau_{\mathrm{c}}},\frac{1}{\tau_q},\frac{1}{\tau_\nu},\frac{1}{\tau_\nu}\right) \tag{3.164}$$

式中："diag"表示对角矩阵；τ_{c} 为守恒量（密度 ρ、x 方向的动量和 y 方向的动量）所对应的量纲为 1 的松弛时间，在应用中可取 $\tau_{\mathrm{c}}=1$；τ_ζ 和 τ_q 为可调松弛参数，它们不影响宏观流体动力学，但与模型的稳定性有关[16]；τ_e 和 τ_ν 分别与体积黏度和动力黏度有关，且有

$$\mu=\rho c_{\mathrm{s}}^2(\tau_\nu-0.5)\delta_t,\quad \mu_{\mathrm{B}}=\rho c_{\mathrm{s}}^2(\tau_e-0.5)\delta_t \tag{3.165}$$

式中，μ 和 μ_{B} 分别为动力黏度和体积黏度。当对角松弛矩阵中的量纲为 1 的松弛时间取

值相同时,多松弛模型退化为单松弛模型。

2) D3Q19 正交多松弛模型

2002 年,d'Humiéres 等[17] 构造了基于 D3Q15 和 D3Q19 的正交多松弛模型。以 D3Q19 模型为例,简单介绍一下三维的正交多松弛模型。根据 Gram-Schmidt 正交化,对 D3Q19 模型可选取如下形式的基向量[17]:

$$
\begin{gathered}
\varphi_{0,\alpha}=1,\quad \varphi_{1,\alpha}=19\ |\boldsymbol{e}_\alpha|^2-30,\quad \varphi_{2,\alpha}=12-\frac{53}{2}|\boldsymbol{e}_\alpha|^2+\frac{21}{2}|\boldsymbol{e}_\alpha|^4,\quad \varphi_{3,\alpha}=e_{\alpha x}\\
\varphi_{4,\alpha}=(5\ |\boldsymbol{e}_\alpha|^2-9)e_{\alpha x},\quad \varphi_{5,\alpha}=e_{\alpha y},\quad \varphi_{6,\alpha}=(5\ |\boldsymbol{e}_\alpha|^2-9)e_{\alpha y},\quad \varphi_{7,\alpha}=e_{\alpha z}\\
\varphi_{8,\alpha}=(5\ |\boldsymbol{e}_\alpha|^2-9)e_{\alpha z},\quad \varphi_{9,\alpha}=3e_{\alpha x}^2-|\boldsymbol{e}_\alpha|^2,\quad \varphi_{10,\alpha}=(3\ |\boldsymbol{e}_\alpha|^2-5)(3e_{\alpha x}^2-|\boldsymbol{e}_\alpha|^2)\\
\varphi_{11,\alpha}=e_{\alpha y}^2-e_{\alpha z}^2,\quad \varphi_{12,\alpha}=(3\ |\boldsymbol{e}_\alpha|^2-5)(e_{\alpha y}^2-e_{\alpha z}^2),\quad \varphi_{13,\alpha}=e_{\alpha x}e_{\alpha y},\quad \varphi_{14,\alpha}=e_{\alpha y}e_{\alpha z}\\
\varphi_{15,\alpha}=e_{\alpha x}e_{\alpha z},\quad \varphi_{16,\alpha}=(e_{\alpha y}^2-e_{\alpha z}^2)e_{\alpha x},\quad \varphi_{17,\alpha}=(e_{\alpha z}^2-e_{\alpha x}^2)e_{\alpha y},\quad \varphi_{18,\alpha}=(e_{\alpha x}^2-e_{\alpha y}^2)e_{\alpha z}
\end{gathered}
\tag{3.166}
$$

式中:$|\boldsymbol{e}_\alpha|^2=e_{\alpha x}^2+e_{\alpha y}^2+e_{\alpha z}^2$。结合式(3.166)及 D3Q19 模型的离散速度配置可得[17]

$$
\boldsymbol{M}=\begin{bmatrix}
1&1&1&1&1&1&1&1&1&1&1&1&1&1&1&1&1&1&1\\
-30&-11&-11&-11&-11&-11&-11&8&8&8&8&8&8&8&8&8&8&8&8\\
12&-4&-4&-4&-4&-4&-4&1&1&1&1&1&1&1&1&1&1&1&1\\
0&1&-1&0&0&0&0&1&-1&1&-1&1&-1&1&-1&0&0&0&0\\
0&-4&4&0&0&0&0&1&-1&1&-1&1&-1&1&-1&0&0&0&0\\
0&0&0&1&-1&0&0&1&1&-1&-1&0&0&0&0&1&-1&1&-1\\
0&0&0&-4&4&0&0&1&1&-1&-1&0&0&0&0&1&-1&1&-1\\
0&0&0&0&0&1&-1&0&0&0&0&1&1&-1&-1&1&1&-1&-1\\
0&0&0&0&0&-4&4&0&0&0&0&1&1&-1&-1&1&1&-1&-1\\
0&2&2&-1&-1&-1&-1&1&1&1&1&1&1&1&1&-2&-2&-2&-2\\
0&-4&-4&2&2&2&2&1&1&1&1&1&1&1&1&-2&-2&-2&-2\\
0&0&0&1&1&-1&-1&1&1&1&1&-1&-1&-1&-1&0&0&0&0\\
0&0&0&-2&-2&2&2&1&1&1&1&-1&-1&-1&-1&0&0&0&0\\
0&0&0&0&0&0&0&1&-1&-1&1&0&0&0&0&0&0&0&0\\
0&0&0&0&0&0&0&0&0&0&0&0&0&0&0&1&-1&-1&1\\
0&0&0&0&0&0&0&0&0&0&0&1&-1&-1&1&0&0&0&0\\
0&0&0&0&0&0&0&1&-1&1&-1&-1&1&-1&1&0&0&0&0\\
0&0&0&0&0&0&0&-1&-1&1&1&0&0&0&0&1&-1&1&-1\\
0&0&0&0&0&0&0&0&0&0&0&1&1&-1&-1&-1&-1&1&1
\end{bmatrix}
$$

类似地,在程序编写中,只需要考虑转换矩阵 $\boldsymbol{M}$ 中的非零项。例如,m_{14} 可写成

$$
m_{14}(\boldsymbol{r},t)=f_{15}(\boldsymbol{r},t)-f_{16}(\boldsymbol{r},t)-f_{17}(\boldsymbol{r},t)+f_{18}(\boldsymbol{r},t)
\tag{3.167}
$$

根据 $\boldsymbol{m}^{\mathrm{eq}}=\boldsymbol{M}\boldsymbol{f}^{\mathrm{eq}}$,平衡态矩函数可选取为

$$
\begin{gathered}
m_0^{\mathrm{eq}}=\rho,\quad m_1^{\mathrm{eq}}=-11\rho+19\rho u^2,\quad m_2^{\mathrm{eq}}=3\rho-5.5\rho u^2,\quad m_3^{\mathrm{eq}}=\rho u_x\\
m_4^{\mathrm{eq}}=-\frac{2}{3}\rho u_x,\quad m_5^{\mathrm{eq}}=\rho u_y,\quad m_6^{\mathrm{eq}}=-\frac{2}{3}\rho u_y,\quad m_7^{\mathrm{eq}}=\rho u_z,\quad m_8^{\mathrm{eq}}=-\frac{2}{3}\rho u_z
\end{gathered}
$$

$$m_9^{\mathrm{eq}}=3\rho u_x^2-\rho u^2,\quad m_{10}^{\mathrm{eq}}=-\frac{1}{2}(3\rho u_x^2-\rho u^2),\quad m_{11}^{\mathrm{eq}}=\rho(u_y^2-u_z^2),\quad m_{12}^{\mathrm{eq}}=-\frac{\rho}{2}(u_y^2-u_z^2)$$

$$m_{13}^{\mathrm{eq}}=\rho u_x u_y,\quad m_{14}^{\mathrm{eq}}=\rho u_y u_z,\quad m_{15}^{\mathrm{eq}}=\rho u_x u_z,\quad m_{16}^{\mathrm{eq}}=0,\quad m_{17}^{\mathrm{eq}}=0,\quad m_{18}^{\mathrm{eq}}=0 \tag{3.168}$$

式中,$u^2=u_x^2+u_y^2+u_z^2$。相应地,该模型的对角松弛矩阵为

$$\boldsymbol{\Lambda}=\mathrm{diag}\left(\frac{1}{\tau_c},\frac{1}{\tau_e},\frac{1}{\tau_\zeta},\frac{1}{\tau_c},\frac{1}{\tau_q},\frac{1}{\tau_c},\frac{1}{\tau_q},\frac{1}{\tau_c},\frac{1}{\tau_q},\frac{1}{\tau_\nu},\frac{1}{\tau_\pi},\frac{1}{\tau_\nu},\frac{1}{\tau_\pi},\frac{1}{\tau_\nu},\frac{1}{\tau_\nu},\frac{1}{\tau_\nu},\frac{1}{\tau_l},\frac{1}{\tau_l},\frac{1}{\tau_l}\right)$$

式中,$\tau_c=1$ 为守恒量对应的量纲为 1 的松弛时间;τ_ζ、τ_q、τ_π、τ_l 为可调松弛参数;τ_e 和 τ_ν 分别与体积黏度和动力黏度有关,且有

$$\mu=\rho c_s^2(\tau_\nu-0.5)\delta_t,\quad \mu_B=\frac{2}{3}\rho c_s^2(\tau_e-0.5)\delta_t \tag{3.169}$$

其中体积黏度的表达式与 D2Q9 模型的表达式稍有不同。

3.6.3 非正交多松弛模型

在正交多松弛模型中,转换矩阵的基向量为 N 个正交向量。实际上,对于多松弛模型,在不影响模型稳定性及宏观流体动力学的前提下,转换矩阵也可以是非正交矩阵,从而衍生出了非正交多松弛模型[18-23]。这类模型主要是通过构造非正交的基向量来简化转换矩阵,即减少转换矩阵中的非零项,从而达到简化算法及程序编写并在一定程度上提高计算效率的目的。以文献[22]中的三维模型为例,介绍非正交多松弛模型的建模思路。

对 D3Q19 模型,可选取如下形式的非正交基向量[22]:

$$\varphi_{0,\alpha}=1,\quad \varphi_{1,\alpha}=e_{\alpha x},\quad \varphi_{2,\alpha}=e_{\alpha y},\quad \varphi_{3,\alpha}=e_{\alpha z} \tag{3.170}$$

$$\varphi_{4,\alpha}=|\boldsymbol{e}_\alpha|^2,\quad \varphi_{5,\alpha}=3e_{\alpha x}^2-|\boldsymbol{e}_\alpha|^2,\quad \varphi_{6,\alpha}=e_{\alpha y}^2-e_{\alpha z}^2 \tag{3.171}$$

$$\varphi_{7,\alpha}=e_{\alpha x}e_{\alpha y},\quad \varphi_{8,\alpha}=e_{\alpha x}e_{\alpha z},\quad \varphi_{9,\alpha}=e_{\alpha y}e_{\alpha z} \tag{3.172}$$

$$\varphi_{10,\alpha}=e_{\alpha x}^2e_{\alpha y},\quad \varphi_{11,\alpha}=e_{\alpha x}e_{\alpha y}^2,\quad \varphi_{12,\alpha}=e_{\alpha x}^2e_{\alpha z},\quad \varphi_{13,\alpha}=e_{\alpha x}e_{\alpha z}^2 \tag{3.173}$$

$$\varphi_{14,\alpha}=e_{\alpha y}^2e_{\alpha z},\quad \varphi_{15,\alpha}=e_{\alpha y}e_{\alpha z}^2,\quad \varphi_{16,\alpha}=e_{\alpha x}^2e_{\alpha y}^2,\quad \varphi_{17,\alpha}=e_{\alpha x}^2e_{\alpha z}^2,\quad \varphi_{18,\alpha}=e_{\alpha y}^2e_{\alpha z}^2 \tag{3.174}$$

式(3.170)中的一阶基向量与守恒量有关;式(3.171)与式(3.172)中的二阶基向量与黏性应力有关;式(3.173)与式(3.174)中的三阶及四阶基向量与 Navier-Stokes 层次的宏观方程无关。

平衡态矩函数可通过 $m_i^{\mathrm{eq}}=\sum\limits_{\alpha=0}^{18}\varphi_{i,\alpha}f_\alpha^{\mathrm{eq}}$ 得到,形式如下:

$$m_0^{\mathrm{eq}}=\rho,\quad m_1^{\mathrm{eq}}=\rho u_x,\quad m_2^{\mathrm{eq}}=\rho u_y,\quad m_3^{\mathrm{eq}}=\rho u_z,\quad m_4^{\mathrm{eq}}=\rho+\rho|\boldsymbol{u}|^2$$

$$m_5^{\mathrm{eq}}=\rho(2u_x^2-u_y^2-u_z^2),\quad m_6^{\mathrm{eq}}=\rho(u_y^2-u_z^2),\quad m_7^{\mathrm{eq}}=\rho u_x u_y,\quad m_8^{\mathrm{eq}}=\rho u_x u_z,\quad m_9^{\mathrm{eq}}=\rho u_y u_z$$

$$m_{10}^{\mathrm{eq}}=\rho c_s^2u_y,\quad m_{11}^{\mathrm{eq}}=\rho c_s^2u_x,\quad m_{12}^{\mathrm{eq}}=\rho c_s^2u_z,\quad m_{13}^{\mathrm{eq}}=\rho c_s^2u_x,\quad m_{14}^{\mathrm{eq}}=\rho c_s^2u_z,\quad m_{15}^{\mathrm{eq}}=\rho c_s^2u_y$$

$$m_{16}^{\mathrm{eq}}=\phi+\frac{3}{2}\rho c_s^2(u_x^2+u_y^2),\quad m_{17}^{\mathrm{eq}}=\phi+\frac{3}{2}\rho c_s^2(u_x^2+u_z^2),\quad m_{18}^{\mathrm{eq}}=\phi+\frac{3}{2}\rho c_s^2(u_y^2+u_z^2) \tag{3.175}$$

式中,$c_s^2=1/3$;$\phi=\rho c_s^4(1-1.5|\boldsymbol{u}|^2)$。注意,$m_{16}^{\mathrm{eq}}$、$m_{17}^{\mathrm{eq}}$ 和 m_{18}^{eq} 不影响 Navier-Stokes 层次的宏观方程,并不一定需要严格按照式(3.175)选取。根据式(3.170)~式(3.174),相应的转换矩阵为

$$
\boldsymbol{M}=\begin{bmatrix}
1&1&1&1&1&1&1&1&1&1&1&1&1&1&1&1&1&1&1\\
0&1&-1&0&0&0&0&1&-1&1&-1&1&-1&1&-1&0&0&0&0\\
0&0&0&1&-1&0&0&1&-1&-1&1&0&0&0&0&1&-1&1&-1\\
0&0&0&0&0&1&-1&0&0&0&0&1&-1&-1&1&1&-1&-1&1\\
0&1&1&1&1&1&1&2&2&2&2&2&2&2&2&2&2&2&2\\
0&2&2&-1&-1&-1&-1&1&1&1&1&1&1&1&1&-2&-2&-2&-2\\
0&0&0&1&1&-1&-1&1&1&1&1&-1&-1&-1&-1&0&0&0&0\\
0&0&0&0&0&0&0&1&1&-1&-1&0&0&0&0&0&0&0&0\\
0&0&0&0&0&0&0&0&0&0&0&1&1&-1&-1&0&0&0&0\\
0&0&0&0&0&0&0&0&0&0&0&0&0&0&0&1&1&-1&-1\\
0&0&0&0&0&0&0&1&-1&-1&1&0&0&0&0&0&0&0&0\\
0&0&0&0&0&0&0&1&-1&1&-1&0&0&0&0&0&0&0&0\\
0&0&0&0&0&0&0&0&0&0&0&1&-1&-1&1&0&0&0&0\\
0&0&0&0&0&0&0&0&0&0&0&1&-1&1&-1&0&0&0&0\\
0&0&0&0&0&0&0&0&0&0&0&0&0&0&0&1&-1&-1&1\\
0&0&0&0&0&0&0&0&0&0&0&0&0&0&0&1&-1&1&-1\\
0&0&0&0&0&0&0&1&1&1&1&0&0&0&0&0&0&0&0\\
0&0&0&0&0&0&0&0&0&0&0&1&1&1&1&0&0&0&0\\
0&0&0&0&0&0&0&0&0&0&0&0&0&0&0&1&1&1&1
\end{bmatrix}
$$

采用正交多松弛碰撞算子时，D3Q19 模型的转换矩阵及其逆矩阵的非零项均为 213 项，而采用上述非正交多松弛碰撞算子时，转换矩阵的非零项减少为 145 项，其逆矩阵的非零项则减少到 96 项[22]。该模型对应的对角松弛矩阵为

$$
\boldsymbol{\Lambda}=\operatorname{diag}\left(\frac{1}{\tau_c},\frac{1}{\tau_c},\frac{1}{\tau_c},\frac{1}{\tau_c},\frac{1}{\tau_e},\frac{1}{\tau_\nu},\frac{1}{\tau_\nu},\frac{1}{\tau_\nu},\frac{1}{\tau_\nu},\frac{1}{\tau_\nu},\frac{1}{\tau_q},\frac{1}{\tau_q},\frac{1}{\tau_q},\frac{1}{\tau_q},\frac{1}{\tau_q},\frac{1}{\tau_q},\frac{1}{\tau_\pi},\frac{1}{\tau_\pi},\frac{1}{\tau_\pi}\right)
$$

式中：$\tau_c=1$ 为守恒量对应的量纲为 1 的松弛时间；τ_q 和 τ_π 为可调松弛参数；τ_e 和 τ_ν 分别与体积黏度和动力黏度有关，其表达式与式(3.169)一致。

3.6.4 多松弛模型的 Chapman-Enskog 展开

与单松弛模型一样，多松弛模型所对应的宏观方程亦可通过 Chapman-Enskog 多尺度展开得到。对式(3.151)做泰勒级数展开可得

$$
\delta_t(\partial_t+\boldsymbol{e}_\alpha\cdot\nabla)f_\alpha+\frac{\delta_t^2}{2}(\partial_t+\boldsymbol{e}_\alpha\cdot\nabla)^2 f_\alpha=-\overline{\Lambda}_{\alpha\beta}(f_\beta-f_\beta^{\text{eq}})\mid_{(r,t)} \tag{3.176}
$$

将式(3.125)~式(3.127)代入式(3.176)，并对比各阶系数可得

$$
K^1:\quad (\partial_{t_1}+\boldsymbol{e}_\alpha\cdot\nabla_1)f_\alpha^{\text{eq}}=-\frac{1}{\delta_t}\overline{\Lambda}_{\alpha\beta}f_\beta^{(1)}\mid_{(r,t)} \tag{3.177}
$$

$$
K^2:\quad \partial_{t_2}f_\alpha^{\text{eq}}+(\partial_{t_1}+\boldsymbol{e}_\alpha\cdot\nabla_1)f_\alpha^{(1)}+\frac{\delta_t}{2}(\partial_{t_1}+\boldsymbol{e}_\alpha\cdot\nabla_1)^2 f_\alpha^{\text{eq}}=-\frac{1}{\delta_t}\overline{\Lambda}_{\alpha\beta}f_\beta^{(2)}\mid_{(r,t)} \tag{3.178}
$$

式(3.177)与式(3.178)左右两端同乘转换矩阵 $\boldsymbol{M}$ 可得

$$
K^1:\quad \boldsymbol{D}_1\boldsymbol{m}^{\text{eq}}=-\frac{1}{\delta_t}\boldsymbol{\Lambda}\boldsymbol{m}^{(1)} \tag{3.179}
$$

$$K^2:\quad \partial_{t_2}\boldsymbol{m}^{\mathrm{eq}}+\boldsymbol{D}_1\boldsymbol{m}^{(1)}+\frac{\delta_t}{2}\boldsymbol{D}_1^2\boldsymbol{m}^{\mathrm{eq}}=-\frac{1}{\delta_t}\boldsymbol{\Lambda m}^{(2)} \tag{3.180}$$

式中，$\boldsymbol{D}_1=\partial_{t_1}\boldsymbol{I}+\boldsymbol{C}_i\partial_{1i}$，其中 $\boldsymbol{C}_i=\boldsymbol{ME}_i\boldsymbol{M}^{-1}$ 与转换矩阵 $\boldsymbol{M}$ 及离散速度配置有关。对 D3Q19 格子，$\boldsymbol{E}_i=\mathrm{diag}(e_{0,i},e_{1,i},...,e_{18,i})$。

将式(3.179)代入式(3.180)可得

$$\partial_{t_2}\boldsymbol{m}^{\mathrm{eq}}+\boldsymbol{D}_1\left(\boldsymbol{I}-\frac{\boldsymbol{\Lambda}}{2}\right)\boldsymbol{m}^{(1)}=-\frac{1}{\delta_t}\boldsymbol{\Lambda m}^{(2)} \tag{3.181}$$

以非正交 D3Q19 多松弛模型为例，根据转换矩阵 $\boldsymbol{M}$ 及相应的逆矩阵 $\boldsymbol{M}^{-1}$，可得 $\boldsymbol{C}_i$ 的表达式，继而根据式(3.179)有[22]

$$\partial_{t_1}\rho+\partial_{1x}m_1^{\mathrm{eq}}+\partial_{1y}m_2^{\mathrm{eq}}+\partial_{1z}m_3^{\mathrm{eq}}=0 \tag{3.182}$$

$$\partial_{t_1}(\rho u_x)+\frac{1}{3}\partial_{1x}(m_4^{\mathrm{eq}}+m_5^{\mathrm{eq}})+\partial_{1y}m_7^{\mathrm{eq}}+\partial_{1z}m_8^{\mathrm{eq}}=0 \tag{3.183}$$

$$\partial_{t_1}(\rho u_y)+\partial_{1x}m_7^{\mathrm{eq}}+\partial_{1y}\left(\frac{1}{3}m_4^{\mathrm{eq}}-\frac{1}{6}m_5^{\mathrm{eq}}+\frac{1}{2}m_6^{\mathrm{eq}}\right)+\partial_{1z}m_9^{\mathrm{eq}}=0 \tag{3.184}$$

$$\partial_{t_1}(\rho u_z)+\partial_{1x}m_8^{\mathrm{eq}}+\partial_{1y}m_9^{\mathrm{eq}}+\partial_{1z}\left(\frac{1}{3}m_4^{\mathrm{eq}}-\frac{1}{6}m_5^{\mathrm{eq}}-\frac{1}{2}m_6^{\mathrm{eq}}\right)=0 \tag{3.185}$$

将式(3.175)代入式(3.182)~式(3.185)可得

$$\partial_{t_1}\rho+\partial_{1x}(\rho u_x)+\partial_{1y}(\rho u_y)+\partial_{1z}(\rho u_z)=0 \tag{3.186}$$

$$\partial_{t_1}(\rho u_x)+\partial_{1x}(p+\rho u_x^2)+\partial_{1y}(\rho u_y u_x)+\partial_{1z}(\rho u_z u_x)=0 \tag{3.187}$$

$$\partial_{t_1}(\rho u_y)+\partial_{1x}(\rho u_x u_y)+\partial_{1y}(p+\rho u_y^2)+\partial_{1z}(\rho u_z u_y)=0 \tag{3.188}$$

$$\partial_{t_1}(\rho u_z)+\partial_{1x}(\rho u_x u_z)+\partial_{1y}(\rho u_y u_z)+\partial_{1z}(p+\rho u_z^2)=0 \tag{3.189}$$

式中，$p=\rho c_s^2$。类似地，根据 $\boldsymbol{C}_i$ 的表达式及式(3.181)可得

$$\partial_{t_2}\rho=0 \tag{3.190}$$

$$\partial_{t_2}(\rho u_x)+\frac{1}{3}\partial_{1x}\left[\left(1-\frac{1}{2\tau_e}\right)m_4^{(1)}+\left(1-\frac{1}{2\tau_\nu}\right)m_5^{(1)}\right]+\partial_{1y}\left(1-\frac{1}{2\tau_\nu}\right)m_7^{(1)}+\partial_{1z}\left(1-\frac{1}{2\tau_\nu}\right)m_8^{(1)}=0 \tag{3.191}$$

$$\partial_{t_2}(\rho u_y)+\partial_{1x}\left(1-\frac{1}{2\tau_\nu}\right)m_7^{(1)}+\partial_{1y}\left[\frac{1}{3}\left(1-\frac{1}{2\tau_e}\right)m_4^{(1)}-\frac{1}{6}\left(1-\frac{1}{2\tau_\nu}\right)m_5^{(1)}+\frac{1}{2}\left(1-\frac{1}{2\tau_\nu}\right)m_6^{(1)}\right]+\partial_{1z}\left(1-\frac{1}{2\tau_\nu}\right)m_9^{(1)}=0 \tag{3.192}$$

$$\partial_{t_2}(\rho u_z)+\partial_{1x}\left(1-\frac{1}{2\tau_\nu}\right)m_8^{(1)}+\partial_{1y}\left(1-\frac{1}{2\tau_\nu}\right)m_9^{(1)}+\partial_{1z}\left[\frac{1}{3}\left(1-\frac{1}{2\tau_e}\right)m_4^{(1)}-\frac{1}{6}\left(1-\frac{1}{2\tau_\nu}\right)m_5^{(1)}-\frac{1}{2}\left(1-\frac{1}{2\tau_\nu}\right)m_6^{(1)}\right]=0 \tag{3.193}$$

此外，根据式(3.179)有

$$\partial_{t_1}m_4^{\mathrm{eq}}+\partial_{1x}(m_1^{\mathrm{eq}}+m_{11}^{\mathrm{eq}}+m_{13}^{\mathrm{eq}})+\partial_{1y}(m_2^{\mathrm{eq}}+m_{10}^{\mathrm{eq}}+m_{15}^{\mathrm{eq}})+\partial_{1z}(m_3^{\mathrm{eq}}+m_{12}^{\mathrm{eq}}+m_{14}^{\mathrm{eq}})=-\frac{1}{\tau_e\delta_t}m_4^{(1)} \tag{3.194}$$

$$\partial_{t_1} m_5^{\mathrm{eq}} + \partial_{1x}(2m_1^{\mathrm{eq}} - m_{11}^{\mathrm{eq}} - m_{13}^{\mathrm{eq}}) + \partial_{1y}(-m_2^{\mathrm{eq}} + 2m_{10}^{\mathrm{eq}} - m_{15}^{\mathrm{eq}}) + \partial_{1z}(-m_3^{\mathrm{eq}} + 2m_{12}^{\mathrm{eq}} - m_{14}^{\mathrm{eq}}) = -\frac{1}{\tau_\nu \delta_t} m_5^{(1)} \tag{3.195}$$

$$\partial_{t_1} m_6^{\mathrm{eq}} + \partial_{1x}(m_{11}^{\mathrm{eq}} - m_{13}^{\mathrm{eq}}) + \partial_{1y}(m_2^{\mathrm{eq}} - m_{15}^{\mathrm{eq}}) + \partial_{1z}(-m_3^{\mathrm{eq}} + m_{14}^{\mathrm{eq}}) = -\frac{1}{\tau_\nu \delta_t} m_6^{(1)} \tag{3.196}$$

$$\partial_{t_1} m_7^{\mathrm{eq}} + \partial_{1x} m_{10}^{\mathrm{eq}} + \partial_{1y} m_{11}^{\mathrm{eq}} = -\frac{1}{\tau_\nu \delta_t} m_7^{(1)} \tag{3.197}$$

$$\partial_{t_1} m_8^{\mathrm{eq}} + \partial_{1x} m_{12}^{\mathrm{eq}} + \partial_{1z} m_{13}^{\mathrm{eq}} = -\frac{1}{\tau_\nu \delta_t} m_8^{(1)} \tag{3.198}$$

$$\partial_{t_1} m_9^{\mathrm{eq}} + \partial_{1y} m_{14}^{\mathrm{eq}} + \partial_{1z} m_{15}^{\mathrm{eq}} = -\frac{1}{\tau_\nu \delta_t} m_9^{(1)} \tag{3.199}$$

将式(3.175)代入式(3.194)~式(3.199),并结合式(3.186)~式(3.189)可得[22]

$$\begin{gathered}
-m_4^{(1)} \approx 2\rho c_s^2 \tau_e \delta_t(\partial_{1x} u_x + \partial_{1y} u_y + \partial_{1z} u_z), \quad -m_5^{(1)} \approx 2\rho c_s^2 \tau_\nu \delta_t(2\partial_{1x} u_x - \partial_{1y} u_y - \partial_{1z} u_z) \\
-m_6^{(1)} \approx 2\rho c_s^2 \tau_\nu \delta_t(\partial_{1y} u_y - \partial_{1z} u_z), \quad -m_7^{(1)} \approx \rho c_s^2 \tau_\nu \delta_t(\partial_{1x} u_y + \partial_{1y} u_x) \\
-m_8^{(1)} \approx \rho c_s^2 \tau_\nu \delta_t(\partial_{1x} u_z + \partial_{1z} u_x), \quad -m_9^{(1)} \approx \rho c_s^2 \tau_\nu \delta_t(\partial_{1y} u_z + \partial_{1z} u_y)
\end{gathered} \tag{3.200}$$

将式(3.200)代入式(3.191)~式(3.193)可得

$$\partial_{t_2}(\rho u_x) = \partial_{1x}\left[\mu_{\mathrm{B}}(\nabla_1 \cdot \boldsymbol{u}) + \frac{2\mu}{3}(2\partial_{1x} u_x - \partial_{1y} u_y - \partial_{1z} u_z)\right] + \partial_{1y}[\mu(\partial_{1y} u_x + \partial_{1x} u_y)] + \partial_{1z}[\mu(\partial_{1z} u_x + \partial_{1x} u_z)] = 0 \tag{3.201}$$

$$\partial_{t_2}(\rho u_y) = \partial_{1x}[\mu(\partial_{1x} u_y + \partial_{1y} u_x)] + \partial_{1y}\left[\mu_{\mathrm{B}}(\nabla_1 \cdot \boldsymbol{u}) + \frac{2\mu}{3}(2\partial_{1y} u_y - \partial_{1x} u_x - \partial_{1z} u_z)\right] + \partial_{1z}[\mu(\partial_{1z} u_y + \partial_{1y} u_z)] = 0 \tag{3.202}$$

$$\partial_{t_2}(\rho u_z) = \partial_{1x}[\mu(\partial_{1x} u_z + \partial_{1z} u_x)] + \partial_{1y}[\mu(\partial_{1y} u_z + \partial_{1z} u_y)] + \partial_{1z}\left[\mu_{\mathrm{B}}(\nabla_1 \cdot \boldsymbol{u}) + \frac{2\mu}{3}(2\partial_{1z} u_z - \partial_{1x} u_x - \partial_{1y} u_y)\right] = 0 \tag{3.203}$$

式中,动力黏度 μ 和体积黏度 μ_{B} 由式(3.169)给出。根据式(3.125)与式(3.126)联立 t_1 和 t_2 时间尺度上的方程可得宏观方程为

$$\partial_t \rho + \nabla \cdot (\rho \boldsymbol{u}) = 0 \tag{3.204}$$

$$\partial_t(\rho \boldsymbol{u}) + \nabla \cdot (\rho \boldsymbol{u}\boldsymbol{u}) = -\nabla p + \nabla \cdot \left\{\mu[\nabla \boldsymbol{u} + (\nabla \boldsymbol{u})^{\mathrm{T}}] - \frac{2}{3}\mu(\nabla \cdot \boldsymbol{u})\boldsymbol{I} + \mu_{\mathrm{B}}(\nabla \cdot \boldsymbol{u})\boldsymbol{I}\right\} \tag{3.205}$$

当 $\tau_e = \tau_\nu$ 时,可得 $\mu_{\mathrm{B}} = 2\mu/3$,此时式(3.205)退化为单松弛模型所对应的宏观动量方程。

3.7 中心矩格子 Boltzmann 模型

中心矩(central-moments-based)格子 Boltzmann 模型最早是由 Geier 等[24]于 2006 年提出的,他们认为伽利略不变性不足是导致单松弛格子 Boltzmann 模型在低黏度工况下出现数值不稳定的重要原因之一(伽利略不变性是指物理规律不随观测的参考系变化而变化)。基于二阶截断形式的平衡态密度分布函数只能严格保证零阶矩~二阶矩的伽利略不变性,而采用对称性更好的离散速度模型和高阶截断形式的平衡态密度分布函数往往

能够改善伽利略不变性。此外，Geier 等认为，尽管多松弛碰撞算子能够提高格子 Boltzmann 方法的数值稳定性，但其增强效果仍有进一步提升的空间，主要是因为多松弛碰撞算子是基于分布函数的原始矩（raw moments）建立的，其中包含了宏观速度，从而在松弛过程中存在一种低阶矩影响高阶矩的“串音”（crosstalk）效应。基于这种考虑，Geier 等[24]提出将松弛过程建立在消除了宏观速度影响的中心矩空间，并约定了一种**级串**（cascaded）松弛顺序，即先松弛低阶矩，然后把松弛后的低阶矩代入高阶矩的松弛方程中，相应的碰撞算子被称作**级串松弛碰撞算子**或**中心矩碰撞算子**。Geier 等通过数值模拟证实了中心矩格子 Boltzmann 模型具有良好的数值稳定性，可以在粗网格上实现很高雷诺数的湍流模拟，使得该模型近年来受到了较多的关注。

3.7.1 早期模型

以 D2Q9 模型为例，我们简单介绍一下 Geier 等[24]所提出的中心矩模型。其推导比较烦琐，但基本原理如下。首先定义一个分布函数矢量

$$\boldsymbol{f}=(r,nw,w,sw,s,se,e,ne,n)^{\mathrm{T}} \tag{3.206}$$

式中：r 代表零速度方向的分布函数；nw 代表西北方向离散速度的分布函数，按逆时针的顺序依次类推。分布函数的中心矩定义如下：

$$\widetilde{k}_{mn}=\sum_{\alpha=0}^{8} f_\alpha\,(e_{\alpha x}-u_x)^m\,(e_{\alpha y}-u_y)^n \tag{3.207}$$

中心矩的平衡态由 Maxwell 平衡态分布的连续（积分）中心矩直接给出。基于标准格子模型的格子速度和声速，可以得到奇数阶次的中心矩平衡态为 0，偶数阶次的中心矩平衡态为

$$\widetilde{k}_{20}^{\mathrm{eq}}=\widetilde{k}_{02}^{\mathrm{eq}}=\rho c_{\mathrm{s}}^2,\quad \widetilde{k}_{22}^{\mathrm{eq}}=\rho c_{\mathrm{s}}^4 \tag{3.208}$$

在中心矩空间执行碰撞步后，分布函数可以重构为

$$\boldsymbol{f}^*=\boldsymbol{f}+\boldsymbol{M}\boldsymbol{k} \tag{3.209}$$

式中：$\boldsymbol{M}$ 为正交转换矩阵；$\boldsymbol{k}$ 为分布函数原始矩相对于其平衡态的偏差所组成的矢量，其求解过程比较烦琐。Geier 等所选定的前三个矩为守恒矩（质量和动量），因此 $k_1=k_2=k_3=0$，其他矩的形式如下（以 k_3、k_4 和 k_8 为例）：

$$k_3=\lfloor \omega_3[\rho(u_x^2+u_y^2)-e-n-s-w-2(se+sw+ne+nw-\rho/3)]/12\rfloor \tag{3.210}$$

$$k_4=\lfloor \omega_4[n+s-e-w+\rho(u_x^2-u_y^2)]/4\rfloor \tag{3.211}$$

$$k_8=\left\lfloor \begin{aligned} &\omega_8/4\{\rho/9-ne-nw-se-sw+2[u_x(ne-nw+se-sw)+\\ &u_y(ne+nw-se-sw)]+4u_xu_y(nw-ne+se-sw)-\\ &u_x^2(n+ne+nw+s+se+sw)+u_y^2(3\rho u_x^2-e-ne-nw-se-sw-w)\}-\\ &2k_3-2u_xk_7-2u_yk_6+4u_xu_yk_5-(3k_3/2-k_4/2)(u_x^2+u_y^2) \end{aligned} \right\rfloor \tag{3.212}$$

式中：$\lfloor\cdot\rfloor$表示一个截断算子；$\omega_{3,\cdots,8}$表示不同矩的量纲为 1 的松弛频率，其中 $\omega_4=\omega_5$ 与黏度系数有关，其他为可调松弛频率。执行碰撞步后，迁移步与单松弛和多松弛模型的迁移步相同。Geier 等发现，当可调松弛频率设定为 1.0 时，该中心矩模型表现出优良的数值性能。然而，早期的中心矩模型不利于应用及推广，主要原因有：

（1）松弛过程过于复杂，求解过程烦琐且缺乏一个简洁通用的表达式，且 Geier 等采用的逆时针离散速度标记法和文献中通用的离散速度标记法不同，其他研究者很难直接

应用这种方法。此外,一些研究者发现 Geier 等提供的附录存在错误[25]。

(2) 与单松弛模型相比,Geier 等的中心矩模型计算量过于庞大。文献[24]中没有提供同样网格下中心矩模型相对于单松弛模型的计算量差异,然而 De Rosis[20]在后来的数值计算中发现中心矩模型的计算量高达单松弛模型的 13~14 倍。

(3) 实际流动问题往往受到各种作用力的影响,Geier 等并未阐明如何在中心矩格子 Boltzmann 模型中添加作用力。

3.7.2 改进的中心矩模型

为了克服早期中心矩模型的缺陷,Fei 和 Luo[26]基于一种广义多松弛框架对中心矩模型重新进行了推导,提出了改进的中心矩模型,引入了简洁的模型表述形式,建立了中心矩模型与前述单松弛和多松弛模型之间的联系,且改进的中心矩模型的计算量仅为单松弛模型的 2 倍左右。

与单松弛和多松弛模型一样,中心矩模型的格子 Boltzmann 方程也可以分为碰撞步和迁移步:其中碰撞步在中心矩空间执行,然后通过重构得到碰撞后的分布函数;迁移步和前述模型完全相同。在不考虑外力作用时,格子 Boltzmann 方程的碰撞步可写成如下的广义多松弛形式:

$$f_\alpha^*(\boldsymbol{r},t)=f_\alpha(\boldsymbol{r},t)-(\boldsymbol{M}^{-1}\boldsymbol{N}^{-1}\boldsymbol{\Lambda}\boldsymbol{N}\boldsymbol{M})(f_\alpha-f_\alpha^{\mathrm{eq}})\big|_{(\boldsymbol{r},t)} \tag{3.213}$$

式中:矩阵 $\boldsymbol{M}$ 和多松弛模型中的转换矩阵一样,用于实现从离散速度空间到原始矩空间的转换;矩阵 $\boldsymbol{N}$ 则实现从原始矩空间到中心矩空间的转换,被称作转移矩阵;$\boldsymbol{\Lambda}$ 为松弛矩阵。矩阵 $\boldsymbol{M}$ 和 $\boldsymbol{N}$ 的具体形式取决于所选取的矩的集合,中心矩模型通常采用一组非正交的基矢量来简化转换矩阵和转移矩阵。此外,为了简化算法,松弛矩阵 $\boldsymbol{\Lambda}$ 可使用块对角矩阵。

根据以上定义可知从离散速度空间 $\boldsymbol{f}$ 到原始矩空间 $\boldsymbol{T}$ 和中心矩空间 $\widetilde{\boldsymbol{T}}$ 存在如下转换关系:

$$\widetilde{\boldsymbol{T}}=\boldsymbol{N}\boldsymbol{T}=\boldsymbol{N}\boldsymbol{M}\boldsymbol{f} \tag{3.214}$$

式中:$\boldsymbol{T}=(T_0,T_1,\cdots,T_{N-2},T_{N-1})^{\mathrm{T}}$;$\widetilde{\boldsymbol{T}}=(\widetilde{T}_0,\widetilde{T}_1,\cdots,\widetilde{T}_{N-2},\widetilde{T}_{N-1})^{\mathrm{T}}$。式(3.213)两端同乘 $\boldsymbol{NM}$ 可得

$$\widetilde{\boldsymbol{T}}^*=\widetilde{\boldsymbol{T}}-\boldsymbol{\Lambda}(\widetilde{\boldsymbol{T}}-\widetilde{\boldsymbol{T}}^{\mathrm{eq}}) \tag{3.215}$$

式中:$\widetilde{\boldsymbol{T}}^{\mathrm{eq}}$为平衡态中心矩;$\widetilde{\boldsymbol{T}}^*$是碰撞后的中心矩。根据式(3.215),程序实施时一般先在中心矩空间执行碰撞步,然后再重构碰撞后的分布函数,即

$$\boldsymbol{f}^*=\boldsymbol{M}^{-1}\boldsymbol{T}^*=\boldsymbol{M}^{-1}\boldsymbol{N}^{-1}\widetilde{\boldsymbol{T}}^* \tag{3.216}$$

之后可在离散速度空间执行迁移步,即式(3.156)。

式(3.214)可以看作是广义的多松弛框架,它建立了不同碰撞算子之间的联系:当转移矩阵为单位矩阵时,中心矩碰撞算子退化为多松弛碰撞算子;当松弛矩阵为对角矩阵且各元素取值相同时,中心矩碰撞算子退化为单松弛碰撞算子。在此基础上,Luo 等[27]建立了单松弛碰撞算子、多松弛碰撞算子和中心矩碰撞算子的统一模型框架。

下面分别介绍二维和三维情况下矩的集合的选取、平衡态中心矩 $\widetilde{\boldsymbol{T}}^{\mathrm{eq}}$的定义以及相应的转换矩阵 $\boldsymbol{M}$、转移矩阵 $\boldsymbol{N}$ 和松弛矩阵 $\boldsymbol{\Lambda}$。

1) D2Q9 中心矩模型

采用 D2Q9 离散速度模型时,需要选择 9 个独立的矩,对应转换矩阵中的 9 个基矢

量。二维中心矩模型通常采用如下形式的原始矩集合

$$\boldsymbol{T}=(k_{00},k_{10},k_{01},k_{20}+k_{02},k_{20}-k_{02},k_{11},k_{21},k_{12},k_{22})^{\mathrm{T}} \tag{3.217}$$

相应地,中心矩集合选取为

$$\widetilde{\boldsymbol{T}}=(\widetilde{k}_{00},\widetilde{k}_{10},\widetilde{k}_{01},\widetilde{k}_{20}+\widetilde{k}_{02},\widetilde{k}_{20}-\widetilde{k}_{02},\widetilde{k}_{11},\widetilde{k}_{21},\widetilde{k}_{12},\widetilde{k}_{22})^{\mathrm{T}} \tag{3.218}$$

式(3.217)和式(3.218)中的原始矩和中心矩分别定义如下:

$$k_{mn}=\langle f_\alpha \mid e_{\alpha x}^m e_{\alpha y}^n\rangle,\quad \widetilde{k}_{mn}=\langle f_\alpha \mid (e_{\alpha x}-u_x)^m(e_{\alpha y}-u_y)^n\rangle \tag{3.219}$$

式中,$\langle\cdot\mid\cdot\rangle$ 表示矢量点乘运算。根据式(3.103)、式(3.214)以及式(3.217)~式(3.219)可得

$$\boldsymbol{M}=\begin{bmatrix}1&1&1&1&1&1&1&1&1\\0&1&0&-1&0&1&-1&-1&1\\0&0&1&0&-1&1&1&-1&-1\\0&1&1&1&1&2&2&2&2\\0&1&-1&1&-1&0&0&0&0\\0&0&0&0&0&1&-1&1&-1\\0&0&0&0&0&1&1&-1&-1\\0&0&0&0&0&1&-1&-1&1\\0&0&0&0&0&1&1&1&1\end{bmatrix} \tag{3.220}$$

$$\boldsymbol{N}=\begin{bmatrix}1&0&0&0&0&0&0&0&0\\-u_x&1&0&0&0&0&0&0&0\\-u_y&0&1&0&0&0&0&0&0\\u_x^2+u_y^2&-2u_x&-2u_y&1&0&0&0&0&0\\u_x^2-u_y^2&-2u_x&2u_y&0&1&0&0&0&0\\u_xu_y&-u_y&-u_x&0&0&1&0&0&0\\-u_x^2u_y&2u_xu_y&u_x^2&-u_y/2&-u_y/2&-2u_x&1&0&0\\-u_y^2u_x&u_y^2&2u_xu_y&-u_x/2&u_x/2&-2u_y&0&1&0\\u_x^2u_y^2&-2u_xu_y^2&-2u_yu_x^2&u_x^2/2+u_y^2/2&u_y^2/2-u_x^2/2&4u_xu_y&-2u_y&-2u_x&1\end{bmatrix} \tag{3.221}$$

从式(3.221)可以看到矩阵 $\boldsymbol{N}$ 是一个下三角矩阵,实际上其逆矩阵 $\boldsymbol{N}^{-1}$ 也是一个下三角矩阵,且与式(3.221)非常相似,具体形式可参阅文献[26]。

在中心矩模型中,平衡态中心矩可选取为 Maxwell 平衡态分布的连续中心矩,即

$$\widetilde{k}_{mn}^{\mathrm{eq}}=\int_{-\infty}^{+\infty}\int_{-\infty}^{+\infty}f^{\mathrm{eq}}(\xi_x-u_x)^m(\xi_y-u_y)^n\mathrm{d}\xi_x\mathrm{d}\xi_y \tag{3.222}$$

根据式(3.222)和式(3.94)可得 D2Q9 模型的平衡态中心矩为

$$\widetilde{\boldsymbol{T}}^{\mathrm{eq}}=\boldsymbol{N}\boldsymbol{T}^{\mathrm{eq}}=\boldsymbol{N}\boldsymbol{M}\boldsymbol{f}^{\mathrm{eq}}=(\rho,0,0,2\rho c_s^2,0,0,0,0,\rho c_s^4)^{\mathrm{T}} \tag{3.223}$$

该模型对应的松弛矩阵为

$$\boldsymbol{\Lambda}=\mathrm{diag}(s_0,s_1,s_1,s_{2\mathrm{B}},s_2,s_2,s_3,s_3,s_4) \tag{3.224}$$

式中:守恒矩的松弛参数可选取为 $s_0=s_1=1$;高阶矩的松弛参数 s_3 和 s_4 不影响宏观流体动力学,可自由选取以提高模型的数值稳定性;s_2 和 $s_{2\mathrm{B}}$ 分别与动力黏度和体积黏度有关

$$\mu=\rho c_s^2(1/s_2-0.5)\delta_t,\quad \mu_{\mathrm{B}}=\rho c_s^2(1/s_{2\mathrm{B}}-0.5)\delta_t \tag{3.225}$$

2）D3Q19 中心矩模型

Fei、Luo 和 Li 在广义多松弛框架下进一步建立了三维的中心矩模型[28]。以 D3Q19 离散速度模型为例，可采用如下形式的原始矩集合：

$$\boldsymbol{T}=(k_{000},k_{100},k_{010},k_{001},\cdots,k_{200},k_{020},k_{002},k_{120},k_{102},k_{210},\cdots,k_{220},k_{202},k_{022})^{\mathrm{T}} \tag{3.226}$$

相应地，中心矩集合可选取为

$$\widetilde{\boldsymbol{T}}=(\widetilde{k}_{000},\widetilde{k}_{100},\widetilde{k}_{010},\widetilde{k}_{001},\cdots,\widetilde{k}_{200},\widetilde{k}_{020},\widetilde{k}_{002},\widetilde{k}_{120},\widetilde{k}_{102},\widetilde{k}_{210},\cdots,\widetilde{k}_{220},\widetilde{k}_{202},\widetilde{k}_{022})^{\mathrm{T}} \tag{3.227}$$

式（3.226）和式（3.227）中的原始矩和中心矩分别定义如下：

$$k_{mnp}=\langle f_\alpha \mid e_{\alpha x}^m e_{\alpha y}^n e_{\alpha z}^p\rangle,\quad \widetilde{k}_{mnp}=\langle f_\alpha \mid (e_{\alpha x}-u_x)^m(e_{\alpha y}-u_y)^n(e_{\alpha z}-u_z)^p\rangle \tag{3.228}$$

转换矩阵 $\boldsymbol{M}$、转移矩阵 $\boldsymbol{N}$ 及其逆矩阵的具体形式可根据式（3.214）及式（3.226）~式（3.228）得到。限于篇幅，这里不再列举，具体可参阅文献[28]。

同样地，平衡态中心矩可选取为 Maxwell 平衡态分布的连续中心矩，即

$$\widetilde{k}_{mnp}^{\mathrm{eq}}=\int_{-\infty}^{+\infty}\int_{-\infty}^{+\infty} f^{\mathrm{eq}}(\xi_x-u_x)^m(\xi_y-u_y)^n(\xi_z-u_z)^p\mathrm{d}\xi_x\mathrm{d}\xi_y\mathrm{d}\xi_z \tag{3.229}$$

据此可得 D3Q19 模型的平衡态中心矩为

$$\widetilde{\boldsymbol{T}}^{\mathrm{eq}}=(\rho,0,0,0,0,0,0,\rho c_{\mathrm{s}}^2,\rho c_{\mathrm{s}}^2,\rho c_{\mathrm{s}}^2,0,0,0,0,0,0,\rho c_{\mathrm{s}}^4,\rho c_{\mathrm{s}}^4,\rho c_{\mathrm{s}}^4)^{\mathrm{T}} \tag{3.230}$$

需要指出的是，为了描述正确的宏观流体动力学，三维中心矩模型还需要独立松弛以下三个中心矩：$\widetilde{k}_{200}-\widetilde{k}_{020}$，$\widetilde{k}_{200}-\widetilde{k}_{002}$ 和 $\widetilde{k}_{200}+\widetilde{k}_{020}+\widetilde{k}_{002}$。如果在式（3.227）中直接包含这三个中心矩，高阶中心矩表述成原始矩的多项式展开在形式上会比较烦琐，对应的转移矩阵 $\boldsymbol{N}$ 及其逆矩阵也比较复杂。为了避免这个问题，式（3.227）采用了另外三个的中心矩：$\widetilde{k}_{200}$、$\widetilde{k}_{020}$ 和 $\widetilde{k}_{002}$，并选取了如下形式的块对角松弛矩阵：

$$\boldsymbol{\Lambda}=\operatorname{diag}\left(s_0,s_1,s_1,s_1,s_\nu,s_\nu,s_\nu,\begin{bmatrix} s_+,s_-,s_- \\ s_-,s_+,s_- \\ s_-,s_-,s_+ \end{bmatrix},s_3,s_3,s_3,s_3,s_3,s_3,s_4,s_4,s_4\right) \tag{3.231}$$

式中，3×3 的块结构用于实现二阶中心矩 $\widetilde{k}_{200}-\widetilde{k}_{020}$、$\widetilde{k}_{200}-\widetilde{k}_{002}$ 和 $\widetilde{k}_{200}+\widetilde{k}_{020}+\widetilde{k}_{002}$ 的独立松弛，其中 $s_+=(s_{2\mathrm{B}}+2s_2)/3$，$s_-=(s_{2\mathrm{B}}-s_2)/3$。类似地，守恒矩的松弛参数可选取为 $s_0=s_1=1$；高阶矩的松弛参数 s_3 和 s_4 可自由选取；s_2 和 $s_{2\mathrm{B}}$ 分别与动力黏度和体积黏度有关，即

$$\mu=\rho c_{\mathrm{s}}^2(1/s_2-0.5)\delta_t,\quad \mu_{\mathrm{B}}=\frac{2}{3}\rho c_{\mathrm{s}}^2(1/s_{2\mathrm{B}}-0.5)\delta_t \tag{3.232}$$

可以看到，式（3.232）与前述三维 D3Q19 多松弛模型的表述式是一致的。

3.8 格子 Boltzmann 方法的作用力格式

流体系统中通常都存在着流体内部或外界环境施加的作用力，如分子间作用力、重力等。外界环境作用在流体上的力可分为体积力和表面力，其中体积力是一种非接触力，通常与质量有关，也称质量力；表面力则是一种接触力，它又可分为垂直于表面的作用力和平行于表面的作用力。例如，剪切力就是平行于表面的作用力。在格子 Boltzmann 方法中，流体内部及外界环境施加的作用力，往往是通过作用力格式嵌入格子 Boltzmann 方程。

3.8.1 He-Shan-Doolen 格式

前面曾提到，单松弛格子 Boltzmann 方程是通过对式（3.99）右端的积分采用矩形法

逼近得到的,这一处理在数学上只有一阶精度。幸运的是,针对碰撞项,矩形法所导致的数值黏性可以吸收到物理黏性中,即式(3.146),从而碰撞项的积分实际上具有二阶精度。但是,当流体系统所受到的作用力随时间或空间变化比较显著时,对作用力项的积分采用矩形法会引入较大的数值误差。为了解决这一问题,He 等[29]提出,碰撞项的积分可以采用矩形法,但作用力项的积分应当采用二阶精度的梯形法,即

$$f_\alpha(\boldsymbol{r}+\boldsymbol{e}_\alpha\delta_t,t+\delta_t)-f_\alpha(\boldsymbol{r},t)=-\frac{1}{\tau}[f_\alpha(\boldsymbol{r},t)-f_\alpha^{\mathrm{eq}}(\boldsymbol{r},t)]+\frac{\delta_t}{2}[G_\alpha(\boldsymbol{r},t)+G_\alpha(\boldsymbol{r}+\boldsymbol{e}_\alpha\delta_t,t+\delta_t)] \tag{3.233}$$

式中,作用力项 G_α 为

$$G_\alpha=\frac{(\boldsymbol{e}_\alpha-\boldsymbol{u})\cdot\boldsymbol{F}}{\rho c_s^2}f_\alpha^{\mathrm{eq}} \tag{3.234}$$

这一表达式源自连续 Boltzmann 方程(3.93)。He 等[29]假定连续 Boltzmann 方程中的作用力项 $\boldsymbol{a}\cdot\nabla_{\xi}f$ 可用 $\boldsymbol{a}\cdot\nabla_{\xi}f^{\mathrm{eq}}$ 近似,并结合式(3.94)得到了 $\boldsymbol{a}\cdot\nabla_{\xi}f^{\mathrm{eq}}$ 的具体表达式,继而导出了离散速度空间的作用力项即式(3.234)。

由于式(3.233)右端含有 $t+\delta_t$ 时刻的作用力项,使得该方程为隐式方程。为此,He 等[29-30]引入了一个新的分布函数

$$\bar{f}_\alpha=f_\alpha-\frac{\delta_t}{2}G_\alpha \tag{3.235}$$

将式(3.235)代入式(3.233)可得

$$\bar{f}_\alpha(\boldsymbol{r}+\boldsymbol{e}_\alpha\delta_t,t+\delta_t)-\bar{f}_\alpha(\boldsymbol{r},t)=-\frac{1}{\tau}[\bar{f}_\alpha(\boldsymbol{r},t)-f_\alpha^{\mathrm{eq}}(\boldsymbol{r},t)]+\delta_t\left(1-\frac{1}{2\tau}\right)G_\alpha(\boldsymbol{r},t) \tag{3.236}$$

与式(3.233)不同,式(3.236)为显式方程。根据式(3.235),对 D2Q9 模型,宏观密度与速度为

$$\rho=\sum_{\alpha=0}^{8}\bar{f}_\alpha,\quad \rho\boldsymbol{u}=\sum_{\alpha=0}^{8}\boldsymbol{e}_\alpha\bar{f}_\alpha+\frac{\delta_t}{2}\boldsymbol{F} \tag{3.237}$$

式(3.236)、式(3.237)及式(3.234)即为 He 等所提出的作用力格式,称为 He-Shan-Doolen 格式。

3.8.2 Guo-Zheng-Shi 格式

2002 年,Guo 等[31]基于 Chapman-Enskog 多尺度展开,研究了不同作用力格式的离散效应,继而提出了一种新的作用力格式,即

$$f_\alpha(\boldsymbol{r}+\boldsymbol{e}_\alpha\delta_t,t+\delta_t)-f_\alpha(\boldsymbol{r},t)=-\frac{1}{\tau}[f_\alpha(\boldsymbol{r},t)-f_\alpha^{\mathrm{eq}}(\boldsymbol{r},t)]+\delta_t\left(1-\frac{1}{2\tau}\right)G_\alpha(\boldsymbol{r},t) \tag{3.238}$$

式中,作用力项 G_α 为

$$G_\alpha=\omega_\alpha\left[\frac{\boldsymbol{e}_\alpha-\boldsymbol{u}}{c_s^2}+\frac{(\boldsymbol{e}_\alpha\cdot\boldsymbol{u})\boldsymbol{e}_\alpha}{c_s^4}\right]\cdot\boldsymbol{F} \tag{3.239}$$

相应的宏观密度与速度定义为[31]

$$\rho=\sum_{\alpha=0}^{8}f_\alpha,\quad \rho\boldsymbol{u}=\sum_{\alpha=0}^{8}\boldsymbol{e}_\alpha f_\alpha+\frac{\delta_t}{2}\boldsymbol{F} \tag{3.240}$$

式(3.238)~式(3.240)即为 Guo-Zheng-Shi 作用力格式。Guo 等[31]通过 Chapman-Enskog 多尺度展开证明了该格式可以恢复正确的宏观方程。

对比上述两种格式可以发现,Guo-Zheng-Shi 格式中的分布函数 f_α 与标准格子 Boltzmann 方程及式(3.233)中的分布函数 f_α 是不同的,它实质上等价于 He-Shan-Doolen 格式中的分布函数 $\bar{f}_\alpha$[32]。此外,这两种格式的作用力项的零阶矩和一阶矩是完全相同的,它们在二阶矩上的差别为 $O(u^3)$。具体地,式(3.239)的二阶矩为

$$\sum_{\alpha=0}^{8} \boldsymbol{e}_\alpha \boldsymbol{e}_\alpha G_\alpha = \boldsymbol{uF}+\boldsymbol{Fu} \tag{3.241}$$

而式(3.234)的二阶矩为

$$\sum_{\alpha=0}^{8} \boldsymbol{e}_\alpha \boldsymbol{e}_\alpha G_\alpha = \boldsymbol{uF}+\boldsymbol{Fu}-\frac{\boldsymbol{F}\cdot\boldsymbol{u}}{c_s^2}\boldsymbol{uu} \tag{3.242}$$

因此,对于低马赫数流动,Guo-Zheng-Shi 格式和 He-Shan-Doolen 格式是等效的。

3.8.3 Exact-Difference-Method 格式

Kupershtokh 等[33-34]提出的一种作用力格式在文献中亦受到了一定的关注,其形式如下:

$$f_\alpha(\boldsymbol{r}+\boldsymbol{e}_\alpha\delta_t, t+\delta_t)-f_\alpha(\boldsymbol{r},t) = -\frac{1}{\tau}[f_\alpha - f_\alpha^{eq}(\rho,\boldsymbol{u})]+f_\alpha^{eq}(\rho,\boldsymbol{u}+\Delta\boldsymbol{u})-f_\alpha^{eq}(\rho,\boldsymbol{u}) \tag{3.243}$$

式中,$\Delta\boldsymbol{u}=\boldsymbol{F}\delta_t/\rho$。与前两种格式不同,式(3.243)中的速度 $\boldsymbol{u}$ 定义为 $\rho\boldsymbol{u}=\sum_{\alpha=0}^{8}\boldsymbol{e}_\alpha f_\alpha$,但流体的宏观速度仍按式(3.240)计算。Kupershtokh 等[33]将该格式命名为精确差分方法(exact-difference-method,EDM)格式,简称 EDM 格式。

与 He-Shan-Doolen 格式类似,Kupershtokh 等亦假定连续 Boltzmann 方程(3.93)中的作用力项 $\boldsymbol{a}\cdot\nabla_{\boldsymbol{\xi}}f$ 可用 $\boldsymbol{a}\cdot\nabla_{\boldsymbol{\xi}}f^{eq}$ 近似,并认为密度近似为常数,于是有

$$\frac{\mathrm{d}f^{eq}(\rho,\boldsymbol{u})}{\mathrm{d}t} = \frac{\partial f^{eq}}{\partial\boldsymbol{u}}\cdot\frac{\mathrm{d}\boldsymbol{u}}{\mathrm{d}t}+\frac{\partial f^{eq}}{\partial\rho}\frac{\mathrm{d}\rho}{\mathrm{d}t}\approx\frac{\partial f^{eq}}{\partial\boldsymbol{u}}\cdot\frac{\mathrm{d}\boldsymbol{u}}{\mathrm{d}t} \tag{3.244}$$

结合式(3.94)可知,$\partial f^{eq}/\partial\boldsymbol{u}=-\partial f^{eq}/\partial\boldsymbol{\xi}=-\nabla_{\boldsymbol{\xi}}f^{eq}$,代入式(3.244)可得

$$\frac{\mathrm{d}f^{eq}(\rho,\boldsymbol{u})}{\mathrm{d}t}\approx\frac{\partial f^{eq}}{\partial\boldsymbol{u}}\cdot\frac{\mathrm{d}\boldsymbol{u}}{\mathrm{d}t} = -\boldsymbol{a}\cdot\nabla_{\boldsymbol{\xi}}f^{eq} \tag{3.245}$$

基于上述结果,Kupershtokh 等[33-34]用 $\mathrm{d}f^{eq}(\boldsymbol{u})/\mathrm{d}t$ 替换了式(3.93)中的作用力项 $-\boldsymbol{a}\cdot\nabla_{\boldsymbol{\xi}}f$。

对 $\mathrm{d}f^{eq}(\boldsymbol{u})/\mathrm{d}t$ 从 t 时刻到 $t+\delta_t$ 时刻积分可得

$$\int_t^{t+\delta_t}\frac{\mathrm{d}f^{eq}(\boldsymbol{u})}{\mathrm{d}t}\mathrm{d}t = f^{eq}(\boldsymbol{u}^{t+\delta_t}) - f^{eq}(\boldsymbol{u}^t) \tag{3.246}$$

式中,$\boldsymbol{u}^{t+\delta_t}$ 为 $t+\delta_t$ 时刻的速度,且有

$$\boldsymbol{u}^{t+\delta_t} = \boldsymbol{u}^t + \int_t^{t+\delta_t}\frac{\mathrm{d}\boldsymbol{u}}{\mathrm{d}t}\mathrm{d}t = \boldsymbol{u}^t + \int_t^{t+\delta_t}\frac{\boldsymbol{F}}{\rho}\mathrm{d}t \tag{3.247}$$

式中,$\boldsymbol{F}/\rho=\boldsymbol{a}=\mathrm{d}\boldsymbol{u}/\mathrm{d}t$ 为加速度。Kupershtokh 等[33-34]假定,从 t 时刻到 $t+\delta_t$ 时刻 $\boldsymbol{F}/\rho$ 近似为常数,于是有

$$\boldsymbol{u}^{t+\delta_t} = \boldsymbol{u}^t + \frac{\boldsymbol{F}}{\rho}\delta_t \tag{3.248}$$

将式(3.248)代入式(3.246),并将其拓展至离散速度空间,即可得式(3.243)中的作用力项。

从 EDM 格式的导出过程可以看到,该格式采用了多个假设,包括密度近似为常数、$\boldsymbol{F}/\rho$ 近似为常数等。因此,EDM 格式并不是一种通用的精确差分格式。实际上,根据 Chapman-Enskog 多尺度展开,EDM 格式在宏观动量方程中所引入的误差项[35]为

$$\boldsymbol{E}_r = -\delta_t^2 \nabla \cdot \left(\frac{\boldsymbol{FF}}{4\rho}\right) \tag{3.249}$$

当密度 ρ 和作用力 $\boldsymbol{F}$ 近似为常数时,EDM 格式的误差项可以忽略。但是,当密度或作用力随空间变化比较显著时,误差项的影响不可忽略[35]。

3.8.4 多松弛模型的作用力格式

针对 D2Q9 多松弛模型,2005 年,McCracken 和 Abraham[36] 提出了相应的作用力格式。类似地,他们对作用力项的积分采用了梯形法,即

$$\begin{aligned} f_\alpha(\boldsymbol{r}+\boldsymbol{e}_\alpha\delta_t, t+\delta_t) - f_\alpha(\boldsymbol{r},t) = & -\overline{\Lambda}_{\alpha\beta}[f_\beta(\boldsymbol{r},t) - f_\beta^{\mathrm{eq}}(\boldsymbol{r},t)] + \\ & \frac{\delta_t}{2}[G_\alpha(\boldsymbol{r},t) + G_\alpha(\boldsymbol{r}+\boldsymbol{e}_\alpha\delta_t, t+\delta_t)] \end{aligned} \tag{3.250}$$

通过引入新的分布函数 $\bar{f}_\alpha = f_\alpha - 0.5\delta_t G_\alpha$,可将隐式方程(3.250)转化为显式方程

$$\begin{aligned} \bar{f}_\alpha(\boldsymbol{r}+\boldsymbol{e}_\alpha\delta_t, t+\delta_t) - \bar{f}_\alpha(\boldsymbol{r},t) = & -\overline{\Lambda}_{\alpha\beta}[\bar{f}_\beta(\boldsymbol{r},t) - f_\beta^{\mathrm{eq}}(\boldsymbol{r},t)] + \\ & \delta_t[G_\alpha(\boldsymbol{r},t) - 0.5\overline{\Lambda}_{\alpha\beta}G_\beta(\boldsymbol{r},t)] \end{aligned} \tag{3.251}$$

相应地,多松弛模型的碰撞步由式(3.157)转化为

$$\boldsymbol{m}^* = \overline{\boldsymbol{m}} - \boldsymbol{\Lambda}(\overline{\boldsymbol{m}} - \boldsymbol{m}^{\mathrm{eq}}) + \delta_t\left(\boldsymbol{I} - \frac{\boldsymbol{\Lambda}}{2}\right)\boldsymbol{S} \tag{3.252}$$

式中,$\boldsymbol{S} = \boldsymbol{MG}$ 为矩空间的作用力项,其中 $\boldsymbol{G} = (G_0, G_1, \cdots, G_8)^{\mathrm{T}}$ 为离散速度空间的作用力项。因此,可通过 $\boldsymbol{S} = \boldsymbol{MG}$ 由离散速度空间的作用力项导出矩空间的作用力项。据此,McCracken 和 Abraham[36] 得到了 D2Q9 多松弛模型的矩空间作用力项

$$\boldsymbol{S} = \begin{bmatrix} 0 \\ 6(u_x F_x + u_y F_y) \\ -6(u_x F_x + u_y F_y) \\ F_x \\ -F_x \\ F_y \\ -F_y \\ 2(u_x F_x - u_y F_y) \\ (u_x F_y + u_y F_x) \end{bmatrix} \tag{3.253}$$

相应的格子 Boltzmann 方程迁移步为

$$\bar{f}_\alpha(\boldsymbol{r}+\boldsymbol{e}_\alpha\delta_t, t+\delta_t) = f_\alpha^*(\boldsymbol{r},t) \tag{3.254}$$

式中,$\boldsymbol{f}^* = \boldsymbol{M}^{-1}\boldsymbol{m}^*$。宏观密度与速度仍由式(3.237)给出。对三维多松弛模型及非正交多松弛模型,可采用类似的方式导出相应的矩空间作用力项。

3.8.5 中心矩模型的作用力格式

Geier 等[24]在提出早期的中心矩格子 Boltzmann 模型时并没有考虑作用力的影响。2009 年,Premnath 和 Banerjee[25]首次讨论了如何在中心矩模型中添加作用力,2014 年,Lycett-Brown 和 Luo [18]将中心矩模型推广到了多相流的模拟,他们采用了单松弛模型中的几种作用力格式并进行了对比。2017 年,Fei 和 Luo[26]基于广义多松弛框架对中心矩模型中已有的作用力格式进行了分析,发现它们均不满足一致性条件:即当中心矩模型退化为单松弛模型或多松弛模型时,相应的作用力格式不能一致性地退化为单松弛或多松弛模型中已广泛认可的正确的作用力格式。为此,他们提出了一种满足一致性条件的中心矩模型作用力格式,并进一步推广到了三维[28]。

这里仅介绍 Fei 和 Luo[26]所提出的一致性作用力格式。以 D2Q9 离散速度模型为例,考虑作用力时,式(3.213)中的广义多松弛框架可改写成

$$f_\alpha^*(\boldsymbol{r},t)=f_\alpha(\boldsymbol{r},t)-(\boldsymbol{M}^{-1}\boldsymbol{N}^{-1}\boldsymbol{\Lambda}\boldsymbol{N}\boldsymbol{M})(\bar{f}_\alpha-f_\alpha^{\mathrm{eq}})\big|_{(\boldsymbol{r},t)}+\frac{\delta_t}{2}\left[R_\alpha\big|_{(\boldsymbol{r},t)}+R_\alpha\big|_{(\boldsymbol{r}+\boldsymbol{e}_\alpha\delta_t,t+\delta_t)}\right] \tag{3.255}$$

通过引入新的分布函数$\bar{f}_\alpha=f_\alpha-0.5\delta_t R_\alpha$,可将式(3.255)转化为

$$f_\alpha^*(\boldsymbol{r},t)=\bar{f}_\alpha(\boldsymbol{r},t)-(\boldsymbol{M}^{-1}\boldsymbol{N}^{-1}\boldsymbol{\Lambda}\boldsymbol{N}\boldsymbol{M})(\bar{f}_\alpha-f_\alpha^{\mathrm{eq}})\big|_{(\boldsymbol{r},t)}+\delta_t\boldsymbol{M}^{-1}\boldsymbol{N}^{-1}\left(\boldsymbol{I}-\frac{\boldsymbol{\Lambda}}{2}\right)\boldsymbol{N}\boldsymbol{M}R_\alpha(\boldsymbol{r},t) \tag{3.256}$$

式(3.256)两端同乘 $\boldsymbol{NM}$ 可得

$$\widetilde{\boldsymbol{T}}^*=\widetilde{\boldsymbol{T}}-\boldsymbol{\Lambda}(\widetilde{\boldsymbol{T}}-\widetilde{\boldsymbol{T}}^{\mathrm{eq}})+\delta_t\left(\boldsymbol{I}-\frac{\boldsymbol{\Lambda}}{2}\right)\boldsymbol{C} \tag{3.257}$$

式中,$\boldsymbol{C}=\boldsymbol{NMR}$ 为中心矩空间的作用力项,其中 $\boldsymbol{R}=(R_0,R_1,\cdots,R_8)^{\mathrm{T}}$ 为离散速度空间的作用力项。在程序实施中并不需要用到离散速度空间作用力项的具体形式,但为了保证一致性,它的中心矩应等于 Boltzmann 方程中作用力项的连续形式中心矩,即

$$\langle R_\alpha\,|\,(e_{\alpha x}-u_x)^m(e_{\alpha y}-u_y)^n\rangle=\int_{-\infty}^{+\infty}\int_{-\infty}^{+\infty}G(\xi_x-u_x)^m(\xi_y-u_y)^n\mathrm{d}\xi_x\mathrm{d}\xi_y \tag{3.258}$$

式中,G 表示 Boltzmann 方程的作用力项。在积分运算时,可采用 He 等[29]所提出的近似,即 $G\approx-\boldsymbol{a}\cdot\nabla_{\xi}f^{\mathrm{eq}}$,其中负号代表将作用力项移至 Boltzmann 方程的右侧。据此可得 D2Q9 模型的中心矩空间作用力项为

$$\boldsymbol{C}=(0,F_x,F_y,0,0,0,c_s^2F_y,c_s^2F_x,0)^{\mathrm{T}} \tag{3.259}$$

相应地,宏观密度和速度的计算与式(3.237)一致。

3.9 小　结

本章沿着 Boltzmann 方程—Maxwell 分布—Boltzmann-BGK 方程—格子 Boltzmann-BGK 方程这一主线对格子 Boltzmann 方法的基础理论进行了详细介绍。Boltzmann 方程是气体动理论的基本方程,Maxwell 分布是 Boltzmann 方程的一个特殊解,它是单组分单原子气体的平衡态分布。Boltzmann 方程与流体力学基本方程之间有着密切的联系,这种联系通过 Maxwell 输运方程得到了清晰展示,它为通过数值求解 Boltzmann 方程来模拟流体的

宏观运动提供了理论基础。但由于 Boltzmann 方程是复杂的微积分方程，它的直接求解通常是不可能的，于是人们便提出了许多近似解法，Boltzmann-BGK 方程应运而生。

本章还对格子 Boltzmann 方法的基本模型和几种常用的作用力格式进行了详细介绍。格子 Boltzmann-BGK 方程可以看作是连续 Boltzmann-BGK 方程的一种特殊离散形式，而多松弛碰撞算子和中心矩碰撞算子的出现主要是为了克服单松弛 BGK 碰撞算子在计算稳定性等方面的不足。需要指出的是，不能因为多松弛碰撞算子和中心矩碰撞算子的出现而无视单松弛碰撞算子的优点，更不能因此否定单松弛格子 Boltzmann 方法的历史意义和研究价值。单松弛格子 Boltzmann 方法简单直观，易于建模，非常有利于发现一些建模的基本规律。对于初学者来说，可以从学习单松弛格子 Boltzmann 方法入手，编写相应的代码，熟悉格子 Boltzmann 方法的程序实施，在此基础上，由易及难，学习并掌握其他碰撞算子的相关理论及程序实施。

习　题

3-1　试证明对离散 Boltzmann-BGK 方程即式(3.97)的时间导数项采用向后差分、空间导数项采用向前差分即可导出格子 Boltzmann-BGK 方程，其中时间和空间离散基于 $\delta_x=e_{\alpha x}\delta_t$、$\delta_y=e_{\alpha y}\delta_t$、$\delta_z=e_{\alpha z}\delta_t$。

3-2　根据 Maxwell 平衡态分布函数式(3.94)导出其三阶矩 $\int \xi_i\xi_j\xi_k f^{\mathrm{eq}}\mathrm{d}\boldsymbol{\xi}$ 的表达式，并分析 D2Q9 模型平衡态分布函数的三阶矩 $\sum_\alpha e_{\alpha i}e_{\alpha j}e_{\alpha k}f_\alpha^{\mathrm{eq}}$ 与之相比有何异同。

3-3　若式(3.171)中的三个向量分别选取为 $\varphi_{4,\alpha}=e_{\alpha x}^2$、$\varphi_{5,\alpha}=e_{\alpha y}^2$、$\varphi_{6,\alpha}=e_{\alpha z}^2$，能否构造相应的非正交多松弛模型？结合 Chapman-Enskog 展开对这一问题进行分析。

3-4　采用 Chapman-Enskog 展开对式(3.233)进行多尺度展开分析，导出其对应的宏观方程，并分析为什么作用力项的积分应当采用梯形法而不能采用矩形法。

3-5　根据式(3.234)以及 D2Q9 模型的转换矩阵和平衡态密度分布函数导出 He-Shan-Doolen 作用力格式在矩空间的表达式。

参考文献

[1] McNamara G R, Zanetti G. Use of the Boltzmann equation to simulate lattice-gas automata[J]. Physical Review Letters, 1988, 61: 2332-2335.

[2] He X Y, Luo L S. A priori derivation of the lattice Boltzmann equation[J]. Physical Review E, 1997, 55(6): R6333-R6336.

[3] He X Y, Luo L S. Theory of the lattice Boltzmann method: From the Boltzmann equation to the lattice Boltzmann equation[J]. Physical Review E, 1997, 56(6): 6811-6817.

[4] 李元香，康立改，陈毓屏．格子气自动机[M]．北京：清华大学出版社，1994.

[5] Bhatnagar P L, Gross E P, Krook M. A model for collision processes in gases. I. Small amplitude processes in charged and neutral one-component systems[J]. Physical Review, 1954, 94(3): 511-525.

[6] 应纯同．气体输运理论及应用[M]．北京：清华大学出版社，1990.

[7] 沈青．稀薄气体动力学[M]．北京：国防工业出版社，2003.

[8] Bird G A. Molecular Gas Dynamics and the Direct Simulation of Gas Flows[M]. Oxford: Clarendon

Press,1994.

[9] Chapman S,Cowling T G. The Mathematical Theory of Non-uniform Gases[M]. 3rd edition. Cambridge: Cambridge University Press,1970.

[10] 郭照立,郑楚光. 格子 Boltzmann 方法的原理及应用[M]. 北京:科学出版社,2009.

[11] Wolf-Gladrow D A. Lattice-Gas Cellular Automata and Lattice Boltzmann Models: An Introduction [M]. Springer Science & Business Media,2000.

[12] Qian Y H,d'Humières D,Lallemand P. Lattice BGK models for Navier-Stokes equation[J]. Europhysics Letters,1992,17(6):479-484.

[13] Lallemand P,Luo L S. Theory of the lattice Boltzmann method:Dispersion,dissipation,isotropy,Galilean invariance,and stability[J]. Physical Review E,2000,61:6546-6562.

[14] Krüger T,Kusumaatmaja H,Kuzmin A,et al. The Lattice Boltzmann Method-Principles and Practice [M]. Springer Nature,2017.

[15] Li Q,Luo K H,He Y L,et al. Coupling lattice Boltzmann model for simulation of thermal flows on standard lattices[J]. Physical Review E,2012,85(1):016710.

[16] Luo L S,Liao W,Chen X W,et al. Numerics of the lattice Boltzmann method:Effects of collision models on the lattice Boltzmann simulations[J]. Physical Review E,2011,83(5):056710.

[17] d'Humières D,Ginzburg I,Krafczyk M,et al. Multiple-relaxation-time lattice Boltzmann models in three dimensions[J]. Philosophical Transactions of the Royal Society of London,Series A,2002,360:437-451.

[18] Lycett-Brown D,Luo K H. Multiphase cascaded lattice Boltzmann method[J]. Computers & Mathematics with Applications,2014,67(2):350-362.

[19] Liu Q,He Y L,Li D,et al. Non-orthogonal multiple-relaxation-time lattice Boltzmann method for incompressible thermal flows[J]. International Journal of Heat and Mass Transfer,2016,102:1334-1344.

[20] De Rosis A. Nonorthogonal central-moments-based lattice Boltzmann scheme in three dimensions [J]. Physical Review E,2017,95(1):013310.

[21] Saito S,De Rosis A,Festuccia A,et al. Color-gradient lattice Boltzmann model with nonorthogonal central moments:Hydrodynamic melt-jet breakup simulations[J]. Physical Review E,2018,98(1):013305.

[22] Li Q,Du D H,Fei L L,et al. Three-dimensional non-orthogonal MRT pseudopotential lattice Boltzmann model for multiphase flows[J]. Computers & Fluids,2019,186:128-140.

[23] Fei L L,Du J Y,Luo K H,et al. Modeling realistic multiphase flows using a non-orthogonal multiple-relaxation-time lattice Boltzmann method[J]. Physics of Fluids,2019,31(4):042105.

[24] Geier M,Greiner A,Korvink J G,Cascaded digital lattice Boltzmann automata for high Reynolds number flow[J]. Physical Review E,2006,73:066705.

[25] Premnath K N,Banerjee S,Incorporating forcing terms in cascaded lattice Boltzmann approach by method of central moments[J]. Physical Review E,2009,80:036702.

[26] Fei L L,K. H. Luo,Consistent forcing scheme in the cascaded lattice Boltzmann method[J]. Physical Review E,2017,96:053307.

[27] Luo K H,Fei L L,Wang G,A unified lattice Boltzmann model and application to multiphase flows [J]. Philosophical Transactions of The Royal Society A:Mathematical Physical and Engineering Sciences, 2021,379:20200397.

[28] Fei L L,Luo K H,Li Q,Three-dimensional cascaded lattice Boltzmann method:Improved implementation and consistent forcing scheme[J]. Physical Review E,2018,97:053309.

[29] He X Y,Shan X W,Doolen G D. Discrete Boltzmann equation model for nonideal gases[J]. Physical Review E,1998,57(1):R13-R16.

[30] He X Y,Chen S Y,Doolen G D. A novel thermal model for the lattice Boltzmann method in incompressible

limit[J]. Journal of Computational Physics, 1998, 146:282-300.

[31] Guo Z L, Zheng C G, Shi B C. Discrete lattice effects on the forcing term in the lattice Boltzmann method [J]. Physical Review E, 2002, 65(4):046308.

[32] Li Q, Luo K H, Kang Q J, et al. Lattice Boltzmann methods for multiphase flow and phase-change heat transfer[J]. Progress in Energy and Combustion Science, 2016, 52:62-105.

[33] Kupershtokh A L, Medvedev D A, Karpov D I. On equations of state in a lattice Boltzmann method [J]. Computers & Mathematics with Applications, 2009, 58(5):965-974.

[34] Kupershtokh A L. New method of incorporating a body force term into the lattice Boltzmann equation [C]. Proc 5th International EHD Workshop, University of Poitiers, Poitiers, France 2004.

[35] Li Q, Zhou P, Yan H J. Revised Chapman-Enskog analysis for a class of forcing schemes in the lattice Boltzmann method[J]. Physical Review E, 2016, 94(4):043313.

[36] McCracken M E, Abraham J. Multiple-relaxation-time lattice-Boltzmann model for multiphase flow [J]. Physical Review E, 2005, 71(3):036701.

第 4 章 单相流动与传热的格子 Boltzmann模型

格子 Boltzmann 方法的基本模型所恢复的宏观方程从形式上来看，是可压缩 Navier-Stokes 方程组。因此，用基本模型去模拟不可压缩流动时，会存在一定的可压缩效应。当这一效应比较明显时，需要对格子 Boltzmann 方法的基本模型做出改进，使其所恢复的宏观方程与不可压缩 Navier-Stokes 方程组相同或非常接近，以消除或降低可压缩效应的影响。这些改进的格子 Boltzmann 模型是本章所要介绍的主要内容之一。

此外，格子 Boltzmann 方法最初只能应用于等温流动，为了扩大格子 Boltzmann 方法的应用范围，人们发展了多种适用于热流体及可压缩流体的格子 Boltzmann 模型。由于种类繁多，对这些模型进行详细的分类及对比超出了本书的范围。为便于叙述，本书大体上将它们分为两类，一类是满足理想气体状态方程（$p=\rho R_g T$）的格子 Boltzmann 模型，如可压缩格子 Boltzmann 模型；另一类模型则不满足理想气体状态方程，如不可压缩热格子 Boltzmann 模型。在本章，将着重介绍应用于不可压缩流动与传热的热格子 Boltzmann 模型以及可压缩格子 Boltzmann 方法中的耦合双分布函数模型。

4.1 不可压缩等温格子 Boltzmann 模型

4.1.1 定常不可压缩模型

1995 年，Zou 等[1] 提出了一个用于定常不可压缩流动的格子 Boltzmann 模型，本书称之为 **Zou-Hou 模型**。他们提出，对定常不可压缩流动，可将 D2Q9 模型的平衡态密度分布函数修改成如下形式：

$$f_\alpha^{\mathrm{eq}}=\omega_\alpha\left[\rho+\frac{\boldsymbol{e}_\alpha\cdot\boldsymbol{u}}{c_s^2}+\frac{(\boldsymbol{e}_\alpha\cdot\boldsymbol{u})^2}{2c_s^4}-\frac{u^2}{2c_s^2}\right] \tag{4.1}$$

相应的宏观密度和速度定义为

$$\rho=\sum_{\alpha=0}^{8}f_\alpha,\quad \boldsymbol{u}=\sum_{\alpha=0}^{8}\boldsymbol{e}_\alpha f_\alpha \tag{4.2}$$

通过 Chapman-Enskog 多尺度展开，可导出 Zou-Hou 模型所对应的宏观方程为

$$\frac{\partial\rho}{\partial t}+\nabla\cdot\boldsymbol{u}=0 \tag{4.3}$$

$$\frac{\partial\boldsymbol{u}}{\partial t}+\nabla\cdot(\boldsymbol{u}\boldsymbol{u})=-\nabla p+\nu\,\nabla\cdot[\nabla\boldsymbol{u}+(\nabla\boldsymbol{u})^{\mathrm{T}}] \tag{4.4}$$

式中，压力 p 和运动黏度 ν 的定义与 D2Q9 模型一样。对定常流动，上述方程组与定常不

可压缩 Navier-Stokes 方程组完全一致。

1996 年，Lin 等[2]指出，Zou-Hou 模型存在一个矛盾，即 Zou-Hou 模型得到的是常密度定常不可压缩 Navier-Stokes 方程组，但在计算压力 p 时，又假定密度是变化的。为此，他们提出了一个改进的用于定常不可压缩流动的格子 Boltzmann 模型。该模型假设流体密度为常数并设 $\rho=1$，平衡态分布函数选取如下：

$$f_\alpha^{\mathrm{eq}}=\omega_\alpha\left[d+\frac{\boldsymbol{e}_\alpha\cdot\boldsymbol{u}}{c_{\mathrm{s}}^2}+\frac{(\boldsymbol{e}_\alpha\cdot\boldsymbol{u})^2}{2c_{\mathrm{s}}^4}-\frac{u^2}{2c_{\mathrm{s}}^2}\right] \tag{4.5}$$

式中，d 是与压力有关的一个参数，且有

$$d=\sum_{\alpha=0}^{8}f_\alpha \tag{4.6}$$

流体宏观速度定义为

$$\boldsymbol{u}=\sum_{\alpha=0}^{8}\boldsymbol{e}_\alpha f_\alpha \tag{4.7}$$

通过 Chapman-Enskog 多尺度展开可以发现，Lin 等[2]的改进模型所恢复的宏观方程组与式(4.3)～式(4.4)相同，但是压力 $p=c_{\mathrm{s}}^2 d$，即压力的计算与密度无关。

4.1.2 He-Luo 模型

前面提到的 Zou-Hou 模型以及 Lin 等的改进模型仅适用于定常不可压缩流动，不能应用于非定常流动。为了解决这一问题，He 和 Luo[3]提出了一个针对一般不可压缩流动的格子 Boltzmann 模型，本书称之为 **He-Luo 模型**。该模型的基本思想是在平衡态分布函数中消除由密度变化所引起的高阶马赫数项。在 He-Luo 模型中，独立的动力学变量是压力 p 而不是密度 ρ，这一点与不可压缩 Navier-Stokes 方程组是一致的。

对不可压缩流体，密度近似为常数，于是密度 ρ 可以分解为 $\rho=\rho_0+\delta\rho$，其中 $\delta\rho$ 为密度的波动。对等温流动，由流体运动的伯努利方程可知 $p+\rho u^2/2=p_{\mathrm{c}}$，其中 p_{c} 为驻点压力，于是有 $\delta p=p_{\mathrm{c}}-p=\rho u^2/2$。此外，流体音速 c_{s} 定义为

$$c_{\mathrm{s}}=\sqrt{\gamma\left(\frac{\partial p}{\partial\rho}\right)_{\mathrm{s}}} \tag{4.8}$$

音速

式中，γ 为比热容比。由式(4.8)可知 $\gamma\delta p=c_{\mathrm{s}}^2\delta\rho$，于是可得[4]

$$\frac{\delta\rho}{\rho}=\frac{\gamma}{2}\left(\frac{u}{c_{\mathrm{s}}}\right)^2=\frac{\gamma}{2}Ma^2 \tag{4.9}$$

因此，平衡态分布函数 f_α^{eq} 可写成如下形式：

$$\begin{aligned}f_\alpha^{\mathrm{eq}}&=\omega_\alpha\rho+\rho_0\left(1+\frac{\delta\rho}{\rho_0}\right)s_\alpha(\boldsymbol{u})\\&=\omega_\alpha\rho+\rho_0 s_\alpha(\boldsymbol{u})+O(Ma^3)\end{aligned} \tag{4.10}$$

式中，

$$s_\alpha(\boldsymbol{u})=\omega_\alpha\left[\frac{\boldsymbol{e}_\alpha\cdot\boldsymbol{u}}{c_{\mathrm{s}}^2}+\frac{(\boldsymbol{e}_\alpha\cdot\boldsymbol{u})^2}{2c_{\mathrm{s}}^4}-\frac{u^2}{2c_{\mathrm{s}}^2}\right] \tag{4.11}$$

略去式(4.10)中的高阶马赫数项，则可得一个新的平衡态分布函数

$$f_\alpha^{\mathrm{eq}}=\omega_\alpha\left\{\rho+\rho_0\left[\frac{\boldsymbol{e}_\alpha\cdot\boldsymbol{u}}{c_{\mathrm{s}}^2}+\frac{(\boldsymbol{e}_\alpha\cdot\boldsymbol{u})^2}{2c_{\mathrm{s}}^4}-\frac{u^2}{2c_{\mathrm{s}}^2}\right]\right\} \tag{4.12}$$

由于不可压缩 Navier-Stokes 方程组使用压力作为独立变量,因此可引入一个压力分布函数 $p_\alpha=c_s^2 f_\alpha$,对应的平衡态分布函数为

$$p_\alpha^{eq}=c_s^2 f_\alpha^{eq}=\omega_\alpha\left\{p+p_0\left[\frac{\boldsymbol{e}_\alpha\cdot\boldsymbol{u}}{c_s^2}+\frac{(\boldsymbol{e}_\alpha\cdot\boldsymbol{u})^2}{2c_s^4}-\frac{u^2}{2c_s^2}\right]\right\} \tag{4.13}$$

式中:$p=c_s^2\rho$;$p_0=c_s^2\rho_0$。由 f_α 的演化方程可导出 p_α 的演化方程为

$$p_\alpha(\boldsymbol{r}+\boldsymbol{e}_\alpha\delta_t,t+\delta_t)-p_\alpha(\boldsymbol{r},t)=-\frac{1}{\tau}\left[p_\alpha(\boldsymbol{r},t)-p_\alpha^{eq}(\boldsymbol{r},t)\right] \tag{4.14}$$

宏观压力和宏观速度定义如下:

$$p=\sum_{\alpha=0}^{8}p_\alpha,\quad \boldsymbol{u}=\frac{1}{\rho_0}\sum_{\alpha=0}^{8}\boldsymbol{e}_\alpha p_\alpha \tag{4.15}$$

式(4.13)~式(4.15)即为 He-Luo 模型。通过 Chapman-Enskog 多尺度展开,可导出 He-Luo 模型所对应的宏观方程为

$$\frac{1}{c_s^2}\frac{\partial P}{\partial t}+\nabla\cdot\boldsymbol{u}=0 \tag{4.16}$$

$$\frac{\partial\boldsymbol{u}}{\partial t}+\nabla\cdot(\boldsymbol{uu})=-\nabla P+\nu\nabla\cdot\left[\nabla\boldsymbol{u}+(\nabla\boldsymbol{u})^{\mathrm{T}}\right] \tag{4.17}$$

式中,$P=p/\rho_0$,运动黏度 ν 仍按之前方式确定。

对定常流动,He-Luo 模型满足不可压缩条件 $\nabla\cdot\boldsymbol{u}=0$,但对非定常流动,正如文献[5]所指出的,为了略去方程(4.16)左端的时间导数项,必须满足附加条件:$t_0\gg L/c_s$。其中,t_0 和 L 分别为流动的特征时间和特征长度。

4.1.3 D2G9 模型

2000 年,Guo 等[6]指出,上述不可压缩模型有一个共同的特点,即 $\sum f_\alpha^{eq}$ 不是常数。因此,这些模型所对应的连续方程必然不满足不可压缩条件。从理论上讲,不可压缩 Navier-Stokes 方程组中的密度为常数,但在 D2Q9 模型中,这一点无法满足。根据压力的表达式 $p=c_s^2\rho$ 可知,流场的压力和该点的密度是一种简单的代数关系,流场中各点的密度变化反映压力的分布,而在实际流动中,往往是通过压力梯度来驱动流体运动。

在格子 Boltzmann 方法的基本模型中,压力并不是独立变量,只能通过密度变化来获得压力梯度。He-Luo 模型虽然采用压力作为独立变量,但可以看到的是,压力与密度实际上仍然存在 $p=c_s^2\rho$ 的关系。因此,如果能够构造一个格子 Boltzmann 模型,使其平衡态分布函数满足 $\sum f_\alpha^{eq}$ 为常数的条件,且用压力取代密度成为真正的独立变量,而密度保持为常数,则可以得到针对一般不可压缩 Navier-Stokes 方程组的模型。基于这一思路,Guo 等[6]提出了一个新的格子 Boltzmann 模型,并命名为 **D2G9 模型**。D2G9 模型仍然使用 D2Q9 格子的离散速度,但其平衡态分布函数选取如下:

$$f_\alpha^{eq}=\begin{cases}\rho_0-4d_0\dfrac{p}{c^2}+\rho_0 s_0(\boldsymbol{u}), & \alpha=0\\[2ex] d_1\dfrac{p}{c^2}+\rho_0 s_\alpha(\boldsymbol{u}), & \alpha=1,2,3,4\\[2ex] d_2\dfrac{p}{c^2}+\rho_0 s_\alpha(\boldsymbol{u}), & \alpha=5,6,7,8\end{cases} \tag{4.18}$$

式中：$s_\alpha(\boldsymbol{u})$ 由式(4.11)给出；d_0、d_1 和 d_2 为模型参数，满足约束条件为

$$\begin{cases} d_1+d_2=d_0 \\ d_1+2d_2=0.5 \end{cases} \tag{4.19}$$

在实际应用中，可选取 $d_0=5/12$、$d_1=1/3$、$d_2=1/12$。由式(4.18)可得

$$\sum_{\alpha=0}^{8} f_\alpha^{\mathrm{eq}}=\rho_0, \qquad \sum_{\alpha=0}^{8} \boldsymbol{e}_\alpha f_\alpha^{\mathrm{eq}}=\rho_0\boldsymbol{u} \tag{4.20}$$

$$\sum_{\alpha=0}^{8} e_{\alpha i}e_{\alpha j}f_\alpha^{\mathrm{eq}}=\rho_0u_iu_j+p\delta_{ij}, \qquad \sum_{\alpha=0}^{8} e_{\alpha i}e_{\alpha j}e_{\alpha k}f_\alpha^{\mathrm{eq}}=\rho_0c_s^2(\delta_{ij}u_k+\delta_{ik}u_j+\delta_{jk}u_i) \tag{4.21}$$

根据 Chapman-Enskog 多尺度展开，可以证明该模型所对应的宏观方程为不可压缩 Navier-Stokes 方程组。其推导过程与 D2Q9 模型的推导类似，但对 D2G9 模型有

$$\begin{aligned}
\sum_{\alpha=0}^{8} e_{\alpha i}e_{\alpha j}f_\alpha^{(1)} &= -\tau\delta_t \sum_{\alpha=0}^{8} e_{\alpha i}e_{\alpha j}\left(\frac{\partial}{\partial t_1}+\boldsymbol{e}_\alpha\cdot\nabla_1\right)f_\alpha^{\mathrm{eq}} \\
&= -\tau\delta_t\left[\frac{\partial}{\partial t_1}(\rho_0u_iu_j+p\delta_{ij})+\frac{\partial}{\partial r_{1k}}\left(\sum_{\alpha=0}^{8} e_{\alpha i}e_{\alpha j}e_{\alpha k}f_\alpha^{\mathrm{eq}}\right)\right] \\
&= -\tau\delta_t\left\{\frac{\partial}{\partial t_1}(\rho_0u_iu_j+p\delta_{ij})+\rho_0c_s^2\left[(\nabla_1\cdot\boldsymbol{u})\delta_{ij}+\frac{\partial u_j}{\partial r_{1i}}+\frac{\partial u_i}{\partial r_{1j}}\right]\right\} \\
&= -\rho_0c_s^2\tau\delta_t\left(\frac{\partial u_j}{\partial r_{1i}}+\frac{\partial u_i}{\partial r_{1j}}\right)+O(Ma^2)
\end{aligned} \tag{4.22}$$

式(4.22)的推导用到了 $\partial p/\partial t=O(Ma^2)$ 和 $\partial u^2/\partial t=O(Ma^2)$。宏观压力的计算公式为

$$p=\rho_0\frac{c^2}{4d_0}\left[\sum_{\alpha=0}^{8} f_\alpha+s_0(\boldsymbol{u})\right] \tag{4.23}$$

D2G9 模型所具有的优点总结如下：

(1) 可导出标准的不可压缩 Navier-Stokes 方程组。在低马赫数假设下，通过 Chapman-Enskog 多尺度展开，可从 D2G9 模型导出一般的不可压缩 Navier-Stokes 方程组。因此，D2G9 模型可以有效地消除 D2Q9 模型所带来的可压缩效应。

(2) 适用于一般不可压缩流动。推导过程针对的是一般的情况，并没有要求流动是定常的，因此 D2G9 模型既可用于定常流动，也可用于非定常流动。

(3) 保留了格子 Boltzmann 方法的基本优点。D2G9 模型的演化方程、平衡态分布函数以及宏观量的计算都与一般的格子 Boltzmann 模型类似，没有增加附加的计算量，保留了格子 Boltzmann 方法所具有的计算量小、并行性好、边界条件易于实施等优点。

4.2 基于温度分布函数的热格子 Boltzmann 模型

用于不可压缩流动与传热的热格子 Boltzmann 模型通常可以分为**双分布函数**(double distribution function, DDF)**模型**和**混合**(hybrid)**模型**两类。其中，混合模型是一种将格子 Boltzmann 方法和其他数值模拟方法结合起来的模型，在这种模型中，流场仍然由格子 Boltzmann 方法求解，而温度场则采用其他数值模拟方法进行求解，如有限差分法、有限容积法等。对混合模型，读者可参阅文献[7]。双分布函数模型则采用两个分布函数，一个是密度分布函数，用于模拟流场；另一个是温度或能量分布函数，用来模拟温度场。双分布函数模型又可细分为温度分布函数模型、热力学能分布函数模型、总能分布函数模型

等。下面,从最早的温度分布函数模型——**被动标量模型**谈起。

1993 年,Shan 和 Chen[8] 提出了一个用于模拟多相/多组分流动的格子 Boltzmann 模型。在该模型中,各组分是否可以混合取决于组分间的相互作用。当相互作用比较弱时,该模型可以用来模拟不同机制驱动的扩散现象。对于有温度变化的流动来说,如果黏性热耗散和可压缩功可以忽略,温度可以看作是一个跟随流体运动的被动标量,并且满足一个简单的对流扩散方程,这一方程与多组分系统中每一组分的扩散方程是相似的,因此可以把温度看作是流体系统的一个附加组分。

基于上述思路,1997 年,Shan[9] 提出使用两组分模型来模拟单组分热流问题:组分Ⅰ模拟流体的运动,组分Ⅱ模拟温度场,这种模型被称为被动标量模型,而组分Ⅱ所对应的分布函数则为温度分布函数,它满足如下形式的热格子 Boltzmann 方程:

$$T_\alpha(\boldsymbol{r}+\boldsymbol{e}_\alpha\delta_t,t+\delta_t)-T_\alpha(\boldsymbol{r},t)=-\frac{1}{\tau_T}[T_\alpha(\boldsymbol{r},t)-T_\alpha^{\text{eq}}(\boldsymbol{r},t)] \tag{4.24}$$

式中:T_α 为温度分布函数;T_α^{eq} 为平衡态温度分布函数;τ_T 为相应的量纲为 1 的松弛时间。采用 D2Q9 格子时,平衡态温度分布函数选取为

$$T_\alpha^{\text{eq}}=\omega_\alpha T\left[1+\frac{\boldsymbol{e}_\alpha\cdot\boldsymbol{u}}{c_s^2}+\frac{(\boldsymbol{e}_\alpha\cdot\boldsymbol{u})^2}{2c_s^4}-\frac{u^2}{2c_s^2}\right] \tag{4.25}$$

式中,$c_s^2=1/3$。式(4.25)的零~二阶矩分别为

$$\sum_{\alpha=0}^{8}T_\alpha^{\text{eq}}=T \tag{4.26}$$

$$\sum_{\alpha=0}^{8}\boldsymbol{e}_\alpha T_\alpha^{\text{eq}}=\boldsymbol{u}T \tag{4.27}$$

$$\sum_{\alpha=0}^{8}\boldsymbol{e}_\alpha\boldsymbol{e}_\alpha T_\alpha^{\text{eq}}=T\boldsymbol{uu}+Tc_s^2\boldsymbol{I} \tag{4.28}$$

对式(4.24)左端做泰勒级数展开可得

$$\delta_t\left(\frac{\partial}{\partial t}+\boldsymbol{e}_\alpha\cdot\nabla\right)T_\alpha+\frac{\delta_t^2}{2}\left(\frac{\partial}{\partial t}+\boldsymbol{e}_\alpha\cdot\nabla\right)^2T_\alpha+O(\delta_t^3)=-\frac{1}{\tau_T}(T_\alpha-T_\alpha^{\text{eq}}) \tag{4.29}$$

对温度分布函数进行多尺度展开有

$$T_\alpha=T_\alpha^{\text{eq}}+KT_\alpha^{(1)}+K^2T_\alpha^{(2)}+\cdots \tag{4.30}$$

将式(3.125)、式(3.126)及式(4.30)代入式(4.29),并对比各阶系数可得

$$K^1:\quad\left(\frac{\partial}{\partial t_1}+\boldsymbol{e}_\alpha\cdot\nabla_1\right)T_\alpha^{\text{eq}}+\frac{1}{\tau_T\delta_t}T_\alpha^{(1)}=0 \tag{4.31}$$

$$K^2:\quad\frac{\partial T_\alpha^{\text{eq}}}{\partial t_2}+\left(\frac{\partial}{\partial t_1}+\boldsymbol{e}_\alpha\cdot\nabla_1\right)T_\alpha^{(1)}+\frac{\delta_t}{2}\left(\frac{\partial}{\partial t_1}+\boldsymbol{e}_\alpha\cdot\nabla_1\right)^2T_\alpha^{\text{eq}}+\frac{1}{\tau_T\delta_t}T_\alpha^{(2)}=0 \tag{4.32}$$

利用式(4.31),式(4.32)可简化为

$$K^2:\quad\frac{\partial T_\alpha^{\text{eq}}}{\partial t_2}+\left(\frac{\partial}{\partial t_1}+\boldsymbol{e}_\alpha\cdot\nabla_1\right)\left(1-\frac{1}{2\tau_T}\right)T_\alpha^{(1)}+\frac{1}{\tau_T\delta_t}T_\alpha^{(2)}=0 \tag{4.33}$$

对式(4.31)和式(4.33)求离散速度的零阶矩可得

$$\frac{\partial T}{\partial t_1}+\nabla_1\cdot(\boldsymbol{u}T)=0 \tag{4.34}$$

$$\frac{\partial T}{\partial t_2}+\nabla_1\cdot\left[\left(1-\frac{1}{2\tau_T}\right)\left(\sum_{\alpha=0}^{8}\boldsymbol{e}_\alpha T_\alpha^{(1)}\right)\right]=0 \tag{4.35}$$

结合式(4.31)、式(4.27)及式(4.28)可得

$$\begin{aligned}\sum_{\alpha=0}^{8} \boldsymbol{e}_\alpha T_\alpha^{(1)} &= -\tau_T \delta_t \sum_{\alpha=0}^{8}\left(\frac{\partial}{\partial t_1}+\boldsymbol{e}_\alpha \cdot \nabla_1\right)\boldsymbol{e}_\alpha T_\alpha^{\mathrm{eq}} \\ &= -\tau_T \delta_t \left[\frac{\partial(\boldsymbol{u}T)}{\partial t_1}+\nabla_1 \cdot (T\boldsymbol{uu}+Tc_s^2\boldsymbol{I})\right] \\ &= -\tau_T \delta_t [c_s^2 \nabla_1 T+\partial_{t_1}(\boldsymbol{u}T)+\nabla_1 \cdot (T\boldsymbol{uu})]\end{aligned} \tag{4.36}$$

将式(4.36)代入式(4.35),并根据式(3.125)和式(3.126)联立 t_1 和 t_2 时间尺度上的方程可得

$$\frac{\partial T}{\partial t}+\nabla \cdot (\boldsymbol{u}T)=\nabla \cdot (\chi \nabla T)+\nabla \cdot \{(\tau_T-0.5)\delta_t K[\partial_{t_1}(\boldsymbol{u}T)+\nabla_1 \cdot (T\boldsymbol{uu})]\} \tag{4.37}$$

式中,$\chi=c_s^2(\tau_T-0.5)\delta_t$ 为热扩散系数。式(4.37)即为被动标量模型所恢复的宏观温度方程,其中右端第二项为误差项。

在许多研究中,式(4.37)右端的误差项被默认可以忽略,继而可以得到关于温度的对流扩散方程。实际上,$\partial_{t_1}(\boldsymbol{u}T)$ 可写成 $\partial_{t_1}(\boldsymbol{u}T)=\boldsymbol{u}\partial_{t_1}T+T\partial_{t_1}\boldsymbol{u}$,其中 $\partial_{t_1}\boldsymbol{u}$ 可通过式(3.134)和式(3.135)得到,即

$$\rho\partial_{t_1}\boldsymbol{u}=-\rho\boldsymbol{u} \cdot \nabla_1\boldsymbol{u}-\nabla_1 p \tag{4.38}$$

式中,$p=\rho c_s^2$。结合式(4.34)、式(4.38)及式(3.125)可得

$$K[\partial_{t_1}(\boldsymbol{u}T)+\nabla_1 \cdot (T\boldsymbol{uu})]=-\frac{T}{\rho}\nabla p=-\frac{Tc_s^2}{\rho}\nabla\rho \tag{4.39}$$

对不可压缩单相热流问题,由于 $\nabla\rho/\rho \sim Ma^2$,误差项可以忽略,式(4.37)应当可写为

$$\frac{\partial T}{\partial t}+\nabla \cdot (\boldsymbol{u}T) \approx \nabla \cdot (\chi\nabla T) \tag{4.40}$$

需要注意的是,在一些研究中,式(4.24)被应用于气液两相流的传热问题,由于气液界面处的密度梯度 $\nabla\rho$ 较大,此时误差项的影响不可忽略。此外,引入作用力项时,式(4.38)右端还将出现与作用力 $\boldsymbol{F}$ 有关的项[10],这是因为作用力 $\boldsymbol{F}$ 将出现在式(3.135)的右端。这意味着,如果作用力随空间变化比较显著,它将通过式(4.37)中的 $\partial_{t_1}(T\boldsymbol{u})$ 给温度场的计算带来一定的数值模拟误差。以气液相变传热问题为例,文献[11]和文献[12]分别展示了这一误差项在二维和三维气液相变传热数值模拟中所造成的误差。

为了消除误差项,一些学者提出了改进的热格子 Boltzmann 模型。事实上,从式(4.36)可以清楚地看到,误差项 $\nabla_1 \cdot (T\boldsymbol{uu})$ 是由式(4.28)右端的 $T\boldsymbol{uu}$ 产生的。因此,为了消除这一项,可将平衡态温度分布函数修改为

$$T_\alpha^{\mathrm{eq}}=\omega_\alpha T\left(1+\frac{\boldsymbol{e}_\alpha \cdot \boldsymbol{u}}{c_s^2}\right) \tag{4.41}$$

修改后的平衡态温度分布函数所对应的二阶矩为 $\sum_{\alpha=0}^{8}\boldsymbol{e}_\alpha\boldsymbol{e}_\alpha T_\alpha^{\mathrm{eq}}=Tc_s^2\boldsymbol{I}$。此外,针对式(4.37)中的误差项 $\partial_{t_1}(T\boldsymbol{u})$,可在热格子 Boltzmann 方程中添加一个修正项[11]

$$T_\alpha(\boldsymbol{r}+\boldsymbol{e}_\alpha\delta_t,t+\delta_t)-T_\alpha(\boldsymbol{r},t)=-\frac{1}{\tau_T}(T_\alpha-T_\alpha^{\mathrm{eq}})+\delta_t\left(1-\frac{1}{2\tau_T}\right)\frac{\omega_\alpha\boldsymbol{e}_\alpha \cdot \partial_t(T\boldsymbol{u})}{c_s^2} \tag{4.42}$$

对单松弛模型,式(4.42)中的时间导数项 $\partial_t(T\boldsymbol{u})$ 需要通过相应的离散格式进行计算,如向后差分格式,而在多松弛格子 Boltzmann 方法框架下,修正项中的时间导数项可通过相

关矩函数的非平衡态部分来表征[11,13]，无须进行差分计算。

4.3 基于热力学能分布函数的热格子 Boltzmann 模型

4.3.1 标准的热力学能分布函数模型

在构造被动标量模型时，需要假设流动中的黏性热耗散和可压缩功可以忽略。因此当这些因素必须考虑时，被动标量模型是不适用的。为了解决这一问题，He 等[14]设计了一个包含黏性热耗散和可压缩功的热格子 Boltzmann 模型。该模型使用两个分布函数：密度分布函数和热力学能分布函数，其中热力学能分布函数是根据密度分布函数确定的。He 等从连续 Boltzmann 方程出发，导出了热力学能分布函数的动理学方程，然后对这两个分布函数的演化方程在时间、空间和速度空间都进行了离散，得到了两个分布函数的格子 Boltzmann 方程，从而构造了基于热力学能分布函数的热格子 Boltzmann 模型，本书称之为**热力学能分布函数模型**。

He 等[14]所设计的热力学能分布函数模型的基本思路是：根据热力学能的定义

$$\rho\varepsilon = \frac{1}{2}\int_{\Re}(\boldsymbol{\xi} - \boldsymbol{u})^2 f \mathrm{d}\boldsymbol{\xi} \tag{4.43}$$

式中，$\Re$ 为分子速度空间。引入热力学能分布函数 $g(\boldsymbol{r},\boldsymbol{\xi},t)=(\boldsymbol{\xi}-\boldsymbol{u})^2 f/2$，于是有

$$\rho\varepsilon = \int_{\Re} g \mathrm{d}\boldsymbol{\xi} \tag{4.44}$$

依据密度分布函数 $f(\boldsymbol{r},\boldsymbol{\xi},t)$ 的连续 Boltzmann 方程（假定系统不受外力影响）

$$\partial_t f+\boldsymbol{\xi}\cdot\nabla f=\Omega_f \tag{4.45}$$

可导出热力学能分布函数的动理学方程为

$$\partial_t g+\boldsymbol{\xi}\cdot\nabla g=\frac{(\boldsymbol{\xi}-\boldsymbol{u})^2}{2}\Omega_f - fq \tag{4.46}$$

式中，$q=(\boldsymbol{\xi}-\boldsymbol{u})\cdot(\partial_t\boldsymbol{u}+\boldsymbol{\xi}\cdot\nabla\boldsymbol{u})$。碰撞项采用 BGK 近似，即 $\Omega_f=-(f-f^{\mathrm{eq}})/\tau_f$，并规定

$$\frac{(\boldsymbol{\xi}-\boldsymbol{u})^2}{2\tau_f}(f-f^{\mathrm{eq}})=\frac{1}{\tau_g}(g-g^{\mathrm{eq}}) \tag{4.47}$$

则有

$$\partial_t f+\boldsymbol{\xi}\cdot\nabla f=-\frac{1}{\tau_f}(f-f^{\mathrm{eq}}) \tag{4.48}$$

$$\partial_t g+\boldsymbol{\xi}\cdot\nabla g=-\frac{1}{\tau_g}(g-g^{\mathrm{eq}})-fq \tag{4.49}$$

式中，

$$f^{\mathrm{eq}}=\frac{\rho}{(2\pi R_{\mathrm{g}}T)^{D/2}}\exp\left[-\frac{(\boldsymbol{\xi}-\boldsymbol{u})^2}{2R_{\mathrm{g}}T}\right] \tag{4.50}$$

$$g^{\mathrm{eq}}=\frac{\rho(\boldsymbol{\xi}-\boldsymbol{u})^2}{2(2\pi R_{\mathrm{g}}T)^{D/2}}\exp\left[-\frac{(\boldsymbol{\xi}-\boldsymbol{u})^2}{2R_{\mathrm{g}}T}\right] \tag{4.51}$$

分别为连续的平衡态密度和热力学能分布函数。经过一些基本的数学运算可得

$$\int_{\Re} f^{\mathrm{eq}}\mathrm{d}\boldsymbol{\xi} = \rho \tag{4.52}$$

$$\int_{\Re} f^{\mathrm{eq}} \xi_i \mathrm{d}\boldsymbol{\xi} = \rho u_i \tag{4.53}$$

$$\int_{\Re} f^{\mathrm{eq}} \xi_i \xi_j \mathrm{d}\boldsymbol{\xi} = \rho u_i u_j + p\delta_{ij} \tag{4.54}$$

$$\int_{\Re} f^{\mathrm{eq}} \xi_i \xi_j \xi_k \mathrm{d}\boldsymbol{\xi} = \rho u_i u_j u_k + p(u_k \delta_{ij} + u_i \delta_{jk} + u_j \delta_{ik}) \tag{4.55}$$

$$\int_{\Re} g^{\mathrm{eq}} \mathrm{d}\boldsymbol{\xi} = \rho \varepsilon \tag{4.56}$$

$$\int_{\Re} g^{\mathrm{eq}} \xi_i \mathrm{d}\boldsymbol{\xi} = \rho \varepsilon u_i \tag{4.57}$$

$$\int_{\Re} g^{\mathrm{eq}} \xi_i \xi_j \mathrm{d}\boldsymbol{\xi} = \rho \varepsilon u_i u_j + p(\varepsilon + R_{\mathrm{g}} T)\delta_{ij} \tag{4.58}$$

式中：$\varepsilon = DR_{\mathrm{g}}T/2$；$p = 2\rho\varepsilon/D$。

通过 Chapman-Enskog 多尺度展开，可导出式(4.48)和式(4.49)所对应的宏观方程，详细推导过程如下。对密度和热力学能分布函数进行多尺度展开：

$$f = f^{\mathrm{eq}} + Kf^{(1)} + K^2 f^{(2)} + \cdots \tag{4.59}$$

$$g = g^{\mathrm{eq}} + Kg^{(1)} + K^2 g^{(2)} + \cdots \tag{4.60}$$

将式(4.59)、式(4.60)、式(3.125)及式(3.126)代入式(4.48)和式(4.49)，整理可得

$$K^1: \quad \left(\frac{\partial}{\partial t_1} + \boldsymbol{\xi} \cdot \nabla_1\right) f^{\mathrm{eq}} + \frac{1}{\tau_f} f_{\alpha}^{(1)} = 0 \tag{4.61}$$

$$K^2: \quad \frac{\partial f^{\mathrm{eq}}}{\partial t_2} + \left(\frac{\partial}{\partial t_1} + \boldsymbol{\xi} \cdot \nabla_1\right) f^{(1)} + \frac{1}{\tau_f} f^{(2)} = 0 \tag{4.62}$$

及

$$K^1: \quad \left(\frac{\partial}{\partial t_1} + \boldsymbol{\xi} \cdot \nabla_1\right) g^{\mathrm{eq}} = -\frac{g^{(1)}}{\tau_g} - f^{\mathrm{eq}} q_1 \tag{4.63}$$

$$K^2: \quad \frac{\partial g^{\mathrm{eq}}}{\partial t_2} + \left(\frac{\partial}{\partial t_1} + \boldsymbol{\xi} \cdot \nabla_1\right) g^{(1)} = -\frac{g^{(2)}}{\tau_g} - f^{(1)} q_1 - f^{\mathrm{eq}} q_2 \tag{4.64}$$

式中，

$$q_1 = (\boldsymbol{\xi} - \boldsymbol{u}) \cdot \left[\frac{\partial \boldsymbol{u}}{\partial t_1} + (\boldsymbol{\xi} \cdot \nabla_1)\boldsymbol{u}\right] + (\boldsymbol{\xi} - \boldsymbol{u})(\boldsymbol{\xi} - \boldsymbol{u}) : \nabla_1 \boldsymbol{u} \tag{4.65}$$

$$q_2 = (\boldsymbol{\xi} - \boldsymbol{u}) \cdot \frac{\partial \boldsymbol{u}}{\partial t_2} \tag{4.66}$$

根据宏观物理量的定义

$$\begin{pmatrix} \rho \\ \rho \boldsymbol{u} \\ \rho \varepsilon \end{pmatrix} = \begin{pmatrix} \int_{\Re} f \mathrm{d}\boldsymbol{\xi} \\ \int_{\Re} \boldsymbol{\xi} f \mathrm{d}\boldsymbol{\xi} \\ \int_{\Re} g \mathrm{d}\boldsymbol{\xi} \end{pmatrix} = \begin{pmatrix} \int_{\Re} f^{\mathrm{eq}} \mathrm{d}\boldsymbol{\xi} \\ \int_{\Re} \boldsymbol{\xi} f^{\mathrm{eq}} \mathrm{d}\boldsymbol{\xi} \\ \int_{\Re} g^{\mathrm{eq}} \mathrm{d}\boldsymbol{\xi} \end{pmatrix} \tag{4.67}$$

可得

$$\int_{\Re} f^{(n)} \mathrm{d}\boldsymbol{\xi} = 0, \quad \int_{\Re} \boldsymbol{\xi} f^{(n)} \mathrm{d}\boldsymbol{\xi} = 0, \quad \int_{\Re} g^{(n)} \mathrm{d}\boldsymbol{\xi} = 0, \quad n > 0 \tag{4.68}$$

式(4.61)～式(4.64)分别对速度空间积分，可得

$$\frac{\partial \rho}{\partial t_1}+\nabla_1 \cdot (\rho \boldsymbol{u})=0 \tag{4.69}$$

$$\frac{\partial \rho}{\partial t_2}=0 \tag{4.70}$$

及

$$\frac{\partial}{\partial t_1}(\rho \varepsilon)+\nabla_1 \cdot (\rho \boldsymbol{u} \varepsilon)=-p \nabla_1 \cdot \boldsymbol{u} \tag{4.71}$$

$$\frac{\partial}{\partial t_2}(\rho \varepsilon)+\int_{\Re}(\boldsymbol{\xi} \cdot \nabla_1) g^{(1)} \mathrm{d} \boldsymbol{\xi}=-\int_{\Re} f^{(1)} q_1 \mathrm{d} \boldsymbol{\xi}-\int_{\Re} f^{\mathrm{eq}} q_2 \mathrm{d} \boldsymbol{\xi} \tag{4.72}$$

式(4.71)的推导用到 $\int_{\Re}(\boldsymbol{\xi}-\boldsymbol{u})(\boldsymbol{\xi}-\boldsymbol{u}) f^{\mathrm{eq}}: \nabla_1 \boldsymbol{u} \mathrm{d} \boldsymbol{\xi}=p \nabla_1 \cdot \boldsymbol{u}$，这说明能量方程中的可压缩功完全取决于 q_1 的第二项。联立式(4.69)和式(4.70)可得连续方程

$$\frac{\partial \rho}{\partial t}+\nabla \cdot (\rho \boldsymbol{u})=0 \tag{4.73}$$

式(4.61)和式(4.62)两边同乘 $\boldsymbol{\xi}$ 再对速度空间积分，可得

$$\frac{\partial}{\partial t_1}(\rho \boldsymbol{u})+\nabla_1 \cdot\left(\int_{\Re} \boldsymbol{\xi} \boldsymbol{\xi} f^{\mathrm{eq}} \mathrm{d} \boldsymbol{\xi}\right)=0 \tag{4.74}$$

$$\frac{\partial}{\partial t_2}(\rho \boldsymbol{u})+\nabla_1 \cdot\left(\int_{\Re} \boldsymbol{\xi} \boldsymbol{\xi} f^{(1)} \mathrm{d} \boldsymbol{\xi}\right)=0 \tag{4.75}$$

结合式(4.54)，式(4.74)可写为

$$\frac{\partial}{\partial t_1}(\rho \boldsymbol{u})+\nabla_1 \cdot (\rho \boldsymbol{u} \boldsymbol{u})=-\nabla_1 p \tag{4.76}$$

由式(4.61)可得

$$\int_{\Re} \boldsymbol{\xi} \boldsymbol{\xi} f^{(1)} \mathrm{d} \boldsymbol{\xi}=-\tau_f\left[\frac{\partial}{\partial t_1}\left(\int_{\Re} \boldsymbol{\xi} \boldsymbol{\xi} f^{\mathrm{eq}} \mathrm{d} \boldsymbol{\xi}\right)+\nabla_1 \cdot\left(\int_{\Re} \boldsymbol{\xi} \boldsymbol{\xi} \boldsymbol{\xi} f^{\mathrm{eq}} \mathrm{d} \boldsymbol{\xi}\right)\right] \tag{4.77}$$

联立式(4.69)和式(4.76)可得

$$\frac{\partial(\rho \boldsymbol{u} \boldsymbol{u})}{\partial t_1}=-\boldsymbol{u} \nabla_1 p-(\boldsymbol{u} \nabla_1 p)^{\mathrm{T}}-\nabla_1 \cdot (\rho \boldsymbol{u} \boldsymbol{u} \boldsymbol{u}) \tag{4.78}$$

根据式(4.71)并注意到 $p=2 \rho \varepsilon / D$，于是有

$$\frac{\partial p}{\partial t_1}=-\nabla_1 \cdot (p \boldsymbol{u})-\frac{2}{D} p \nabla_1 \cdot \boldsymbol{u} \tag{4.79}$$

利用上述结果并将式(4.54)和式(4.55)代入式(4.77)可得

$$\int_{\Re} \boldsymbol{\xi} \boldsymbol{\xi} f^{(1)} \mathrm{d} \boldsymbol{\xi}=-\tau_f\left[p \nabla_1 \boldsymbol{u}+p(\nabla_1 \boldsymbol{u})^{\mathrm{T}}-\frac{2}{D} p(\nabla_1 \cdot \boldsymbol{u}) \boldsymbol{I}\right] \tag{4.80}$$

联立式(4.75)、式(4.76)及式(4.80)可得宏观动量方程

$$\frac{\partial(\rho \boldsymbol{u})}{\partial t}+\nabla \cdot (\rho \boldsymbol{u} \boldsymbol{u})=-\nabla p+\nabla \cdot \boldsymbol{\Pi} \tag{4.81}$$

式中：$\boldsymbol{\Pi}$ 为黏性应力张量，

$$\boldsymbol{\Pi}=\mu\left[\nabla \boldsymbol{u}+(\nabla \boldsymbol{u})^{\mathrm{T}}-\frac{2}{D}(\nabla \cdot \boldsymbol{u}) \boldsymbol{I}\right] \tag{4.82}$$

$\mu=\tau_f p$ 为动力黏度；$\boldsymbol{I}$ 为单位二阶张量。

下面推导能量方程,根据式(4.67)有

$$\int_{\Re} f^{eq} q_2 \mathrm{d}\boldsymbol{\xi} = 0 \tag{4.83}$$

式(4.83)表明 q_2 不影响能量方程。利用式(4.68)可得

$$\int_{\Re} f^{(1)} (\boldsymbol{\xi} - \boldsymbol{u}) \cdot \left[\frac{\partial \boldsymbol{u}}{\partial t_1} + (\boldsymbol{u} \cdot \nabla_1) \boldsymbol{u}\right] \mathrm{d}\boldsymbol{\xi} = 0 \tag{4.84}$$

$$\int_{\Re} [f^{(1)} (\boldsymbol{\xi} - \boldsymbol{u})(\boldsymbol{\xi} - \boldsymbol{u}) : \nabla_1 \boldsymbol{u}] \mathrm{d}\boldsymbol{\xi} = \int_{\Re} (f^{(1)} \boldsymbol{\xi\xi} : \nabla_1 \boldsymbol{u}) \mathrm{d}\boldsymbol{\xi} \tag{4.85}$$

联立式(4.65)、式(4.84)、式(4.85)及式(4.80)可得

$$\int_{\Re} f^{(1)} q_1 \mathrm{d}\boldsymbol{\xi} = - \boldsymbol{\Pi}_1 : \nabla_1 \boldsymbol{u} \tag{4.86}$$

式(4.86)表明黏性热耗散完全取决于 q_1 的第二项。由式(4.63)可得

$$\int_{\Re} (\boldsymbol{\xi} \cdot \nabla_1) g^{(1)} \mathrm{d}\boldsymbol{\xi} = - \int_{\Re} \nabla_1 \cdot \left\{\tau_g \left[\frac{\partial}{\partial t_1}(\boldsymbol{\xi} g^{eq}) + \nabla_1 \cdot (\boldsymbol{\xi\xi} g^{eq}) + f^{eq} \boldsymbol{\xi} q_1\right]\right\} \mathrm{d}\boldsymbol{\xi} \tag{4.87}$$

利用式(4.53)~式(4.55)可得

$$\int_{\Re} f^{eq} \boldsymbol{\xi} (\boldsymbol{\xi} - \boldsymbol{u}) \cdot \left(\frac{\partial \boldsymbol{u}}{\partial t_1} + \boldsymbol{u} \cdot \nabla_1 \boldsymbol{u}\right) \mathrm{d}\boldsymbol{\xi} = p\left(\frac{\partial \boldsymbol{u}}{\partial t_1} + \boldsymbol{u} \cdot \nabla_1 \boldsymbol{u}\right) \tag{4.88}$$

$$\int_{\Re} [f^{eq} \boldsymbol{\xi} (\boldsymbol{\xi} - \boldsymbol{u})(\boldsymbol{\xi} - \boldsymbol{u}) : \nabla_1 \boldsymbol{u}] \mathrm{d}\boldsymbol{\xi} = p\boldsymbol{u}(\nabla_1 \cdot \boldsymbol{u}) \tag{4.89}$$

联立式(4.69)和式(4.76)可得

$$\frac{\partial \boldsymbol{u}}{\partial t_1} = -\boldsymbol{u} \cdot \nabla_1 \boldsymbol{u} - \frac{1}{\rho} \nabla_1 p \tag{4.90}$$

将式(4.90)代入式(4.88),并注意到 $p=\rho R_g T$,可得

$$\int_{\Re} f^{eq} \boldsymbol{\xi} (\boldsymbol{\xi} - \boldsymbol{u}) \cdot \left(\frac{\partial \boldsymbol{u}}{\partial t_1} + \boldsymbol{u} \cdot \nabla_1 \boldsymbol{u}\right) \mathrm{d}\boldsymbol{\xi} = - R_g T \nabla_1 p \tag{4.91}$$

根据式(4.89)和式(4.91)可得

$$\int_{\Re} f^{eq} \boldsymbol{\xi} q_1 \mathrm{d}\boldsymbol{\xi} = - R_g T \nabla_1 p + p\boldsymbol{u}(\nabla_1 \cdot \boldsymbol{u}) \tag{4.92}$$

联立式(4.57)、式(4.71)和式(4.90)可得

$$\frac{\partial}{\partial t_1}\left(\int_{\Re} \boldsymbol{\xi} g^{eq} \mathrm{d}\boldsymbol{\xi}\right) = - \boldsymbol{u} \nabla_1 \cdot (\rho \boldsymbol{u} \varepsilon) - p\boldsymbol{u}(\nabla_1 \cdot \boldsymbol{u}) - \rho \varepsilon \boldsymbol{u} \cdot \nabla_1 \boldsymbol{u} - \varepsilon \nabla_1 p \tag{4.93}$$

根据式(4.58)有

$$\nabla_1 \cdot \left(\int_{\Re} \boldsymbol{\xi\xi} g^{eq} \mathrm{d}\boldsymbol{\xi}\right) = \nabla_1 \cdot (\rho \varepsilon \boldsymbol{uu}) + \left[(\varepsilon + R_g T) \nabla_1 p + \frac{D+2}{2} p R_g \nabla_1 T\right] \tag{4.94}$$

将式(4.92)~式(4.94)代入式(4.87)可得

$$\int_{\Re} (\boldsymbol{\xi} \cdot \nabla_1) g^{(1)} \mathrm{d}\boldsymbol{\xi} = -\nabla_1 \cdot \left(\frac{D+2}{2} p R_g \tau_g \nabla_1 T\right) \tag{4.95}$$

联立式(4.83)、式(4.86)和式(4.95)可得

$$\frac{\partial}{\partial t_2}(\rho \varepsilon) = \nabla_1 \cdot \left(\frac{D+2}{2} p R_g \tau_g \nabla_1 T\right) + \boldsymbol{\Pi}_1 : \nabla_1 \boldsymbol{u} \tag{4.96}$$

联立式(4.71)和式(4.96),可得宏观能量方程

$$\frac{\partial}{\partial t}(\rho \varepsilon) + \nabla \cdot (\rho \boldsymbol{u} \varepsilon) = \nabla \cdot (\lambda \nabla T) + \boldsymbol{\Pi} : \nabla \boldsymbol{u} - p \nabla \cdot \boldsymbol{u} \tag{4.97}$$

式中，$\lambda=(D+2)\tau_g pR_g/2$ 为导热系数。模型的普朗特数为 $Pr=\mu c_p/\lambda=\tau_f/\tau_g$。

如果系统存在外力，可根据含外力项的连续 Boltzmann 方程对模型分布函数的演化方程做适当调整，

$$\frac{\partial f}{\partial t}+\boldsymbol{\xi}\cdot\nabla f+\boldsymbol{a}\cdot\nabla_{\boldsymbol{\xi}}f=\Omega_f \tag{4.98}$$

对外力项 $\boldsymbol{a}\cdot\nabla_{\boldsymbol{\xi}}f$ 可做如下处理：

$$\boldsymbol{a}\cdot\nabla_{\boldsymbol{\xi}}f\approx\boldsymbol{a}\cdot\nabla_{\boldsymbol{\xi}}f^{\mathrm{eq}}=-\frac{\boldsymbol{a}\cdot(\boldsymbol{\xi}-\boldsymbol{u})}{R_g T}f^{\mathrm{eq}} \tag{4.99}$$

于是，式(4.48)可修改为

$$\frac{\partial f}{\partial t}+\boldsymbol{\xi}\cdot\nabla f=-\frac{1}{\tau_f}(f-f^{\mathrm{eq}})+G \tag{4.100}$$

式中，

$$G=\frac{\boldsymbol{a}\cdot(\boldsymbol{\xi}-\boldsymbol{u})}{R_g T}f^{\mathrm{eq}} \tag{4.101}$$

对式(4.101)求矩，并利用式(4.52)~式(4.55)可得

$$\int_{\Re}G\mathrm{d}\boldsymbol{\xi}=0 \tag{4.102}$$

$$\int_{\Re}G\boldsymbol{\xi}\mathrm{d}\boldsymbol{\xi}=\rho\boldsymbol{a} \tag{4.103}$$

$$\int_{\Re}G\boldsymbol{\xi}\boldsymbol{\xi}\mathrm{d}\boldsymbol{\xi}=\rho\boldsymbol{u}\boldsymbol{a}+\rho\boldsymbol{a}\boldsymbol{u} \tag{4.104}$$

这与 Guo 等[15]通过分析外力项在 Chapman-Enskog 展开中所产生的影响而提出的外力项约束条件是一致的。

为了便于进行数值计算，He 等[14]基于 D2Q9 格子对速度空间进行了离散，并采用二阶精度的数值积分格式对式(4.49)和式(4.100)进行了时间和空间离散，继而得到了基于热力学能分布函数的热格子 Boltzmann 模型，相应的分布函数演化方程为

$$f_\alpha(\boldsymbol{r}+\boldsymbol{e}_\alpha\delta_t,t+\delta_t)-f_\alpha(\boldsymbol{r},t)=-\omega_f(f_\alpha-f_\alpha^{\mathrm{eq}})+\tau_f\,\omega_f G_\alpha \tag{4.105}$$

$$g_\alpha(\boldsymbol{r}+\boldsymbol{e}_\alpha\delta_t,t+\delta_t)-g_\alpha(\boldsymbol{r},t)=-\omega_g(g_\alpha-g_\alpha^{\mathrm{eq}})-\tau_g\omega_g\bar{f}_\alpha(\boldsymbol{r},t)q_\alpha(\boldsymbol{r},t) \tag{4.106}$$

式中，

$$\omega_f=\frac{\delta_t}{\tau_f+0.5\delta_t},\quad \omega_g=\frac{\delta_t}{\tau_g+0.5\delta_t},\quad G_\alpha=\frac{\boldsymbol{a}\cdot(\boldsymbol{e}_\alpha-\boldsymbol{u})}{R_g T}f_\alpha^{\mathrm{eq}} \tag{4.107}$$

$$\bar{f}_\alpha=\omega_f\left(\frac{\tau_f f_\alpha}{\delta_t}+0.5f_\alpha^{\mathrm{eq}}+\tau_f F_\alpha\right),\quad q_\alpha=(\boldsymbol{e}_\alpha-\boldsymbol{u})\cdot[\partial_t\boldsymbol{u}+(\boldsymbol{e}_\alpha\cdot\nabla)\boldsymbol{u}] \tag{4.108}$$

在数值计算中，$\partial_t\boldsymbol{u}+(\boldsymbol{e}_\alpha\cdot\nabla)\boldsymbol{u}$ 可离散为 $[\boldsymbol{u}(\boldsymbol{r}+\boldsymbol{e}_\alpha\delta_t,t+\delta_t)-\boldsymbol{u}(\boldsymbol{r},t)]/\delta_t$，由于 $\boldsymbol{u}(\boldsymbol{r}+\boldsymbol{e}_\alpha\delta_t,t+\delta_t)$ 为新时层的宏观速度，因此在实际应用中常用 $\boldsymbol{u}(\boldsymbol{r}+\boldsymbol{e}_\alpha\delta_t,t)$ 代替。

对二维问题，采用 D2Q9 格子时，平衡态分布函数可选取为

$$f_\alpha^{\mathrm{eq}}=\rho\omega_\alpha\left[1+3\frac{\boldsymbol{e}_\alpha\cdot\boldsymbol{u}}{c^2}+4.5\frac{(\boldsymbol{e}_\alpha\cdot\boldsymbol{u})^2}{c^4}-1.5\frac{u^2}{c^2}\right] \tag{4.109}$$

$$g_0^{\mathrm{eq}}=-\frac{2\rho\varepsilon}{3}\frac{u^2}{c^2} \tag{4.110}$$

$$g_{1,2,3,4}^{\text{eq}}=\frac{\rho\varepsilon}{9}\left[1.5+1.5\frac{\boldsymbol{e}_\alpha\cdot\boldsymbol{u}}{c^2}+4.5\frac{(\boldsymbol{e}_\alpha\cdot\boldsymbol{u})^2}{c^4}-1.5\frac{u^2}{c^2}\right] \tag{4.111}$$

$$g_{5,6,7,8}^{\text{eq}}=\frac{\rho\varepsilon}{36}\left[3+6\frac{\boldsymbol{e}_\alpha\cdot\boldsymbol{u}}{c^2}+4.5\frac{(\boldsymbol{e}_\alpha\cdot\boldsymbol{u})^2}{c^4}-1.5\frac{u^2}{c^2}\right] \tag{4.112}$$

式中，$c=\sqrt{3R_gT_0}$。在标准格子 Boltzmann 方法框架下，为了保证粒子在格线上运动，必须满足 $c=\delta_x/\delta_t$，而空间步长和时间步长通常取为 $\delta_x=\delta_t=1$，于是可选取 $R_g=1/(3T_0)$ 以保证 $c=1$，从而压力 $p=\rho c^2/3=\rho R_gT_0$，这意味着用于具体数值计算的模型并不能保证理想气体状态方程，由此将导致实际求解的宏观方程发生变化。首先，黏性应力张量的形式将发生变化，由于 $p=p(\rho)$，根据连续方程可得

$$\frac{\partial p}{\partial t}+\nabla\cdot(p\boldsymbol{u})=0 \tag{4.113}$$

因此式(4.79)不再成立，黏性应力张量将变为

$$\boldsymbol{\Pi}=\mu[\nabla\boldsymbol{u}+(\nabla\boldsymbol{u})^{\mathrm{T}}] \tag{4.114}$$

式中，μ 依然等于 $\tau_f p$，不过此时的压力 p 等于 ρR_gT_0。其次，模型所对应的能量方程将出现偏差项。当 $p=\rho R_gT_0$ 时，式(4.91)和式(4.92)中的温度 T 应由 T_0 替换，这将导致能量方程变为[16]

$$\frac{\partial}{\partial t}(\rho\varepsilon)+\nabla\cdot(\rho\boldsymbol{u}\varepsilon)=\nabla\cdot(\lambda\nabla T)+\boldsymbol{\Pi}:\nabla\boldsymbol{u}-p\nabla\cdot\boldsymbol{u}+\nabla\cdot[\tau_gR_g(T-T_0)\nabla p] \tag{4.115}$$

式(4.115)右端最后一项为偏差项。

在上述热力学能分布函数模型中，流体的宏观量计算如下：

$$\rho=\sum_{\alpha=0}^{8}f_\alpha \tag{4.116}$$

$$\rho\boldsymbol{u}=\sum_{\alpha=0}^{8}\boldsymbol{e}_\alpha f_\alpha+\frac{\rho\boldsymbol{a}}{2}\delta_t \tag{4.117}$$

$$\rho\varepsilon=\sum_{\alpha=0}^{8}g_\alpha-\delta_t\sum_{\alpha=0}^{8}\bar{f}_\alpha q_\alpha \tag{4.118}$$

热力学能分布函数模型的主要优点在于物理背景明确、具有良好的数值稳定性、普朗特数不受限制及较好的通用性。主要缺点在于模型的计算量较大，在每一个时间步长都需要计算 $\bar{f}_\alpha$ 和 q_α，特别是 q_α 的计算，不仅增加了计算量，还可能引入额外的数值误差。此外，模型所恢复的宏观方程存在偏差项。

4.3.2 简化的热力学能分布函数模型

结合不可压缩等温流动的 He-Luo 模型，He 等[14]对标准的热力学能分布函数模型进行了简化，提出了用于不可压缩流动与传热的热力学能分布函数模型。在该模型中，平衡态热力学能分布函数与标准热力学能分布函数模型的形式一致，但用于模拟流场的平衡态分布函数根据 He-Luo 不可压缩模型取为

$$f_\alpha^{\text{eq}}=\omega_\alpha\rho_0\left[\frac{3p}{\rho_0c^2}+\frac{\boldsymbol{e}_\alpha\cdot\boldsymbol{u}}{c_s^2}+\frac{(\boldsymbol{e}_\alpha\cdot\boldsymbol{u})^2}{2c_s^4}-\frac{u^2}{2c_s^2}\right] \tag{4.119}$$

式中：ρ_0 为常数；压力 p 的计算格式为

$$p=c_s^2\sum_{\alpha=0}^{8}f_\alpha \tag{4.120}$$

同时,式(4.106)中的 q_α 被拆分为 $q_\alpha=q_{\alpha,c}+q_{\alpha,d}$,且有

$$q_{\alpha,c}=(\boldsymbol{e}_\alpha-\boldsymbol{u})\cdot[\partial_t\boldsymbol{u}+(\boldsymbol{u}\cdot\nabla)\boldsymbol{u}] \tag{4.121}$$

$$q_{\alpha,d}=(\boldsymbol{e}_\alpha-\boldsymbol{u})(\boldsymbol{e}_\alpha-\boldsymbol{u}):\nabla\boldsymbol{u} \tag{4.122}$$

由于 $q_{\alpha,d}$ 只与黏性热耗散和可压缩功有关,因此在不可压缩模型中可以去掉,而 $q_{\alpha,c}$ 则由于与热传导项有关被保留了下来。因此,由 q_α 所带来的计算量较大、引入额外数值误差的问题仍然存在。

为此,Peng 等[17]在 He 等的基础上提出了一个简化的热力学能分布函数模型。简化模型的平衡态分布函数依然采用标准热力学能分布函数模型中的形式,不同之处在于简化模型直接舍弃了式(4.106)中的 q_α,从而有

$$f_\alpha(\boldsymbol{r}+\boldsymbol{e}_\alpha\delta_t,t+\delta_t)-f_\alpha(\boldsymbol{r},t)=-\frac{1}{\tau_f}[f_\alpha(\boldsymbol{r},t)-f_\alpha^{\mathrm{eq}}(\boldsymbol{r},t)]+\delta_t G_\alpha \tag{4.123}$$

$$g_\alpha(\boldsymbol{r}+\boldsymbol{e}_\alpha\delta_t,t+\delta_t)-g_\alpha(\boldsymbol{r},t)=-\frac{1}{\tau_g}[g_\alpha(\boldsymbol{r},t)-g_\alpha^{\mathrm{eq}}(\boldsymbol{r},t)] \tag{4.124}$$

对于不可压缩热流问题来说,该模型与 He 等[14]的模型相比不但简单而且计算效率要高。

随后,Li 等[16]研究发现,Peng 等直接舍弃 q_α 的做法在理论上存在不足。事实上,从前面的 Chapman-Enskog 多尺度展开可以看到,与 $q_{\alpha,c}$ 相关的项为式(4.92)右端的第一项,这一项在合成式(4.95)时与式(4.94)中对应的项相消。通过分析可以发现,与之相消的项是由式(4.58)中的 $pR_gT\delta_{ij}$ 产生的。因此,舍弃 $q_{\alpha,c}$ 时,应当对平衡态热力学能分布函数所满足的约束条件做相应的调整,即此时平衡态热力学能分布函数所满足的约束条件应调整为

$$\sum_{\alpha=0}^{8}g_\alpha^{\mathrm{eq}}=\rho\varepsilon \tag{4.125}$$

$$\sum_{\alpha=0}^{8}e_{\alpha i}g_\alpha^{\mathrm{eq}}=\rho\varepsilon u_i \tag{4.126}$$

$$\sum_{\alpha=0}^{8}e_{\alpha i}e_{\alpha j}g_\alpha^{\mathrm{eq}}=\rho\varepsilon u_iu_j+p\varepsilon\delta_{ij} \tag{4.127}$$

于是平衡态热力学能分布函数可选取为

$$g_\alpha^{\mathrm{eq}}=\omega_\alpha\rho\varepsilon\left[1+\frac{\boldsymbol{e}_\alpha\cdot\boldsymbol{u}}{c_s^2}+\frac{(\boldsymbol{e}_\alpha\cdot\boldsymbol{u})^2}{2c_s^4}-\frac{u^2}{2c_s^2}\right] \tag{4.128}$$

宏观能量方程的推导如下:对时间导数、空间导数和分布函数均采用多尺度展开,代入式(4.124),并对比各阶系数,可得

$$\left(\frac{\partial}{\partial t_1}+\boldsymbol{e}_\alpha\cdot\nabla_1\right)g_\alpha^{\mathrm{eq}}+\frac{1}{\tau_g\delta_t}g_\alpha^{(1)}=0 \tag{4.129}$$

$$\frac{\partial g_\alpha^{\mathrm{eq}}}{\partial t_2}+\left(\frac{\partial}{\partial t_1}+\boldsymbol{e}_\alpha\cdot\nabla_1\right)\left(1-\frac{1}{2\tau_g}\right)g_\alpha^{(1)}+\frac{1}{\tau_g\delta_t}g_\alpha^{(2)}=0 \tag{4.130}$$

式(4.129)和式(4.130)对 α 求和,可得

$$\frac{\partial}{\partial t_1}(\rho\varepsilon)+\nabla_1\cdot(\rho\boldsymbol{u}\varepsilon)=0 \tag{4.131}$$

$$\frac{\partial}{\partial t_2}(\rho\varepsilon)+\sum_{\alpha=0}^{8}\left[\frac{\partial}{\partial t_1}+(\boldsymbol{e}_\alpha\cdot\nabla_1)\right]\left(1-\frac{1}{2\tau_g}\right)g_\alpha^{(1)}=0 \tag{4.132}$$

利用式(4.129)并注意到 $\sum_{\alpha=0}^{8} g_\alpha^{(1)}=0$,可得

$$\frac{\partial}{\partial t_2}(\rho\varepsilon)=\nabla_1\cdot\left\{\delta_t\left(\tau_g-\frac{1}{2}\right)\left[\frac{\partial}{\partial t_1}\left(\sum_{\alpha=0}^{8}\boldsymbol{e}_\alpha g_\alpha^{\mathrm{eq}}\right)+\nabla_1\cdot\left(\sum_{\alpha=0}^{8}\boldsymbol{e}_\alpha\boldsymbol{e}_\alpha g_\alpha^{\mathrm{eq}}\right)\right]\right\} \tag{4.133}$$

由式(4.126)有

$$\frac{\partial}{\partial t_1}\left(\sum_{\alpha=0}^{8}\boldsymbol{e}_\alpha g_\alpha^{\mathrm{eq}}\right)=\boldsymbol{u}\frac{\partial}{\partial t_1}(\rho\varepsilon)+\rho\varepsilon\frac{\partial\boldsymbol{u}}{\partial t_1} \tag{4.134}$$

将式(4.90)和式(4.131)代入式(4.134)可得

$$\frac{\partial}{\partial t_1}\left(\sum_{\alpha=0}^{8}\boldsymbol{e}_\alpha g_\alpha^{\mathrm{eq}}\right)=-\boldsymbol{u}\nabla_1\cdot(\rho\boldsymbol{u}\varepsilon)-\rho\varepsilon\boldsymbol{u}\cdot\nabla_1\boldsymbol{u}-\varepsilon\nabla_1 p \tag{4.135}$$

同时,由式(4.127)有

$$\nabla_1\cdot\left(\sum_{\alpha=0}^{8}\boldsymbol{e}_\alpha\boldsymbol{e}_\alpha g_\alpha^{\mathrm{eq}}\right)=\nabla_1\cdot(\rho\varepsilon\boldsymbol{u}\boldsymbol{u})+\varepsilon\nabla_1 p+p\nabla_1\varepsilon \tag{4.136}$$

将式(4.135)和式(4.136)代入式(4.133)可得

$$\frac{\partial}{\partial t_2}(\rho\varepsilon)=\nabla_1\cdot\left[\delta_t\left(\tau_g-\frac{1}{2}\right)p\nabla_1\varepsilon\right] \tag{4.137}$$

联立式(4.131)和式(4.137),并注意到 $\varepsilon=DR_{\mathrm{g}}T/2$,可得如下形式的能量方程:

$$\frac{\partial}{\partial t}(\rho\varepsilon)+\nabla\cdot(\rho\boldsymbol{u}\varepsilon)=\nabla\cdot(\lambda\nabla T) \tag{4.138}$$

式中,导热系数 $\lambda=(\tau_g-0.5)Dp R_g\delta_t/2$。

忽略黏性热耗散和可压缩功时,式(4.138)与式(2.61)是一致的。需要强调的是,上述推导过程是建立在比定容热容 c_V 为常数的基础上。当 c_V 随温度变化时,$\nabla\varepsilon\neq c_V\nabla T$,此时式(4.138)的右端不能写成 $\nabla\cdot(\lambda\nabla T)$ 的形式。

比定压热容和比定容热容

4.4 基于总能分布函数的热格子 Boltzmann 模型

前面我们提到,He 等[14]所设计的热力学能分布函数模型虽然能够包含黏性热耗散和可压缩功,但仍然存在不足之处。为了解决这一问题,Guo 等[18]提出了一个**总能分布函数模型**。与 He 等从热力学能出发不同,Guo 等根据总能的定义

$$\rho E=\rho\left(\varepsilon+\frac{u^2}{2}\right)=\int_{\Re}\frac{1}{2}\xi^2 f\mathrm{d}\boldsymbol{\xi} \tag{4.139}$$

引入了一个总能分布函数

$$h=\frac{1}{2}\xi^2 f \tag{4.140}$$

式中,$\xi^2=\boldsymbol{\xi}\cdot\boldsymbol{\xi}$。类似地,总能分布函数的动理学方程亦可从连续 Boltzmann 方程(4.98)导出。对式(4.98)两端同乘 $\xi^2/2$ 可得

$$\partial_t h+\boldsymbol{\xi}\cdot\nabla h+\boldsymbol{a}\cdot(\nabla_{\boldsymbol{\xi}}h-\boldsymbol{\xi}f)=\Omega_h \tag{4.141}$$

式中,$\Omega_h=\Omega_f\xi^2/2$。

Guo 等[18]指出,构建总能分布函数动理学模型的关键在于确定 Ω_h 的具体形式,为此他们将总能碰撞算子 Ω_h 分成如下两部分:

$$\Omega_h=\Omega_i+\Omega_m \tag{4.142}$$

式中,$\Omega_i=(\boldsymbol{\xi}-\boldsymbol{u})^2\Omega_f/2$ 为热力学能部分,而

$$\Omega_m=\Omega_h-\Omega_i=\left[\frac{\xi^2}{2}-\frac{(\boldsymbol{\xi}-\boldsymbol{u})^2}{2}\right]\Omega_f=Z\Omega_f \tag{4.143}$$

则为机械能部分,且有 $Z=\boldsymbol{\xi}\cdot\boldsymbol{u}-u^2/2$。采用 BGK 近似时,$\Omega_f=-(f-f^{\mathrm{eq}})/\tau_f$,于是有

$$\Omega_m=Z\Omega_f=-\frac{Z}{\tau_f}(f-f^{\mathrm{eq}}) \tag{4.144}$$

对于热力学能部分 Ω_i,注意到$(\boldsymbol{\xi}-\boldsymbol{u})^2\Omega_f/2$ 亦出现在式(4.46)中,因此可取 $\Omega_i=-(g-g^{\mathrm{eq}})/\tau_g$,但这样就会引入一个热力学能分布函数,为此 Guo 等采用$(h-Zf)$取代了 g,故有

$$\Omega_i=-\frac{1}{\tau_h}[(h-h^{\mathrm{eq}})-Z(f-f^{\mathrm{eq}})] \tag{4.145}$$

将式(4.144)和式(4.145)代入式(4.142),可得

$$\Omega_h=-\frac{1}{\tau_h}(h-h^{\mathrm{eq}})+\frac{Z}{\tau_{hf}}(f-f^{\mathrm{eq}}) \tag{4.146}$$

式中,

$$\frac{1}{\tau_{hf}}=\frac{1}{\tau_h}-\frac{1}{\tau_f} \tag{4.147}$$

最终可得

$$\partial_t f+\boldsymbol{\xi}\cdot\nabla f+\boldsymbol{a}\cdot\nabla_{\boldsymbol{\xi}} f=-\frac{1}{\tau_f}(f-f^{\mathrm{eq}}) \tag{4.148}$$

$$\partial_t h+\boldsymbol{\xi}\cdot\nabla h+\boldsymbol{a}\cdot\nabla_{\boldsymbol{\xi}} h=-\frac{1}{\tau_h}(h-h^{\mathrm{eq}})+\frac{Z}{\tau_{hf}}(f-f^{\mathrm{eq}})+f\boldsymbol{\xi}\cdot\boldsymbol{a} \tag{4.149}$$

式中,f^{eq}由式(4.50)给出,而 h^{eq}则计算如下:

$$h^{\mathrm{eq}}=\frac{\rho\xi^2}{2(2\pi R_{\mathrm{g}}T)^{D/2}}\exp\left[-\frac{(\boldsymbol{\xi}-\boldsymbol{u})^2}{2R_{\mathrm{g}}T}\right]=\frac{\xi^2}{2}f^{\mathrm{eq}} \tag{4.150}$$

通过 Chapman-Enskog 多尺度展开,可导出相应的宏观方程为(详见文献[18]的附录 A)

$$\frac{\partial\rho}{\partial t}+\nabla\cdot(\rho\boldsymbol{u})=0 \tag{4.151}$$

$$\frac{\partial(\rho\boldsymbol{u})}{\partial t}+\nabla\cdot(\rho\boldsymbol{u}\boldsymbol{u})=-\nabla p+\nabla\cdot\boldsymbol{\Pi}+\rho\boldsymbol{a} \tag{4.152}$$

$$\frac{\partial}{\partial t}(\rho E)+\nabla\cdot[(\rho E+p)\boldsymbol{u}]=\nabla\cdot(\lambda\nabla T)+\nabla\cdot(\boldsymbol{u}\cdot\boldsymbol{\Pi})+\rho\boldsymbol{u}\cdot\boldsymbol{a} \tag{4.153}$$

式中:$\boldsymbol{\Pi}=\mu[\nabla\boldsymbol{u}+(\nabla\boldsymbol{u})^{\mathrm{T}}-2(\nabla\cdot\boldsymbol{u})\boldsymbol{I}/D]$;$\mu=\tau_f p$;$\lambda=(D+2)R_{\mathrm{g}}\tau_h p/2$;$p=\rho R_{\mathrm{g}}T$。

为了获得相应的离散模型,文献[18]先对平衡态分布函数进行了 Hermite 展开

$$f^{\mathrm{eq}}=\omega(\boldsymbol{\xi},T)\sum_{n=0}^{\infty}\frac{\boldsymbol{A}^{(n)}(\boldsymbol{x},t)}{n!}\boldsymbol{H}^{(n)}(\hat{\boldsymbol{\xi}}) \tag{4.154}$$

$$h^{\mathrm{eq}}=\omega(\boldsymbol{\xi},T)\sum_{n=0}^{\infty}\frac{\boldsymbol{B}^{(n)}(\boldsymbol{x},t)}{n!}\boldsymbol{H}^{(n)}(\hat{\boldsymbol{\xi}}) \tag{4.155}$$

式中：$\hat{\boldsymbol{\xi}}=\boldsymbol{\xi}/\sqrt{R_g T}$；$\omega(\boldsymbol{\xi},T)$ 为

$$\omega(\boldsymbol{\xi},T)=\frac{1}{(2\pi R_g T)^{D/2}}\exp\left(-\frac{\xi^2}{2R_g T}\right) \tag{4.156}$$

式(4.154)和式(4.155)中的 $\boldsymbol{H}^{(n)}(\hat{\boldsymbol{\xi}})$ 是 n 阶的 Hermite 张量多项式，其定义为[19]

$$\boldsymbol{H}^{(n)}(\hat{\boldsymbol{\xi}})=\frac{(-1)^n}{\omega(\hat{\boldsymbol{\xi}},T)}\nabla_{\hat{\xi}}^{n}\omega(\hat{\boldsymbol{\xi}},T) \tag{4.157}$$

零～三阶的 Hermite 张量多项式为[19-20]

$$\boldsymbol{H}^{(0)}(\hat{\boldsymbol{\xi}})=1,\quad \boldsymbol{H}_i^{(1)}(\hat{\boldsymbol{\xi}})=\hat{\xi}_i \tag{4.158}$$

$$\boldsymbol{H}_{ij}^{(2)}(\hat{\boldsymbol{\xi}})=\hat{\xi}_i\hat{\xi}_j-\delta_{ij},\quad \boldsymbol{H}_{ijk}^{(3)}(\hat{\boldsymbol{\xi}})=\hat{\xi}_i\hat{\xi}_j\hat{\xi}_k-(\hat{\xi}_i\delta_{jk}+\hat{\xi}_j\delta_{ik}+\hat{\xi}_k\delta_{ij}) \tag{4.159}$$

式(4.154)和式(4.155)中的展开系数 $\boldsymbol{A}^{(n)}$ 和 $\boldsymbol{B}^{(n)}$ 定义如下：

$$\boldsymbol{A}^{(n)}=\int_{\Re} f^{\mathrm{eq}}\boldsymbol{H}^{(n)}(\hat{\boldsymbol{\xi}})\,\mathrm{d}\boldsymbol{\xi} \tag{4.160}$$

$$\boldsymbol{B}^{(n)}=\int_{\Re} h^{\mathrm{eq}}\boldsymbol{H}^{(n)}(\hat{\boldsymbol{\xi}})\,\mathrm{d}\boldsymbol{\xi} \tag{4.161}$$

前几个 $\boldsymbol{A}^{(n)}$ 分别为

$$\boldsymbol{A}^{(0)}=\rho,\quad \boldsymbol{A}_i^{(1)}=\rho\hat{u}_i,\quad \boldsymbol{A}_{ij}^{(2)}=\rho\hat{u}_i\hat{u}_j,\quad \boldsymbol{A}_{ijk}^{(3)}=\rho\hat{u}_i\hat{u}_j\hat{u}_k \tag{4.162}$$

根据 Hermite 张量多项式及相应的展开系数，可得到连续平衡态分布函数的 Hermite 多项式展开。例如，对平衡态密度分布函数，根据式(4.158)、式(4.159)及式(4.162)可得其三阶展开为

$$f^{\mathrm{eq},3}(T)=\omega(\boldsymbol{\xi},T)\rho\left\{1+\frac{\boldsymbol{\xi}\cdot\boldsymbol{u}}{R_g T}+\frac{1}{2}\left(\frac{\boldsymbol{\xi}\cdot\boldsymbol{u}}{R_g T}\right)^2-\frac{u^2}{2R_g T}+\frac{\boldsymbol{\xi}\cdot\boldsymbol{u}}{6R_g T}\left[\left(\frac{\boldsymbol{\xi}\cdot\boldsymbol{u}}{R_g T}\right)^2-\frac{3u^2}{R_g T}\right]\right\} \tag{4.163}$$

同理，根据需要对 h^{eq} 取其二阶展开，可得

$$\begin{aligned}h^{\mathrm{eq},2}(T)=\omega(\boldsymbol{\xi},T)&\left\{\rho E+(p+\rho E)\frac{\boldsymbol{\xi}\cdot\boldsymbol{u}}{R_g T}+\frac{p}{2}\left(\frac{\xi^2}{R_g T}-D\right)+\right.\\&\left.\left(p+\frac{\rho E}{2}\right)\left[\left(\frac{\boldsymbol{\xi}\cdot\boldsymbol{u}}{R_g T}\right)^2-\frac{u^2}{R_g T}\right]\right\}\end{aligned} \tag{4.164}$$

对低马赫数流动，式(4.163)中的三阶速度项可以忽略，于是有

$$f^{\mathrm{eq},2}(T)=\omega(\boldsymbol{\xi},T)\rho\left[1+\frac{\boldsymbol{\xi}\cdot\boldsymbol{u}}{R_g T}+\frac{1}{2}\left(\frac{\boldsymbol{\xi}\cdot\boldsymbol{u}}{R_g T}\right)^2-\frac{u^2}{2R_g T}\right] \tag{4.165}$$

$$h^{\mathrm{eq},2}(T)=\omega(\boldsymbol{\xi},T)p\left[\frac{\boldsymbol{\xi}\cdot\boldsymbol{u}}{R_g T}+\left(\frac{\boldsymbol{\xi}\cdot\boldsymbol{u}}{R_g T}\right)^2-\frac{u^2}{R_g T}+\frac{1}{2}\left(\frac{\xi^2}{R_g T}-D\right)\right]+Ef^{\mathrm{eq},2}(T) \tag{4.166}$$

对外力项 $\boldsymbol{a}\cdot\nabla_{\boldsymbol{\xi}}f$，可做如下处理[19]：

$$\begin{aligned}\boldsymbol{a}\cdot\nabla_{\boldsymbol{\xi}}f&\approx\boldsymbol{a}\cdot\nabla_{\boldsymbol{\xi}}f^{\mathrm{eq}}=\left(\frac{\boldsymbol{a}}{\sqrt{R_g T}}\cdot\nabla_{\hat{\xi}}\right)f^{\mathrm{eq}}\\&=\frac{\boldsymbol{a}}{\sqrt{R_g T}}\cdot\sum_{n=0}^{\infty}\frac{\boldsymbol{A}^{(n)}(\boldsymbol{x},t)}{n!}\nabla_{\hat{\xi}}[\omega(\boldsymbol{\xi},T)\boldsymbol{H}^{(n)}(\hat{\boldsymbol{\xi}})]\\&=\frac{\boldsymbol{a}}{\sqrt{R_g T}}\cdot\sum_{n=0}^{\infty}\frac{\boldsymbol{A}^{(n)}(\boldsymbol{x},t)}{n!}\frac{\omega(\boldsymbol{\xi},T)}{\omega(\hat{\boldsymbol{\xi}},T)}(-1)^n\nabla_{\hat{\xi}}[\nabla_{\hat{\xi}}^{n}\omega(\hat{\boldsymbol{\xi}},T)]\end{aligned}$$

$$= -\omega(\boldsymbol{\xi},T)\frac{\boldsymbol{a}}{\sqrt{R_g T}}\cdot\sum_{n=0}^{\infty}\frac{\boldsymbol{A}^{(n)}(\boldsymbol{x},t)}{n!}\boldsymbol{H}^{(n+1)}(\hat{\boldsymbol{\xi}}) \tag{4.167}$$

取前两阶可得

$$\boldsymbol{a}\cdot\nabla_{\xi}f=-\omega(\boldsymbol{\xi},T)\rho\left[\frac{\boldsymbol{\xi}\cdot\boldsymbol{a}}{R_gT}+\frac{(\boldsymbol{\xi}\cdot\boldsymbol{a})(\boldsymbol{\xi}\cdot\boldsymbol{u})}{(R_gT)^2}-\frac{\boldsymbol{a}\cdot\boldsymbol{u}}{R_gT}\right] \tag{4.168}$$

类似地，对 $\boldsymbol{a}\cdot\nabla_{\xi}h$ 取其一阶可得

$$\boldsymbol{a}\cdot\nabla_{\xi}h=-\omega(\boldsymbol{\xi},T)\rho E\frac{\boldsymbol{\xi}\cdot\boldsymbol{a}}{R_gT} \tag{4.169}$$

与热力学能分布函数模型一样，文献[17]亦采用 D2Q9 模型的离散速度和二阶数值积分格式对式(4.148)和式(4.149)进行离散，继而得到如下形式的离散模型：

$$f_\alpha(\boldsymbol{r}+\boldsymbol{e}_\alpha\delta_t,t+\delta_t)-f_\alpha(\boldsymbol{r},t)=-\omega_f(f_\alpha-f_\alpha^{\mathrm{eq}})+\delta_t\left(1-\frac{\omega_f}{2}\right)G_\alpha \tag{4.170}$$

$$h_\alpha(\boldsymbol{r}+\boldsymbol{e}_\alpha\delta_t,t+\delta_t)-h_\alpha(\boldsymbol{r},t)=-\omega_h(h_\alpha-h_\alpha^{\mathrm{eq}})+\delta_t\left(1-\frac{\omega_h}{2}\right)H_\alpha+(\omega_h-\omega_f)Z_\alpha\left(f_\alpha-f_\alpha^{\mathrm{eq}}+\frac{\delta_t}{2}G_\alpha\right) \tag{4.171}$$

式中：$\omega_f=\delta_t/(\tau_f+0.5\delta_t)$；$\omega_h=\delta_t/(\tau_h+0.5\delta_t)$；离散的平衡态分布函数为

$$f_\alpha^{\mathrm{eq}}=\rho\omega_\alpha\left[1+\frac{\boldsymbol{e}_\alpha\cdot\boldsymbol{u}}{R_gT_0}+\frac{1}{2}\left(\frac{\boldsymbol{e}_\alpha\cdot\boldsymbol{u}}{R_gT_0}\right)^2-\frac{u^2}{2R_gT_0}\right] \tag{4.172}$$

$$h_\alpha^{\mathrm{eq}}=\omega_\alpha p_0\left[\frac{\boldsymbol{e}_\alpha\cdot\boldsymbol{u}}{R_gT_0}+\left(\frac{\boldsymbol{e}_\alpha\cdot\boldsymbol{u}}{R_gT_0}\right)^2-\frac{u^2}{R_gT_0}+\frac{1}{2}\left(\frac{e_\alpha^2}{R_gT_0}-D\right)\right]+Ef_\alpha^{\mathrm{eq}} \tag{4.173}$$

外力项 G_α 和 H_α 分别为

$$G_\alpha=\omega_\alpha\rho\left[\frac{\boldsymbol{e}_\alpha\cdot\boldsymbol{a}}{R_gT_0}+\frac{(\boldsymbol{e}_\alpha\cdot\boldsymbol{a})(\boldsymbol{e}_\alpha\cdot\boldsymbol{u})}{(R_gT_0)^2}-\frac{\boldsymbol{a}\cdot\boldsymbol{u}}{R_gT_0}\right] \tag{4.174}$$

$$H_\alpha=\omega_\alpha\rho E\frac{\boldsymbol{e}_\alpha\cdot\boldsymbol{u}}{R_gT_0}+f_\alpha^{\mathrm{eq}}\boldsymbol{e}_\alpha\cdot\boldsymbol{a} \tag{4.175}$$

该模型宏观物理量的计算公式为

$$\rho=\sum_{\alpha=0}^{8}f_\alpha \tag{4.176}$$

$$\rho\boldsymbol{u}=\sum_{\alpha=0}^{8}\boldsymbol{e}_\alpha f_\alpha+\frac{\delta_t}{2}\rho\boldsymbol{a} \tag{4.177}$$

$$\rho E=\sum_{\alpha=0}^{8}h_\alpha+\frac{\delta_t}{2}\rho\boldsymbol{u}\cdot\boldsymbol{a} \tag{4.178}$$

离散模型所对应的宏观方程为

$$\frac{\partial\rho}{\partial t}+\nabla\cdot(\rho\boldsymbol{u})=0 \tag{4.179}$$

$$\frac{\partial(\rho\boldsymbol{u})}{\partial t}+\nabla\cdot(\rho\boldsymbol{u}\boldsymbol{u})=-\nabla p_0+\nabla\cdot\widetilde{\boldsymbol{\Pi}}+\rho\boldsymbol{a} \tag{4.180}$$

$$\frac{\partial}{\partial t}(\rho E)+\nabla\cdot[(\rho E+p_0)\boldsymbol{u}]=\nabla\cdot(\lambda\nabla T)+\nabla\cdot(\boldsymbol{u}\cdot\widetilde{\boldsymbol{\Pi}})+\rho\boldsymbol{u}\cdot\boldsymbol{a} \tag{4.181}$$

式中：$p_0=\rho R_gT_0$；$\widetilde{\boldsymbol{\Pi}}=\mu[\nabla\boldsymbol{u}+(\nabla\boldsymbol{u})^{\mathrm{T}}]$；$\mu=\tau_fp_0$；$\lambda=(D+2)R_g\tau_hp_0/2$。

对比热力学能分布函数模型和总能分布函数模型可得出如下结论。

两个模型的相同点：

(1) 从连续 Boltzmann 方程出发，物理背景明确；推导离散的平衡态分布函数时用到了低马赫数假设；所恢复的宏观能量方程包含可压缩功和黏性热耗散。

(2) 采用 D2Q9 格子；为了解决动量方程和能量方程中黏度系数不一致的问题[10]，均采用了二阶数值积分。

两个模型的不同点：

(1) 采用的能量分布函数不同，分别为热力学能分布函数和总能分布函数。

(2) 总能分布函数模型的演化方程中没有热力学能分布函数模型中那样复杂的梯度项，因此计算更简单，数值稳定性更好。

4.5 可压缩流动的耦合双分布函数模型

由流体运动所引起的可压缩性通常采用马赫数 Ma 来度量。当 $Ma \leqslant 0.3$ 时，可认为流体是不可压缩的；当 $Ma > 0.3$ 时，流体逐渐表现出比较明显的可压缩性[21]。常温常压下空气的声速约为 340 m/s，那么速度小于 110 m/s 的空气流动，就可认为是不可压缩的；而速度大于 110 m/s 的空气流动，如研究在空气中高速飞行的物体，就必须认为空气是可压缩的。

利用 $Ma = 0.3$ 作为分界线的划分方法在空气动力学中被广泛采用，但需要注意的是，这里所说的可压缩性是指由于流体运动所引起的密度的较大变化，而在内燃机、脉管制冷机、热声制冷机、斯特林制冷机等工程应用中，流体流动的马赫数很小，甚至接近于零，但流场中的密度却有明显的变化。这是由于流体经过压缩和膨胀热力过程，流体温度和密度发生了较大的变化。这类流动虽然马赫数很小，但流动的可压缩性必须考虑，才能正确反映所经历的热力过程的能量传递和转换，这种由于热力过程而导致的气体密度发生明显变化的流动称为**低马赫数可压缩流动**[22]。

激波

与不可压缩流动相比，流体可压缩性的存在，使得流体运动变得更加复杂，这是因为：首先，流体的密度不再是常数，密度的变化不仅会引起流体热状态的变化，同时它又会反过来影响流体的力学状态，此时连续方程和动量方程必须与能量方程耦合求解；其次，连续方程变为非线性方程，使得求解困难；此外，在某些情况下，可能产生物理量的间断面，这些间断面通常称为**激波**，流体质点经过激波时，熵、密度、压力、温度、速度等都将产生一个急剧（跳跃）的变化，从而对数值计算方法提出了很高的要求。

前向台阶激波反射的 LBM 模拟

可压缩流动的格子 Boltzmann 模型大体上可分为三类：多速度模型、比热容比可调模型以及耦合双分布函数模型。多速度模型是等温格子 Boltzmann 模型的直接推广，即在等温模型的基础上增加若干新的离散速度方向，同时在平衡态密度分布函数中加入高阶速度项。在一般的多速度模型中，比热容比与模型的空间维数有关，从而通常不具有真实的物理意义。为了解决这一问题，多个比热容比可调模型应运而生。考虑到双分布函数格子 Boltzmann 方法在调节比热容比和普朗特数上的优势及其良好的数值稳定性，将双分布函数格子 Boltzmann 方法拓展到了可压缩流动，提出了满足理想气体状态方程、适用于可压缩流动的耦合双分布函数格子 Boltzmann 模型。本节对这一模型进行介绍。

4.5.1 基本原理

按照两个分布函数是否耦合可以将双分布函数格子 Boltzmann 模型分为两类:一类是非耦合的双分布函数模型,即密度分布函数影响温度或能量分布函数,而温度或能量分布函数并不影响密度分布函数,反映在流动中即为流场影响温度场,而温度场并不反作用于流场;另一类则为耦合的双分布函数模型,比较常见是针对不可压缩流动与传热问题,通过 Boussinesq 假设耦合起来的双分布函数模型,即通过在密度分布函数的演化方程中添加一个与温度有关的作用力项,从而将温度场对流场的影响考虑进去。

相对于不可压缩模型来说,要构造可压缩流动的双分布函数模型需考虑以下几个问题。

(1) 基于 D2Q9 格子的双分布函数模型所恢复的宏观动量方程为

$$\partial_t(\rho\boldsymbol{u})+\nabla\cdot(\rho\boldsymbol{u}\boldsymbol{u})=-\nabla p+\nabla\cdot\left\{\rho\nu[\nabla\boldsymbol{u}+(\nabla\boldsymbol{u})^{\mathrm{T}}]-\frac{\nu}{c_s^2}\nabla\cdot(\rho\boldsymbol{u}\boldsymbol{u}\boldsymbol{u})\right\} \tag{4.182}$$

前面我们曾提到,构造等温不可压缩格子 Boltzmann 模型主要是对 D2Q9 模型做出改进使其所恢复的宏观方程是不可压缩 Navier-Stokes 方程组。同理,要构造可压缩流动的格子 Boltzmann 模型,就必须采取措施使得模型所恢复的宏观方程是可压缩 Navier-Stokes 方程组。通过分析可知,要消除式(4.182)右端的偏差项,需保证格子张量零~六阶都是各向同性的(参见附录 B),而 D2Q9 格子至多只能满足零~五阶格子张量是各向同性的,因此必须选取对称性更好的离散速度配置。比较常见的做法是像多速度模型那样增加必要的离散速度方向,并在平衡态密度分布函数中加入更高阶的速度项。

(2) 需要满足理想气体状态方程,即 $p=\rho R_g T$。在一般的双分布函数模型中,$p=\rho c^2/3$,其中格子速度 c 为常数,于是压力 p 只与密度有关,因此模型是解耦的。此外,对不可压缩流动,比定压热容 c_p 可近似认为与比定容热容 c_V 相等,因此不需要考虑比热容比的问题,但在构造可压缩模型时,这一问题需着重处理。

(3) 采用双分布函数比采用单分布函数多了一个确定平衡态能量分布函数的过程,这势必增加建模和编程的工作量。如果能找到一个合理的关系式,将平衡态能量分布函数和平衡态密度分布函数联系起来,则可以方便不少。

从上述问题出发,我们通过将双分布函数模型和多速度模型结合起来,构造了满足理想气体状态方程、适用于可压缩流动的**耦合双分布函数格子 Boltzmann 模型**[23]。下面介绍其基本原理。

为了恢复可压缩形式的动量方程,耦合双分布函数模型中的密度分布函数应当满足下列约束条件:

$$\sum_{\alpha=0}^{m-1} f_\alpha^{\mathrm{eq}}=\rho \tag{4.183}$$

$$\sum_{\alpha=0}^{m-1} f_\alpha^{\mathrm{eq}} e_{\alpha i}=\rho u_i \tag{4.184}$$

$$\sum_{\alpha=0}^{m-1} f_\alpha^{\mathrm{eq}} e_{\alpha i} e_{\alpha j}=\rho u_i u_j+p\delta_{ij} \tag{4.185}$$

$$\sum_{\alpha=0}^{m-1} f_\alpha^{\mathrm{eq}} e_{\alpha i} e_{\alpha j} e_{\alpha k}=\rho u_i u_j u_k+p(u_k\delta_{ij}+u_j\delta_{ik}+u_i\delta_{jk}) \tag{4.186}$$

$$\sum_{\alpha=0}^{m-1} f_{\alpha}^{\mathrm{eq}} e_{\alpha}^{2}=\rho u^{2}+Dp \tag{4.187}$$

$$\sum_{\alpha=0}^{m-1} f_{\alpha}^{\mathrm{eq}} e_{\alpha}^{2} e_{\alpha i}=[\rho u^{2}+(D+2)p]u_{i} \tag{4.188}$$

式中:m 为离散速度的个数;$p=\rho R_{\mathrm{g}}T$。应当指出的是,满足上述约束条件的平衡态密度分布函数可以独立地构成一个单分布的多速度格子 Boltzmann 模型,它可以恢复标准的可压缩连续方程和动量方程,并能保证 Euler 层次上的能量方程,故可以看作是一个比热容比为 $\gamma=(D+2)/D$ 的可压缩 Euler 模型,其宏观压力可计算如下:

$$p=\frac{1}{D}\left(\sum_{\alpha=0}^{m-1} f_{\alpha} e_{\alpha}^{2}-\rho u^{2}\right) \tag{4.189}$$

对可压缩流动的格子 Boltzmann 模型来说,比热容比的调节是一个非常重要的问题。一般模型通常只能保证 D 个自由度,因此必须采取措施增加自由度以调节比热容比。根据文献[18,24],适合单原子、双原子和多原子气体的密度分布函数的通用表达式为 $\hat{f}=\hat{f}(\boldsymbol{r},\boldsymbol{\xi},\boldsymbol{\eta},t)$,其中 $\boldsymbol{\eta}$ 为与内自由度有关的含有 d 个($d=b-D$)分量的速度矢量。此时,连续 Boltzmann-BGK 方程仍然成立:

$$\partial_{t} \hat{f}+(\boldsymbol{\xi}\cdot\nabla)\hat{f}=-\frac{1}{\tau_{f}}(\hat{f}-\hat{f}^{\mathrm{eq}}) \tag{4.190}$$

式中,

$$\hat{f}^{\mathrm{eq}}=\frac{\rho}{(2\pi R_{\mathrm{g}}T)^{(D+d)/2}}\exp\left[-\frac{(\boldsymbol{\xi}-\boldsymbol{u})^{2}+\eta^{2}}{2R_{\mathrm{g}}T}\right] \tag{4.191}$$

相对应的热力学能分布函数及其平衡态分布函数应定义如下:

$$\hat{g}=\frac{(\boldsymbol{\xi}-\boldsymbol{u})^{2}+\eta^{2}}{2}\hat{f},\quad \hat{g}^{\mathrm{eq}}=\frac{(\boldsymbol{\xi}-\boldsymbol{u})^{2}+\eta^{2}}{2}\hat{f}^{\mathrm{eq}} \tag{4.192}$$

根据宏观量的定义

$$\rho=\int_{\Re}\int_{\Re_{\eta}}\hat{f}\mathrm{d}\boldsymbol{\eta}\mathrm{d}\boldsymbol{\xi},\quad \rho\boldsymbol{u}=\int_{\Re}\int_{\Re_{\eta}}\boldsymbol{\xi}\hat{f}\mathrm{d}\boldsymbol{\eta}\mathrm{d}\boldsymbol{\xi},\quad \rho\frac{b}{2}R_{\mathrm{g}}T=\int_{\Re}\int_{\Re_{\eta}}\hat{g}\mathrm{d}\boldsymbol{\eta}\mathrm{d}\boldsymbol{\xi} \tag{4.193}$$

式中,$\Re_{\eta}$ 表示 $\boldsymbol{\eta}$ 对应的速度空间。取 $f=\int_{\Re_{\eta}}\hat{f}\mathrm{d}\boldsymbol{\eta}$、$g=\int_{\Re_{\eta}}\hat{g}\mathrm{d}\boldsymbol{\eta}$,则依据 Boltzmann 方程可得

$$\partial_{t} f+(\boldsymbol{\xi}\cdot\nabla)f=-\frac{1}{\tau_{f}}(f-f^{\mathrm{eq}}) \tag{4.194}$$

$$\partial_{t} g+(\boldsymbol{\xi}\cdot\nabla)g=-\frac{1}{\tau_{g}}(g-g^{\mathrm{eq}})-fq \tag{4.195}$$

式中,

$$f^{\mathrm{eq}}=\int_{\Re_{\eta}}\hat{f}^{\mathrm{eq}}\mathrm{d}\boldsymbol{\eta}=\frac{\rho}{(2\pi R_{\mathrm{g}}T)^{D/2}}\exp\left[-\frac{(\boldsymbol{\xi}-\boldsymbol{u})^{2}}{2R_{\mathrm{g}}T}\right] \tag{4.196}$$

$$g^{\mathrm{eq}}=\int_{\Re_{\eta}}\hat{g}^{\mathrm{eq}}\mathrm{d}\boldsymbol{\eta}=\frac{\rho[(\boldsymbol{\xi}-\boldsymbol{u})^{2}+(b-D)R_{\mathrm{g}}T]}{2(2\pi R_{\mathrm{g}}T)^{D/2}}\exp\left[-\frac{(\boldsymbol{\xi}-\boldsymbol{u})^{2}}{2R_{\mathrm{g}}T}\right] \tag{4.197}$$

上述两式的推导过程中用到数学积分:$\int_{-\infty}^{+\infty}\exp(-\zeta x^{2})\mathrm{d}x=\sqrt{\pi/\zeta}$ 和 $\int_{-\infty}^{+\infty}x^{2}\exp(-\zeta x^{2})\mathrm{d}x=\sqrt{\pi/\zeta^{3}}/2$。由式(4.194)和式(4.196)可知,对单分布函数模型来说,即使从基本的统计理论出发引入与内自由度有关的量,也并未对密度分布函数的演化方程及其平衡态形式造成

任何影响,因此必须采取其他措施才能调节单分布函数模型的比热容比。对双分布函数模型来说,尽管平衡态密度分布函数的形式没有发生变化,但对比式(4.51)和式(4.197)可知平衡态能量分布函数发生了变化,这一变化正是由内自由度的相关量引起的。

式(4.197)对速度空间积分可得通用热力学能分布函数所满足的统计关系为

$$\int_{\Re} g^{\mathrm{eq}} \mathrm{d}\boldsymbol{\xi} = \rho \frac{b}{2} R_{\mathrm{g}} T \tag{4.198}$$

$$\int_{\Re} g^{\mathrm{eq}} \xi_i \mathrm{d}\boldsymbol{\xi} = \rho \frac{b}{2} R_{\mathrm{g}} T u_i \tag{4.199}$$

$$\int_{\Re} g^{\mathrm{eq}} \xi_i \xi_j \mathrm{d}\boldsymbol{\xi} = \rho \frac{b}{2} R_{\mathrm{g}} T u_i u_j + p \frac{b+2}{2} R_{\mathrm{g}} T \delta_{ij} \tag{4.200}$$

如果读者对双分布函数模型的 Chapman-Enskog 多尺度展开过程比较熟悉,完全可以脱离上述分析,直接依据应恢复的宏观能量方程反推出平衡态分布函数应满足的约束条件。这也是格子 Boltzmann 方法中比较常见的建模思路。

同理,对**总能分布函数**有

$$h^{\mathrm{eq}} = \frac{\rho[\xi^2 + (b-D) R_{\mathrm{g}} T]}{2(2\pi R_{\mathrm{g}} T)^{D/2}} \exp\left[-\frac{(\boldsymbol{\xi}-\boldsymbol{u})^2}{2R_{\mathrm{g}} T}\right] \tag{4.201}$$

式(4.201)对速度空间积分可得

$$\int_{\Re} h^{\mathrm{eq}} \mathrm{d}\boldsymbol{\xi} = \rho E \tag{4.202}$$

$$\int_{\Re} h^{\mathrm{eq}} \xi_i \mathrm{d}\boldsymbol{\xi} = (\rho E + p) u_i \tag{4.203}$$

$$\int_{\Re} h^{\mathrm{eq}} \xi_i \xi_j \mathrm{d}\boldsymbol{\xi} = (\rho E + 2p) u_i u_j + p(E + R_{\mathrm{g}} T) \delta_{ij} \tag{4.204}$$

式中,$\rho E = b\rho R_{\mathrm{g}} T/2 + \rho u^2/2$ 为适用于单原子、双原子和多原子气体的单位体积总能表达式。相应地,离散的总能分布函数应满足如下约束条件:

$$\sum_{\alpha=0}^{m-1} h_\alpha^{\mathrm{eq}} = \rho E \tag{4.205}$$

$$\sum_{\alpha=0}^{m-1} e_{\alpha i} h_\alpha^{\mathrm{eq}} = (\rho E + p) u_i \tag{4.206}$$

$$\sum_{\alpha=0}^{m-1} e_{\alpha i} e_{\alpha j} h_\alpha^{\mathrm{eq}} = (\rho E + 2p) u_i u_j + p(E + R_{\mathrm{g}} T) \delta_{ij} \tag{4.207}$$

不考虑外力项时,密度分布函数 f_α 和总能分布函数 h_α 的演化方程分别为

$$\partial_t f_\alpha + (\boldsymbol{e}_\alpha \cdot \nabla) f_\alpha = -\frac{1}{\tau_f}(f_\alpha - f_\alpha^{\mathrm{eq}}) \tag{4.208}$$

$$\partial_t h_\alpha + (\boldsymbol{e}_\alpha \cdot \nabla) h_\alpha = -\frac{1}{\tau_h}(h_\alpha - h_\alpha^{\mathrm{eq}}) + \frac{1}{\tau_{hf}}(\boldsymbol{e}_\alpha \cdot \boldsymbol{u})(f_\alpha - f_\alpha^{\mathrm{eq}}) \tag{4.209}$$

式(4.208)和式(4.209)所对应的宏观方程可通过 Chapman-Enskog 多尺度展开导出。能量方程的推导可参阅文献[23],动量方程的推导与 3.5.2 节相似,所不同的是,$\sum_{\alpha=0}^{m-1} \boldsymbol{e}_\alpha \boldsymbol{e}_\alpha f_\alpha^{(1)}$ 推导如下:

$$\sum_{\alpha=0}^{m-1} \boldsymbol{e}_\alpha \boldsymbol{e}_\alpha f_\alpha^{(1)} = -\tau_f \sum_{\alpha=0}^{m-1} (\partial_{t_1} + \boldsymbol{e}_\alpha \cdot \nabla_1) \boldsymbol{e}_\alpha \boldsymbol{e}_\alpha f_\alpha^{\mathrm{eq}}$$

$$= -\tau_f \left[\partial_{t_1}(\rho \boldsymbol{u}\boldsymbol{u} + p\boldsymbol{I}) + \nabla_1 \cdot \left(\sum_{\alpha=0}^{m-1} \boldsymbol{e}_\alpha \boldsymbol{e}_\alpha \boldsymbol{e}_\alpha f_\alpha^{\mathrm{eq}} \right) \right] \tag{4.210}$$

且有

$$\partial_{t_1}(\rho \boldsymbol{u}\boldsymbol{u}) = -\boldsymbol{u}\nabla_1 p - (\boldsymbol{u}\nabla_1 p)^{\mathrm{T}} - \nabla_1 \cdot (\rho \boldsymbol{u}\boldsymbol{u}\boldsymbol{u}) \tag{4.211}$$

$$\partial_{t_1} p = -\nabla_1 \cdot (p\boldsymbol{u}) - (\gamma - 1) p \nabla_1 \cdot \boldsymbol{u} \tag{4.212}$$

结合式(4.210)~式(4.212)及式(4.186)可导出宏观动量方程中的黏性应力张量为

$$\boldsymbol{\Pi} = \mu \left[\nabla \boldsymbol{u} + (\nabla \boldsymbol{u})^{\mathrm{T}} - \frac{2}{D}(\nabla \cdot \boldsymbol{u})\boldsymbol{I} \right] + \mu_{\mathrm{B}}(\nabla \cdot \boldsymbol{u})\boldsymbol{I} \tag{4.213}$$

式中,$\mu_{\mathrm{B}} = 2(1/D - 1/b)\mu$ 为体积黏度。

由于f_α^{eq}和 h_α^{eq} 的约束条件不同,通常需要分别独立确定。如果能找到一个合理的关系式将 h_α^{eq} 和f_α^{eq}联系起来,可以方便不少。考虑到连续的平衡态总能分布函数和 Maxwell 分布函数存在如下关系[23]:

$$h^{\mathrm{eq}} = \frac{\rho[\xi^2 + (b-D)RT]}{2(2\pi R_{\mathrm{g}} T)^{D/2}} \exp\left[-\frac{(\boldsymbol{\xi} - \boldsymbol{u})^2}{2R_{\mathrm{g}} T} \right] = \frac{1}{2}[\xi^2 + (b-D)R_{\mathrm{g}} T] f^{\mathrm{eq}} \tag{4.214}$$

那么离散的平衡态分布函数 h_α^{eq} 和 f_α^{eq}应该也可以由某种类似的关系式联系起来。为此,对式(4.214)右端做如下重组:

$$\frac{1}{2}[\xi^2 + (b-D)R_{\mathrm{g}} T] f^{\mathrm{eq}} = \frac{1}{2}[(\boldsymbol{\xi} - \boldsymbol{u}) \cdot (\boldsymbol{\xi} - \boldsymbol{u}) + (b-D)R_{\mathrm{g}} T + u^2] f^{\mathrm{eq}} + (\boldsymbol{\xi} - \boldsymbol{u}) \cdot \boldsymbol{u} f^{\mathrm{eq}} \tag{4.215}$$

式(4.215)右端第一项分别乘 1、ξ_i 和 $\xi_i\xi_j$,然后再对 $\boldsymbol{\xi}$ 积分有

$$\int_{\Re} \frac{1}{2}[(\boldsymbol{\xi} - \boldsymbol{u})^2 + (b-D)R_{\mathrm{g}} T + u^2] f^{\mathrm{eq}} \mathrm{d}\boldsymbol{\xi} = E \int_{\Re} f^{\mathrm{eq}} \mathrm{d}\boldsymbol{\xi} \tag{4.216}$$

$$\int_{\Re} \frac{1}{2}[(\boldsymbol{\xi} - \boldsymbol{u})^2 + (b-D)R_{\mathrm{g}} T + u^2] \xi_i f^{\mathrm{eq}} \mathrm{d}\boldsymbol{\xi} = E \int_{\Re} \xi_i f^{\mathrm{eq}} \mathrm{d}\boldsymbol{\xi} \tag{4.217}$$

$$\int_{\Re} \frac{1}{2}[(\boldsymbol{\xi} - \boldsymbol{u})^2 + (b-D)R_{\mathrm{g}} T + u^2] \xi_i\xi_j f^{\mathrm{eq}} \mathrm{d}\boldsymbol{\xi} = E \int_{\Re} \xi_i\xi_j f^{\mathrm{eq}} \mathrm{d}\boldsymbol{\xi} + pR_{\mathrm{g}} T\delta_{ij} \tag{4.218}$$

式(4.215)~式(4.218)表明,对 Navier-Stokes 层次的宏观方程来说,如下形式的 h^{eq}定义式与式(4.214)是完全等价的,即

$$h^{\mathrm{eq}} = [E + (\boldsymbol{\xi} - \boldsymbol{u}) \cdot \boldsymbol{u}] f^{\mathrm{eq}} + \omega(\boldsymbol{\xi}, T) \tag{4.219}$$

式中,$\omega(\boldsymbol{\xi}, T)$满足

$$\int_{\Re} \omega(\boldsymbol{\xi}, T) \mathrm{d}\boldsymbol{\xi} = 0, \quad \int_{\Re} \omega(\boldsymbol{\xi}, T) \boldsymbol{\xi} \mathrm{d}\boldsymbol{\xi} = 0, \quad \int_{\Re} \omega(\boldsymbol{\xi}, T) \xi_i\xi_j \mathrm{d}\boldsymbol{\xi} = pR_{\mathrm{g}} T\delta_{ij} \tag{4.220}$$

基于这一思路,我们在文献[23]中提出了如下关系式:

$$h_\alpha^{\mathrm{eq}} = [E + (\boldsymbol{e}_\alpha - \boldsymbol{u}) \cdot \boldsymbol{u}] f_\alpha^{\mathrm{eq}} + \omega_\alpha \frac{p}{c^2} R_{\mathrm{g}} T \tag{4.221}$$

式中,c 为格子特征速度。对式(4.221)两边分别同乘 1、$e_{\alpha i}$和 $e_{\alpha i}e_{\alpha j}$,再对 α 求和,分别可得

$$\sum_{\alpha=0}^{m-1} h_\alpha^{\mathrm{eq}} = \sum_{\alpha=0}^{m-1} \left[E f_\alpha^{\mathrm{eq}} + (\boldsymbol{e}_\alpha - \boldsymbol{u}) \cdot \boldsymbol{u} f_\alpha^{\mathrm{eq}} + \omega_\alpha \frac{p}{c^2} R_{\mathrm{g}} T \right]$$

$$= \rho E + \sum_{\alpha=0}^{m-1} \omega_\alpha \frac{p}{c^2} R_{\mathrm{g}} T \tag{4.222}$$

$$\sum_{\alpha=0}^{m-1} e_{\alpha i}h_\alpha^{\mathrm{eq}} = \sum_{\alpha=0}^{m-1} e_{\alpha i}\left[Ef_\alpha^{\mathrm{eq}}+(\boldsymbol{e}_\alpha-\boldsymbol{u})\cdot\boldsymbol{u}f_\alpha^{\mathrm{eq}}+\omega_\alpha\frac{p}{c^2}R_{\mathrm{g}}T\right]$$
$$=\rho Eu_i+pu_i+\sum_{\alpha=0}^{m-1}\omega_\alpha e_{\alpha i}\frac{p}{c^2}R_{\mathrm{g}}T \tag{4.223}$$

$$\sum_{\alpha=0}^{m-1} e_{\alpha i}e_{\alpha j}h_\alpha^{\mathrm{eq}} = \sum_{\alpha=0}^{m-1} e_{\alpha i}e_{\alpha j}\left[Ef_\alpha^{\mathrm{eq}}+(\boldsymbol{e}_\alpha-\boldsymbol{u})\cdot\boldsymbol{u}f_\alpha^{\mathrm{eq}}+\omega_\alpha\frac{p}{c^2}R_{\mathrm{g}}T\right]$$
$$=\rho Eu_iu_j+pE\delta_{ij}+2pu_iu_j+\sum_{\alpha=0}^{m-1}e_{\alpha i}e_{\alpha j}\omega_\alpha\frac{p}{c^2}R_{\mathrm{g}}T \tag{4.224}$$

将上述式子与式(4.205)~式(4.207)对比,可知 ω_α 需满足的约束条件为

$$\sum_{\alpha=0}^{m-1}\omega_\alpha=0,\quad \sum_{\alpha=0}^{m-1}\omega_\alpha e_{\alpha i}=0,\quad \sum_{\alpha=0}^{m-1}\omega_\alpha e_{\alpha i}e_{\alpha j}=c^2\delta_{ij} \tag{4.225}$$

于是,在确定平衡态密度分布函数 f_α^{eq} 之后,可以比较方便地根据式(4.221)得到 h_α^{eq}。

4.5.2 二维模型

根据前面的分析,对二维问题,D2Q9 格子的离散速度并不能满足可压缩耦合双分布函数模型的需要。为了保证零~六阶格子张量是各向同性的,通常可采用 12 个或 13 个离散速度的格子(D2Q12 或 D2Q13,如图 4.1 所示),其中 D2Q12 的粒子速度配置为

$$\boldsymbol{e}_\alpha=c\begin{cases}\{\cos[(\alpha-1)\pi/2],\sin[(\alpha-1)\pi/2]\}, & \alpha=1,2,3,4\\ \sqrt{2}\{\cos[(2\alpha-1)\pi/2],\sin[(2\alpha-1)\pi/2]\}, & \alpha=5,6,7,8\\ 2\{\cos[(\alpha-9)\pi/2],\sin[(\alpha-9)\pi/2]\}, & \alpha=9,10,11,12\end{cases} \tag{4.226}$$

式中:$c=\sqrt{R_{\mathrm{g}}T_{\mathrm{c}}}$ 为格子特征速度;T_{c} 为特征温度。在 D2Q12 格子原点上布置一个静止粒子则为 D2Q13 格子。

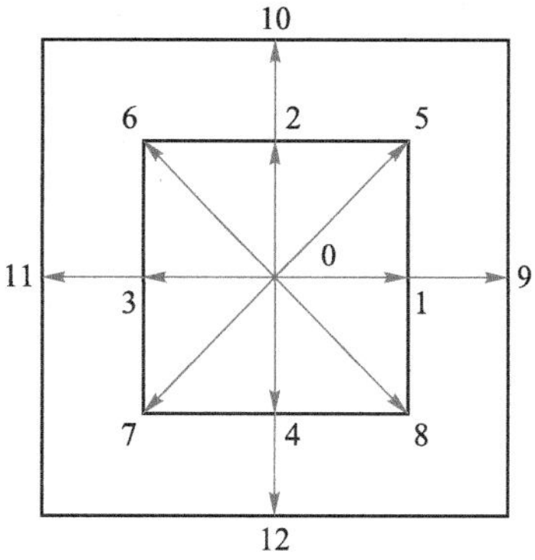

图 4.1 D2Q12 格子

格子 Boltzmann 方法中构造平衡态分布函数的方法通常可分为如下两大类:

(1) 一类是从 Maxwell 分布函数出发构造平衡态分布函数。这种方法是最基本也是最常见的方法,它通常又可以分为两种:一种是对 Maxwell 分布函数做泰勒级数展开形成截断形式的 Maxwell 分布函数,并采用待定系数法或 Gauss-Hermite 求积方法根据相关约束条件确定离散的平衡态分布函数;另一种则是对 Maxwell 分布函数做 Hermite 展开,而后通过 Gauss-Hermite 求积方法确定离散的平衡态分布函数。这两种方法对等温模型来说基本上没有区别,不过对热模型来说,得到的结果稍有不同[25]。

(2) 第二大类方法与 Maxwell 分布函数没有直接联系。例如,有的可压缩模型采用克罗内克 δ 函数作为离散的平衡态分布函数;又如 Qu 等[25]提出用圆函数或其他类似的分布函数取代 Maxwell 分布函数,而后采用拉格朗日插值多项式将连续的分布函数分配到各个离散速度方向上。第二大类方法的提出主要是为了克服第一类方法往往只能适用于低中马赫数流动的缺陷。

在文献[23]中,我们基于截断形式的 Maxwell 分布函数及 Qu 等[25]的圆函数分布,给

出了两个二维的耦合双分布函数模型，相应的平衡态密度分布函数分别为

$$
\begin{aligned}
f_{1,2,3,4}^{\mathrm{eq}} &= \frac{\rho}{12}\left[(5-4\bar{p})+(8-12\bar{p})(\bar{\boldsymbol{e}}_\alpha\cdot\bar{\boldsymbol{u}})-2\bar{u}^2+2(\bar{\boldsymbol{e}}_\alpha\cdot\bar{\boldsymbol{u}})^2-6(\bar{\boldsymbol{e}}_\alpha\cdot\bar{\boldsymbol{u}})\bar{u}^2+4(\bar{\boldsymbol{e}}_\alpha\cdot\bar{\boldsymbol{u}})^3\right]\\
f_{5,6,7,8}^{\mathrm{eq}} &= \frac{\rho}{8}\left[(2\bar{p}-1)+2\bar{p}(\bar{\boldsymbol{e}}_\alpha\cdot\bar{\boldsymbol{u}})-\bar{u}^2+(\bar{\boldsymbol{e}}_\alpha\cdot\bar{\boldsymbol{u}})^2-(\bar{\boldsymbol{e}}_\alpha\cdot\bar{\boldsymbol{u}})\bar{u}^2+(\bar{\boldsymbol{e}}_\alpha\cdot\bar{\boldsymbol{u}})^3\right]\\
f_{9,10,11,12}^{\mathrm{eq}} &= \frac{\rho}{48}\left[2(2\bar{p}-1)+2(3\bar{p}-1)(\bar{\boldsymbol{e}}_\alpha\cdot\bar{\boldsymbol{u}})+2\bar{u}^2+(\bar{\boldsymbol{e}}_\alpha\cdot\bar{\boldsymbol{u}})^2+0.5(\bar{\boldsymbol{e}}_\alpha\cdot\bar{\boldsymbol{u}})^3\right]
\end{aligned}
\tag{4.227}
$$

及

$$
\begin{aligned}
f_0^{\mathrm{eq}} &= \frac{\rho}{4}\left[\bar{u}_x^4+5\bar{p}^2-10\bar{p}+4+4\bar{u}_x^2\bar{u}_y^2+\bar{u}_y^4+(10\bar{p}-5)(\bar{u}_x^2+\bar{u}_y^2)\right]\\
f_1^{\mathrm{eq}} &= -\frac{\rho}{6}(-4\bar{u}_x^2+3\bar{p}^2+\bar{u}_x^4-4\bar{p}+3\bar{p}\bar{u}_y^2+3\bar{u}_x\bar{u}_y^2+9\bar{p}\bar{u}_x^2+6\bar{p}\bar{u}_x+3\bar{u}_x^2\bar{u}_y^2-4\bar{u}_x+\bar{u}_x^3)\\
f_2^{\mathrm{eq}} &= -\frac{\rho}{6}(-4\bar{u}_y^2+3\bar{p}^2+\bar{u}_y^4-4\bar{p}+3\bar{p}\bar{u}_x^2+3\bar{u}_x^2\bar{u}_y+9\bar{p}\bar{u}_y^2+6\bar{p}\bar{u}_y+3\bar{u}_x^2\bar{u}_y^2-4\bar{u}_y+\bar{u}_y^3)\\
f_3^{\mathrm{eq}} &= -\frac{\rho}{6}(-4\bar{u}_x^2+3\bar{p}^2+\bar{u}_x^4-4\bar{p}+3\bar{p}\bar{u}_y^2-3\bar{u}_x\bar{u}_y^2+9\bar{p}\bar{u}_x^2-6\bar{p}\bar{u}_x+3\bar{u}_x^2\bar{u}_y^2+4\bar{u}_x-\bar{u}_x^3)\\
f_4^{\mathrm{eq}} &= -\frac{\rho}{6}(-4\bar{u}_y^2+3\bar{p}^2+\bar{u}_y^4-4\bar{p}+3\bar{p}\bar{u}_x^2-3\bar{u}_x^2\bar{u}_y+9\bar{p}\bar{u}_y^2-6\bar{p}\bar{u}_y+3\bar{u}_x^2\bar{u}_y^2+4\bar{u}_y-\bar{u}_y^3)\\
f_5^{\mathrm{eq}} &= \frac{\rho}{4}(\bar{u}_x\bar{u}_y^2+\bar{u}_x\bar{u}_y+\bar{p}\bar{u}_x+\bar{p}\bar{u}_y+\bar{u}_x^2\bar{u}_y+0.5\bar{p}^2+\bar{u}_x^2\bar{u}_y^2+\bar{p}\bar{u}_x^2+\bar{p}\bar{u}_y^2)\\
f_6^{\mathrm{eq}} &= \frac{\rho}{4}(-\bar{u}_x\bar{u}_y^2-\bar{u}_x\bar{u}_y-\bar{p}\bar{u}_x+\bar{p}\bar{u}_y+\bar{u}_x^2\bar{u}_y+0.5\bar{p}^2+\bar{u}_x^2\bar{u}_y^2+\bar{p}\bar{u}_x^2+\bar{p}\bar{u}_y^2)\\
f_7^{\mathrm{eq}} &= \frac{\rho}{4}(-\bar{u}_x\bar{u}_y^2+\bar{u}_x\bar{u}_y-\bar{p}\bar{u}_x-\bar{p}\bar{u}_y-\bar{u}_x^2\bar{u}_y+0.5\bar{p}^2+\bar{u}_x^2\bar{u}_y^2+\bar{p}\bar{u}_x^2+\bar{p}\bar{u}_y^2)\\
f_8^{\mathrm{eq}} &= \frac{\rho}{4}(\bar{u}_x\bar{u}_y^2-\bar{u}_x\bar{u}_y+\bar{p}\bar{u}_x-\bar{p}\bar{u}_y-\bar{u}_x^2\bar{u}_y+0.5\bar{p}^2+\bar{u}_x^2\bar{u}_y^2+\bar{p}\bar{u}_x^2+\bar{p}\bar{u}_y^2)\\
f_9^{\mathrm{eq}} &= \frac{\rho}{24}(-2\bar{u}_x+\bar{u}_x^4-\bar{p}-\bar{u}_x^2+6\bar{p}\bar{u}_x^2+1.5\bar{p}^2+2\bar{u}_x^3+6\bar{p}\bar{u}_x)\\
f_{10}^{\mathrm{eq}} &= \frac{\rho}{24}(-2\bar{u}_y+\bar{u}_y^4-\bar{p}-\bar{u}_y^2+6\bar{p}\bar{u}_y^2+1.5\bar{p}^2+2\bar{u}_y^3+6\bar{p}\bar{u}_y)\\
f_{11}^{\mathrm{eq}} &= \frac{\rho}{24}(2\bar{u}_x+\bar{u}_x^4-\bar{p}-\bar{u}_x^2+6\bar{p}\bar{u}_x^2+1.5\bar{p}^2-2\bar{u}_x^3-6\bar{p}\bar{u}_x)\\
f_{12}^{\mathrm{eq}} &= \frac{\rho}{24}(2\bar{u}_y+\bar{u}_y^4-\bar{p}-\bar{u}_y^2+6\bar{p}\bar{u}_y^2+1.5\bar{p}^2-2\bar{u}_y^3-6\bar{p}\bar{u}_y)
\end{aligned}
\tag{4.228}
$$

式中：量纲为 1 的量 $\bar{p}=p/(\rho c^2)$；$\bar{\boldsymbol{e}}_\alpha=\boldsymbol{e}_\alpha/c$；$\bar{u}_x=u_x/c$；$\bar{u}_y=u_y/c$。

需要注意的是，式(4.227)针对的是 D2Q12 格子，采用 D2Q13 或其他对称性更好的离散速度模型也是可行的，而式(4.228)采用 D2Q13 格子则是沿用了文献[25]的做法。针对式(4.228)，由于格子的对称性，同一亚格子各个方向上的平衡态分布函数亦具有对称性[26]。例如，$f_2^{\mathrm{eq}}(\bar{u}_x,\bar{u}_y)=f_1^{\mathrm{eq}}(\bar{u}_y,\bar{u}_x)$，即由于方向 2 和方向 1 关于 $y=x$ 对称，于是互换 f_1^{eq} 中的下标 x 和 y 便可得到 f_2^{eq}；又如方向 3 和方向 1 关于 $x=0$ 对称，则可得 $f_3^{\mathrm{eq}}(\bar{u}_x,\bar{u}_y)=$

$f_1^{eq}(-\bar{u}_x,\bar{u}_y)$。

在确定平衡态密度分布函数之后，可通过式(4.221)得到 h_α^{eq}。选取的参数 ω_α 为[23]

$$\omega_0=0,\quad \omega_{1,2,3,4}=-\frac{1}{3},\quad \omega_{5,6,7,8}=\frac{1}{4},\quad \omega_{9,10,11,12}=\frac{1}{12} \tag{4.229}$$

由于 ω_0 为 0，故上述参数亦适用于 D2Q12 格子模型。流体密度、速度、温度和压力的计算格式为

$$\rho=\sum_{\alpha=0}^{12} f_\alpha,\quad \rho\boldsymbol{u}=\sum_{\alpha=0}^{12}\boldsymbol{e}_\alpha f_\alpha \tag{4.230}$$

$$T=2\frac{\left(\sum_{\alpha=0}^{12} h_\alpha-\frac{1}{2}\rho u^2\right)}{\rho bR_g},\quad p=\rho R_g T \tag{4.231}$$

在具体的数值计算过程中，特征温度 T_c 的选取非常重要，它在一定程度上关乎着模型的数值稳定性，对存在激波或间断的可压缩流动来说则更加重要。通常情况下，T_c 应不小于流场中滞止温度的最大值，即[23,25]

$$T_c\geqslant\max\left[T\left(1+\frac{\gamma-1}{2}Ma^2\right)\right] \tag{4.232}$$

当然，并不是说不按照式(4.232)选取，模型就一定会失稳。对于具体问题需要具体分析，可结合模型的数值稳定性以及 T_c 对时间步长的影响选取一个适当的值。

4.5.3 三维模型

对三维流动，我们提出采用**球面函数**取代 Maxwell 分布函数，其定义为[26]

$$f^{eq}=\begin{cases}\dfrac{\rho}{4\pi r^2}, & \|\boldsymbol{\xi}-\boldsymbol{u}\|=\|\boldsymbol{r}\|=r\\ 0, & \text{else}\end{cases} \tag{4.233}$$

式中，$\boldsymbol{r}=\boldsymbol{\xi}-\boldsymbol{u}$（即 $\boldsymbol{\xi}=\boldsymbol{u}+\boldsymbol{r}$）为特异速度，$r$ 为其模。

式(4.233)表明相空间中的所有粒子均集中于一个球面上（图 4.2），因此对整个空间的积分就转化为对球面的积分，即

$$d\boldsymbol{\xi}=r^2\sin\theta d\theta d\varphi \tag{4.234}$$

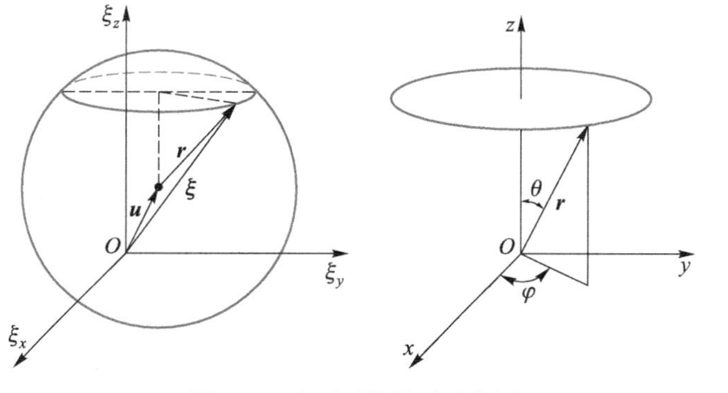

图 4.2 球面函数模型示意图

对式(4.233)求矩可得

$$\int_{\Re} f^{\mathrm{eq}} \mathrm{d}\boldsymbol{\xi} = \int_0^{2\pi}\int_0^{\pi} \frac{\rho}{4\pi r^2} r^2 \sin\theta \mathrm{d}\theta \mathrm{d}\varphi = \rho \tag{4.235}$$

$$\int_{\Re} f^{\mathrm{eq}} \xi_i \mathrm{d}\boldsymbol{\xi} = \int_0^{2\pi}\int_0^{\pi} \frac{\rho}{4\pi r^2} (\boldsymbol{u}+\boldsymbol{r})_i r^2 \sin\theta \mathrm{d}\theta \mathrm{d}\varphi = \rho u_i \tag{4.236}$$

$$\begin{aligned}\int_{\Re} f^{\mathrm{eq}} \xi_i \xi_j \mathrm{d}\boldsymbol{\xi} &= \int_0^{2\pi}\int_0^{\pi} \frac{\rho}{4\pi r^2} (\boldsymbol{u}+\boldsymbol{r})_i (\boldsymbol{u}+\boldsymbol{r})_j r^2 \sin\theta \mathrm{d}\theta \mathrm{d}\varphi \\ &= \rho u_i u_j + \int_0^{2\pi}\int_0^{\pi} \frac{\rho}{4\pi} r_i r_i \delta_{ij} \sin\theta \mathrm{d}\theta \mathrm{d}\varphi \\ &= \rho u_i u_j + \delta_{ij} \int_0^{2\pi}\int_0^{\pi} \frac{\rho}{4\pi} (r\sin\theta\cos\varphi)^2 \sin\theta \mathrm{d}\theta \mathrm{d}\varphi \\ &= \rho u_i u_j + \frac{1}{3}\rho r^2 \delta_{ij}\end{aligned} \tag{4.237}$$

$$\begin{aligned}\int_{\Re} f^{\mathrm{eq}} \xi_i \xi_j \xi_k \mathrm{d}\boldsymbol{\xi} &= \int_0^{2\pi}\int_0^{\pi} \frac{\rho}{4\pi r^2} (\boldsymbol{u}+\boldsymbol{r})_i (\boldsymbol{u}+\boldsymbol{r})_j (\boldsymbol{u}+\boldsymbol{r})_k r^2 \sin\theta \mathrm{d}\theta \mathrm{d}\varphi \\ &= \rho u_i u_j u_k + \int_0^{2\pi}\int_0^{\pi} \frac{\rho}{4\pi} (u_i r_j r_j \delta_{jk} + u_j r_i r_i \delta_{ik} + u_k r_i r_i \delta_{ij}) \sin\theta \mathrm{d}\theta \mathrm{d}\varphi \\ &= \rho u_i u_j u_k + \frac{1}{3}\rho r^2 (u_i \delta_{jk} + u_j \delta_{ik} + u_k \delta_{ij})\end{aligned} \tag{4.238}$$

为了满足相关的统计关系式，$\rho r^2/3$ 须等于 p，对理想气体则有 $r=\sqrt{3R_g T}$。构造用于黏性可压缩流动的耦合双分布函数模型，至少需要三阶的拉格朗日插值多项式做分配函数。于是，可采用如下形式的多项式

$$\begin{aligned}P(x,y,z) = & a_0+a_1x+a_2y+a_3z+a_4x^2+a_5y^2+a_6z^2+a_7xy+a_8xz+ \\ & a_9yz+a_{10}x^3+a_{11}y^3+a_{12}z^3+a_{13}x^2y+a_{14}x^2z+a_{15}y^2z+a_{16}xy^2+ \\ & a_{17}xz^2+a_{18}z^2y+a_{19}x^4+a_{20}y^4+a_{21}z^4+a_{22}x^2y^2+a_{23}x^2z^2+a_{24}y^2z^2\end{aligned} \tag{4.239}$$

相应的格子为三维 25 速的格子（D3Q25，如图 4.3 所示），其粒子速度配置为[26]

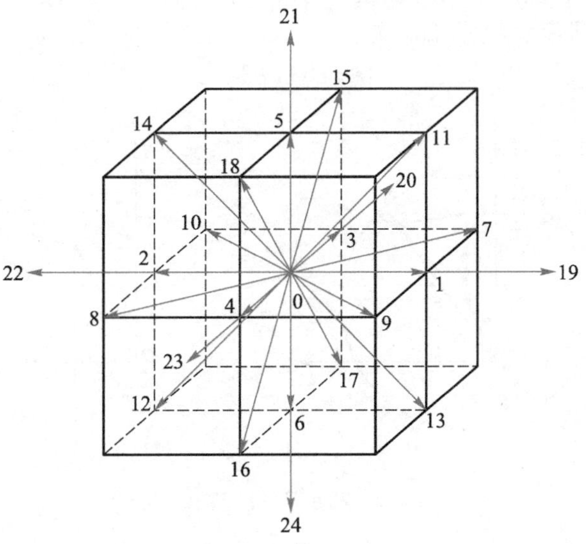

图 4.3 D3Q25 格子

$$e_{\alpha x}=c\{0,1,0,0,-1,0,0,1,1,0,-1,-1,0,-1,-1,0,1,1,0,2,0,0,-2,0,0\}$$
$$e_{\alpha y}=c\{0,0,1,0,0,-1,0,1,0,1,1,0,-1,-1,0,-1,-1,0,1,0,2,0,0,-2,0\}$$
$$e_{\alpha z}=c\{0,0,0,1,0,0,-1,0,1,1,0,1,1,0,-1,-1,0,-1,-1,0,0,2,0,0,-2\}$$
(4.240)

根据式(4.239),分配函数 $\phi_\alpha(x,y,z)$ 可写成

$$\phi_\alpha(x,y,z)=\boldsymbol{a}_\alpha \boldsymbol{t} \tag{4.241}$$

式中,

$$\boldsymbol{a}_\alpha=[a_{\alpha,0},a_{\alpha,1},\cdots,a_{\alpha,11},a_{\alpha,12},\cdots,a_{\alpha,24}],\quad \boldsymbol{t}=[1,x,\cdots,y^3,z^3,\cdots,y^2z^2]^{\mathrm{T}} \tag{4.242}$$

根据拉格朗日插值多项式的 δ 性质,可得[26]

$$\phi_0(x,y,z)=1+x^2y^2+x^2z^2+y^2z^2-\frac{5}{4}x^2-\frac{5}{4}y^2-\frac{5}{4}z^2+\frac{1}{4}x^4+\frac{1}{4}y^4+\frac{1}{4}z^4$$

$$\phi_1(x,y,z)=\frac{2}{3}x-\frac{1}{2}xy^2-\frac{1}{2}xz^2-\frac{1}{2}x^2y^2-\frac{1}{2}x^2z^2+\frac{2}{3}x^2-\frac{1}{6}x^3-\frac{1}{6}x^4$$

$$\phi_2(x,y,z)=\frac{2}{3}y-\frac{1}{2}yx^2-\frac{1}{2}yz^2-\frac{1}{2}y^2x^2-\frac{1}{2}y^2z^2+\frac{2}{3}y^2-\frac{1}{6}y^3-\frac{1}{6}y^4$$

$$\phi_3(x,y,z)=\frac{2}{3}z-\frac{1}{2}zx^2-\frac{1}{2}zy^2-\frac{1}{2}z^2x^2-\frac{1}{2}z^2y^2+\frac{2}{3}z^2-\frac{1}{6}z^3-\frac{1}{6}z^4$$

$$\phi_4(x,y,z)=-\frac{2}{3}x+\frac{1}{2}xy^2+\frac{1}{2}xz^2-\frac{1}{2}x^2y^2-\frac{1}{2}x^2z^2+\frac{2}{3}x^2+\frac{1}{6}x^3-\frac{1}{6}x^4$$

$$\phi_5(x,y,z)=-\frac{2}{3}y+\frac{1}{2}yx^2+\frac{1}{2}yz^2-\frac{1}{2}y^2x^2-\frac{1}{2}y^2z^2+\frac{2}{3}y^2+\frac{1}{6}y^3-\frac{1}{6}y^4$$

$$\phi_6(x,y,z)=-\frac{2}{3}z+\frac{1}{2}zx^2+\frac{1}{2}zy^2-\frac{1}{2}z^2x^2-\frac{1}{2}z^2y^2+\frac{2}{3}z^2+\frac{1}{6}z^3-\frac{1}{6}z^4$$

$$\phi_7(x,y,z)=\frac{1}{4}(xy+x^2y+xy^2+x^2y^2),\quad \phi_8(x,y,z)=\frac{1}{4}(xz+x^2z+xz^2+x^2z^2)$$

$$\phi_9(x,y,z)=\frac{1}{4}(yz+y^2z+yz^2+y^2z^2),\quad \phi_{10}(x,y,z)=\frac{1}{4}(-xy+x^2y-xy^2+x^2y^2)$$

$$\phi_{11}(x,y,z)=\frac{1}{4}(-xz+x^2z-xz^2+x^2z^2),\quad \phi_{12}(x,y,z)=\frac{1}{4}(-yz+y^2z-yz^2+y^2z^2)$$

$$\phi_{13}(x,y,z)=\frac{1}{4}(xy-x^2y-xy^2+x^2y^2),\quad \phi_{14}(x,y,z)=\frac{1}{4}(xz-x^2z-xz^2+x^2z^2)$$

$$\phi_{15}(x,y,z)=\frac{1}{4}(yz-y^2z-yz^2+y^2z^2),\quad \phi_{16}(x,y,z)=\frac{1}{4}(-xy-x^2y+xy^2+x^2y^2)$$

$$\phi_{17}(x,y,z)=\frac{1}{4}(-xz-x^2z+xz^2+x^2z^2),\quad \phi_{18}(x,y,z)=\frac{1}{4}(-yz-y^2z+yz^2+y^2z^2)$$

$$\phi_{19}(x,y,z)=\frac{1}{24}(-2x-x^2+2x^3+x^4),\quad \phi_{20}(x,y,z)=\frac{1}{24}(-2y-y^2+2y^3+y^4)$$

$$\phi_{21}(x,y,z)=\frac{1}{24}(-2z-z^2+2z^3+z^4),\quad \phi_{22}(x,y,z)=\frac{1}{24}(2x-x^2-2x^3+x^4)$$

$$\phi_{23}(x,y,z)=\frac{1}{24}(2y-y^2-2y^3+y^4),\quad \phi_{24}(x,y,z)=\frac{1}{24}(2z-z^2-2z^3+z^4) \tag{4.243}$$

确定分配函数之后,可通过如下积分得到平衡态密度分布函数:

$$f_\alpha^{\mathrm{eq}}=\int_{\Re}\phi_\alpha(\boldsymbol{\xi})f^{\mathrm{eq}}\mathrm{d}\boldsymbol{\xi}=\int_0^{2\pi}\int_0^{\pi}\frac{\rho}{4\pi}\phi_\alpha(\xi_x,\xi_y,\xi_z)\sin\theta\mathrm{d}\theta\mathrm{d}\phi \tag{4.244}$$

式中,ξ_x、ξ_y 和 ξ_z 分别为 $u_x+r\sin\theta\cos\varphi$、$u_y+r\sin\theta\sin\varphi$ 和 $u_z+r\cos\theta$。平衡态总能分布函数则可通过式(4.221)从平衡态密度分布函数直接得到。对 D3Q25 格子,系数 ω_α 确定如下[26]:

$$\omega_0=0,\quad \omega_{1,\cdots,6}=-\frac{1}{3},\quad \omega_{7,\cdots,18}=\frac{1}{8},\quad \omega_{19,\cdots,24}=\frac{1}{12} \tag{4.245}$$

最后,对耦合双分布函数模型进行总结。其耦合性可做如下理解:能量分布函数的演化方程中存在的密度分布函数项,以及平衡态能量分布函数中存在的密度和速度项体现了流场对温度场的影响;而温度场对流场的影响则是通过理想气体状态方程 $p=p(\rho,T)$ 实现的,由于平衡态密度分布函数中含有温度或压力项,于是温度场可通过状态方程影响平衡态密度分布函数,从而反作用于流场。

对可压缩流动,假定压力给定,那么连续方程和动量方程四个标量方程含有四个未知量(密度及三个速度分量),方程组封闭,密度场、速度场可以确定,而后温度场亦可以确定。于是,可以通过状态方程由密度和温度确定压力场,并开始新一轮的演化。

与单分布函数可压缩模型相比,耦合双分布函数可压缩模型具有如下优势:

(1) 从恢复同一层次的宏观方程来看,如果确定平衡态分布函数的方法相同,则由于双分布函数对约束条件的降阶,双分布函数模型的平衡态分布函数通常比单分布函数模型的简单。

(2) 从调节比热容比和普朗特数来看,双分布函数模型具有较大的优势。在双分布函数模型中,通过在平衡态能量分布函数中引入一个与比热容比相关的参数便可调节比热容比,这比单分布函数模型中通过采用势能或者多能级来调节比热容比方便不少。

4.6 小　　结

本章介绍了用于单相流动与传热的格子 Boltzmann 模型。单相流动模型方面,为了消除或降低格子 Boltzmann 方法基本模型的可压缩效应,产生了 Zou-Hou 模型、Lin 模型、He-Luo 模型、D2G9 模型等。前两个模型在理论上只能应用于定常流动;He-Luo 模型本质上是求解不可压缩 Navier-Stokes 方程组的人工压缩方法,要忽略可压缩性所造成的影响,需要增加附加条件,从而限制了模型的应用范围;D2G9 模型能够有效地消除基本模型中的可压缩效应,既可用于定常问题,也可用于非定常问题。

热模型方面,双分布函数模型取得了很大的成功。其中,被动标量模型(温度分布函数模型)广泛应用于不考虑黏性热耗散和可压缩功的传热问题中,而总能分布函数模型的提出则是在热力学能分布函数模型的基础上更进一步地发展和完善了双分布函数格子 Boltzmann 方法。本章还介绍了用于高马赫数可压缩流动的耦合双分布函数格子 Boltzmann 模型。对于高马赫可压缩流动来说,在有限网格的基础上,通常需要采用高阶的激波捕捉格式去求解微分形式的 Boltzmann-BGK 方程以便更好地捕捉激波或间断(参见第 9 章有限差分格子 Boltzmann 方法)。由于 Boltzmann-BGK 方程比 Navier-Stokes 方程组简单,与同样采用激波捕捉格式的传统有限差分法相比,格子 Boltzmann 方法的算法较为简单。此外,Boltzmann-BGK 方程中不存在二阶导数项,从而格子 Boltzmann 方法更容易在任意贴体坐标系中实施。

习　题

4-1　列出 He-Luo 模型平衡态密度分布函数式(4.12)的零阶~三阶矩,并与 D2G9 模型的相关结果进行对比。

4-2　在热力学能分布函数模型中,外力项的处理如式(4.99)所示,而在总能分布函数模型中,外力项由式(4.168)给出,试分析这两种处理的主要差异。

4-3　当比定容热容 c_V 为变量时,平衡态热力学能分布函数式(4.128)将无法恢复正确的热传导项,此时需对 g_α^{eq} 进行修改。若将 g_α^{eq} 的二阶矩修改为 $\sum_\alpha e_{\alpha i}e_{\alpha j}g_\alpha^{\mathrm{eq}}=\rho\varepsilon u_i u_j+pc_V^0 T\delta_{ij}$,能否恢复正确的热传导项? 其中 c_V^0 为参考比定容热容,其值为常数。结合 Chapman-Enskog 展开对这一问题进行分析。

4-4　与式(4.153)中的黏性应力张量不同,式(4.181)中的黏性应力张量为 $\widehat{\boldsymbol{\Pi}}=\mu[\nabla\boldsymbol{u}+(\nabla\boldsymbol{u})^{\mathrm{T}}]$,试分析产生这一改变的原因。

4-5　式(4.235)~式(4.238)列出了球面分布函数的零阶~三阶矩,试给出球面分布函数的四阶矩,并与 Maxwell 分布函数的四阶矩进行对比。

4-6　结合附录 D 中顶盖驱动流的程序以及本章所学习的热格子 Boltzmann 模型,编制程序模拟封闭方腔自然对流问题(该问题的描述可参考 10.1 节)。

参 考 文 献

[1] Zou Q S, Hou S L, Chen S Y, et al. A improved incompressible lattice Boltzmann model for time-independent flows[J]. Journal of Statistical Physics,1995,81(1):35-48.

[2] Lin Z F,Fang H P,Tao R B. Improved lattice Boltzmann model for incompressible two-dimensional steady flows[J]. Physical Review E,1996,54(6):6323-6330.

[3] He X Y,Luo L S. Lattice Boltzmann model for the incompressible Navier-Stokes equation[J]. Journal of Statistical Physics,1997,88(3):927-944.

[4] 董曾南,章梓雄. 非粘性流体力学[M]. 北京:清华大学出版社,2003.

[5] 郭照立,郑楚光,李青,等. 流体动力学的格子 Boltzmann 方法[M]. 武汉:湖北科学技术出版社,2002.

[6] Guo Z L,Shi B C,Wang N C. Lattice BGK model for incompressible Navier-Stokes equation[J]. Journal of Computational Physics,2000,165(1):288-306.

[7] Lallemand P,Luo L S. Theory of the lattice Boltzmann method:Acoustic and thermal properties in two and three dimensions[J]. Physical Review E,2003,68(3):036706.

[8] Shan X W,Chen H D. Lattice Boltzmann model for simulating flows with multiple phases and components [J]. Physical Review E,1993,47:1815-1820.

[9] Shan X W. Simulation of Rayleigh-Bénard convection using a lattice Boltzmann method[J]. Physical Review E,1997,55(3):2780-2788.

[10] Li Q,Luo K H. Effect of the forcing term in the pseudopotential lattice Boltzmann modeling of thermal flows[J]. Physical Review E,2014,89(5):053022.

[11] Li Q, Zhou P, Yan H J. Improved thermal lattice Boltzmann model for simulation of liquid-vapor phase change[J]. Physical Review E, 2017, 96(6): 063303.

[12] Li Q, Yu Y, Luo K H. Improved three-dimensional thermal multiphase lattice Boltzmann model for liquid-vapor phase change[J]. Physical Review E, 2022, 105(2): 025308.

[13] Huang R Z, Wu H Y. A modified multiple-relaxation-time lattice Boltzmann model for convection-diffusion equation[J]. Journal of Computational Physics, 2014, 274: 50-63.

[14] He X Y, Chen S Y, Doolen G D. A novel thermal model for the lattice Boltzmann method in incompressible limit[J]. Journal of Computational Physics, 1998, 146(1): 282-300.

[15] Guo Z L, Zheng C G, Shi B C. Discrete lattice effects on the forcing term in the lattice Boltzmann method [J]. Physical Review E, 2002, 65(4): 046308.

[16] Li Q, He Y L, Wang Y, et al. An improved thermal lattice Boltzmann model for flows without viscous heat dissipation and compression work[J]. International Journal of Modern Physics C, 2008, 19(1): 125-150.

[17] Peng Y, Shu C, Chew Y T. Simplified thermal lattice Boltzmann model for incompressible thermal flows [J]. Physical Review E, 2003, 68(2): 026701.

[18] Guo Z L, Zheng C G, Shi B C, et al. Thermal lattice Boltzmann equation for low Mach number flows: Decoupling model[J]. Physical Review E, 2007, 75(3): 036704.

[19] Shan X W, Yuan X F, Chen H D. Kinetic theory representation of hydrodynamics: A way beyond the Navier-Stokes equation[J]. Journal of Fluid Mechanics, 2006, 550: 413-441.

[20] 应纯同. 气体输运理论及应用[M]. 北京:清华大学出版社, 1990.

[21] Thompson P A. Compressible Fluid Dynamics[M]. New York: Hemisphere Publishing Corporation, 1972.

[22] 陶文铨, 何雅玲. 跨尺度和可压缩交变流动与换热中的科学问题及其现代数值模拟方法研究[M]. 香山科学会议第240次学术研讨会. 北京. 2004.

[23] Li Q, He Y L, Wang Y, et al. Coupled double-distribution-function lattice Boltzmann method for the compressible Navier-Stokes equations[J]. Physical Review E, 2007, 76(5): 056705.

[24] Prendergast K H, Xu K. Numerical hydrodynamics from gas-kinetic theory[J]. Journal of Computational Physics, 1993, 109(1): 53-66.

[25] Qu K, Shu C, Chew Y T. Alternative method to construct equilibrium distribution functions in lattice-Boltzmann method simulation of inviscid compressible flows at high Mach number[J]. Physical Review E, 2007, 75(3): 036706.

[26] Li Q, He Y L, Wang Y, et al. Three-dimensional non-free-parameter lattice-Boltzmann model and its application to inviscid compressible flows[J]. Physics Letters A, 2009, 373(25): 2101-2108.

第 5 章　气液两相流格子Boltzmann模型

多相流是指两种或两种以上不同相态的物质共存并具有明确分界面的多相流体流动，它广泛存在于自然界及许多工业领域中，如雨水、雪花等在大气中的运动、江河湖海表面的波浪运动、压缩空气雾化、二氧化碳驱油等。多相流通常可分为气液两相流、气固两相流、液固两相流、气液固三相流、液液固三相流等[1]。此外，还可根据参与流动的各相组分对多相流进行分类。例如，气液两相流可分为单组分气液两相流和双组分气液两相流。水和水蒸气的混合流动即为单组分气液两相流，而水和空气的混合流动则为双组分气液两相流。与单相流相比，多相流的主要特点是流动中各相之间存在着界面，且该界面随着流动在不断地变化。在工程实际中，界面的产生与运动是非常复杂的物理过程，且往往伴随着组分扩散、相变、化学反应等过程。以流体体积法（volume of fluid，VOF）、水平集法（level-set）等方法为代表的界面捕捉或界面跟踪方法，虽然可以描述少数大的界面，但难以捕获或跟踪大量细小的、分散的界面[2]。对复杂结构中的多相流动，如多孔介质孔隙尺度气液两相流，界面捕捉或界面跟踪方法往往难以胜任。

从物理本质来看，多相流中界面的产生与运动是微观分子间相互作用的宏观体现。如果这些相互作用能够得到正确的描述，就可以从底层刻画这类复杂的流动。分子动力学模拟方法可以从分子尺度描述这些相互作用，但其模拟过程所涉及的计算量非常大，限制了它所能应用的几何区域尺寸。格子 Boltzmann 方法的介观物理背景和粒子特性，使其能够非常方便地描述流体内部的相互作用，从而对多相/多组分流动等复杂流动的模拟具有明显的优势。本章将介绍气液两相流的格子 Boltzmann 模型，涉及固相的流固两相流模型及固液相变格子 Boltzmann 模型将在下一章进行介绍。依据描述相互作用的方式，气液两相流格子 Boltzmann 模型主要可分为自由能模型、伪势模型、颜色梯度模型和相场模型四类。在介绍这四类气液两相流模型之前，我们先对稠密流体的平均场动理论（kinetic mean-field theory）做一个简单的介绍，方便读者了解相关模型的建模基础和模型本质。

流体体积法和水平集法简介

5.1　稠密流体的平均场动理论

在讨论 Boltzmann 方程时，一般认为气体是稀疏的，并且假定分子本身的大小与分子间的距离相比可以忽略不计。因此经典的 Boltzmann 方程仅适用于稀薄气体，不能用于稠密气体或液体。为了解决这一问题，Enskog 将气体动理论拓展到了稠密气体[3]，其主要贡献是在动理论中考虑了分子的有限体积效应和碰撞输运（collisional transfer）的影响，但由于依然采用刚性球分子模型和分子混沌假设，从而忽略了分子间的长程相互作用。事实上，对于稠密气体，即使它是由刚性球分子组成的，相邻分子的速度之间仍然可能存在某种关联,这是因为它们彼此间刚刚发生过相互作用,或者它们与同一相邻分子刚刚发生过相互作用[3]。为此,一些学者对 Enskog 理论进行了修正,形成了以 Enskog-Vlasov 方程为基础的稠密流体平均场动理论[4-5],但相应的碰撞项较为复杂,影响了它的实际应用。

2002 年,He 和 Doolen[6]通过结合动理论和平均场理论建立了一个简化的稠密流体动理学方程。依据动理论中的 BBGKY 级次理论(hierarchy theory),单粒子分布函数的动理学方程形式如下[6]:

$$\frac{\partial f}{\partial t}+\boldsymbol{\xi}_1\cdot\frac{\partial f}{\partial \boldsymbol{r}_1}+\boldsymbol{a}\cdot\frac{\partial f}{\partial \boldsymbol{\xi}_1}=\iint_{\mathbb{Z}\,\Re}\frac{\partial f^{(2)}}{\partial \boldsymbol{\xi}_1}\cdot\frac{\partial V(r_{12})}{\partial \boldsymbol{r}_1}\mathrm{d}\boldsymbol{\xi}_2\mathrm{d}\boldsymbol{r}_2 \tag{5.1}$$

式中:$f(\boldsymbol{\xi}_1,\boldsymbol{r}_1,t)$为单粒子分布函数;$f^{(2)}(\boldsymbol{\xi}_1,\boldsymbol{r}_1,\boldsymbol{\xi}_2,\boldsymbol{r}_2,t)$为两粒子分布函数;$\boldsymbol{a}$ 为外力加速度;$\mathbb{Z}$ 表示物理空间;$\Re$ 表示分子速度空间;$V(r_{12})$为成对分子间势,其中 $r_{12}=|\boldsymbol{r}_1-\boldsymbol{r}_2|$。式(5.1)右端的物理空间积分域 $\mathbb{Z}$ 可以分为两部分[6]:$\{\mathbb{Z}_1:|\boldsymbol{r}_1-\boldsymbol{r}_2|\leqslant d\}$和$\{\mathbb{Z}_2:|\boldsymbol{r}_1-\boldsymbol{r}_2|>d\}$,$d$ 为分子有效直径。相应地,式(5.1)右端可写为

$$\iint_{\mathbb{Z}\,\Re}\frac{\partial f^{(2)}}{\partial \boldsymbol{\xi}_1}\cdot\frac{\partial V(r_{12})}{\partial \boldsymbol{r}_1}\mathrm{d}\boldsymbol{\xi}_2\mathrm{d}\boldsymbol{r}_2=\underbrace{\iint_{\mathbb{Z}_1\Re}\frac{\partial f^{(2)}}{\partial \boldsymbol{\xi}_1}\cdot\frac{\partial V(r_{12})}{\partial \boldsymbol{r}_1}\mathrm{d}\boldsymbol{\xi}_2\mathrm{d}\boldsymbol{r}_2}_{J_1}+\underbrace{\iint_{\mathbb{Z}_2\Re}\frac{\partial f^{(2)}}{\partial \boldsymbol{\xi}_1}\cdot\frac{\partial V(r_{12})}{\partial \boldsymbol{r}_1}\mathrm{d}\boldsymbol{\xi}_2\mathrm{d}\boldsymbol{r}_2}_{J_2} \tag{5.2}$$

式(5.2)右端第一个积分项 J_1 描述的是分子间的短程相互作用,而第二个积分项 J_2 则表征分子间的长程相互作用。其中,J_1 可通过 Enskog 理论描述[3]为

$$\begin{aligned}J_1&=\iint_{\mathbb{Z}_1\Re}\frac{\partial f^{(2)}}{\partial \boldsymbol{\xi}_1}\cdot\frac{\partial V(r_{12})}{\partial \boldsymbol{r}_1}\mathrm{d}\boldsymbol{\xi}_2\mathrm{d}\boldsymbol{r}_2\\&=\chi\Omega_0-b\rho\chi f^{\mathrm{eq}}\left\{(\boldsymbol{\xi}_1-\boldsymbol{u})\cdot\left[\nabla\ln(\rho^2\chi T)+\frac{3}{5}\left(|\boldsymbol{C}|^2-\frac{5}{2}\right)\nabla\ln T\right]+\right.\\&\left.\frac{2}{5}\left[2\boldsymbol{CC}:\nabla\boldsymbol{u}+\left(|\boldsymbol{C}|^2-\frac{5}{2}\right)\nabla\cdot\boldsymbol{u}\right]\right\}\end{aligned} \tag{5.3}$$

式中:Ω_0 为不考虑分子体积时的常规碰撞项;$\boldsymbol{C}=(\boldsymbol{\xi}_1-\boldsymbol{u})/\sqrt{2R_gT}$;$\boldsymbol{u}$ 为宏观速度;R_g 为气体常数;f^{eq}为平衡态分布函数;$b\rho$ 为分子的协体积[7];χ 为碰撞概率的量乘因子,它反映的是气体稠密性所引起的碰撞概率的增加。对稀薄气体,χ 等于 1;对稠密气体,χ 大于1,且随着密度增大,χ 也增大。

式(5.2)右端第二个积分项 J_2 在 Enskog 理论中是被忽略的,但许多研究特别是 van der Waals 理论证实了分子间的长程相互作用对稠密气体有重要影响。为了模化积分项 J_2,两粒子分布函数可写为[7]

$$f^{(2)}(\boldsymbol{\xi}_1,\boldsymbol{r}_1,\boldsymbol{\xi}_2,\boldsymbol{r}_2,t)=f(\boldsymbol{\xi}_1,\boldsymbol{r}_1,t)f(\boldsymbol{\xi}_2,\boldsymbol{r}_2,t)+\hbar(\boldsymbol{\xi}_1,\boldsymbol{r}_1,\boldsymbol{\xi}_2,\boldsymbol{r}_2,t) \tag{5.4}$$

式中,$\hbar(\boldsymbol{\xi}_1,\boldsymbol{r}_1,\boldsymbol{\xi}_2,\boldsymbol{r}_2,t)$为两粒子关联函数。He 和 Doolen[6]认为,在 $\mathbb{Z}_2$区域即$|\boldsymbol{r}_2-\boldsymbol{r}_1|>d$区域,对大部分流体 $\hbar$ 近似为零。根据这一近似可得

$$J_2=\iint_{\mathbb{Z}_2\Re}\frac{\partial f^{(2)}}{\partial \boldsymbol{\xi}_1}\cdot\frac{\partial V(r_{12})}{\partial \boldsymbol{r}_1}\mathrm{d}\boldsymbol{\xi}_2\mathrm{d}\boldsymbol{r}_2=\frac{\partial}{\partial \boldsymbol{r}_1}\left[\int_{\mathbb{Z}_2}\rho(\boldsymbol{r}_2)V(r_{12})\mathrm{d}\boldsymbol{r}_2\right]\cdot\frac{\partial f(\boldsymbol{\xi}_1,\boldsymbol{r}_1,t)}{\partial \boldsymbol{\xi}_1} \tag{5.5}$$

式(5.5)右端中括号内的积分项正是分子间引力势的平均场近似[8],即

$$V_{\mathrm{m}}=\int_{\mathbb{Z}_2}\rho(\boldsymbol{r}_2)V(r_{12})\mathrm{d}\boldsymbol{r}_2 \tag{5.6}$$

假定密度 $\rho(\boldsymbol{r}_2)$ 为缓变量，根据泰勒级数展开可得[6]

$$\rho(\boldsymbol{r}_2)=\rho(\boldsymbol{r}_1)+\boldsymbol{r}_{21}\cdot\nabla\rho+\frac{1}{2}\boldsymbol{r}_{21}\boldsymbol{r}_{21}:\nabla\nabla\rho+\cdots \tag{5.7}$$

式中，$\boldsymbol{r}_{21}=\boldsymbol{r}_2-\boldsymbol{r}_1$。将式(5.7)代入式(5.6)可得

$$V_{\mathrm{m}}\approx-2a\rho-\kappa\nabla^2\rho \tag{5.8}$$

式中，系数 a 和 κ 分别为[8]

$$a=-\frac{1}{2}\int_{r>d}V(r)\,\mathrm{d}\boldsymbol{r},\quad \kappa=-\frac{1}{6}\int_{r>d}r^2V(r)\,\mathrm{d}\boldsymbol{r} \tag{5.9}$$

系数 a 将出现在状态方程中，而系数 κ 则与表面张力的大小有关。这两个系数通常被假定为常数[6]。将式(5.8)代入式(5.5)可得

$$J_2=\iint_{\mathbb{Z}_2\Re}\frac{\partial f^{(2)}}{\partial\boldsymbol{\xi}_1}\cdot\frac{\partial V(r_{12})}{\partial\boldsymbol{r}_1}\mathrm{d}\boldsymbol{\xi}_2\mathrm{d}\boldsymbol{r}_2=\nabla V_{\mathrm{m}}\cdot\nabla_{\xi_1}f \tag{5.10}$$

相应地，式(5.1)可写成

$$\frac{\partial f}{\partial t}+\boldsymbol{\xi}\cdot\nabla f+\boldsymbol{a}\cdot\nabla_{\xi}f=J_1+\nabla V_{\mathrm{m}}\cdot\nabla_{\xi}f \tag{5.11}$$

为了公式的简洁，式中粒子速度 $\boldsymbol{\xi}_1$ 已由 $\boldsymbol{\xi}$ 替换。

通过 Chapman-Enskog 多尺度展开，可从式(5.11)导出如下形式的连续性方程和动量方程[6]：

$$\frac{\partial\rho}{\partial t}+\nabla\cdot(\rho\boldsymbol{u})=0 \tag{5.12}$$

$$\frac{\partial(\rho\boldsymbol{u})}{\partial t}+\nabla\cdot(\rho\boldsymbol{u}\boldsymbol{u})=\rho\boldsymbol{a}-\rho\nabla V_{\mathrm{m}}-\nabla[\rho R_{\mathrm{g}}T(1+b\rho\chi)]+\nabla\cdot\boldsymbol{\Pi} \tag{5.13}$$

式中，$\boldsymbol{\Pi}$ 为黏性应力张量。根据式(5.8)，式(5.13)右端第二项和第三项可做如下整理：

$$\rho\nabla V_{\mathrm{m}}+\nabla[\rho R_{\mathrm{g}}T(1+b\rho\chi)]=\nabla[\rho R_{\mathrm{g}}T(1+b\rho\chi)-a\rho^2]-\kappa\rho\nabla\nabla^2\rho \tag{5.14}$$

于是可得状态方程

$$p_{\mathrm{EOS}}=\rho R_{\mathrm{g}}T(1+b\rho\chi)-a\rho^2 \tag{5.15}$$

当 χ 取 $1/(1-b\rho)$ 时，式(5.15)即为 van der Waals 状态方程；当 χ 取 $(1-0.5\eta)/(1-\eta)^3$ 时，式(5.15)为 Carnahan-Starling 状态方程[9]，其中 $\eta=b\rho/4$。式(5.14)右端第二项 $\kappa\rho\nabla\nabla^2\rho$ 与表面张力有关。从上述理论推导可以看到，状态方程中的 $b\rho\chi$ 源自分子间的短程相互作用(分子有效体积的排斥效应)，而 $-a\rho^2$ 则来自分子间的长程相互作用(吸引力)，这与 van der Waals 理论是一致的。此外，表面张力亦起源于分子间的长程相互作用。

van der Waals 状态方程

式(5.14)的右端项还有另外两种表达形式，分别是化学势形式和压力张量形式。定义 $E_{\mathrm{f}}(\rho)$ 为体相区的自由能密度(bulk free-energy density)，则化学势与状态方程可根据下列式子得到[10-12]：

$$\mu_{\mathrm{c}}=E_{\mathrm{f}}'(\rho)-\kappa\nabla^2\rho,\quad p_{\mathrm{EOS}}=\rho E_{\mathrm{f}}'(\rho)-E_{\mathrm{f}}(\rho) \tag{5.16}$$

式中：μ_{c} 为化学势；$E_{\mathrm{f}}'(\rho)=\mathrm{d}E_{\mathrm{f}}(\rho)/\mathrm{d}\rho$。注意到

$$\nabla[\rho E_{\mathrm{f}}'(\rho)]=\rho\nabla E_{\mathrm{f}}'(\rho)+E_{\mathrm{f}}'(\rho)\nabla\rho=\rho\nabla E_{\mathrm{f}}'(\rho)+\nabla E_{\mathrm{f}}(\rho) \tag{5.17}$$

因此，式(5.14)右端可写成如下化学势形式：

$$\nabla p_{\mathrm{EOS}}-\kappa\rho\nabla\nabla^2\rho=\rho\nabla\mu_{\mathrm{c}} \tag{5.18}$$

可以证明，van der Waals 状态方程对应的自由能密度为[10-11]

$$E_{\mathrm{f}}=\rho R_{\mathrm{g}}T\ln\left(\frac{\rho}{1-b\rho}\right)-a\rho^2 \tag{5.19}$$

对其他状态方程,相应的自由能密度及化学势可参阅文献[13]。

下面介绍压力张量形式,根据张量运算(参见附录 A),$\kappa\rho\nabla\nabla^2\rho$ 可展开为

$$\begin{aligned}
\kappa\rho\partial_i\partial_j\partial_j\rho &=\kappa\partial_i(\rho\partial_j\partial_j\rho)-\kappa(\partial_i\rho)(\partial_j\partial_j\rho)\\
&=\kappa\partial_i(\rho\partial_j\partial_j\rho)-\{\kappa\partial_j[(\partial_i\rho)(\partial_j\rho)]-\kappa(\partial_j\rho)(\partial_j\partial_i\rho)\}\\
&=\kappa\partial_i(\rho\partial_j\partial_j\rho)-\left\{\kappa\partial_j[(\partial_i\rho)(\partial_j\rho)]-\frac{\kappa}{2}[\partial_i(\partial_j\rho)^2]\right\}\\
&=\partial_j\left\{\left[\frac{\kappa}{2}(\partial_j\rho)^2+\kappa\rho\partial_j\partial_j\rho\right]\delta_{ij}-\kappa(\partial_i\rho)(\partial_j\rho)\right\}
\end{aligned} \tag{5.20}$$

式(5.20)的推导用到$\partial_j\partial_i\rho=\partial_i\partial_j\rho$。根据式(5.20),式(5.14)右端亦可写为

$$\nabla p_{\mathrm{EOS}}-\kappa\rho\nabla\nabla^2\rho=\nabla\cdot\boldsymbol{P}=\nabla\cdot\left[\left(p_{\mathrm{EOS}}-\frac{\kappa}{2}|\nabla\rho|^2-\kappa\rho\nabla^2\rho\right)\boldsymbol{I}+\kappa\nabla\rho\nabla\rho\right] \tag{5.21}$$

式中:$\boldsymbol{I}$ 为二阶单位张量;$\boldsymbol{P}$ 为压力张量,即

$$\boldsymbol{P}=\left(p_{\mathrm{EOS}}-\frac{\kappa}{2}|\nabla\rho|^2-\kappa\rho\nabla^2\rho\right)\boldsymbol{I}+\kappa\nabla\rho\nabla\rho \tag{5.22}$$

表面张力

可以看到,分子间相互作用的宏观体现主要包括两部分:一个是状态方程,它既与分子间的短程相互作用有关,也与分子间的长程相互作用有关;另一个是表面张力,它源自分子间的长程相互作用。这两部分在宏观方程中可以统一成压力张量形式,即式(5.21),或化学势形式,即式(5.18)。实际上,结合 Navier-Stokes 方程组和式(5.18)或式(5.21)就可以描述等温单组分气液两相流,这一点可从文献[14]得到证实。但采用状态方程时,密度场和速度场耦合,不便于实施以压力为基本求解变量的数值模拟算法,因此这种处理在传统方法中并不常用,比较多的还是采用流体体积法、水平集法等界面捕捉或界面跟踪方法。

在格子 Boltzmann 方法框架下,根据稠密流体的平均场动理论,气液两相流的建模思路至少有三种。一种是直接从动理学方程(5.11)出发构建相应的格子 Boltzmann 模型,但由于存在较多的梯度项需要离散,这种建模思路在格子 Boltzmann 方法领域并没有得到有效的数值实施。另一种是依据式(5.14)的压力张量形式或化学势形式,将其纳入格子 Boltzmann 模型当中,从而实现对状态方程和表面张力的描述。第三种则是结合格子 Boltzmann 方法的介观特性,直接刻画粒子间的相互作用,这种相互作用可以描述宏观层面的相分离和表面张力效应。下面即将介绍的自由能模型和伪势模型对应的正是第二、第三种建模思路。

5.2 自由能格子 Boltzmann 模型

自由能格子 Boltzmann 模型最早是由 Swift 等[10-11]于 1995 年提出的,他们从自由能泛函出发,构造了与热力学理论一致的气液两相流格子 Boltzmann 模型。其基本思路是根据自由能泛函所导出的热力学压力张量来设计模型的平衡态密度分布函数,从而实现对相分离及表面张力效应的描述。

5.2.1 热力学理论及 Swift 自由能模型

对于包含相界面的单组分等温气液两相流体系,自由能泛函可以表示为[8,15-17]

$$\mathcal{F}=\int_{\Omega_V}\varepsilon(\rho,\nabla\rho)\mathrm{d}\Omega_V=\int_{\Omega_V}\left[E_{\mathrm{f}}(\rho)+\frac{\kappa}{2}|\nabla\rho|^2\right]\mathrm{d}\Omega_V \tag{5.23}$$

式中,Ω_V 为系统占据的空间;$\varepsilon(\rho,\nabla\rho)$ 为自由能密度函数,它包括体相区的自由能密度 $E_{\mathrm{f}}(\rho)$ 以及梯度自由能密度 $0.5\kappa|\nabla\rho|^2$(κ 与表面张力的大小有关)。相应的化学势为

$$\mu_{\mathrm{c}}=\frac{\delta\mathcal{F}}{\delta\rho} \tag{5.24}$$

式中,δ 表示变分运算。根据式(5.23)及变分运算[15]可导出如下形式的化学势:

$$\mu_{\mathrm{c}}=E_{\mathrm{f}}'(\rho)-\kappa\nabla^2\rho \tag{5.25}$$

在此基础上,可以定义一个非局部压力[18]

$$p=\rho\mu_{\mathrm{c}}-\varepsilon(\rho,\nabla\rho) \tag{5.26}$$

将式(5.25)和 $\varepsilon(\rho,\nabla\rho)$ 的表达式代入式(5.26)可得

$$p=p_{\mathrm{EOS}}-\kappa\rho\nabla^2\rho-\frac{\kappa}{2}|\nabla\rho|^2 \tag{5.27}$$

式中,$p_{\mathrm{EOS}}=\rho E_{\mathrm{f}}'(\rho)-E_{\mathrm{f}}(\rho)$ 为状态方程。需要注意的是,文献[10]中非局部压力的定义有误,不能导出正确的表达式。结合式(5.27),热力学压力张量可以定义为[18-20]

$$\boldsymbol{P}=p\boldsymbol{I}+\frac{\partial\varepsilon(\rho,\nabla\rho)}{\partial(\nabla\rho)}\nabla\rho=\left(p_{\mathrm{EOS}}-\kappa\rho\nabla^2\rho-\frac{\kappa}{2}|\nabla\rho|^2\right)\boldsymbol{I}+\kappa\nabla\rho\nabla\rho \tag{5.28}$$

对比式(5.28)和式(5.22)可以看到,从热力学理论所导出的压力张量与平均场动理论所给出的压力张量是一致的。

Swift 等[10]采用的是不含作用力项的格子 Boltzmann 方程,为了将热力学压力张量纳入格子 Boltzmann 模型,他们基于 D2Q7 格子定义了如下形式的平衡态密度分布函数:

$$f_0^{\mathrm{eq}}=A_0+C_0|\boldsymbol{u}|^2 \tag{5.29}$$

$$f_\alpha^{\mathrm{eq}}=A+B(\boldsymbol{e}_\alpha\cdot\boldsymbol{u})+C|\boldsymbol{u}|^2+D(\boldsymbol{e}_\alpha\cdot\boldsymbol{u})^2+\boldsymbol{G}:\boldsymbol{e}_\alpha\boldsymbol{e}_\alpha,\quad \alpha\neq 0 \tag{5.30}$$

上述平衡态分布函数需满足下列约束条件:

$$\sum_{\alpha=0}^{6}f_\alpha^{\mathrm{eq}}=\rho,\quad \sum_{\alpha=0}^{6}\boldsymbol{e}_\alpha f_\alpha^{\mathrm{eq}}=\rho\boldsymbol{u},\quad \sum_{\alpha=0}^{6}\boldsymbol{e}_\alpha\boldsymbol{e}_\alpha f_\alpha^{\mathrm{eq}}=\boldsymbol{P}+\rho\boldsymbol{u}\boldsymbol{u} \tag{5.31}$$

根据约束条件可以确定平衡态密度分布函数的参数[10]

$$\begin{gathered}A_0=\rho-6A,\quad C_0=-\rho\\ A=\frac{1}{3}(p_{\mathrm{EOS}}-\kappa\rho\nabla^2\rho),\quad B=\frac{\rho}{3},\quad C=-\frac{\rho}{6},\quad D=\frac{2\rho}{3}\\ G_{xx}=-G_{yy}=\frac{\kappa}{3}\left[\left(\frac{\partial\rho}{\partial x}\right)^2-\left(\frac{\partial\rho}{\partial y}\right)^2\right],\quad G_{xy}=\frac{2\kappa}{3}\frac{\partial\rho}{\partial x}\frac{\partial\rho}{\partial y}\end{gathered} \tag{5.32}$$

对于 D2Q9 格子模型,相应的平衡态密度分布函数可参阅文献[21]。通过 Chapman-Enskog 多尺度展开,Swift 等[11]导出了自由能模型所对应的宏观方程,即

$$\frac{\partial\rho}{\partial t}+\nabla\cdot(\rho\boldsymbol{u})=0 \tag{5.33}$$

$$\begin{aligned}\frac{\partial(\rho\boldsymbol{u})}{\partial t}+\nabla\cdot(\rho\boldsymbol{u}\boldsymbol{u})=&-\nabla\cdot\boldsymbol{P}+\nabla\cdot[\nu\nabla(\rho\boldsymbol{u})]+\nabla[\lambda\nabla\cdot(\rho\boldsymbol{u})]-\\&\left(\tau-\frac{1}{2}\right)\frac{\mathrm{d}p_{\mathrm{EOS}}}{\mathrm{d}\rho}\delta_t\nabla\cdot[\boldsymbol{u}\nabla\rho+(\nabla\rho)\boldsymbol{u}]\end{aligned} \tag{5.34}$$

式中,ν 和 λ 分别为

$$\nu=\frac{1}{4}\left(\tau-\frac{1}{2}\right)\delta_t, \quad \lambda=\left(\tau-\frac{1}{2}\right)\left(\frac{1}{2}-\frac{\mathrm{d}p_{\mathrm{EOS}}}{\mathrm{d}\rho}\right)\delta_t \tag{5.35}$$

式(5.34)右端最后一项为误差项,该项在单相区为零,在界面处由于密度梯度很大,会带来较大的数值误差。此外,动量方程中的黏性项是以$\nabla\cdot[\nu\nabla(\rho\boldsymbol{u})]$的形式出现,这将导致模型丧失伽利略不变性。

实际上,从式(5.34)可以看到,压力张量的散度$\nabla\cdot\boldsymbol{P}$可以通过作用力项的形式嵌入格子 Boltzmann 模型,从而无须修改平衡态密度分布函数的二阶矩。但是,压力张量的散度含有密度的三阶导数$\nabla\nabla^2\rho$,会增加数值实施的复杂程度。为了避免这一问题,Swift 等选择修改平衡态密度分布函数的二阶矩来引入压力张量,但带来了另一些问题,包括动量方程存在误差项、破坏了伽利略不变性等。针对这些问题,一些学者提出了改进的模型,我们将在 5.2.3 节进行介绍。

5.2.2 热力学一致性

作为从热力学理论出发而构建的模型,自由能模型的主要优势在于它在理论上能够满足热力学一致性,即理论上相平衡时的气液共存密度与 Maxwell 等面积法则所给出的结果是一致的。下面以图 5.1 所示的平直气液相界面为例介绍自由能模型的热力学一致性。当图中所示的气液两相处于相平衡时,气相和液相具有相同的压力,即 $p_{\mathrm{EOS}}(\rho_{\mathrm{L}})=p_{\mathrm{EOS}}(\rho_{\mathrm{G}})=p_{\mathrm{b}}$,这里的 ρ_{L} 和 ρ_{G} 分别为液相和气相的密度。由于速度 $\boldsymbol{u}$ 为零,根据动量方程,此时压力张量的分量 P_{xx} 沿 x 方向应为常数,即 $\mathrm{d}P_{xx}/\mathrm{d}x=0$。

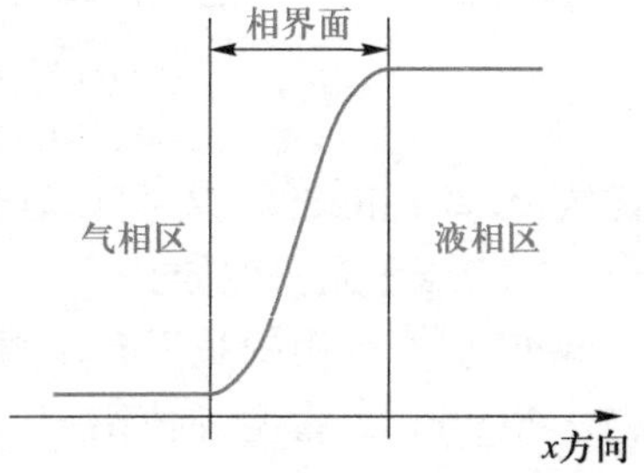

图 5.1 平直气液相界面的密度变化示意图

根据上述分析及压力张量的表达式可得

$$P_{xx}=p_{\mathrm{EOS}}(\rho)-\kappa\rho\frac{\mathrm{d}^2\rho}{\mathrm{d}x^2}+\frac{\kappa}{2}\left(\frac{\mathrm{d}\rho}{\mathrm{d}x}\right)^2=p_{\mathrm{EOS}}(\rho_{\mathrm{L}})=p_{\mathrm{EOS}}(\rho_{\mathrm{G}})=p_{\mathrm{b}} \tag{5.36}$$

其中二阶导数 $\mathrm{d}^2\rho/\mathrm{d}x^2$ 可做如下转换:

$$\frac{\mathrm{d}^2\rho}{\mathrm{d}x^2}=\frac{1}{2}\frac{\mathrm{d}}{\mathrm{d}\rho}\left[\left(\frac{\mathrm{d}\rho}{\mathrm{d}x}\right)^2\right] \tag{5.37}$$

于是式(5.36)可写成

$$P_{xx}=p_{\mathrm{EOS}}(\rho)-\kappa\frac{\rho^2}{2}\frac{\mathrm{d}}{\mathrm{d}\rho}\left[\frac{1}{\rho}\left(\frac{\mathrm{d}\rho}{\mathrm{d}x}\right)^2\right]=p_{\mathrm{b}} \tag{5.38}$$

对式(5.38)稍做整理可得

$$\frac{2}{\kappa}[p_{\mathrm{EOS}}(\rho)-p_{\mathrm{b}}]\frac{1}{\rho^2}=\frac{\mathrm{d}}{\mathrm{d}\rho}\left[\frac{1}{\rho}\left(\frac{\mathrm{d}\rho}{\mathrm{d}x}\right)^2\right] \tag{5.39}$$

式(5.39)两端同乘 $\mathrm{d}\rho$,并从 ρ_{G} 到 ρ_{L} 积分可得(κ 为常数且 $\mathrm{d}\rho/\mathrm{d}x$ 在单相区为零)

$$\int_{\rho_{\mathrm{G}}}^{\rho_{\mathrm{L}}}[p_{\mathrm{EOS}}(\rho)-p_{\mathrm{b}}]\frac{1}{\rho^2}\mathrm{d}\rho=0 \tag{5.40}$$

式(5.40)即为自由能模型在连续层面的**力学稳定性条件**(mechanical stability condition),

它与 Maxwell 等面积法则的表达式是一致的,这表明自由能模型在理论上能够满足热力学一致性。

5.2.3 改进的自由能模型

在 5.2.1 节我们曾提到,Swift 自由能模型的主要缺点在于破坏了伽利略不变性。这是因为黏性应力项的密度出现在第一次导数中[2],即$\nabla\cdot[\nu\nabla(\rho\boldsymbol{u})]$,而在 Navier-Stokes 方程中,相应项为$\nabla\cdot(\nu\rho\nabla\boldsymbol{u})$。在单相区,密度为常数,影响不大,但在气液界面处有

$$\nabla\cdot[\nu\nabla(\rho\boldsymbol{u})]=\nabla\cdot(\nu\rho\nabla\boldsymbol{u})+\nabla\cdot(\nu\boldsymbol{u}\nabla\rho) \tag{5.41}$$

由于界面处的密度梯度较大,式(5.41)右端第二项将引入较大的数值误差,导致一些非物理的现象。例如,一些研究[22-23]表明,在匀速运动的流场中,Swift 自由能模型会导致圆形的液滴或气泡逐渐变成椭圆形。

针对 Swift 自由能模型所存在的问题,一些学者提出了改进的自由能模型。这些模型主要可分为三类:

(1) 第一类自由能模型仍然沿用 Swift 等的思路,通过改变平衡态分布函数的二阶矩来引入热力学压力张量,但增加一些修正项来改进或恢复模型的伽利略不变性[22-23]。

(2) 第二类自由能模型舍弃了 Swift 等的思路,采用作用力项将压力张量的散度$\nabla\cdot\boldsymbol{P}$引入格子 Boltzmann 模型,从而避免了对模型伽利略不变性的影响。

(3) 第三类自由能模型则采用化学势形式,即$\nabla\cdot\boldsymbol{P}=\rho\nabla\mu_c$。尽管化学势形式与压力张量形式在数学上是等价的,但它们在数值模拟中的表现不尽相同。对于任意两个变量a和b,等式$\nabla(ab)=a\nabla b+b\nabla a$在数学上是成立的,但在数值模拟中,受限于离散格式的精度,式子左右两边可能会存在一定的偏差,且数值稳定性可能不同。例如,a和b的乘积可能变化比较平缓,而a或b可能变化比较剧烈。

第一类自由能模型由于涉及较多的修正项,在实际应用中已较少被采用。这里简单介绍一下第二类和第三类自由能模型。2006 年,Wagner 和 Li[24] 提出通过作用力项将压力张量的散度引入自由能格子 Boltzmann 模型。由于格子 Boltzmann 模型本身会产生压力项$-\nabla(\rho c_s^2)$,因此他们添加的作用力为

$$\boldsymbol{F}=-\nabla\cdot\boldsymbol{P}+\nabla(\rho c_s^2) \tag{5.42}$$

式(5.42)右端第二项用于抵消格子 Boltzmann 模型本身所产生的压力项。在数值实施中,需要对$\nabla\rho$、$\nabla^2\rho$以及压力张量的散度进行离散。为了将自由能模型拓展到较大密度比的气液两相流,Mazloomi 等[25]采用了一种多项式形式的状态方程,并通过熵格子 Boltzmann 方法增强了模型的数值稳定性。需要注意的是,无论是文献[24]还是文献[25],所采用的作用力格式均存在离散格子效应。例如,Mazloomi 等[25]采用的是 EDM 作用力格式。在 3.7 节我们对 EDM 格式进行了详细的介绍,并给出了它所带来的宏观方程偏差项。

2006 年,Lee 和 Fischer[26] 研究发现,采用化学势形式可以有效降低相界面附近产生的虚假速度(spurious currents)。针对等式$\nabla p_{\mathrm{EOS}}=\rho\nabla\mu_0$,式中$\mu_0=E_f'(\rho)$,Lee 和 Fischer[26] 通过对离散格式进行泰勒级数展开发现

$$\nabla p_{\mathrm{EOS}}\big|_{(r,t)}-\rho\nabla\mu_0\big|_{(r,t)}=\sum_{\alpha=1}^{8}\frac{\omega_\alpha}{6c_s^2\delta_t}\left[\left(\frac{\partial\mu_0}{\partial\rho}\right)(\delta_t\boldsymbol{e}_\alpha\cdot\nabla\rho)(\delta_t\boldsymbol{e}_\alpha\cdot\nabla)^2\rho\right]\Bigg|_{(r,t)} \tag{5.43}$$

在气液界面处,由于密度梯度较大,式(5.43)右端会引入较大的数值误差,从而导致压力张量形式在数值模拟中产生较大的虚假速度。Wen 等[13] 导出了多个状态方程所对应的

自由能密度和化学势，研究了采用不同状态方程时自由能模型的数值表现，并构建了适用于大密度比气液两相流的化学势形式自由能模型[27]。此外，Qiao 等[28]采用 Beam-Warming 格式构造了基于多松弛碰撞算子的化学势形式自由能模型。

2011 年，Guo 等[29]发现，采用作用力项的自由能模型在离散层面存在力的不平衡，从而导致化学势梯度不为零。随后，Lou 和 Guo[30]基于 Lax-Wendroff 迁移格式，提出了一个改进的自由能模型，相应的演化方程如下：

$$f_\alpha^*(\boldsymbol{r},t)=f_\alpha(\boldsymbol{r},t)-\frac{1}{\tau}[f_\alpha(\boldsymbol{r},t)-f_\alpha^{\mathrm{eq}}(\boldsymbol{r},t)]+\delta_t\left(1-\frac{1}{2\tau}\right)\frac{(\boldsymbol{e}_\alpha-\boldsymbol{u})\cdot\boldsymbol{F}}{\rho c_s^2}f_\alpha^{\mathrm{eq}} \tag{5.44}$$

$$f_\alpha(\boldsymbol{r},t+\delta_t)=a_0 f_\alpha^*(\boldsymbol{r},t)+a_1 f_\alpha^*(\boldsymbol{r}+\boldsymbol{e}_\alpha\delta_t,t)+a_{-1}f_\alpha^*(\boldsymbol{r}-\boldsymbol{e}_\alpha\delta_t,t) \tag{5.45}$$

式中：$\boldsymbol{F}=-\rho\nabla\mu_c+\nabla(\rho c_s^2)$；$a_0=1-A^2$；$a_1=A(A-1)/2$；$a_{-1}=A(A+1)/2$，其中 A 为 Courant-Friedrichs-Lewy 数。Lou 和 Guo[30]发现，尽管化学势形式的自由能模型在理论上自动满足热力学一致性，但在离散层面存在力的不平衡，导致数值模拟中的气液共存密度会偏离 Maxwell 等面积法所给出的结果。

从数值误差的角度来看，离散层面力的不平衡是由相应的离散误差项引起的，它主要包含两部分。首先，格子 Boltzmann 方法所采用的梯度离散算子会引入一部分离散误差。例如，基于 D2Q9 格子，对 $\nabla\rho$ 有

$$\nabla\rho^{(\mathrm{discrete})}=\frac{1}{c_s^2\delta_t}\sum_{\alpha=0}^{8}\omega_\alpha\rho(\boldsymbol{r}+\boldsymbol{e}_\alpha\delta_t)\boldsymbol{e}_\alpha \tag{5.46}$$

根据泰勒级数展开可得

$$\rho(\boldsymbol{r}+\boldsymbol{e}_\alpha\delta_t)=\rho(\boldsymbol{r})+e_{\alpha k}\delta_t\partial_k\rho(\boldsymbol{r})+\frac{\delta_t^2}{2}e_{\alpha k}e_{\alpha l}\partial_k\partial_l\rho(\boldsymbol{r})+\frac{\delta_t^3}{6}e_{\alpha k}e_{\alpha l}e_{\alpha m}\partial_k\partial_l\partial_m\rho(\boldsymbol{r})+\cdots \tag{5.47}$$

将式(5.47)代入式(5.46)可得

$$\nabla\rho^{(\mathrm{discrete})}=\nabla\rho+\frac{\delta_t^2}{6}\nabla\nabla^2\rho+\cdots \tag{5.48}$$

式(5.48)右端第二项即为梯度离散算子所引入的三阶误差项。另一方面，通过 Chapman-Enskog 多尺度展开，可导出三阶展开时所恢复的宏观动量方程为[31]

$$\frac{\partial(\rho\boldsymbol{u})}{\partial t}+\nabla\cdot(\rho\boldsymbol{u}\boldsymbol{u})=-\nabla(\rho c_s^2)+\boldsymbol{F}+\nabla\cdot\boldsymbol{\Pi}+\frac{\delta_t^2}{12}\nabla\cdot\nabla\boldsymbol{F} \tag{5.49}$$

式中，$\rho\boldsymbol{u}=\sum_{\alpha=0}^{8}\boldsymbol{e}_\alpha f_\alpha+0.5\delta_t\boldsymbol{F}$。式(5.49)右端最后一项即为作用力格式所引入的三阶误差项。对第二类和第三类自由能模型，由于作用力项中含有 $\nabla(\rho c_s^2)$，随着密度比的增大，三阶误差项的影响不可忽略。它们的存在会增大数值模拟中的虚假速度，并使得模型所给出的气液共存密度与 Maxwell 等面积法的结果不一致。基于上述分析，Li 等[32]提出了一种改进格式用于实现化学势形式自由能模型的热力学一致性，研究结果如图 5.2 所示。从图中可以看到，随着对比温度的降低（即密度比增大），标准的化学势形式自由能模型的数值模拟结果逐渐偏离 Maxwell 等面积法的结果，而改进格式的结果则吻合得非常好。

5.2.4 接触角格式

接触角是表征液体在固体表面润湿性的重要参数。当气液界面与固体表面接触时，气、液、固三相交会处的气液界面切线方向与固液界面之间的夹角即为接触角，如图 5.3

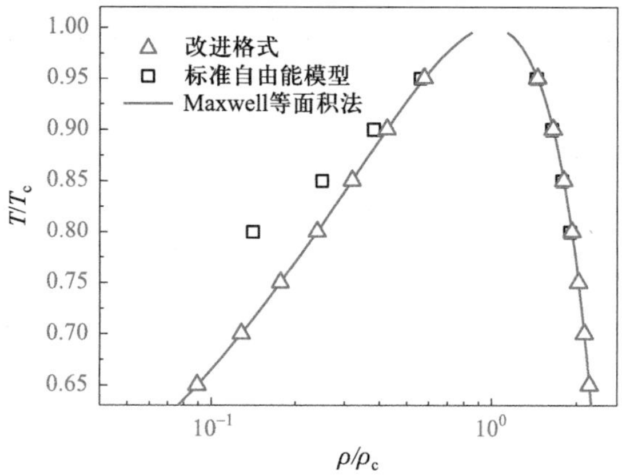

图 5.2 标准和改进化学势形式自由能模型的对比[32]

所示。接触角的大小与界面张力有关。当液滴在水平固体表面静止不动时,所形成的接触角与各界面张力之间符合杨氏方程(Young's equation):

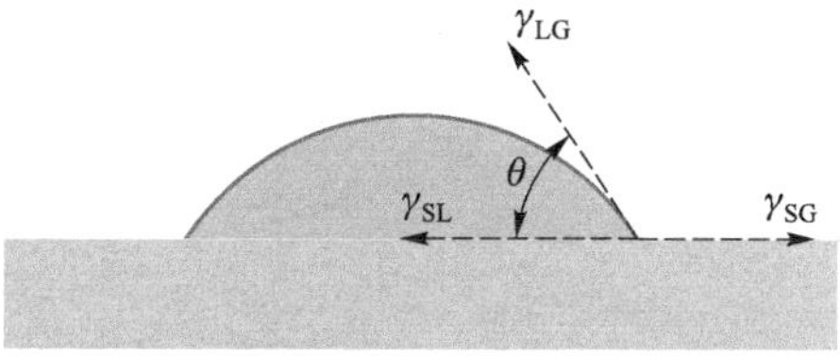

图 5.3 接触角示意图

$$\cos\,\theta=\frac{\gamma_{SG}-\gamma_{SL}}{\gamma_{LG}} \tag{5.50}$$

式中,γ_{LG}、γ_{SL}和 γ_{SG}分别为气液、固液和固气界面的界面张力。

亲水、疏水表面及荷叶效应

2004 年,Briant 等[33]研究了自由能模型在湿润性表面上的实施,并提出了相应的接触角格式。根据 Cahn[34]的假设,流固之间的相互作用在系统原有自由能密度的基础上增加了一项表面自由能密度

$$\mathcal{F}=\int_{\Omega_V}\left[E_f(\rho)+\frac{\kappa}{2}|\nabla\rho|^2\right]\mathrm{d}\Omega_V+\int_S\mathcal{H}(\rho_w)\,\mathrm{d}S \tag{5.51}$$

式中:S 为系统所占据空间的表面;$\mathcal{H}(\rho_w)$为表面自由能密度函数,它与表面密度 ρ_w 有关。一个比较简单的形式为 $\mathcal{H}(\rho_w)=-\omega\rho_w$,其中 ω 为常数。Briant 等[33]提出可在固体表面实施如下形式的边界条件:

$$\kappa\boldsymbol{n}\cdot\nabla\rho=\frac{\mathrm{d}\mathcal{H}(\rho_w)}{\mathrm{d}\rho_w}=-\omega \tag{5.52}$$

式中,$\boldsymbol{n}$ 为指向流体内部的固体表面法向矢量。

为了获得解析形式的界面张力,Briant 等[33]采用的体相区自由能密度函数为

$$E_f(\rho)=p_c\,(\bar{\rho}+1)^2(\bar{\rho}^2-2\bar{\rho}+3-2\beta\bar{T}) \tag{5.53}$$

式中:$\bar{\rho}=(\rho-\rho_c)/\rho_c$;$\bar{T}=(T-T_c)/T_c$;$\beta$ 为常数;p_c、ρ_c 和 T_c 分别为临界压力、临界密度和临界温度。根据式(5.53)及 $\mathcal{H}(\rho_w)$的表达式,固气界面和固液界面的界面张力分别为[33]

$$\gamma_{SG}=-\omega\rho_c+\frac{\gamma_{LG}}{2}-\frac{\gamma_{LG}}{2}(1-\Omega)^{3/2} \tag{5.54}$$

$$\gamma_{SL}=-\omega\rho_c+\frac{\gamma_{LG}}{2}-\frac{\gamma_{LG}}{2}(1+\Omega)^{3/2} \tag{5.55}$$

式中，$\Omega=\omega/(\beta\bar{T}\sqrt{2\kappa p_c})$。将式(5.54)和式(5.55)代入式(5.50)可得

$$\cos\theta=\frac{1}{2}[(1+\Omega)^{3/2}-(1-\Omega)^{3/2}] \tag{5.56}$$

当接触角 θ 给定时，可通过式(5.56)求得 Ω 的值，继而得到 ω 的值，最终通过式(5.52)实施相应的湿润性边界条件。

需要注意的是，上述 Ω 的表达式是建立在式(5.53)的基础上，对其他形式的自由能密度函数不一定适用。针对 van der Waals 状态方程，Huang 等[21]提出可近似采用如下形式的 Ω：

$$\Omega=\left(\frac{2\rho_c}{\rho_L-\rho_G}\right)^2\frac{\omega}{\sqrt{\sigma_{LG}}} \tag{5.57}$$

此外，Wen 等[13]提出可以通过给定固体节点的化学势来实施相应的接触角；Yu 等[35]提出了一种相容性的接触角格式来实施化学势形式自由能模型的湿润性边界条件，部分模拟结果如图 5.4 所示。

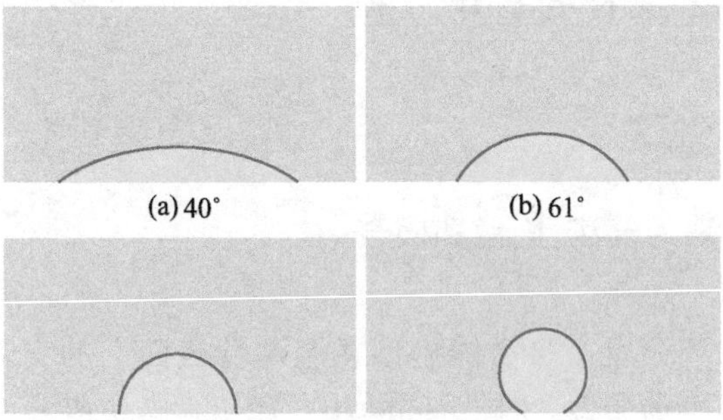

图 5.4 基于化学势形式自由能模型的接触角模拟[35]

5.3 伪势格子 Boltzmann 模型

无论是压力张量形式的自由能模型，还是化学势形式的自由能模型，都是从分子间相互作用的宏观体现出发，进行相应的梯度离散，这种模拟方式属于“**自顶向下**”[36]。本节将介绍的伪势模型则是通过一个伪势相互作用力来描述流体内部的相互作用。实际上，伪势相互作用力在宏观层面所对应的也是压力张量。不同的是，伪势模型并不是通过对压力张量或压力张量的散度逐项进行离散，而是“**自底向上**”通过一个简单的伪势相互作用力来描述流体内部的相互作用，继而反映宏观层面相对复杂的压力张量，实现对相分离和表面张力效应的描述。

5.3.1 Shan-Chen 模型

1993 年，Shan 和 Chen[37-38]提出了一种基于伪势的相互作用力来描述流体内部的相互作用，具体形式如下：

$$\boldsymbol{F}(\boldsymbol{r},t)=-G\psi(\boldsymbol{r})\sum_{\alpha=0}^{8}w_\alpha\psi(\boldsymbol{r}+\boldsymbol{e}_\alpha\delta_t)\boldsymbol{e}_\alpha \tag{5.58}$$

式中：$\psi(\boldsymbol{r})$ 为空间位置 $\boldsymbol{r}$ 处的伪势；$\psi(\boldsymbol{r}+\boldsymbol{e}_\alpha\delta_t)$ 为 $\boldsymbol{e}_\alpha$ 方向相邻节点的伪势；G 表示相互作用强度；w_α 为权系数[39]。在早期的伪势模型中，是通过改变平衡态密度分布函数中的速度将伪势相互作用力嵌入格子 Boltzmann 方程中，

$$f_\alpha(\boldsymbol{r}+\boldsymbol{e}_\alpha\delta_t,t+\delta_t)-f_\alpha(\boldsymbol{r},t)=-\frac{1}{\tau}[f_\alpha-f_\alpha^{\mathrm{eq}}(\rho,\boldsymbol{u}^{\mathrm{eq}})] \tag{5.59}$$

式中，速度 $\boldsymbol{u}^{\mathrm{eq}}=\left(\sum_{\alpha=0}^{8}f_\alpha\boldsymbol{e}_\alpha+\tau\delta_t\boldsymbol{F}\right)/\rho$。同时，真实的流体速度定义如下[40]：

$$\rho\boldsymbol{u}=\sum_{\alpha=0}^{8}f_\alpha\boldsymbol{e}_\alpha+\frac{\delta_t}{2}\boldsymbol{F} \tag{5.60}$$

这一作用力格式通常被称为 Shan-Chen 或 Shan-Doolen 格式，该格式已被证明会在宏观动量方程中引入误差项，且误差项与松弛因子 τ 有关，从而导致早期的伪势模型所给出的气液共存密度依赖于流体的运动黏度[41]。

对于指数形式的伪势，即 $\psi(\rho)=\psi_0\exp(-\rho_0/\rho)$[38]，相关研究已表明[42-43]，应当采用 Guo 等[44]提出的作用力格式：

$$f_\alpha(\boldsymbol{r}+\boldsymbol{e}_\alpha\delta_t,t+\delta_t)-f_\alpha(\boldsymbol{r},t)=-\frac{1}{\tau}[f_\alpha-f_\alpha^{\mathrm{eq}}(\rho,\boldsymbol{u})]+\delta_t\left(1-\frac{1}{2\tau}\right)F_\alpha(\boldsymbol{r},t) \tag{5.61}$$

式中，

$$F_\alpha(\boldsymbol{r},t)=\omega_\alpha\left[\frac{\boldsymbol{e}_\alpha\cdot\boldsymbol{F}}{c_s^2}+\frac{(\boldsymbol{uF}+\boldsymbol{Fu}):(\boldsymbol{e}_\alpha\boldsymbol{e}_\alpha-c_s^2\boldsymbol{I})}{2c_s^4}\right] \tag{5.62}$$

通过 Chapman-Enskog 展开可导出伪势模型所对应的宏观动量方程为

$$\frac{\partial(\rho\boldsymbol{u})}{\partial t}+\nabla\cdot(\rho\boldsymbol{uu})=-\nabla\cdot\boldsymbol{P}+\nabla\cdot\boldsymbol{\Pi} \tag{5.63}$$

式中，$\nabla\cdot\boldsymbol{P}=\nabla\cdot(\rho c_s^2\boldsymbol{I})-\boldsymbol{F}$。

需要注意的是，式(5.63)是通过二阶 Chapman-Enskog 展开导出的。He 和 Doolen[6]基于这一结果定义了一种连续形式的压力张量，但由于二阶 Chapman-Enskog 展开无法反映作用力格式所引入的三阶项，即式(5.49)右端最后一项，从而导致连续形式的压力张量不能正确地描述伪势模型的相分离特性。2008 年，Shan[45]强调，对伪势模型的分析应当基于离散形式的压力张量。2016 年，Huang 和 Wu[31]发现，当连续形式的压力张量纳入高阶项之后，它实际上与离散形式的压力张量是一致的。具体地，伪势模型离散形式的压力张量可定义为[45-46]

$$\boldsymbol{P}=\left(\rho c_s^2+\frac{Gc^2\psi^2}{2}\right)\boldsymbol{I}+\frac{G\psi}{2}\sum_{\alpha=0}^{8}w_\alpha[\psi(\boldsymbol{r}+\boldsymbol{e}_\alpha\delta_t)-\psi(\boldsymbol{r})]\boldsymbol{e}_\alpha\boldsymbol{e}_\alpha \tag{5.64}$$

根据泰勒级数展开有

$$\psi(\boldsymbol{r}+\boldsymbol{e}_\alpha\delta_t)=\psi(\boldsymbol{r})+\delta_t\boldsymbol{e}_\alpha\cdot\nabla\psi(\boldsymbol{r})+\frac{\delta_t^2}{2}\boldsymbol{e}_\alpha\boldsymbol{e}_\alpha:\nabla\nabla\psi(\boldsymbol{r})+\cdots \tag{5.65}$$

将式(5.65)代入式(5.64)并结合格子张量运算（参见附录 B）可得[46]

$$\boldsymbol{P}\approx\left(\rho c_s^2+\frac{Gc^2}{2}\psi^2+\frac{Gc^4}{12}\psi\nabla^2\psi\right)\boldsymbol{I}+\frac{Gc^4}{6}\psi\nabla\nabla\psi \tag{5.66}$$

式(5.66)即为伪势模型所对应的压力张量，尽管它在形式上与热力学压力张量有所不同，但它同样具备了描述相分离和表面张力的能力。实际上，伪势模型最重要的特征就是“自底向

上”通过一个简单的伪势相互作用力来描述流体内部的相互作用,继而反映宏观层面相对复杂的压力张量。正如 Succi 在展望 2038 年的格子 Boltzmann 方法时所指出的[47]:“This simple variant opens up a vast scenario of applications involving multiphase and multicomponent flows”。

在 5.2.2 节,我们推导了自由能模型的力学稳定性条件。根据式(5.66),可对伪势模型做类似分析。结合图 5.1 和式(5.66),x 方向压力张量的分量为

$$P_{xx}=p_{\mathrm{EOS}}(\rho)+\frac{Gc^4}{4}\psi\frac{\mathrm{d}^2\psi}{\mathrm{d}x^2},\quad p_{\mathrm{EOS}}(\rho)=\rho c_s^2+\frac{Gc^2}{2}\psi^2 \tag{5.67}$$

式中,$p_{\mathrm{EOS}}(\rho)$ 为伪势模型的状态方程。根据下列关系式[45,48]:

$$\frac{\mathrm{d}^2\psi}{\mathrm{d}x^2}=\frac{1}{2}\frac{\mathrm{d}}{\mathrm{d}\psi}\left(\frac{\mathrm{d}\psi}{\mathrm{d}x}\right)^2,\quad \frac{\mathrm{d}}{\mathrm{d}\psi}=\frac{1}{\psi'}\frac{\mathrm{d}}{\mathrm{d}\rho},\quad \left(\frac{\mathrm{d}\psi}{\mathrm{d}x}\right)^2=\psi'^2\left(\frac{\mathrm{d}\rho}{\mathrm{d}x}\right)^2 \tag{5.68}$$

可将 $\psi\mathrm{d}^2\psi/\mathrm{d}x^2$ 转换为

$$\psi\frac{\mathrm{d}^2\psi}{\mathrm{d}x^2}=\frac{1}{2}\frac{\psi}{\psi'}\frac{\mathrm{d}}{\mathrm{d}\rho}\left[\psi'^2\left(\frac{\mathrm{d}\rho}{\mathrm{d}x}\right)^2\right] \tag{5.69}$$

式中,$\psi'=\mathrm{d}\psi/\mathrm{d}\rho$。结合式(5.67)和式(5.69),可得伪势模型的力学稳定性条件为

$$\int_{\rho_G}^{\rho_L}\left[p_{\mathrm{EOS}}(\rho)-p_b\right]\frac{\psi'}{\psi}\mathrm{d}\rho=0 \tag{5.70}$$

当伪势取指数形式时,即 $\psi(\rho)=\psi_0\exp(-\rho_0/\rho)$,求导可得 $\psi'=\rho_0\psi/\rho^2$。由于 ρ_0 为常数,此时式(5.70)与 Maxwell 等面积法则式(5.40)是一致的,但状态方程被固定为

$$p_{\mathrm{EOS}}(\rho)=\rho c_s^2+\frac{Gc^2}{2}\psi_0^2\exp(-2\rho_0/\rho) \tag{5.71}$$

此外,指数形式的伪势往往难以胜任大密度比气液两相流的模拟。2002 年,He 和 Doolen[6] 指出,为了与热力学中的状态方程取得一致,应当采用平方根形式的伪势,即

$$\psi(\rho)=\sqrt{\frac{2(p_{\mathrm{EOS}}-\rho c_s^2)}{Gc^2}} \tag{5.72}$$

式中,p_{EOS} 为热力学状态方程,如 van der Waals 状态方程、Carnahan-Starling 状态方程等[9]。

可以看到,指数形式的伪势虽然能够自动满足 Maxwell 等面积法则,但无法取得状态方程的一致性;而平方根形式的伪势虽然能够取得状态方程的一致性,但此时式(5.70)与式(5.40)不同,从而导致气液共存密度与 Maxwell 等面积法则的结果不一致。这一问题在文献中通常被称为伪势模型的**热力学不一致性**(thermodynamic inconsistency)。此外,在 5.1 节曾提到分子间的相互作用可分为短程和长程相互作用,而伪势模型采用的相互作用力,即式(5.58),实质上是单参数相互作用力,从而导致表面张力的大小依赖于状态方程(气液密度比),不能自由调节。

针对上述问题,一些学者开展了相关研究,我们将在 5.3.2 节和 5.3.3 节进行介绍。在四类气液两相流格子 Boltzmann 模型中,伪势模型是相对简单且应用比较广泛的一类[48-49]。这主要是因为伪势模型直接对粒子间的相互作用进行描述,反映了多相流体动力学的物理本质[2],从而获得了比较广泛的应用。一个简单的相互作用力,结合数十行格子 Boltzmann 方法的代码,就可以模拟非常复杂的多相流动,这是采用任何其他算法时难以想象的。伪势模型的简单性必然是付出了相应的“代价”而获得的,这些代价可以理解为它所存在的问题,例如热力学不一致性、表面张力不能自由调节等。目前,这些问题已在一定程度上得到了解决或改善。

5.3.2 实现热力学一致性的 Li-Luo-Li 方法

2012 年,Li 等[50]提出,采用平方根形式的伪势时,可以通过调整力学稳定性条件来实现伪势模型的热力学一致性。具体地,他们通过构造新的作用力格式将伪势模型的力学稳定性条件从式(5.70)调整成如下形式 :

$$\int_{\rho_G}^{\rho_L}\left[p_{\mathrm{EOS}}(\rho)-p_{\mathrm{b}}\right]\frac{\psi'}{\psi^{1+\varepsilon}}\mathrm{d}\rho=0 \tag{5.73}$$

式中,ε 为大于零的常数。以 Carnahan-Starling 状态方程为例,他们对比了式(5.73)的解析解和 Maxwell 等面积法则的结果,发现在区间 $\varepsilon\in(1,2)$ 存在着一个 ε 的值($\varepsilon\approx1.7$),此时式(5.73)所给出的气液共存密度与 Maxwell 等面积法则的结果基本一致。基于这一发现,Li 等[50]结合式(5.61)提出了一个新的作用力项

$$F_\alpha(\boldsymbol{r},t)=\omega_\alpha\left[\frac{\boldsymbol{e}_\alpha\cdot\boldsymbol{F}}{c_s^2}+\frac{(\boldsymbol{u}_{\mathrm{new}}\boldsymbol{F}+\boldsymbol{F}\boldsymbol{u}_{\mathrm{new}}):(\boldsymbol{e}_\alpha\boldsymbol{e}_\alpha-c_s^2\boldsymbol{I})}{2c_s^4}\right] \tag{5.74}$$

式中,$\boldsymbol{u}_{\mathrm{new}}=\boldsymbol{u}+\sigma\boldsymbol{F}/[(\tau-0.5)\delta_t\psi^2]$,常数 σ 用来调节力学稳定性条件即式(5.73)中的 ε,且 $\sigma=-\varepsilon/(16G)$。这里的 $\boldsymbol{u}_{\mathrm{new}}$ 与 $\boldsymbol{u}$ 相比所增加的额外项会改变作用力项的二阶矩[48,50],从而达到调整力学稳定性条件继而实现伪势模型热力学一致性的目的。平衡态密度分布函数中的速度及真实流体的速度仍为 $\boldsymbol{u}$,由式(5.60)给出。与原有格式相比,新格式以较小的改动实现了力学稳定性条件的调整,没有增加复杂的计算,保留了伪势模型的优势和特色。

在此基础上,Li 等[46]构造了基于多松弛碰撞算子的大密度比伪势模型,该模型采用如下形式的矩空间作用力项:

$$\boldsymbol{S}=\begin{bmatrix}0\\ 6\boldsymbol{F}\cdot\boldsymbol{u}+\dfrac{12\sigma\,|\boldsymbol{F}|^2}{\psi^2\delta_t(\tau_e-0.5)}\\ -6\boldsymbol{F}\cdot\boldsymbol{u}-\dfrac{12\sigma\,|\boldsymbol{F}|^2}{\psi^2\delta_t(\tau_\zeta-0.5)}\\ F_x\\ -F_x\\ F_y\\ -F_y\\ 2(u_xF_x-u_yF_y)\\ (u_xF_y+u_yF_x)\end{bmatrix} \tag{5.75}$$

式中:$|\boldsymbol{F}|^2=(F_x^2+F_y^2)$;流体速度 $\boldsymbol{u}$ 由式(5.60)定义。与单松弛碰撞算子相比,多松弛碰撞算子增强了模型的数值稳定性。该模型的建立实现了伪势格子 Boltzmann 方法对较高雷诺数的大密度比气液两相流的数值模拟,部分模拟结果如图 5.5 所示。通过 Chapman-Enskog 多尺度展开,可以导出式(5.75)所对应的宏观压力张量为

$$\boldsymbol{P}_{\mathrm{new}}\approx\boldsymbol{P}+2G^2c^4\sigma\,|\nabla\psi|^2\boldsymbol{I} \tag{5.76}$$

推导中用到 $|\boldsymbol{F}|^2\approx G^2c^4\psi^2\,|\nabla\psi|^2$。式中,$\boldsymbol{P}$ 由式(5.66)给出,$c=1$,$\sigma=-\varepsilon/(16G)$。当 $\sigma=0$ 时,式(5.75)退化为标准的多松弛碰撞算子的作用力格式。对于平方根形式的伪势,参

数 G 不再具有相互作用强度的物理意义，它的主要作用是确保根号中的项非负[9]。在数值模拟中，常取 $G=-1$。需要强调的是，这一调节方法并不仅限于 Carnahan-Starling 状态方程，它也适用于 van der Waals 状态方程、Peng-Robinson 状态方程等热力学中其他形式的状态方程[9]。

基于大密度比伪势模型的液滴撞击模拟

上述通过调整力学稳定性条件来实现伪势模型热力学一致性的方法已在许多文献中应用或推广。其中，Zhang 等[51]、Xu 等[52]和 Li 等[53]分别基于 D3Q19 格子、D3Q15 格子和非正交多松弛碰撞算子将式(5.75)拓展到了三维；Lycett-Brown 和 Luo[54]根据这一方法构造了适用于大密度比的 Cascaded 伪势格子 Boltzmann 模型；Hou 等[55-56]和 Zheng 等[57]将该方法推广到了多组分系统。在应用方面，该方法已被应用到沸腾[58-62]、冷凝[57,63-64]、燃料电池[55,65]、液滴动力学[66-68]、空化[69-71]、缔合流体[72]等。

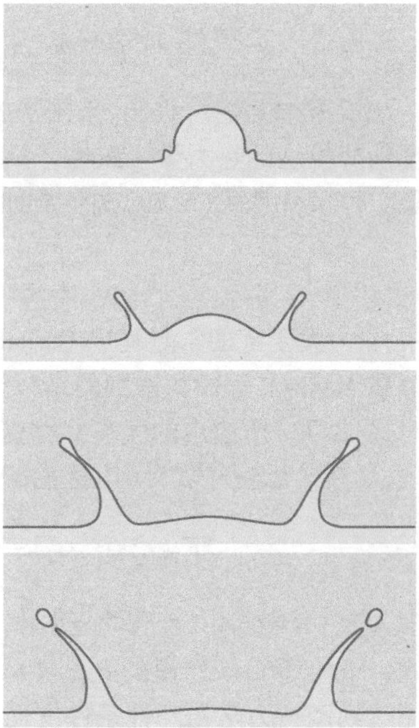

图 5.5 基于大密度比伪势模型的液滴撞击液膜的数值模拟[46]

5.3.3 表面张力

前面提到，伪势模型的另一个缺点是表面张力依赖于气液密度比，不能自由调节。这一问题的根源在于式(5.58)是单参数相互作用力(只有参数 G)。对于平直气液界面，可导出如下形式的伪势模型表面张力[48]

$$\gamma=-\frac{Gc^4}{6}\int_{-\infty}^{+\infty}\left(\frac{\mathrm{d}\psi}{\mathrm{d}n}\right)\mathrm{d}\psi=-\frac{Gc^4}{6}\int_{\rho_G}^{\rho_L}\psi'^2\left(\frac{\mathrm{d}\rho}{\mathrm{d}n}\right)\mathrm{d}\rho \tag{5.77}$$

式中：$\psi'=\mathrm{d}\psi/\mathrm{d}\rho$；$n$ 为气液界面的法线方向。从形式上看，式(5.77)表明伪势模型的表面张力可以通过相互作用强度 G 来进行调节，但采用指数形式的伪势时，相互作用强度 G 影响气液密度比，因此改变表面张力的大小会同时改变气液密度比。相应地，采用平方根形式的伪势时，ψ'^2 正比于 $1/G$，从而改变 G 的大小对式(5.77)没有影响。为了解决这一问题，Sbragaglia 等[73]提出了一种多范围(multi-range)伪势相互作用力

$$\boldsymbol{F}(\boldsymbol{r},t)=-\psi(\boldsymbol{r})\sum_{\alpha=0}^{8}w_\alpha\left[G_1\psi(\boldsymbol{r}+\boldsymbol{e}_\alpha\delta_t)+G_2\psi(\boldsymbol{r}+2\boldsymbol{e}_\alpha\delta_t)\right]\boldsymbol{e}_\alpha \tag{5.78}$$

式中，G_1 和 G_2 为可调参数。当 $G_1=G$，$G_2=0$ 时，式(5.78)退化为式(5.58)。Sbragaglia 等[73]认为式(5.78)所对应的压力张量为

$$\boldsymbol{P}=\left[\rho c_s^2+\frac{A_1}{2}c^2\psi^2+A_2\left(\frac{c^4}{12}|\nabla\psi|^2+\frac{c^4}{6}\psi\,\nabla^2\psi\right)\right]\boldsymbol{I}-\frac{A_2c^4}{6}\nabla\psi\,\nabla\psi \tag{5.79}$$

式中：$A_1=G_1+2G_2$；$A_2=G_1+8G_2$。Huang 等[41]对多范围伪势相互作用力进行了研究，他们发现，虽然可通过调节 A_2/A_1 的大小对表面张力进行调节，但仍然会改变伪势模型的气液密度比。

2013 年，Li 和 Luo[74]指出，式(5.79)是连续形式的压力张量，而根据 Shan 和 Chen[37-38,45]的研究，对伪势模型的分析应当基于离散层面的压力张量。为此，Li 和

Luo[74]导出了多范围伪势相互作用力所对应的离散形式压力张量。他们发现，当参数 A_2 减小时，与表面张力有关的项中有一项会减小但另一项会增大，从而导致表面张力的可调范围较窄。此外，采用多范围伪势相互作用力时，表面张力的调节会影响力学稳定性条件，因此调节表面张力时仍然会改变气液密度比。为此，Li 和 Luo[74]提出了另一种调节方法，即在多松弛碰撞算子的碰撞步添加一个源项

$$\boldsymbol{m}^{*}=\boldsymbol{m}-\boldsymbol{\Lambda}(\boldsymbol{m}-\boldsymbol{m}^{\mathrm{eq}})+\delta_t\left(\boldsymbol{I}-\frac{\boldsymbol{\Lambda}}{2}\right)\boldsymbol{S}+\delta_t\boldsymbol{C}_{\mathrm{s}} \tag{5.80}$$

式中：$\boldsymbol{m}$ 为矩函数；$\boldsymbol{m}^{\mathrm{eq}}$为平衡态矩函数；$\boldsymbol{m}^{*}$为碰撞后的矩函数；$\boldsymbol{\Lambda}$ 与松弛因子有关；$\boldsymbol{I}$ 为二阶单位张量；$\boldsymbol{S}$ 为作用力项；$\boldsymbol{C}_{\mathrm{s}}$ 为源项，对 D2Q9 格子有

$$\boldsymbol{C}_{\mathrm{s}}=\begin{bmatrix}0\\ 1.5\tau_e^{-1}(Q_{xx}+Q_{yy})\\ -1.5\tau_\zeta^{-1}(Q_{xx}+Q_{yy})\\ 0\\ 0\\ 0\\ 0\\ -\tau_v^{-1}(Q_{xx}-Q_{yy})\\ -\tau_v^{-1}Q_{xy}\end{bmatrix} \tag{5.81}$$

式中，Q_{xx}、Q_{yy}和 Q_{xy}为二阶张量 $\boldsymbol{Q}$ 的分量，其定义为

$$\boldsymbol{Q}=\kappa\frac{G}{2}\psi(\boldsymbol{r})\sum_{\alpha=0}^{8}w_\alpha[\psi(\boldsymbol{r}+\boldsymbol{e}_\alpha\delta_t)-\psi(\boldsymbol{r})]\boldsymbol{e}_\alpha\boldsymbol{e}_\alpha \tag{5.82}$$

式中，参数 κ 用来调节表面张力的大小。基于上述方法，Li 和 Luo[74]对伪势模型的表面张力进行较大范围的调节，且调节表面张力时气液密度比基本不变。该方法已被 Xu 等[52]、Ammar 等[75]、Fei 等[76]分别基于 D3Q15 格子、D3Q19 格子、非正交多松弛碰撞算子拓展至三维。此外，Huang 和 Wu[31]基于三阶 Chapman-Enskog 多尺度分析，通过结合式(5.75)和式(5.81)，提出了调节伪势模型力学稳定性条件和表面张力的多松弛碰撞算子统一格式，并开展了消除伪势模型高阶速度项的研究[77]。在此基础上，Wu 等[78]对伪势模型的作用力项进行了四阶 Chapman-Enskog 多尺度分析，并对多松弛碰撞算子中松弛因子的选取提出了相应的建议。

5.3.4 接触角格式

对气液两相流的模拟，当存在气液界面与固体表面的接触时，需要引入接触角格式来实施相应的湿润性边界。伪势模型中最早的接触角格式是由 Martys 和 Chen[79]提出的，他们借鉴流-流相互作用力的概念和表达式，提出了一种流固相互作用力来实施接触角

$$\boldsymbol{F}_{\mathrm{ads}}(\boldsymbol{r},t)=-G_{\mathrm{w}}\rho(\boldsymbol{r})\sum_{\alpha=0}^{8}w_\alpha s(\boldsymbol{r}+\boldsymbol{e}_\alpha\delta_t)\boldsymbol{e}_\alpha \tag{5.83}$$

式中：$s(\boldsymbol{r}+\boldsymbol{e}_\alpha\delta_t)$为开关函数，当 $\boldsymbol{r}+\boldsymbol{e}_\alpha\delta_t$ 处在固体区域时取 1，处在流体区域时取 0；G_{w} 用来调节接触角的大小。此外，一些学者[80-81]提出用伪势取代式(5.83)求和符号前的密度，继而得到如下形式的流固作用力格式：

$$\boldsymbol{F}_{\mathrm{ads}}(\boldsymbol{r},t) = -G_{\mathrm{w}}\psi(\boldsymbol{r})\sum_{\alpha=0}^{8} w_{\alpha}s(\boldsymbol{r}+\boldsymbol{e}_{\alpha}\delta_t)\boldsymbol{e}_{\alpha} \tag{5.84}$$

流固作用力格式因其概念清晰,在伪势模型中得到了比较广泛的应用,但它产生的虚假速度通常比较大[82]。

伪势模型中另一种常用的接触角格式是 Benzi 等[83]于 2006 年提出的虚拟密度格式。该格式针对固体区域设置一个虚拟密度 ρ_{w}(其值为常数),从而可以在流体边界节点处直接实施流-流相互作用力,即伪势相互作用力式(5.58)。虚拟密度格式的数值实施非常简单,且产生的虚假速度较小,往往比流固作用力格式产生的虚假速度小一个数量级,因而在处理复杂结构表面的接触角时具有明显的优势,但该格式会在固体表面形成一层非物理密度层(图 5.6),影响数值模拟的准确性。

此外,Ding 和 Spelt[85]针对相场方法所提出的几何接触角格式近年来也被应用到伪势格子 Boltzmann 模型[86-87]。对于平直固体表面,几何接触角格式在固体表面下布置一层虚拟层,且虚拟层的密度计算如下:

$$\rho_{i,0}=\rho_{i,2}+\tan\left(\frac{\pi}{2}-\theta_{\mathrm{a}}\right)\left|\rho_{i+1,1}-\rho_{i-1,1}\right| \tag{5.85}$$

式中:下标 i 表示 x 方向的坐标;下标 0 表示固体表面下的虚拟层。几何接触角格式的优点在于可以通过式(5.85)中的 θ_{a} 预先给定接触角的大小,且准确度较高(数值模拟得到的接触角与给定的接触角偏差不大)。2015 年,Liu 和 Ding[88]将几何接触角格式拓展到了二维曲线表面。这里简要介绍一下二维曲线表面几何接触角格式的思路,详细内容可参阅文献[88]。如图 5.7 所示,点 S 是任意一个与流体边界节点相邻的固体节点,$\boldsymbol{n}_S$ 是该节点附近固体表面的法向矢量,θ 是接触角,$\boldsymbol{l}_1$ 和 $\boldsymbol{l}_2$ 是接触角给定时两个可能的气液界面切向矢量,这两个切向矢量分别交流体区域的网格线于点 D_1 和点 D_2。

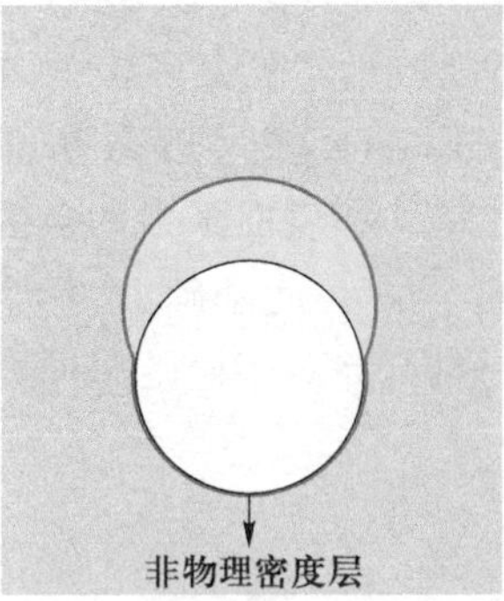

图 5.6 虚拟密度格式所形成的非物理密度层[84]

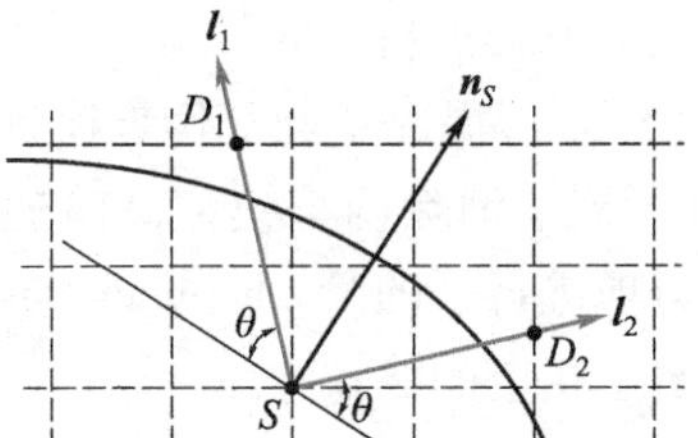

图 5.7 二维曲线表面几何接触角格式的示意图

实施几何接触角格式的关键在于根据流体节点的信息构造出相邻固体节点的虚拟密度。从这一点来看,几何接触角格式本质上也是一种虚拟密度格式。不同的是,Benzi 等[83]的虚拟密度格式采用的是全局密度(常数 ρ_{w}),而几何接触角格式则是构造出固体节点的局部密度。根据 Liu 和 Ding[88]的研究,图 5.7 中固体节点 S 的虚拟密度近似等于点 D_1或点 D_2的密度,且有

$$\rho(S)=\begin{cases}\max(\rho_{D_1},\rho_{D_2}), & \theta\leqslant\pi/2\\ \min(\rho_{D_1},\rho_{D_2}), & \theta>\pi/2\end{cases} \tag{5.86}$$

因此,首先应当求得点 S 附近固体表面的法向矢量,即 $\boldsymbol{n}_S$,然后依据 $\boldsymbol{n}_S$ 及给定的接触角 θ,通过矢量旋转(vector rotation)得到 $\boldsymbol{l}_1$ 和 $\boldsymbol{l}_2$,继而找到点 D_1 和点 D_2,之后通过插值格式依据相邻流体节点的密度得到点 D_1 和点 D_2 的密度[88],并根据式(5.86)确定点 S 的虚拟密度。

除了可以预先给定接触角的大小,几何接触角格式产生的虚假速度也比较小,而且不存在非物理密度层。实际上,虚拟密度格式产生非物理密度层的根源在于每个流体边界节点的密度是不同的,而所有固体节点的虚拟密度却固定为 ρ_{w}。因此,为了降低或消除非物理密度层的影响,固体节点的密度应当具有局部性。几何接触角格式具有这一特征,但其数值实施较为复杂(平直表面除外),从而限制了它的应用。

为此,Li 等[84]通过结合虚拟密度格式的全局性和几何接触角格式的局部性,提出了一种改进的虚拟密度格式

$$\rho_{\mathrm{w}}(\boldsymbol{r})=\begin{cases}\varphi\rho_{\mathrm{ave}}(\boldsymbol{r}), & \theta\leqslant\theta_{\mathrm{s}}\\ \rho_{\mathrm{ave}}(\boldsymbol{r})-\Delta\rho, & \theta>\theta_{\mathrm{s}}\end{cases} \tag{5.87}$$

式(5.87)分段选取的依据详见文献[84]。式中:$\boldsymbol{r}$ 为固体节点的空间位置,它至少有一个相邻节点在流体区域;$\varphi\geqslant1$ 和 $\Delta\rho>0$ 为常数,它们体现了该格式的全局性;$\rho_{\mathrm{ave}}(\boldsymbol{r})$ 是固体节点附近流体节点的平均密度,对 D2Q9 格子有

$$\rho_{\mathrm{ave}}(\boldsymbol{r})=\frac{\sum_{\alpha=0}^{8}w_{\alpha}\rho(\boldsymbol{r}+\boldsymbol{e}_{\alpha}\delta_t)s_{\mathrm{w}}(\boldsymbol{r}+\boldsymbol{e}_{\alpha}\delta_t)}{\sum_{\alpha=0}^{8}w_{\alpha}s_{\mathrm{w}}(\boldsymbol{r}+\boldsymbol{e}_{\alpha}\delta_t)} \tag{5.88}$$

式中,$s_{\mathrm{w}}(\boldsymbol{r}+\boldsymbol{e}_{\alpha}\delta_t)$ 为开关函数,对流体区域取 1,固体区域取 0。当 $\varphi=1$ 时,$\rho_{\mathrm{w}}(\boldsymbol{r})=\rho_{\mathrm{ave}}(\boldsymbol{r})$,相应的理论接触角为 $\theta_{\mathrm{s}}=90°$。该格式设置了一个限制器,即 $\rho_G\leqslant\rho_w(\boldsymbol{r})\leqslant\rho_{\mathrm{L}}$。当式(5.87)的计算结果大于液相密度时,取 $\rho_{\mathrm{w}}(\boldsymbol{r})=\rho_{\mathrm{L}}$。类似地,当式(5.87)的计算结果小于气相密度时,取 $\rho_{\mathrm{w}}(\boldsymbol{r})=\rho_{\mathrm{G}}$。这一改进格式保留了原有虚拟密度格式简单易实施的优点,对复杂表面具有良好的适应性。同时,该格式结合了一定的局部性,有效地消除了非物理密度层对数值模拟准确性的影响[84]。此外,它产生的虚假速度较小,比流固作用力格式所产生的虚假速度小一个数量级以上。图 5.8 给出了改进格式对三维曲线表面接触角的模拟结果。

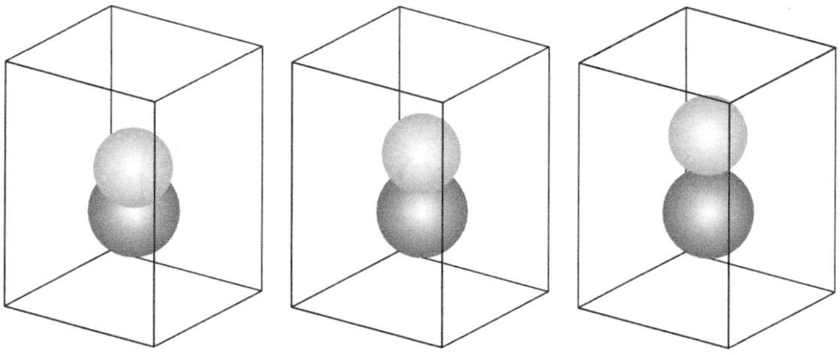

(a) $\theta\approx53°$　(b) $\theta\approx88°$　(c) $\theta\approx145°$

图 5.8　基于改进虚拟密度格式的三维曲线表面接触角模拟[84]

5.3.5 多组分伪势模型简介

除了应用于单组分两相流动，伪势模型在多组分流动中也得到了广泛的应用。对多组分系统，伪势相互作用力可写成如下形式[37,40,48]：

$$\boldsymbol{F}_k(\boldsymbol{r},t)=-\psi_k(\boldsymbol{r})\sum_{\bar{k}=1}^{n}G_{k\bar{k}}\sum_{\alpha=0}^{m-1}w_\alpha\psi_{\bar{k}}(\boldsymbol{r}+\boldsymbol{e}_\alpha\delta_t)\boldsymbol{e}_\alpha \tag{5.89}$$

式中：$G_{k\bar{k}}$为相互作用强度；n 为组分的个数；m 为离散速度的个数，对 D3Q19 格子（$m=19$），它包含离散速度为零的方向（即 $\alpha=0$）以及 18 个非零离散速度方向；下标 k 和 $\bar{k}$ 表示组分，其中 k 是自由指标，$\bar{k}$ 是哑指标。当 $\bar{k}\neq k$ 时，$G_{k\bar{k}}=G_{\bar{k}k}$，当 $\bar{k}=k$ 时，G_{kk}表征组分内部的相互作用强度。以组分 A 和组分 B 构成的两组分系统为例，组分 A 的作用力为

$$\boldsymbol{F}_{\mathrm{A}}(\boldsymbol{r},t)=-\psi_{\mathrm{A}}(\boldsymbol{r})\left[G_{\mathrm{AA}}\sum_{\alpha=0}^{m-1}w_\alpha\psi_{\mathrm{A}}(\boldsymbol{r}+\boldsymbol{e}_\alpha\delta_t)\boldsymbol{e}_\alpha+G_{\mathrm{AB}}\sum_{\alpha=0}^{m-1}w_\alpha\psi_{\mathrm{B}}(\boldsymbol{r}+\boldsymbol{e}_\alpha\delta_t)\boldsymbol{e}_\alpha\right] \tag{5.90}$$

式中，G_{AB}为组分 A 和组分 B 之间的相互作用强度，且 $G_{\mathrm{AB}}=G_{\mathrm{BA}}$。若两个组分均为理想气体，则 $G_{\mathrm{AA}}=G_{\mathrm{BB}}=0$。在伪势函数的选取上，对理想气体，通常可取 $\psi_k(\boldsymbol{r})=\rho_k(\boldsymbol{r})$；对实际气体，可选取指数形式[37,40]：或平方根形式的伪势[57,63,89]，此时模型通常被称为**多组分多相伪势模型**。

早期的多组分伪势模型采用如下形式的演化方程[37,40]：

$$f_\alpha^k(\boldsymbol{r}+\boldsymbol{e}_\alpha\delta_t,t+\delta_t)-f_\alpha^k(\boldsymbol{r},t)=-\frac{1}{\tau_k}[f_\alpha^k-f_\alpha^{k,\mathrm{eq}}(\rho_k,\boldsymbol{u}_k^{\mathrm{eq}})] \tag{5.91}$$

式中：f_α^k为组分 k 的密度分布函数；$f_\alpha^{k,\mathrm{eq}}$为平衡态分布函数；$\rho_k=\sum_{\alpha=0}^{m-1}f_\alpha^k$ 为组分 k 的密度；平衡态分布函数中的速度 $\boldsymbol{u}_k^{\mathrm{eq}}$ 由 $\rho_k\boldsymbol{u}_k^{\mathrm{eq}}=\rho_k\boldsymbol{u}'+\tau_k\boldsymbol{F}_k$[40] 定义。为了确保在没有作用力时系统碰撞过程的动量守恒，$\boldsymbol{u}'$需满足[40]：

$$\sum_{k=1}^{n}\sum_{\alpha=0}^{m-1}\left\{\frac{\boldsymbol{e}_\alpha}{\tau_k}[f_\alpha^k-f_\alpha^{k,\mathrm{eq}}(\rho_k,\boldsymbol{u}')]\right\}=0 \tag{5.92}$$

根据式(5.92)可得

$$\boldsymbol{u}'=\frac{\sum_{k=1}^{n}\dfrac{\sum_{\alpha=0}^{m-1}\boldsymbol{e}_\alpha f_\alpha^k}{\tau_k}}{\sum_{k=1}^{n}\dfrac{\rho_k}{\tau_k}} \tag{5.93}$$

混合流体的密度和速度分别定义如下[40]：

$$\rho=\sum_{k=1}^{n}\sum_{\alpha=0}^{m-1}f_\alpha^k,\quad \rho\boldsymbol{u}=\sum_{k=1}^{n}\left(\sum_{\alpha=0}^{m-1}\boldsymbol{e}_\alpha f_\alpha^k+\frac{\delta_t}{2}\boldsymbol{F}_k\right) \tag{5.94}$$

由于平衡态分布函数中的速度 $\boldsymbol{u}_k^{\mathrm{eq}}$ 依赖于松弛因子 τ_k，导致早期的多组分伪势模型所给出的结果往往依赖于运动黏度（例如，表面张力随运动黏度变化）。

为此，Porter 等[90]将 He-Shan-Doolen 作用力格式[91]拓展到了多组分伪势模型

$$f_\alpha^k(\boldsymbol{r}+\boldsymbol{e}_\alpha\delta_t,t+\delta_t)-f_\alpha^k(\boldsymbol{r},t)=-\frac{1}{\tau_k}[f_\alpha^k-f_\alpha^{k,\mathrm{eq}}(\rho_k,\boldsymbol{u}^{\mathrm{eq}})]+\left(1-\frac{1}{2\tau_k}\right)\delta_tF_\alpha(\boldsymbol{r},t) \tag{5.95}$$

式中，作用力项 $F_\alpha(\boldsymbol{r},t)$ 为

$$F_\alpha(\boldsymbol{r},t)=\frac{(\boldsymbol{e}_\alpha-\boldsymbol{u}^{\mathrm{eq}})\cdot\boldsymbol{F}_k}{\rho_k c_s^2}f_\alpha^{k,\mathrm{eq}}(\rho_k,\boldsymbol{u}^{\mathrm{eq}}) \tag{5.96}$$

与早期伪势模型的作用力格式不同，这里的平衡态分布函数及作用力项中的速度不是 $\boldsymbol{u}_k^{\mathrm{eq}}$，而是 $\boldsymbol{u}^{\mathrm{eq}}=\boldsymbol{u}'$，其中 $\boldsymbol{u}'$ 由式(5.93)给出。混合流体的密度和速度仍由式(5.94)定义。根据上述思路，Porter 等[90]亦构造了基于多松弛碰撞算子的多组分伪势模型，其中平衡态分布函数及作用力项中的速度 $\boldsymbol{u}^{\mathrm{eq}}$ 满足如下关系式：

$$\sum_{k=1}^{n}s_c^k\sum_{\alpha=0}^{m-1}\boldsymbol{e}_\alpha\left[f_\alpha^k-f_\alpha^{k,\mathrm{eq}}(\rho_k,\boldsymbol{u}^{\mathrm{eq}})\right]=0 \tag{5.97}$$

式中，s_c^k 表示与动量有关的两个矩函数的弛豫频率。根据式(5.97)可得

$$\boldsymbol{u}^{\mathrm{eq}}=\frac{\sum_{k=1}^{n}s_c^k\sum_{\alpha=0}^{m-1}\boldsymbol{e}_\alpha f_\alpha^k}{\sum_{k=1}^{n}s_c^k\rho_k} \tag{5.98}$$

当 $s_c^k=1/\tau_k$ 时，式(5.98)退化为式(5.93)。需要注意的是，文献[90]是用 $\rho_k\boldsymbol{u}_k$ 表示式(5.93)和式(5.98)中的 $\sum_{\alpha=0}^{m-1}\boldsymbol{e}_\alpha f_\alpha^k$，但在其他地方又定义了 $\rho_k\boldsymbol{u}_k$ 的另一种表达式。为了避免混淆，这里保留了原有形式。

理论上，多组分伪势模型应当遵守这样一个原则，即当多组分模型退化为单组分模型时，相应的作用力格式应退化为单组分模型的作用力格式。可以看到，Porter 等[90]所提出的作用力格式实际上并不符合这一原则。以式(5.95)为例，当多组分退化为单组分时，由 $\boldsymbol{u}^{\mathrm{eq}}=\boldsymbol{u}'$ 及式(5.93)可得

$$\boldsymbol{u}^{\mathrm{eq}}=\frac{1}{\rho}\sum_{\alpha=0}^{m-1}\boldsymbol{e}_\alpha f_\alpha \tag{5.99}$$

相应地，流体的真实速度为

$$\boldsymbol{u}=\frac{1}{\rho}\sum_{\alpha=0}^{m-1}\left(\boldsymbol{e}_\alpha f_\alpha+\frac{\delta_t}{2}\boldsymbol{F}\right) \tag{5.100}$$

显然，平衡态分布函数及作用力项中的速度与流体的真实速度并不相同，这与 He-Shan-Doolen 作用力格式[91]只有一个速度不相符。

文献中也有一些学者[92-93]针对多组分伪势模型采用了统一速度的作用力格式，但并未对格式是否满足 Shan 和 Chen[37,40]所提出的约束条件做相应的讨论。从式(5.91)和式(5.92)来看，平衡态分布函数中的速度是否与 $\boldsymbol{F}_k$ 有关并不影响 Shan 和 Chen[37,40]所提出的约束条件，即混合流体的碰撞过程在没有作用力时应当满足动量守恒。因此，式(5.95)和式(5.96)中的速度 $\boldsymbol{u}^{\mathrm{eq}}$ 理论上可以定义为

$$\boldsymbol{u}^{\mathrm{eq}}=\boldsymbol{u}'+\frac{\delta_t}{2\rho}\sum_{k=1}^{n}\boldsymbol{F}_k \tag{5.101}$$

当两个组分的松弛因子相同时，式(5.101)所给出的 $\boldsymbol{u}^{\mathrm{eq}}$ 与式(5.94)的速度 $\boldsymbol{u}$ 是一致的。类似地，对多松弛碰撞算子，式(5.98)中的 $\boldsymbol{u}^{\mathrm{eq}}$ 可定义为

$$\boldsymbol{u}^{\mathrm{eq}}=\frac{\sum_{k=1}^{n}s_c^k\sum_{\alpha=0}^{m-1}\boldsymbol{e}_\alpha f_\alpha^k}{\sum_{k=1}^{n}s_c^k\rho_k}+\frac{\delta_t}{2\rho}\sum_{k=1}^{n}\boldsymbol{F}_k \tag{5.102}$$

根据这一定义,式(5.97)在没有作用力时仍然是成立的。当各组分的 s_c^k 相同时,$\boldsymbol{u}^{eq}$ 与 $\boldsymbol{u}$ 完全一致。对多松弛碰撞算子,s_c^k 相同并不影响每个组分的运动黏度。也就是说,对多组分伪势模型可以采用统一速度的作用力格式,但应当注意相应的实施条件。

5.4 颜色梯度格子 Boltzmann 模型

颜色梯度模型是格子 Boltzmann 方法中最早出现的多相/多组分模型,且早期的颜色梯度模型[94]是基于不混溶流体(immiscible fluids)的格子气自动机模型[95]建立的。因此,颜色梯度模型在一定程度上体现了格子 Boltzmann 方法对格子气自动机的继承和发展。该模型的特点是采用不同颜色的分布函数来区分不同相态或不同组分的流体[2],流体之间的相互作用则通过颜色梯度来实现,并根据颜色梯度来调整流体粒子的运动趋势。

5.4.1 基本原理

早期的颜色梯度模型[94]保留了一些格子气自动机的特征。例如,采用的碰撞项是基于格子气自动机碰撞规则的算子。这类碰撞算子在计算效率和数值稳定性方面存在明显不足[2]。之后的颜色梯度模型往往采用单松弛碰撞算子,即 BGK 碰撞算子。实际上,模拟气液两相流的关键在于构建描述相分离和表面张力效应的机制。在颜色梯度模型中,这两个机制分别是“**重新标色算法**(recoloring algorithm)”和“**扰动算子**(perturbation operator)”。

颜色梯度模型通常采用两个不同颜色的分布函数f_α^R和f_α^B来分别表示红色流体和蓝色流体,它们的演化方程为

$$f_\alpha^k(\boldsymbol{r}+\boldsymbol{e}_\alpha\delta_t,t+\delta_t)-f_\alpha^k(\boldsymbol{r},t)=\Omega_\alpha^k(\boldsymbol{r},t) \tag{5.103}$$

上标 k 表示 R 或 B,Ω_α^k 是碰撞项,它通常可表示为[96-97]

$$\Omega_\alpha^k=(\Omega_\alpha^k)^{(3)}[(\Omega_\alpha^k)^{(1)}+(\Omega_\alpha^k)^{(2)}] \tag{5.104}$$

式中:$(\Omega_\alpha^k)^{(1)}$ 为单相碰撞算子;$(\Omega_\alpha^k)^{(2)}$ 为扰动算子;$(\Omega_\alpha^k)^{(3)}$ 表征用于描述相分离的重新标色算法。采用 BGK 碰撞算子时,$(\Omega_\alpha^k)^{(1)}$ 可表示为

$$(\Omega_\alpha^k)^{(1)}=-\frac{1}{\tau}[f_\alpha^k(\boldsymbol{r},t)-f_\alpha^{k,\mathrm{eq}}(\boldsymbol{r},t)] \tag{5.105}$$

式中:τ 为量纲为 1 的松弛因子;$f_\alpha^{k,\mathrm{eq}}$ 为平衡态分布函数。相应的密度和速度计算如下:

$$\rho_R=\sum_{\alpha=0}^{m-1}f_\alpha^R,\quad \rho_B=\sum_{\alpha=0}^{m-1}f_\alpha^B,\quad \rho=\sum_{k=R}^{B}\rho_k,\quad \rho\boldsymbol{u}=\sum_{k=R}^{B}\sum_{\alpha=0}^{m-1}\boldsymbol{e}_\alpha f_\alpha^k \tag{5.106}$$

式中:ρ_R 和 ρ_B 分别为红色流体和蓝色流体的密度;ρ 为混合流体的密度;$\boldsymbol{u}$ 为混合流体的速度。

早期的颜色梯度模型只能适用于密度比为 1 的情形,首个密度比可调的颜色梯度模型是由 Grunau 等[98]基于 D2Q7 格子构造的。考虑到 D2Q9 格子更常用,这里仅讨论基于 D2Q9 格子的平衡态密度分布函数[97]

$$f_\alpha^{k,\mathrm{eq}}=\rho_k\left\{\phi_\alpha^k+\omega_\alpha\left[\frac{3}{c^2}(\boldsymbol{e}_\alpha\cdot\boldsymbol{u})+\frac{9}{2c^4}(\boldsymbol{e}_\alpha\cdot\boldsymbol{u})^2-\frac{3}{2c^2}|\boldsymbol{u}|^2\right]\right\} \tag{5.107}$$

式中,ϕ_α^k 为

$$\phi_\alpha^k=\begin{cases}\lambda_k, & \alpha=0\\(1-\lambda_k)/5, & \alpha=1,2,3,4\\(1-\lambda_k)/20, & \alpha=5,6,7,8\end{cases}\tag{5.108}$$

式中,参数 $0<\lambda_k<1$ 与状态方程有关。根据式(5.107)的二阶矩可得相应的状态方程为[97]

$$p_k=\frac{3}{5}\rho_k(1-\lambda_k)\tag{5.109}$$

由于相平衡时两相压力相等,因此密度比可定义为 $\rho_{R,0}/\rho_{B,0}=(1-\lambda_B)/(1-\lambda_R)$,于是密度比可通过 λ_B 和 λ_R 进行调节。例如,选取 $\lambda_B=0.2$ 和 $\lambda_R=0.92$ 可得密度比为 $\rho_{R,0}/\rho_{B,0}=10$[99]。可以看到,上述模型是通过改变平衡态密度分布函数的二阶矩来实现密度比可调,因此它与早期的自由能模型一样,会丧失伽利略不变性。关于这一点,将在 5.4.3 节进行讨论。

在早期的颜色梯度模型中,重新标色过程需要求解一个极大化问题[94,98],以驱使流体流向相同颜色的区域,从而达到相分离的目的[2]。但这一标色过程的算法比较复杂,且计算量较大,严重限制了模型的应用。为此,一些学者提出了公式化的简化算法。1995 年,d'Ortona 等[100]提出了如下形式的重新标色算法:

$$f_\alpha^{R,+}=\frac{\rho_R}{\rho}f_\alpha^*+\beta\frac{\rho_R\rho_B}{\rho^2}\cos(\varphi_\alpha),\quad f_\alpha^{B,+}=\frac{\rho_B}{\rho}f_\alpha^*-\beta\frac{\rho_R\rho_B}{\rho^2}\cos(\varphi_\alpha)\tag{5.110}$$

式中:β 为自由参数;φ_α 为颜色梯度 $\boldsymbol{C}$ 与离散速度方向 $\boldsymbol{e}_\alpha$ 之间的夹角,即 $\cos(\varphi_\alpha)=(\boldsymbol{e}_\alpha\cdot\boldsymbol{C})/(|\boldsymbol{C}||\boldsymbol{e}_\alpha|)$;$f_\alpha^*=\sum_{k=R}^{B}f_\alpha^{*,k}$ 是碰撞后总的分布函数,其中 $f_\alpha^{*,k}$ 为

$$f_\alpha^{*,k}(\boldsymbol{r},t)=f_\alpha^k(\boldsymbol{r},t)+(\Omega_\alpha^k)^{(1)}(\boldsymbol{r},t)+(\Omega_\alpha^k)^{(2)}(\boldsymbol{r},t)\tag{5.111}$$

颜色梯度 $\boldsymbol{C}$ 有多种定义方式[99],一种常见的定义为[97-98,101]

$$\boldsymbol{C}(\boldsymbol{r},t)=\nabla[\rho_R(\boldsymbol{r},t)-\rho_B(\boldsymbol{r},t)]\tag{5.112}$$

式中,红色流体和蓝色流体的密度由式(5.106)给出。根据式(5.110)计算 $f_\alpha^{R,+}$ 和 $f_\alpha^{B,+}$ 之后,即可执行格子 Boltzmann 方程的迁移步

$$f_\alpha^k(\boldsymbol{r}+\boldsymbol{e}_\alpha\delta_t,t+\delta_t)=f_\alpha^{k,+}(\boldsymbol{r},t)\tag{5.113}$$

上述简化算法避免了求解一个极大化问题,有效地提高了颜色梯度模型的计算效率。为了确保不会出现负的分布函数,Latva-Kokko 和 Rothman[101]随后对 d'Ortona 等[100]的算法进行了改进:

$$f_\alpha^{R,+}=\frac{\rho_R}{\rho}f_\alpha^*+\beta\frac{\rho_R\rho_B}{\rho^2}\cos(\varphi_\alpha)f_\alpha^{eq}(\rho,\boldsymbol{u}=0)\tag{5.114}$$

$$f_\alpha^{B,+}=\frac{\rho_B}{\rho}f_\alpha^*-\beta\frac{\rho_R\rho_B}{\rho^2}\cos(\varphi_\alpha)f_\alpha^{eq}(\rho,\boldsymbol{u}=0)\tag{5.115}$$

式中,f_α^{eq} 只与混合流体有关。当 f_α^{eq} 取标准的平衡态密度分布函数时,$f_\alpha^{eq}(\rho,\boldsymbol{u}=0)=\rho\omega_\alpha$。相应地,$\beta$ 的选取范围为 $0<\beta<1$。一般来说,β 越小,界面越宽,界面处的虚假速度越小[21]。随后,Halliday 等[102]提出用 $\rho\omega_\alpha|\boldsymbol{e}_\alpha|$ 替换上述算法中的 $f_\alpha^{eq}(\rho,\boldsymbol{u}=0)$,形成了如下形式的算法:

$$f_\alpha^{R,+}=\frac{\rho_R}{\rho}f_\alpha^*+\beta\frac{\rho_R\rho_B}{\rho}\omega_\alpha\cos(\varphi_\alpha)|\boldsymbol{e}_\alpha|\tag{5.116}$$

$$f_\alpha^{B,+}=\frac{\rho_B}{\rho}f_\alpha^*-\beta\frac{\rho_R\rho_B}{\rho}\omega_\alpha\cos(\varphi_\alpha)\,|\boldsymbol{e}_\alpha| \tag{5.117}$$

此外,针对密度比可调的颜色梯度模型,Leclaire 等[103]提出用 $\sum_{k=R}^{B}f_\alpha^{k,\mathrm{eq}}(\rho_k,\lambda_k,\boldsymbol{u}=0)$ 取代式(5.114)和式(5.115)中的 $f_\alpha^{\mathrm{eq}}(\rho,\boldsymbol{u}=0)$,其中 $f_\alpha^{k,\mathrm{eq}}$ 由式(5.107)给出。

重新标色算法汇总

可以看到,与前面两类模型一样,颜色梯度模型也有描述相分离的机制。不同的是,在颜色梯度模型中,相分离是通过重新标色算法实现的,而在自由能模型和伪势模型中,相分离是通过状态方程实现的。此外,前两类模型可引入热力学状态方程,因此它们不但适用于等温气液两相流,也适用于气液相变。

在颜色梯度模型中,式(5.105)中的松弛因子 τ 也有不同的形式,一种常见的形式为[98]

$$\tau=\begin{cases}\tau_R, & \rho^N>\delta\\ g^R(\rho^N), & \delta\geqslant\rho^N>0\\ g^B(\rho^N), & 0\geqslant\rho^N\geqslant-\delta\\ \tau_B, & \rho^N<-\delta\end{cases} \tag{5.118}$$

式中:τ_R 和 τ_B 分别为红色流体和蓝色流体的松弛因子;δ 为可调参数,通常取 0.98 或 0.99;函数 g^R 和 g^B 的表达式可参阅文献[97,98];颜色函数 ρ^N 定义如下[98]:

$$\rho^N=\frac{\rho_R(\boldsymbol{r},t)-\rho_B(\boldsymbol{r},t)}{\rho_R(\boldsymbol{r},t)+\rho_B(\boldsymbol{r},t)},\quad -1\leqslant\rho^N\leqslant1 \tag{5.119}$$

Grunau 等[98]提出式(5.118)的主要目的在于确保流体黏度在界面处光滑过渡。此外,Tölke 等[96]提出了一种线性插值形式的松弛因子

$$\tau=\frac{1}{2}(1+\rho^N)\tau_R+\frac{1}{2}(1-\rho^N)\tau_B \tag{5.120}$$

5.4.2 表面张力

颜色梯度模型中描述表面张力的扰动算子最早是由 Gunstensen 等[94]引入的,其形式如下:

$$(\Omega_\alpha^k)^{(2)}=A\,|\boldsymbol{C}|\cos(\varphi_\alpha) \tag{5.121}$$

式中:φ_α 为颜色梯度 $\boldsymbol{C}$ 与离散速度方向 $\boldsymbol{e}_\alpha$ 之间的夹角;参数 A 控制表面张力的大小。随后,Grunau 等[98]基于 D2Q7 格子提出了如下形式的扰动算子:

$$(\Omega_\alpha^k)^{(2)}=\frac{A_k}{2}|\boldsymbol{C}|\left[\frac{(\boldsymbol{e}_\alpha\cdot\boldsymbol{C})^2}{|\boldsymbol{C}|^2}-\frac{1}{2}\right] \tag{5.122}$$

2007 年,Reis 和 Phillips[97]发现上述扰动算子不能给出正确的宏观动量方程。为此,他们基于 D2Q9 格子对式(5.122)进行了改进:

$$(\Omega_\alpha^k)^{(2)}=\frac{A_k}{2}|\boldsymbol{C}|\left[\omega_\alpha\frac{(\boldsymbol{e}_\alpha\cdot\boldsymbol{C})^2}{|\boldsymbol{C}|^2}-d_\alpha\right] \tag{5.123}$$

式中,系数 d_α 满足下列关系式:

$$\sum_{\alpha=0}^{8}d_\alpha=\frac{1}{3},\quad \sum_{\alpha=0}^{8}\boldsymbol{e}_\alpha d_\alpha=0,\quad \sum_{\alpha=0}^{8}\boldsymbol{e}_\alpha\boldsymbol{e}_\alpha d_\alpha=\frac{1}{3}\boldsymbol{I} \tag{5.124}$$

对 D2Q9 格子,d_α 可选取为 $d_0=-4/27$、$d_{1-4}=2/27$、$d_{5-8}=5/108$[97]。通过 Chapman-Enskog

多尺度展开,可以导出扰动算子在宏观动量方程中所恢复的项为

$$\nabla\cdot\mathcal{G}=\nabla\cdot\left[-\tau\sum_{k=R}^{B}\sum_{\alpha=0}^{8}\boldsymbol{e}_{\alpha}\boldsymbol{e}_{\alpha}\left(\Omega_{\alpha}^{k}\right)^{(2)}\right] \tag{5.125}$$

结合格子张量运算(参见附录 B)可得

$$\begin{aligned}\mathcal{G}&=-\tau\sum_{k=R}^{B}\sum_{\alpha=0}^{8}\boldsymbol{e}_{\alpha}\boldsymbol{e}_{\alpha}\left(\Omega_{\alpha}^{k}\right)^{(2)}\\&=-\frac{\tau(A_R+A_B)}{2}|\boldsymbol{C}|\left[\frac{1}{|\boldsymbol{C}|^2}\sum_{\alpha=0}^{8}\omega_{\alpha}(\boldsymbol{e}_{\alpha}\cdot\boldsymbol{C})^2\boldsymbol{e}_{\alpha}\boldsymbol{e}_{\alpha}-\sum_{\alpha=0}^{8}d_{\alpha}\boldsymbol{e}_{\alpha}\boldsymbol{e}_{\alpha}\right]\\&=-\frac{\tau(A_R+A_B)}{2}|\boldsymbol{C}|\left[\frac{1}{9|\boldsymbol{C}|^2}(|\boldsymbol{C}|^2\boldsymbol{I}+2\boldsymbol{C}\boldsymbol{C})-\frac{1}{3}\boldsymbol{I}\right]\end{aligned} \tag{5.126}$$

因此,二阶张量 $\mathcal{G}$ 可写成

$$\mathcal{G}=\frac{\tau(A_{\mathrm{R}}+A_{\mathrm{B}})}{9|\boldsymbol{C}|}(|\boldsymbol{C}|^2\boldsymbol{I}-\boldsymbol{C}\boldsymbol{C}) \tag{5.127}$$

随后,Liu 等[104]将 Reis 和 Phillips 所提出的扰动算子拓展到了三维,他们采用的颜色梯度为 $\boldsymbol{C}=\nabla\rho^N$,其中 ρ^N 由式(5.119)给出。为了建立表面张力与参数 A_k 之间的关系式,Liu 等[104]将动量方程中与表面张力相关的项表示为

$$\boldsymbol{F}_{\mathrm{s}}=\nabla\cdot\left[\frac{\sigma_{\mathrm{s}}}{2|\boldsymbol{C}|}(|\boldsymbol{C}|^2\boldsymbol{I}-\boldsymbol{C}\boldsymbol{C})\right] \tag{5.128}$$

式中,σ_{s} 为表面张力。结合 $\boldsymbol{F}_{\mathrm{s}}=\nabla\cdot\mathcal{G}$ 与式(5.128)和式(5.127),可得 $\sigma_{\mathrm{s}}=2(A_R+A_B)\tau/9$[104]。需要注意的是,这里的 σ_{s} 与松弛因子 τ 有关。当红色流体和蓝色流体的松弛因子不同时,根据式(5.118)或式(5.120),τ 在界面处是变化的。为了解除 σ_{s} 对 τ 的依赖,可在式(5.123)右端乘以 $1/\tau$[105]。

除了采用扰动算子来描述表面张力之外,也有学者提出通过作用力项将表面张力引入颜色梯度模型。2003 年,Lishchuk 等[106]从界面曲率出发开展了这方面的研究,

$$K=-\nabla_{\mathrm{s}}\cdot\boldsymbol{n},\quad \nabla_{\mathrm{s}}=(\boldsymbol{I}-\boldsymbol{n}\boldsymbol{n})\cdot\nabla \tag{5.129}$$

式中:K 为界面曲率;∇_{s}为表面梯度算子;$\boldsymbol{n}=-\nabla\rho^N/|\nabla\rho^N|$。对二维问题,由式(5.129)可得[106]

$$K=n_x n_y(\partial_x n_y+\partial_y n_x)-n_y^2\partial_x n_x-n_x^2\partial_y n_y \tag{5.130}$$

式中,n_x 和 n_y 为 $\boldsymbol{n}$ 在 x 方向和 y 方向的分量。相应的界面力可定义为[102]

$$\boldsymbol{F}=-\frac{1}{2}\sigma_{\mathrm{s}}K\nabla\rho^N \tag{5.131}$$

在此基础上,可以通过作用力格式将式(5.131)引入格子 Boltzmann 方程。上述描述颜色梯度模型表面张力的方式在文献中也得到了比较广泛的应用[99,107-110]。

实际上,从式(5.125)可以看到,扰动算子是通过它的二阶矩来描述表面张力,而 Lishchuk 等的方式则是通过作用力项的一阶矩将表面张力纳入颜色梯度模型。也就是说,采用扰动算子本质上等价于自由能模型修改平衡态密度分布函数的二阶矩,而 Lishchuk 等的方式则与采用作用力项纳入压力张量的自由能模型是一致的。需要注意的是,Halliday 等[102]认为,作用力项的引入会导致颜色梯度模型不满足运动学条件(kinematic condition)$\mathrm{D}\rho^N/\mathrm{D}t=0$。对于不改变宏观速度定义的作用力格式,Halliday 等[102]导出了如下关系式

$$\frac{\mathrm{D}\rho^N}{\mathrm{D}t} = -\frac{1}{2\rho}\boldsymbol{F}\cdot\nabla\rho^N \tag{5.132}$$

在界面处，由于 $\boldsymbol{F}$ 和 $\nabla\rho^N$ 不为零，导致 $\mathrm{D}\rho^N/\mathrm{D}t \neq 0$。针对这一问题，Hollis 等[111]提出了一些改进的措施。

5.4.3 伽利略不变性

在 5.4.1 节我们曾提到，通过修改平衡态分布函数的二阶矩来实现密度比可调将导致颜色梯度模型丧失伽利略不变性。以 D2Q9 模型为例，其平衡态分布函数的二阶矩为

$$\sum_{\alpha=0}^{8} \boldsymbol{e}_\alpha \boldsymbol{e}_\alpha f_\alpha^{\mathrm{eq}} = \rho c_s^2 \boldsymbol{I} + \rho\boldsymbol{uu} \tag{5.133}$$

修改平衡态分布函数的二阶矩通常是指将式(5.133)修改为

$$\sum_{\alpha=0}^{8} \boldsymbol{e}_\alpha \boldsymbol{e}_\alpha f_\alpha^{\mathrm{eq}} = \mathcal{P} + \rho\boldsymbol{uu} \tag{5.134}$$

式中，$\mathcal{P}$为二阶张量。对 Swift 自由能模型，$\mathcal{P}$即为压力张量，而对密度比可调的颜色梯度模型，$\mathcal{P}=p_k\boldsymbol{I}$，其中状态方程 p_k 由式(5.109)给出。通过 Chapman-Enskog 多尺度展开可以导出式(5.134)所造成的宏观方程的偏差项，这里我们给出一个通用的结果：

$$\frac{\partial(\rho\boldsymbol{u})}{\partial t} + \nabla\cdot(\rho\boldsymbol{uu}) = -\nabla\cdot(\mathcal{D}+\rho c_s^2\boldsymbol{I}) + \nabla\cdot\boldsymbol{\Pi} - \nabla\cdot\{(\tau-0.5)[\boldsymbol{u}\nabla\cdot\mathcal{D} + (\boldsymbol{u}\nabla\cdot\mathcal{D})^{\mathrm{T}} + \partial_\rho\mathcal{D}\nabla\cdot(\rho\boldsymbol{u})]\} \tag{5.135}$$

式中：$\boldsymbol{\Pi}$ 为标准格子 Boltzmann 模型所产生的黏性应力张量；$\mathcal{D}=\mathcal{P}-\rho c_s^2\boldsymbol{I}$；$\partial_\rho\mathcal{D}=\partial\mathcal{D}/\partial\rho$。式(5.135)右端第二行即为修改平衡态分布函数二阶矩所导致的偏差项。当 $\mathcal{D}=0$ 时，式(5.135)退化为标准格子 Boltzmann 模型所对应的宏观动量方程。

对于密度比可调的颜色梯度模型，根据 $\mathcal{D}^k=p_k\boldsymbol{I}-\rho_k c_s^2\boldsymbol{I}$ 可得

$$\mathcal{D}^k=\rho_k[(c_s^k)^2-c_s^2]\boldsymbol{I},\quad \partial_{\rho_k}\mathcal{D}^k=[(c_s^k)^2-c_s^2]\boldsymbol{I} \tag{5.136}$$

式中，$(c_s^k)^2$ 可由状态方程(5.109)得到。结合式(5.135)和式(5.136)可得

$$\boldsymbol{\varepsilon}^k=-\nabla\cdot\{(\tau-0.5)[(c_s^k)^2-c_s^2][\boldsymbol{u}\nabla\rho_k+(\boldsymbol{u}\nabla\rho_k)^{\mathrm{T}}+\nabla\cdot(\rho_k\boldsymbol{u})\boldsymbol{I}]\} \tag{5.137}$$

式(5.137)的推导用到了关系式$\nabla\cdot(a\boldsymbol{I})=\nabla a$。式(5.137)即为密度比可调的颜色梯度模型所对应的误差项，它与 2013 年 Huang 等[112]的分析是一致的。这些误差项的存在会带来数值误差，并导致模型丧失伽利略不变性。

实际上，产生上述误差项的根源在于平衡态分布函数的二阶矩和三阶矩是相互关联的，只改变二阶矩而不对三阶矩做相应的调整导致了上述误差项的出现。对于密度比可调的颜色梯度模型，其平衡态分布函数的二阶矩和三阶矩分别为

$$\sum_{\alpha=0}^{8} e_{\alpha i}e_{\alpha j}f_\alpha^{k,\mathrm{eq}} = \rho_k(c_s^k)^2\delta_{ij} + \rho_k u_i u_j \tag{5.138}$$

$$\sum_{\alpha=0}^{8} e_{\alpha i}e_{\alpha j}e_{\alpha l}f_\alpha^{k,\mathrm{eq}} = \rho_k c_s^2(u_i\delta_{jl} + u_j\delta_{il} + u_l\delta_{ij}) \tag{5.139}$$

式中，$c_s^2=1/3$。为了消除式(5.137)中的误差项，平衡态分布函数的三阶矩应调整为

$$\sum_{\alpha=0}^{8} e_{\alpha i}e_{\alpha j}e_{\alpha l}f_\alpha^{k,\mathrm{eq}} = \rho_k(c_s^k)^2(u_i\delta_{jl} + u_j\delta_{il} + u_l\delta_{ij}) \tag{5.140}$$

实现这一调整除改动平衡态分布函数之外，还应采用对称性更好的格子(零至六阶格子张

量各向同性)[113],而 D2Q9 格子因为对称性不够至多只能满足如下关系式[114]:

$$\sum_{\alpha=0}^{8} e_{\alpha i}e_{\alpha j}e_{\alpha l}f_{\alpha}^{k,\mathrm{eq}}=\begin{cases}\rho_k c_s^2(u_i\delta_{jl}+u_j\delta_{il}+u_l\delta_{ij}), & i=j=l\\ \rho_k(c_s^k)^2(u_i\delta_{jl}+u_j\delta_{il}+u_l\delta_{ij}), & \text{otherwise}\end{cases} \tag{5.141}$$

式(5.141)可以消除式(5.137)中与 $\boldsymbol{u}\nabla\rho_k$ 和 $(\boldsymbol{u}\nabla\rho_k)^{\mathrm{T}}$ 有关的误差项,但与$\nabla\cdot(\rho_k\boldsymbol{u})\boldsymbol{I}$ 有关的误差项仍需在格子 Boltzmann 方程中引入修正项予以消除[114]。

上述思路是 Li 等[114]为构造低马赫数可压缩格子 Boltzmann 模型所提出的。Ba 等[99]和 Wen 等[105]将这一思路分别拓展到了二维和三维颜色梯度模型。这里以二维多松弛颜色梯度模型为例,介绍相应的修正项,三维模型可参阅文献[105]。引入修正项时,多松弛碰撞算子的碰撞步可写成[99,105]

$$\boldsymbol{m}^{*,k}=\boldsymbol{m}^k-\boldsymbol{\Lambda}(\boldsymbol{m}^k-\boldsymbol{m}^{k,\mathrm{eq}})+\delta_t\left(\boldsymbol{I}-\frac{\boldsymbol{\Lambda}}{2}\right)\boldsymbol{N}^k \tag{5.142}$$

式中:$\boldsymbol{\Lambda}$ 与松弛因子有关;$\boldsymbol{N}^k$ 为修正项,具体形式为

$$\boldsymbol{N}^k=(0,3N_1^k,0,0,0,0,0,N_7^k,0)^{\mathrm{T}} \tag{5.143}$$

式中,N_1^k 和 N_7^k 分别为

$$N_1^k=[1-3\,(c_s^k)^2][\partial_x(\rho_k u_x)+\partial_y(\rho_k u_y)] \tag{5.144}$$

$$N_7^k=[1-3\,(c_s^k)^2][\partial_x(\rho_k u_x)-\partial_y(\rho_k u_y)] \tag{5.145}$$

梯度项$\partial_x(\rho_k u_x)$和$\partial_y(\rho_k u_y)$的离散采用与颜色梯度一样的离散格式。相应地,流体运动黏度的表达式应调整为 $\nu_k=(c_s^k)^2(\tau_k-0.5)\delta_t$。Ba 等[99]和 Wen 等[105]通过模拟通道中的分层两相流、瑞利-泰勒非稳定性、液滴撞击等问题展示了原有颜色梯度模型在数值模拟中所造成的误差(部分模拟结果如图 5.9 所示),并证明了改进模型的有效性。

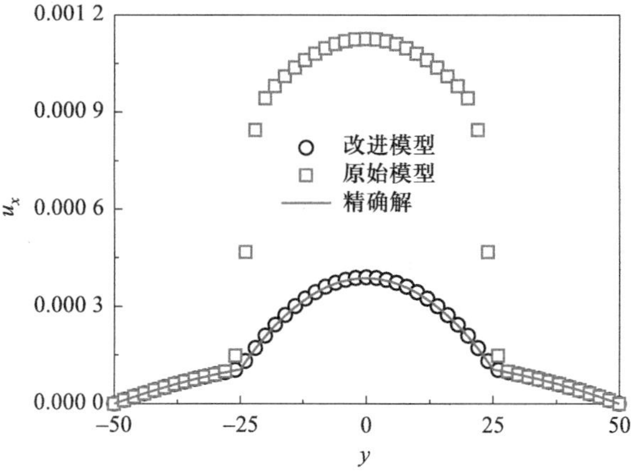

图 5.9 基于原始和改进颜色梯度模型的分层两相流模拟[105]

5.4.4 接触角格式

颜色梯度模型中常用的一种接触角格式是由 Latva-Kokko 和 Rothman[115]于 2005 年提出的。该格式在文献中也被称为虚拟密度或虚密度(fictitious-density)格式,它与伪势模型的虚拟密度格式一样,具有简单易实施的优点,因而得到了比较广泛的应用[21,116-117],但其缺点是会在固体表面形成一层非物理的密度层[110,118]。

为此,Leclaire 等[118]提出了一种新的思路来实施颜色梯度模型的接触角,他们通过预

估-修正(prediction-correction)的方式对流体边界节点处颜色梯度的方向进行先预估后修正,从而使其与给定的接触角相匹配。首先,对任意一个与流体节点相邻的固体节点,可通过计算其相邻流体节点的平均密度得到一个虚拟的固体节点密度。在此基础上,对流体边界节点可根据相应的离散格式计算出颜色梯度 $\boldsymbol{C}^*$ 及颜色梯度的方向[118]

$$\boldsymbol{n}^* = \frac{\boldsymbol{C}^*}{|\boldsymbol{C}^*|} \tag{5.146}$$

式(5.146)即为预估的颜色梯度方向。在修正步,需要找到与接触角相匹配的颜色梯度方向 $\boldsymbol{n}_c$。对二维问题,当接触角给定时,存在两个可能的气液界面法向矢量 $\boldsymbol{n}_1$ 和 $\boldsymbol{n}_2$,如图5.10所示,图中 $\boldsymbol{n}_s$ 为三相交汇处固液界面的法向矢量。Leclaire 等[118] 提出可通过如下方式确定 $\boldsymbol{n}_c$:

$$\boldsymbol{n}_c = \begin{cases} \boldsymbol{n}_1, & D_1 < D_2 \\ \boldsymbol{n}_2, & \text{otherwise} \end{cases} \tag{5.147}$$

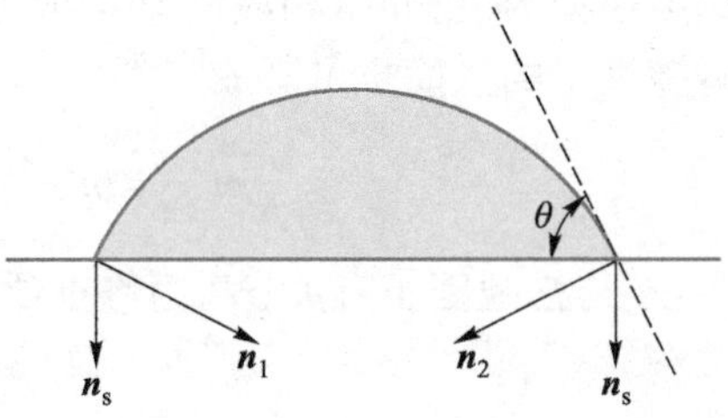

图 5.10 三相交汇处气液界面法向矢量示意图

式中:$D_1 = |\boldsymbol{n}_1 - \boldsymbol{n}^*|$;$D_2 = |\boldsymbol{n}_2 - \boldsymbol{n}^*|$。通过式(5.147)确定 $\boldsymbol{n}_c$ 之后,可对颜色梯度进行修正,即

$$\boldsymbol{C} = \boldsymbol{n}_c |\boldsymbol{C}^*| \tag{5.148}$$

在 Leclaire 等[118]的研究中,气液界面的法向矢量 $\boldsymbol{n}_1$ 和 $\boldsymbol{n}_2$ 是通过线性最小二乘法插值得到的,他们发现这一处理在数值模拟中会造成非对称效应,即液滴的质心位置与预期位置相比会偏离1~2个格子。为了消除非对称效应,Xu 等[109]采用矢量旋转来计算 $\boldsymbol{n}_1$ 和 $\boldsymbol{n}_2$:

$$\begin{cases} \boldsymbol{n}_1 = (n_{s,x}\cos\theta - n_{s,y}\sin\theta, n_{s,x}\sin\theta + n_{s,y}\cos\theta) \\ \boldsymbol{n}_1 = (n_{s,x}\cos\theta + n_{s,y}\sin\theta, -n_{s,x}\sin\theta + n_{s,y}\cos\theta) \end{cases} \tag{5.149}$$

式中,$n_{s,x}$和 $n_{s,y}$为 $\boldsymbol{n}_s$ 分别在 x 方向和 y 方向的分量。

对于三维问题,三相接触线所对应的气液界面法线方向理论上有无穷多个。因此,式(5.147)和式(5.149)将不再适用。为了解决这一问题,2017年 Leclaire 等[119]提出了一种迭代的方法来寻找与接触角相匹配的颜色梯度方向 $\boldsymbol{n}_c$,即

$$\boldsymbol{n}_c^{(0)} = \boldsymbol{n}^* \tag{5.150}$$

$$\boldsymbol{n}_c^{(1)} = \boldsymbol{n}^* - 0.5(\boldsymbol{n}^* + \boldsymbol{n}_s) \tag{5.151}$$

$$\boldsymbol{n}_c^{(m)} = \frac{\boldsymbol{n}_c^{(m-2)} g(\boldsymbol{n}_c^{(m-1)}) - \boldsymbol{n}_c^{(m-1)} g(\boldsymbol{n}_c^{(m-2)})}{g(\boldsymbol{n}_c^{(m-1)}) - g(\boldsymbol{n}_c^{(m-2)})} \tag{5.152}$$

式中,

$$g(\boldsymbol{n}_c) = \boldsymbol{n}_c \cdot \boldsymbol{n}_s - |\boldsymbol{n}_c| \cos\theta \tag{5.153}$$

为了节省计算时间,Leclaire 等[119]设定 $m=2$,即只计算到 $\boldsymbol{n}_c^{(2)}$。针对 $\boldsymbol{n}_s$ 的计算他们提出了一种平滑算子(smoothing operator)。此外,Akai 等[110]提出了一种 $\boldsymbol{n}^*$ 和 $\boldsymbol{n}_s$ 的线性组合格式来实施三维颜色梯度模型的接触角,具体形式可参阅文献[110]。

上述接触角格式与5.3.4节提到的几何接触角格式有许多相似之处。例如,它们都可以预先给定接触角的大小,且都不会产生非物理的密度层。此外,二维曲线表面的几何

接触角格式也需要进行矢量旋转。这两个格式的主要区别在于,几何接触角格式是根据给定的接触角构造出相应的固体节点的虚拟密度,而颜色梯度模型的预估-修正格式只需要修正边界附近流体节点的颜色梯度方向。需要注意的是,在预估-修正的过程中,颜色梯度的模是保持不变的,即 $|\boldsymbol{C}|=|\boldsymbol{C}^*|$。这一处理是否会对格式的数值表现造成影响(特别是接触角偏离90°较远时)有待进一步的研究。

5.5 相场格子 Boltzmann 模型

相场格子 Boltzmann 模型是通过结合格子 Boltzmann 方法和相场方法而建立的。与 VOF、水平集法等方法类似,相场方法也属于界面捕捉方法,它的主要特征是相界面是由一个变化迅速但连续的相场变量所描述[120],这个变量被称为**序参数**(order parameter)。1958 年,Cahn 和 Hilliard[16-17]基于热力学并考虑非均匀性体系的扩散界面,构造了相场理论中著名的 Cahn-Hilliard 方程。相场方法早期主要应用于凝固动力学、断裂动力学等学科,在 20 世纪后期[121-122]被引入流体力学学科。近年来,该方法已被广泛应用于复杂流体流动、固态相变以及固液相变、图像处理等多个领域。

5.5.1 相场理论

与自由能模型类似,相场模型也是从自由能泛函出发[15-16],即

$$\mathcal{F}=\int_{\Omega_V}\left[E_\phi(\phi)+\frac{\kappa}{2}|\nabla\phi|^2\right]\mathrm{d}\Omega_V \tag{5.154}$$

式中:Ω_V 为系统占据的空间;$E_\phi(\phi)$ 为体相区的自由能密度;$\kappa|\nabla\phi|^2/2$ 为梯度自由能密度,其中 κ 与表面张力的大小有关。根据式(5.154),相应的化学势为

$$\mu_\phi=\frac{\delta\mathcal{F}}{\delta\phi}=E'_\phi(\phi)-\kappa\nabla^2\phi \tag{5.155}$$

式中,$E'_\phi(\phi)=\mathrm{d}E_\phi(\phi)/\mathrm{d}\phi$。在相场模型中,$E_\phi(\phi)$ 通常选取为双阱形式[12],即

$$E_\phi(\phi)=\beta(\phi-\phi_\mathrm{L})^2(\phi-\phi_\mathrm{G})^2 \tag{5.156}$$

式中:ϕ_L 和 ϕ_G 分别为液相和气相所对应的序参数;参数 β 与式(5.155)中的参数 κ 一起控制表面张力的大小及相界面的宽度。将式(5.156)代入式(5.155)可得

$$\mu_\phi=4\beta(\phi-\phi_\mathrm{L})(\phi-\phi_\mathrm{G})(\phi-\phi_\mathrm{m})-\kappa\nabla^2\phi \tag{5.157}$$

式中,$\phi_\mathrm{m}=(\phi_\mathrm{L}+\phi_\mathrm{G})/2$。van der Waals[123]认为,系统平衡时界面的分布应使得自由能泛函具有最小的极值。由式(5.155)可知,此时 $\mu_\phi=0$。因此 $\mu_\phi=0$ 可视为系统平衡时序参数的控制方程。在此基础上,Cahn 和 Hilliard[16-17]假定一般状态下界面处的扩散通量与化学势梯度 $\nabla\mu_\phi$ 成正比,即[124]

$$\frac{\partial\phi}{\partial t}=\nabla\cdot(M_\phi\nabla\mu_\phi) \tag{5.158}$$

式中,M_ϕ 为迁移系数(mobility)。相应地,包含对流项的 Cahn-Hilliard 方程可写为[125]

$$\frac{\partial\phi}{\partial t}+\boldsymbol{u}\cdot\nabla\phi=\nabla\cdot(M_\phi\nabla\mu_\phi) \tag{5.159}$$

对于不可压缩流动,连续性方程和动量方程分别为

$$\nabla\cdot\boldsymbol{u}=0 \tag{5.160}$$

$$\rho\left(\frac{\partial \boldsymbol{u}}{\partial t}+\boldsymbol{u}\cdot\nabla\boldsymbol{u}\right)=-\nabla p+\nabla\cdot\boldsymbol{\Pi}+\boldsymbol{F}_s \tag{5.161}$$

式中，$\boldsymbol{F}_s$ 为表面张力项，常用的表达形式有 $\boldsymbol{F}_s=\mu_\phi\nabla\phi$[125]、$\boldsymbol{F}_s=-\phi\nabla\mu_\phi$[122]等。

式(5.159)~式(5.161)在文献中通常被称为耦合的 Navier-Stokes-Cahn-Hilliard 方程组，即相场理论中不可压缩气液两相流的控制方程组，其中密度由序参数确定，即

$$\rho=\rho_G+\frac{\phi-\phi_G}{\phi_L-\phi_G}(\rho_L-\rho_G) \tag{5.162}$$

式中，ρ_L 和 ρ_G 分别为液相和气相的密度。根据式(5.162)和式(5.159)并结合 $\nabla\cdot\boldsymbol{u}=0$ 可得[126]

$$\frac{\partial\rho}{\partial t}+\nabla\cdot(\rho\boldsymbol{u})=\frac{\partial\rho}{\partial t}+\boldsymbol{u}\cdot\nabla\rho=\frac{\mathrm{d}\rho}{\mathrm{d}\phi}\nabla\cdot(M_\phi\nabla\mu_\phi) \tag{5.163}$$

式(5.163)表明，在满足不可压缩条件($\nabla\cdot\boldsymbol{u}=0$)的前提下，相场模型在界面处不满足质量守恒。结合式(5.163)和高斯散度定理可得系统总体质量的变化为

$$\begin{aligned}\frac{\mathrm{D}}{\mathrm{D}t}\int_{\Omega_V}\rho\,\mathrm{d}\Omega_V&=\int_{\Omega_V}\left(\frac{\mathrm{D}\rho}{\mathrm{D}t}+\rho\nabla\cdot\boldsymbol{u}\right)\mathrm{d}\Omega_V\\&=\frac{\mathrm{d}\rho}{\mathrm{d}\phi}\int_{\Omega_V}[\nabla\cdot(M_\phi\nabla\mu_\phi)]\,\mathrm{d}\Omega_V\\&=\frac{\mathrm{d}\rho}{\mathrm{d}\phi}\int_A(M_\phi\nabla\mu_\phi\cdot\boldsymbol{n})\,\mathrm{d}A\end{aligned} \tag{5.164}$$

式中：A 为空间 Ω_V 的表面；$\boldsymbol{n}$ 为其法向矢量。因此，当边界条件设定为 $\nabla\mu_\phi\cdot\boldsymbol{n}=0$ 时，系统的总体质量是守恒的[126]。如果界面处满足质量守恒[$\partial_t\rho+\nabla\cdot(\rho\boldsymbol{u})=0$]，则速度散度无法满足不可压缩条件，相应的模型通常被称为准不可压缩(quasi-incompressible)相场模型[127]。除了 Cahn-Hilliard 方程，Allen-Cahn 方程在相场格子 Boltzmann 模型中也得到了广泛的应用[128-131]。Chai 等[131]基于相场格子 Boltzmann 模型研究了局部和非局部 Allen-Cahn 方程在质量守恒、数值稳定性等方面的表现，相关结果可参阅文献[131]。

与颜色梯度模型一样，相场模型中的表面张力也可以通过相关式子直接给定。对平直气液界面，系统平衡时序参数的界面分布可通过求解 $\mu_\phi=0$ 得到。以 $\phi_L=\phi^*$ 和 $\phi_G=-\phi^*$ 为例，有 $\mu_\phi=4\beta\phi(\phi^2-\phi^{*2})-\kappa\nabla^2\phi$，于是可得[125]

$$\phi(x)=\phi^*\tanh\left(\frac{x}{\sqrt{2}w}\right) \tag{5.165}$$

式中：x 为界面法线方向；$w=(1/2\phi^*)\sqrt{\kappa/\beta}$。结合式(5.155)与 $\mu_\phi=0$ 可得

$$\frac{\mathrm{d}E_\phi(\phi)}{\mathrm{d}\phi}=\kappa\frac{\mathrm{d}^2\phi}{\mathrm{d}x^2}=\frac{\kappa}{2}\frac{\mathrm{d}}{\mathrm{d}\phi}\left(\frac{\mathrm{d}\phi}{\mathrm{d}x}\right)^2 \tag{5.166}$$

当 $\phi=\pm\phi^*$ 时，$E_\phi(\phi)=0$ 且 $\mathrm{d}\phi/\mathrm{d}x=0$。因此，对式(5.166)两边积分可得

$$E_\phi(\phi)=\frac{\kappa}{2}\left(\frac{\mathrm{d}\phi}{\mathrm{d}x}\right)^2 \tag{5.167}$$

当系统处于平衡状态时，表面张力应等于自由能密度函数沿整个界面的积分[125]。因此，根据式(5.167)和式(5.154)，表面张力表达式为

$$\sigma_s=\int_{-\infty}^{+\infty}\left[E_\phi(\phi)+\frac{\kappa}{2}\left(\frac{\mathrm{d}\phi}{\mathrm{d}x}\right)^2\right]\mathrm{d}x=\int_{-\infty}^{+\infty}\kappa\left(\frac{\mathrm{d}\phi}{\mathrm{d}x}\right)^2\mathrm{d}x \tag{5.168}$$

将式(5.165)代入式(5.168)可得[48]

$$\sigma_s = \int_{-\infty}^{+\infty} \kappa \left(\frac{\mathrm{d}\phi}{\mathrm{d}x}\right)^2 \mathrm{d}x = \kappa \phi^{*2} \frac{4}{3} \frac{1}{\sqrt{2}w} = \frac{4\phi^{*3}}{3}\sqrt{2\kappa\beta} \tag{5.169}$$

式(5.169)表明表面张力的大小可通过参数 κ 和 β 进行调节。此外,根据式(5.165)可以定义相应的界面宽度。文献中通常有两种定义方式[48]。一种定义方式为 $W=2\sqrt{2}w$[132],对应 $\phi=-(\tanh 1)\phi^*$ 到 $\phi=(\tanh 1)\phi^*$;另一种定义方式为 $W=2\sqrt{2}w\tanh^{-1}(0.9)\approx 4.164w$[125],对应 $\phi=-0.9\phi^*$ 到 $\phi=0.9\phi^*$,如图 5.11 所示。在数值模拟中,往往是给定表面张力的大小和界面宽度来确定参数 κ 和 β 的值。

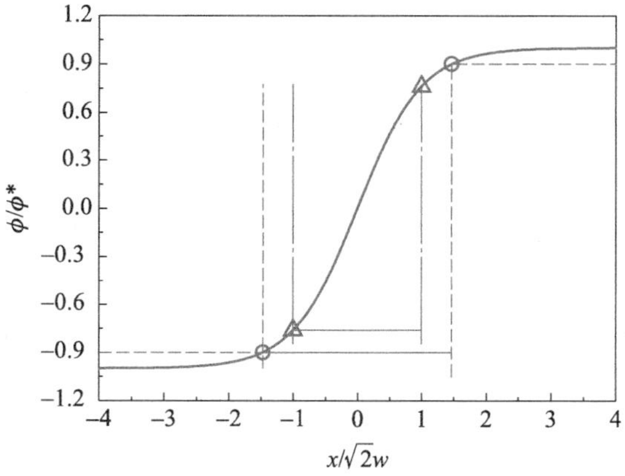

图 5.11 相场模型序参数界面分布示意图[48]

可以看到,相场模型的基础理论是独立于格子 Boltzmann 方法而存在的。因此,针对相场模型,格子 Boltzmann 方法主要是作为其控制方程组的一种求解器,这也使得相场格子 Boltzmann 模型与前面两类模型有比较明显的区别。从相分离和表面张力效应的描述来看,伪势模型和颜色梯度模型利用了格子 Boltzmann 方法的介观背景和粒子特性,体现了介观方法在多相/多组分流动模拟方面的特色。另一方面,从相界面的描述来看,本章所介绍的四类气液两相流格子 Boltzmann 模型实际上都可归入扩散界面法。在数值模拟中,这四类模型都需要维持一定的界面宽度(通常是 4~5 个格子)。太窄的界面将导致界面附近的虚假速度激增,并影响模型的数值稳定性,而太宽的界面则需要配套更多的计算网格,导致计算量增加。

5.5.2 早期模型

1999 年,He 等[133]提出了一个用于不可压缩气液两相流的格子 Boltzmann 模型,他们引入一个指标函数 ϕ(index function)来捕捉相界面的位置。该指标函数所对应的界面捕捉方程与 Cahn-Hilliard 方程相似,He 等的模型也因此被视为第一个相场格子 Boltzmann 模型。具体地,他们提出了如下形式的演化方程[133]:

$$f_\alpha(\boldsymbol{r}+\boldsymbol{e}_\alpha\delta_t, t+\delta_t)-f_\alpha(\boldsymbol{r},t) = -\frac{1}{\tau}(f_\alpha - f_\alpha^{\mathrm{eq}}) - \delta_t\left(1-\frac{1}{2\tau}\right)\frac{\varpi_\alpha(\boldsymbol{u})}{c_s^2}(\boldsymbol{e}_\alpha-\boldsymbol{u})\cdot\nabla\varphi(\phi) \tag{5.170}$$

$$g_\alpha(\boldsymbol{r}+\boldsymbol{e}_\alpha\delta_t, t+\delta_t)-g_\alpha(\boldsymbol{r},t) = -\frac{1}{\tau}(g_\alpha - g_\alpha^{\mathrm{eq}}) + \delta_t\left(1-\frac{1}{2\tau}\right)(\boldsymbol{e}_\alpha-\boldsymbol{u})\cdot\boldsymbol{Q} \tag{5.171}$$

式中：f_α 为指标分布函数；g_α 为压力分布函数；f_α^{eq} 和 g_α^{eq} 为平衡态分布函数；$\varphi(\phi)=p_{EOS}(\phi)-\phi c_s^2$；$\boldsymbol{\mathcal{Q}}=\varpi_\alpha(\boldsymbol{u})\boldsymbol{F}_s-[\varpi_\alpha(\boldsymbol{u})-\varpi_\alpha(0)]\nabla(p-\rho c_s^2)$；$\boldsymbol{F}_s$ 为表面张力项；$\varpi_\alpha(\boldsymbol{u})$ 为

$$\varpi_\alpha(\boldsymbol{u})=\omega_\alpha\left[1+\frac{(\boldsymbol{e}_\alpha\cdot\boldsymbol{u})}{c_s^2}+\frac{(\boldsymbol{e}_\alpha\cdot\boldsymbol{u})^2}{2c_s^4}-\frac{|\boldsymbol{u}|^2}{2c_s^2}\right] \tag{5.172}$$

通过 Chapman-Enskog 多尺度展开，可以导出式(5.170)和式(5.171)所恢复的宏观方程为[48]

$$\frac{1}{\rho c_s^2}\left(\frac{\partial p}{\partial t}+\boldsymbol{u}\cdot\nabla p\right)+\nabla\cdot\boldsymbol{u}=0 \tag{5.173}$$

$$\frac{\partial(\rho\boldsymbol{u})}{\partial t}+\nabla\cdot(\rho\boldsymbol{u}\boldsymbol{u})=-\nabla p+\nabla\cdot\boldsymbol{\Pi}+\boldsymbol{F}_s \tag{5.174}$$

$$\frac{\partial\phi}{\partial t}+\nabla\cdot(\phi\boldsymbol{u})=\nabla\cdot\left[\lambda_\phi\left(\nabla p_{EOS}(\phi)-\frac{\phi}{\rho}\nabla p\right)\right] \tag{5.175}$$

式中：λ_ϕ 与松弛因子有关；密度 ρ 由指标函数 ϕ 确定。式(5.175)是指标函数的控制方程，作为界面捕捉方程，它与 Cahn-Hilliard 方程的作用相同。基于该模型，He 等成功地模拟了不同尺度相界面的产生、破碎、卷吸等界面动力学行为。

随后，Inamuro 等[134]提出了一个适用于大密度比不可压缩气液两相流的格子 Boltzmann 模型。类似地，他们采用两个分布函数，一个是序参数分布函数，用于捕捉界面；另一个是速度分布函数，用于恢复不含压力梯度的动量方程。该模型每一步都需要求解压力的泊松方程，从而对速度场进行修正，其恢复的宏观方程为[134]

$$\frac{\partial\phi}{\partial t}+\boldsymbol{u}\cdot\nabla\phi=M_\phi\nabla\cdot(\nabla\cdot\boldsymbol{P}) \tag{5.176}$$

$$\nabla\cdot\boldsymbol{u}=0 \tag{5.177}$$

$$\frac{\partial\boldsymbol{u}}{\partial t}+\boldsymbol{u}\cdot\nabla\boldsymbol{u}=-\frac{1}{\rho}\nabla p+\frac{1}{\rho}\nabla\cdot\boldsymbol{\Pi}+\nabla\cdot\left[\frac{k_g}{\rho}(\nabla\rho\nabla\rho-|\nabla\rho|^2\boldsymbol{I})\right] \tag{5.178}$$

式中：$\boldsymbol{P}$ 为热力学压力张量；k_g 与表面张力的大小有关。Inamuro 等[134]通过数值模拟证明了该模型适用于大密度比气液两相流，但由于每一步都需要求解泊松方程，导致它丧失了格子 Boltzmann 方法的基本优点。

为了增强不可压缩气液两相流模型的数值稳定性，Lee 和 Lin[12]提出了一个三段式的求解格式用以求解相关的演化方程。他们采用密度作为序参数，相应的密度分布函数和压力分布函数分别满足如下形式的离散 Boltzmann 方程[12]

$$\frac{\partial f_\alpha}{\partial t}+\boldsymbol{e}_\alpha\cdot\nabla f_\alpha=-\frac{1}{\tau}(f_\alpha-f_\alpha^{eq})+\frac{\varpi_\alpha(\boldsymbol{u})}{c_s^2}(\boldsymbol{e}_\alpha-\boldsymbol{u})\cdot[\nabla(\rho c_s^2)-\rho\nabla(\mu_\phi-\kappa\nabla^2\rho)] \tag{5.179}$$

$$\frac{\partial g_\alpha}{\partial t}+\boldsymbol{e}_\alpha\cdot\nabla g_\alpha=-\frac{1}{\tau}(g_\alpha-g_\alpha^{eq})+\frac{1}{c_s^2}(\boldsymbol{e}_\alpha-\boldsymbol{u})\cdot\boldsymbol{\mathcal{Q}} \tag{5.180}$$

式中：μ_ϕ 为化学势；$\boldsymbol{\mathcal{Q}}=\varpi_\alpha(\boldsymbol{u})\boldsymbol{F}_s+[\varpi_\alpha(\boldsymbol{u})-\varpi_\alpha(0)]\nabla(\rho c_s^2)$。该模型恢复的宏观方程为

$$\frac{\partial\rho}{\partial t}+\nabla\cdot(\rho\boldsymbol{u})=\nabla\cdot[(\mu/c_s^2)\nabla\mu_\phi] \tag{5.181}$$

$$\frac{1}{\rho c_s^2}\frac{\partial p}{\partial t}+\nabla\cdot\boldsymbol{u}=0 \tag{5.182}$$

$$\frac{\partial(\rho\boldsymbol{u})}{\partial t}+\nabla\cdot(\rho\boldsymbol{u}\boldsymbol{u})=-\nabla p+\nabla\cdot\boldsymbol{\Pi}+\boldsymbol{F}_{\mathrm{s}} \tag{5.183}$$

式中:μ 为动力黏度;μ/c_{s}^2 称为界面捕捉方程的迁移系数。式(5.181)表明,该模型在界面处不满足质量守恒。为了增强模型的数值稳定性,Lee 和 Lin[12] 针对式(5.179)和式(5.180)提出了一个三段式的求解格式。具体地,他们将求解过程分为迁移前的碰撞步、迁移步和迁移后的碰撞步,并且在迁移前和迁移后,采用不同的梯度离散格式,即在迁移前采用偏差分格式或混合差分格式,而在迁移后采用中心差分格式

$$\delta_t\boldsymbol{e}_\alpha\cdot\nabla^{\mathrm{BD}}\varphi\big|_{(\boldsymbol{r},t)}=\frac{1}{2}[-\varphi(\boldsymbol{r}+2\boldsymbol{e}_\alpha\delta_t)+4\varphi(\boldsymbol{r}+\boldsymbol{e}_\alpha\delta_t)-3\varphi(\boldsymbol{r})] \tag{5.184}$$

$$\delta_t\boldsymbol{e}_\alpha\cdot\nabla^{\mathrm{CD}}\varphi\big|_{(\boldsymbol{r},t)}=\frac{1}{2}[\varphi(\boldsymbol{r}+\boldsymbol{e}_\alpha\delta_t)-\varphi(\boldsymbol{r}-\boldsymbol{e}_\alpha\delta_t)] \tag{5.185}$$

式中,上标“BD”和“CD”分别表示偏差分和中心差分。混合差分格式是中心差分和偏差分的组合格式:$\delta_t\boldsymbol{e}_\alpha\cdot\nabla^{\mathrm{MD}}\varphi=(\delta_t\boldsymbol{e}_\alpha\cdot\nabla^{\mathrm{CD}}\varphi+\delta_t\boldsymbol{e}_\alpha\cdot\nabla^{\mathrm{BD}}\varphi)/2$。基于这一模型,Lee 和 Lin[12] 在格子 Boltzmann 方法领域率先实现了对较高雷诺数的大密度比气液两相流的数值模拟。该模型的主要缺点在于三段式的求解格式使得计算量较大。此外,混合差分格式破坏了模型的伽利略不变性并且导致系统的总体质量不守恒[135]。

5.5.3 大密度比模型

由于早期的模型无法恢复正确的 Cahn-Hilliard 方程,Zheng 等[132] 提出对序参数分布函数的演化方程进行修正:

$$f_\alpha(\boldsymbol{r}+\boldsymbol{e}_\alpha\delta_t,t+\delta_t)-f_\alpha(\boldsymbol{r},t)=-\frac{1}{\tau_\phi}(f_\alpha-f_\alpha^{\mathrm{eq}})+(1-q)[f_\alpha(\boldsymbol{r}+\boldsymbol{e}_\alpha\delta_t,t)-f_\alpha(\boldsymbol{r},t)] \tag{5.186}$$

式中,$q=1/(\tau_\phi+0.5)$。式(5.186)右端与 q 有关的项用于消除界面捕捉方程中所产生的偏差项。此外,Zheng 等[132] 引入了一个粒子分布函数来描述平均密度 $n=(n_{\mathrm{A}}+n_{\mathrm{B}})/2$,其中 n_{A} 和 n_{B} 分别是 A 流体和 B 流体的密度,对应的演化方程为

$$g_\alpha(\boldsymbol{r}+\boldsymbol{e}_\alpha\delta_t,t+\delta_t)-g_\alpha(\boldsymbol{r},t)=-\frac{1}{\tau_n}(g_\alpha-g_\alpha^{\mathrm{eq}})+\delta_t F_\alpha \tag{5.187}$$

式中,F_α 为作用力项。该模型恢复的宏观方程为

$$\frac{\partial\phi}{\partial t}+\nabla\cdot(\phi\boldsymbol{u})=\nabla\cdot(M_\phi\nabla\mu_\phi) \tag{5.188}$$

$$\frac{\partial n}{\partial t}+\nabla\cdot(n\boldsymbol{u})=0 \tag{5.189}$$

$$\frac{\partial(n\boldsymbol{u})}{\partial t}+\nabla\cdot(n\boldsymbol{u}\boldsymbol{u})=-\nabla(p+\phi\mu_\phi)+\nabla\cdot\boldsymbol{\Pi}+\boldsymbol{F}_{\mathrm{s}} \tag{5.190}$$

式中,$\boldsymbol{F}_{\mathrm{s}}=\mu_\phi\nabla\phi$ 为表面张力项。当$\nabla\cdot\boldsymbol{u}=0$ 时,式(5.188)与 Cahn-Hilliard 方程是一致的。但是该模型采用的是平均密度 n,且 n 并不由序参数 ϕ 决定,这一点与式(5.162)是不同的,从而导致该模型实际上是一个密度匹配的二元流模型,而不是真正意义上的大密度比气液两相流模型。

2010 年,Fakhari 和 Rahimian[136] 指出了 Zheng 等[132] 模型所存在的上述问题。他们采用该模型研究了平均密度相等的两个算例:$n_{\mathrm{A}}=1\ 000$ 和 $n_{\mathrm{B}}=1$ 及 $n_{\mathrm{A}}=501$ 和 $n_{\mathrm{B}}=500$。数

值模拟显示这两个算例的结果完全相同。为此,Fakhari 和 Rahimian[136]提出了一个改进模型,对静态气泡可以适用于大密度比,但对动态气液两相流所能适用的密度比仍然较小。同年,Lee 和 Liu[137]在 Lee 和 Lin[12]的基础上提出了一个修正模型,他们用序参数分布函数取代了原来的密度分布函数,并采用如下形式的演化方程:

$$\frac{\partial h_\alpha}{\partial t}+\boldsymbol{e}_\alpha\cdot\nabla h_\alpha=-\frac{1}{\tau}(h_\alpha-h_\alpha^{\mathrm{eq}})+M_\phi\varpi_\alpha(\boldsymbol{u})\nabla^2\mu_\phi+$$
$$(\boldsymbol{e}_\alpha-\boldsymbol{u})\cdot\left[\nabla\phi-\frac{\phi}{\rho c_{\mathrm{s}}^2}(\nabla p+\phi\nabla\mu_\phi)\right]\varpi_\alpha(\boldsymbol{u}) \tag{5.191}$$

式中:h_α 为序参数分布函数;$\varpi_\alpha(\boldsymbol{u})$ 由式(5.172)给出。通过 Chapman-Enskog 多尺度展开,Lee 和 Liu[137]证明当$\nabla\cdot\boldsymbol{u}=0$ 时式(5.191)所恢复的界面捕捉方程与 Cahn-Hilliard 方程是一致的。为了增强模型的数值稳定性,他们沿用了 Lee 和 Lin[12]所提出的三段式求解格式。

2015 年,Wang 等[138-139]基于通量求解器(flux solver)提出了一个大密度比相场格子 Boltzmann 模型,该模型的控制方程为

$$\frac{\partial p}{\partial t}+\nabla\cdot\left(\sum_{\alpha=0}^{m-1}\boldsymbol{e}_\alpha f_\alpha^{\mathrm{eq}}\right)+\boldsymbol{u}\cdot\nabla(p-\rho c_{\mathrm{s}}^2)=0 \tag{5.192}$$

$$\frac{\partial(\rho c_{\mathrm{s}}^2\boldsymbol{u})}{\partial t}+\nabla\cdot(\boldsymbol{\Pi}-\mu\boldsymbol{\Pi}^{\mathrm{e}})-\boldsymbol{F}_{\mathrm{s}}c_{\mathrm{s}}^2=0 \tag{5.193}$$

式中:m 为离散速度的个数;$\boldsymbol{\Pi}$ 为动量通量张量,它通过f_α^{eq}和f_α^{neq}的二阶矩定义,且有$f_\alpha^{\mathrm{neq}}=-\tau\delta_t(\partial_t+\boldsymbol{e}_\alpha\cdot\nabla)f_\alpha^{\mathrm{eq}}$;$\boldsymbol{\Pi}^{\mathrm{e}}$ 则定义为[138]

$$\boldsymbol{\Pi}^{\mathrm{e}}=\frac{1}{\rho}[\boldsymbol{u}\nabla\rho+(\boldsymbol{u}\nabla\rho)^{\mathrm{T}}](p-\rho c_{\mathrm{s}}^2) \tag{5.194}$$

基于该模型,Wang 等[139]实现了大密度比液滴撞击液膜的数值模拟,相应的密度比为 1 000,雷诺数为 2 000,部分模拟结果如图 5.12 所示。为了增强模型的数值稳定性,他们采用有限差分格式求解 Cahn-Hilliard 方程。赵宁和王东红[140]认为,Cahn-Hilliard 方程的求解过程不涉及泊松方程,因此可以采用传统的离散方法直接求解,且有利于发挥高阶离散格式在数值稳定性上的优势。

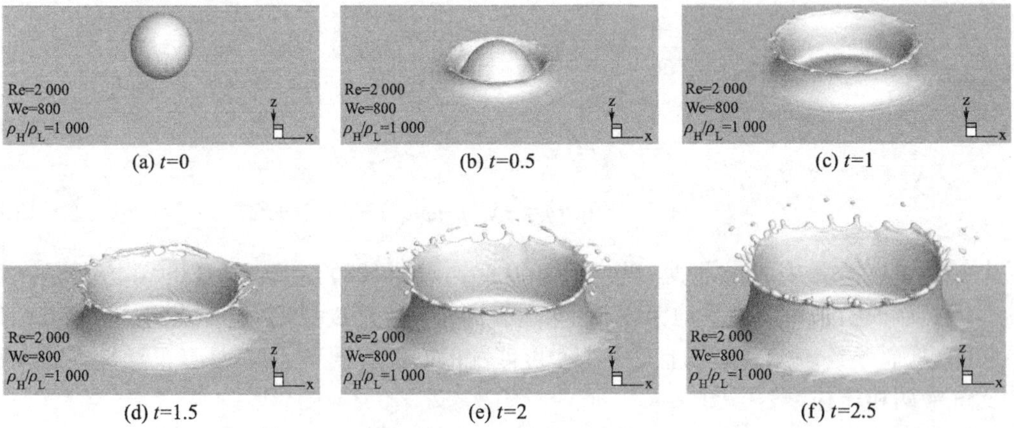

图 5.12 基于通量求解器相场模型的液滴撞击液膜的数值模拟(t 为量纲为 1 的时间[139])

2017 年，Fakhari 和 Bolster[141] 提出了一个基于 Allen-Cahn 方程的大密度比相场格子 Boltzmann 模型，其中序参数分布函数的演化方程为

$$h_\alpha(\boldsymbol{r}+\boldsymbol{e}_\alpha\delta_t,t+\delta_t)-h_\alpha(\boldsymbol{r},t)=-\frac{1}{\tau_\phi+0.5}(h_\alpha-h_\alpha^{\mathrm{eq}}) \tag{5.195}$$

式中，平衡态分布函数 h_α^{eq} 定义为

$$h_\alpha^{\mathrm{eq}}=\phi\,\varpi_\alpha(\boldsymbol{u})+\omega_\alpha\frac{M_\phi}{c_s^2}\left[\frac{4}{W}\phi(1-\phi)\right](\boldsymbol{e}_\alpha\cdot\boldsymbol{n}) \tag{5.196}$$

其中，W 为界面宽度，$\boldsymbol{n}=\nabla\phi/|\nabla\phi|$ 为界面法向矢量。随后，Liang 等[142] 指出式(5.195)所恢复的界面捕捉方程存在偏差项。为此，他们提出了一个改进模型，相应的序参数分布函数的演化方程为

$$f_\alpha(\boldsymbol{r}+\boldsymbol{e}_\alpha\delta_t,t+\delta_t)-f_\alpha(\boldsymbol{r},t)=-\frac{1}{\tau}(f_\alpha-f_\alpha^{\mathrm{eq}})+\delta_t\left(1-\frac{1}{2\tau}\right)\frac{\omega_\alpha\boldsymbol{e}_\alpha}{c_s^2}\cdot[\partial_t(\phi\boldsymbol{u})+c_s^2\lambda\boldsymbol{n}] \tag{5.197}$$

式中：$f_\alpha^{\mathrm{eq}}=\omega_\alpha\phi[1+(\boldsymbol{e}_\alpha\cdot\boldsymbol{u}/c_s^2)]$；$\lambda=4\phi(1-\phi)/W$。对式(5.197)进行 Chapman-Enskog 多尺度展开可以导出如下形式的 Allen-Cahn 方程：

$$\frac{\partial\phi}{\partial t}+\nabla\cdot(\phi\boldsymbol{u})=\nabla\cdot[M_\phi(\nabla\phi-\lambda\boldsymbol{n})] \tag{5.198}$$

基于该模型，Liang 等[142] 模拟了大密度比气液两相流。

需要注意的是，大部分相场格子 Boltzmann 模型所恢复的动量方程是守恒形式的，而 Navier-Stokes 方程，即式(5.161)，是非守恒形式的[143]，且存在如下关系式：

$$\frac{\partial(\rho\boldsymbol{u})}{\partial t}+\nabla\cdot(\rho\boldsymbol{u}\boldsymbol{u})=\rho\left(\frac{\partial\boldsymbol{u}}{\partial t}+\boldsymbol{u}\cdot\nabla\boldsymbol{u}\right)+\boldsymbol{u}\left[\frac{\partial\rho}{\partial t}+\nabla\cdot(\rho\boldsymbol{u})\right] \tag{5.199}$$

当且仅当 $\partial_t\rho+\nabla\cdot(\rho\boldsymbol{u})=0$ 时，$\partial_t(\rho\boldsymbol{u})+\nabla\cdot(\rho\boldsymbol{u}\boldsymbol{u})$ 与 $\rho(\partial_t\boldsymbol{u}+\boldsymbol{u}\cdot\nabla\boldsymbol{u})$ 才是相等的，而相场模型在界面处往往不满足质量守恒，这将导致动量方程存在偏差项，即动量不守恒。2012 年，Li 等[126] 指出了相场格子 Boltzmann 模型所存在的这一问题，并研究了式(5.199)右端第二项所引起的附加界面力。他们发现，当速度或雷诺数逐渐增大时，附加界面力在数值模拟中所造成的误差会越来越大[126]。为了确保相场模型的数值模拟精度，这一附加界面力应当予以消除。

5.5.4 接触角格式

相场模型的接触角格式主要可分为表面自由能格式和几何接触角格式。实际上，通过研究不同模型的接触角格式可以发现，部分接触角格式对几类模型都是适用的，其中最典型的代表就是几何接触角格式。这一格式最早是由 Ding 和 Spelt[85] 提出的，其基本思路是依据接触角的大小和流体边界节点的信息构造出固体节点的虚拟序参数，相关介绍参见 5.3.4 节。这里主要介绍一下相场模型的表面自由能格式，该格式是基于 Cahn[34] 的假设建立的，即流固之间的相互作用在系统原有自由能密度的基础上增加一项表面自由能密度

$$\mathcal{F}_\phi=\int_{\Omega_V}\left[E_\phi(\phi)+\frac{\kappa}{2}|\nabla\phi|^2\right]\mathrm{d}\Omega_V+\int_S\mathcal{H}(\phi_w)\mathrm{d}S \tag{5.200}$$

式中：$\mathcal{H}(\phi_w)$ 为表面自由能密度；ϕ_w 为表面序参数。根据 $\mathcal{H}(\phi_w)$ 的具体形式，表面自由能格式可分为线性格式、二次方格式、三次方格式等[144-146]。线性格式与 5.2.4 节所介绍

的自由能模型的接触角格式类似,即表面自由能密度选取为$\mathcal{H}(\phi_w)=-\omega\phi_w$,其中 ω 为常数。此时,式(5.56)仍然适用。不同的是,式中的 Ω 应为[147]

$$\Omega=\frac{4\omega}{(\phi_L-\phi_G)^2\sqrt{2\kappa\beta}} \tag{5.201}$$

类似地,对二次方格式和三次方格式,$\mathcal{H}(\phi_w)$分别选取为二次多项式和三次多项式。以三次方格式为例,$\mathcal{H}(\phi_w)$形式为[137]

$$\mathcal{H}(\phi_w)=\omega\left(\frac{1}{2}\phi_w^2-\frac{1}{3}\phi_w^3\right) \tag{5.202}$$

根据式(5.200)和式(5.202)可以导出各界面张力,将相应的结果代入杨氏方程可得[137]

$$\cos\theta=-\frac{\omega}{\sqrt{2\kappa\beta}} \tag{5.203}$$

根据接触角的大小可以确定 ω 的值,之后可通过如下关系式实施固体表面的湿润性边界条件:

$$\kappa\boldsymbol{n}_w\cdot\nabla\phi=\frac{\mathrm{d}\mathcal{H}(\phi_w)}{\mathrm{d}\phi_w}=\omega\phi_w(1-\phi_w) \tag{5.204}$$

式中,$\boldsymbol{n}_w$ 为指向流体内部的固体表面法向矢量。对于平直固体表面,假定 z 方向为固体表面的法线方向,边界处的一阶导数$(\partial\phi/\partial z)_{z=0}$可通过式(5.204)确定,相应的二阶导数可采用偏差分格式

$$\left.\frac{\partial^2\phi}{\partial z^2}\right|_{z=0}\approx\frac{1}{2}\left(-3\left.\frac{\partial\phi}{\partial z}\right|_{z=0}+4\left.\frac{\partial\phi}{\partial z}\right|_{z=1}-\left.\frac{\partial\phi}{\partial z}\right|_{z=2}\right) \tag{5.205}$$

此外,对化学势应当实施无通量(no flux)边界条件以确保系统的总体质量守恒,即$\nabla\mu_\phi\cdot\boldsymbol{n}=0$。

2017年,Fakhari 和 Bolster[141]将式(5.204)拓展到了曲线表面,他们提出了两种处理方案,分别为方案一和方案二,如图5.13所示。图中,点 S 是任意一个与流体边界节点相邻的固体节点,$\boldsymbol{n}_w$ 是该节点附近固体表面的法向矢量,它与固体表面相交于点 W。方案一选取的点 P 是法向矢量方向上的一个点,且点 P 到点 W 的距离与点 S 到点 W 的距离相等,即 $h=|\boldsymbol{r}_P-\boldsymbol{r}_W|=|\boldsymbol{r}_S-\boldsymbol{r}_W|$。结合式(5.204)和图5.13(a)可得

$$\boldsymbol{n}_w\cdot\nabla\phi=\frac{\partial\phi}{\partial\hat{n}_w}=\frac{\phi_P-\phi_S}{2h}=\frac{\omega}{\kappa}\phi_W(1-\phi_W) \tag{5.206}$$

在此基础上,Fakhari 和 Bolster 假定 $\phi_W=(\phi_P+\phi_S)/2$,于是有[141]

$$\phi_S=\frac{1}{a}\left[1+a-\sqrt{(1+a)^2-4a\phi_P}\right]-\phi_P,\quad a=\frac{\omega}{\kappa}h\neq0\quad(\theta\neq90°) \tag{5.207}$$

式(5.207)即为 Fakhari 和 Bolster 提出的方案一,式中 ϕ_P 的值通过插值确定。当 $a=0$,接触角 $\theta=90°$,此时 $\phi_S=\phi_P$。在确定固体节点的序参数之后,流体边界节点处的序参数梯度 $\nabla\phi$ 及二阶导数$\nabla^2\phi$ 可通过相应的离散格式得到。

方案一主要存在的问题在于通过插值格式确定 ϕ_P 的值时可能要用到 ϕ_S 的值,从而式(5.207)的求解需要用到迭代。为了解决这一问题,Fakhari 和 Bolster[141]提出可采用上一时层的 ϕ_S 更新 ϕ_P,或采用方案二,如图5.13b所示。其中点 P 是固体表面法向矢量与流体区域网格线的交点,根据具体情况的不同,它可能是法向矢量与 x 方向网格线的交

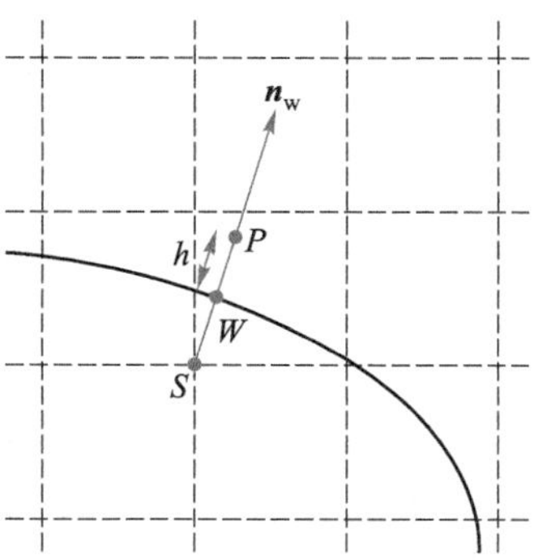

(a) 方案一

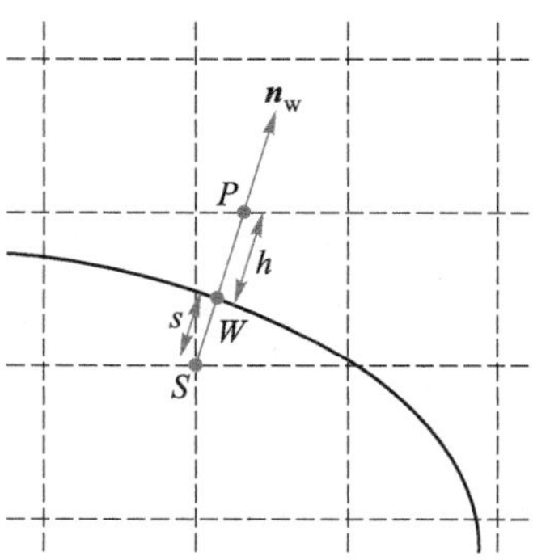

(b) 方案二

图 5.13 表面自由能格式的曲线边界处理

点，也可能是法向矢量与 y 方向网格线的交点。无论哪种情况，在通过插值确定 ϕ_P 的值时都不再需要点 S 的信息。对方案二，Fakhari 和 Bolster[141] 提出可通过如下公式来确定 ϕ_S 的值：

$$\phi_S=\frac{s+h}{2ah}\left[1+a-\sqrt{(1+a)^2-4a\phi_P}\right]-\frac{s}{h}\phi_P,\quad a=\frac{\omega}{\kappa}h\neq 0\quad(\theta\neq 90^\circ)\tag{5.208}$$

式中：$s=|\boldsymbol{r}_S-\boldsymbol{r}_W|$；$h=|\boldsymbol{r}_P-\boldsymbol{r}_W|$。在实际应用中，$\phi_P$ 的值可以根据相邻流体节点的序参数通过线性插值或二次多项式插值获得。类似地，曲线表面的几何接触角格式也需要用到相应的插值[88]。

5.6 小结

自 1991 年第一个两相流格子 Boltzmann 模型诞生以来，气液两相流格子 Boltzmann 方法已历经 30 多年的发展，形成了以自由能模型、伪势模型、颜色梯度模型和相场模型为主体的模型体系。本章对这四类气液两相流格子 Boltzmann 模型逐一进行了介绍，包括它们的基本原理、早期模型、改进模型、新的发展、接触角格式等。上述四类模型有其各自的优势与特色。自由能模型从自由能泛函出发，因而在理论上能够满足状态方程的一致性和热力学一致性；伪势模型“自底向上”通过一个伪势相互作用力直接对流体内部的相互作用进行描述，反映了多相流体动力学的物理本质；颜色梯度模型结合了格子 Boltzmann 方法的介观特性，通过公式化的重新标色算法实现相分离，模型的消融性较小；相场模型则基于相场理论，通过求解序参数的控制方程，对两相界面进行描述。在实际应用中，应当根据所研究问题的具体情况选取相对合适的模型。需要强调的是，这四类格子 Boltzmann 模型并不存在一类模型对另一类模型具有压倒性的优势。实际上，这一点也适用于格子 Boltzmann 方法与 VOF、水平集法等传统方法。作为一种介观方法，格子 Boltzmann 方法为探索多相/多组分流动的机制与规律提供了一种可选的途径，它的发展目标并不是要取代 VOF、水平集法等传统方法，而是针对这些方法难以胜任的领域发挥格子 Boltzmann 方法的优势和特色。

习 题

5-1 当χ取$(1-0.5\eta)/(1-\eta)^3$且$\eta=b\rho/4$时，式(5.15)即为 Carnahan-Starling 状态方程。试根据式(5.16)中状态方程与自由能密度的关系导出 Carnahan-Starling 状态方程所对应的自由能密度和化学势。

5-2 结合泰勒级数展开[参考式(5.65)]分析伪势相互作用力的数学本质。

5-3 基于伪势模型的状态方程，从式(5.73)出发导出$\varepsilon=1$和$\varepsilon=2$时力学稳定性条件的具体表达式。

5-4 式(5.107)为密度比可调的颜色梯度模型的平衡态密度分布函数，当密度比为1时，可否按照式(5.108)选取ϕ_α^k？若可行，请给出具体的参数。

5-5 相场模型序参数界面分布的解析表达式可通过求解$\mu_\phi=0$得到，其中式(5.165)是基于$\mu_\phi=4\beta\phi(\phi^2-\phi^{*2})-\kappa\nabla^2\phi$导出的，试结合式(5.157)导出其对应的界面分布解析表达式。

5-6 根据式(5.70)和式(5.71)数值求解伪势模型的气液共存密度(基于指数形式的伪势函数)，并结合附录E中伪势模型的相分离程序对G取不同值时的结果进行验证。

5-7 结合附录E中伪势模型的相关程序以及本章所学习的接触角格式，编制程序对平直表面的湿润性问题进行模拟(无滑移边界可采用半步长反弹格式，参考8.2.4节)。

参考文献

[1] 林宗虎．变幻流动的科学：多相流体力学[M]. 北京：清华大学出版社，2000.

[2] 郭照立，郑楚光．格子 Boltzmann 方法的原理及应用[M]. 北京：科学出版社，2009.

[3] Chapman S, Cowling T G. The Mathematical Theory of Non-uniform Gases[M]. 3rd edition. Cambridge: Cambridge University Press, 1970.

[4] Karkheck J, Stell G. Kinetic mean-field theories[J]. Journal of Chemical Physics, 1981, 75(3): 1475-1487.

[5] Frezzotti A, Gibelli L, Lorenzani S. Mean field kinetic theory description of evaporation of a fluid into vacuum[J]. Physics of Fluids, 2005, 17(1): 012102.

[6] He X Y, Doolen G D. Thermodynamic foundations of kinetic theory and lattice Boltzmann models for multiphase flows[J]. Journal of Statistical Physics, 2002, 107(1-2): 309-328.

[7] 应纯同．气体输运理论及应用[M]. 北京：清华大学出版社，1990.

[8] Rowlinson J S, Widom B. Molecular Theory of Capillarity[M]. Oxford: Clarendon Press, 1982.

[9] Yuan P, Schaefer L. Equations of state in a lattice Boltzmann model[J]. Physics of Fluids, 2006, 18: 042101.

[10] Swift M R, Osborn W R, Yeomans J M. Lattice Boltzmann simulation of nonideal fluids[J]. Physical Review Letters, 1995, 75(5): 830-833.

[11] Swift M R, Orlandini E, Osborn W R, et al. Lattice Boltzmann simulations of liquid-gas and binary fluid systems[J]. Physical Review E, 1996, 54(5): 5041-5052.

[12] Lee T, Lin C L. A stable discretization of the lattice Boltzmann equation for simulation of incompressible two-phase flows at high density ratio[J]. Journal of Computational Physics, 2005, 206(1): 16-47.

[13] Wen B H, Zhou X, He B, et al. Chemical-potential-based lattice Boltzmann method for nonideal fluids

[J]. Physical Review E,2017,95(6):063305.

[14] Nadiga B T,Zaleski S. Investigations of a two-phase fluid model[J]. European Journal of Mechanics-B/Fluids,1995,15:885-896.

[15] Penrose O,Fife P C. Thermodynamically consistent models of phase-field type for the kinetic of phase transitions[J]. Physica D:Nonlinear Phenomena,1990,43(1):44-62.

[16] Cahn J W,Hilliard J E. Free energy of a nonuniform system. I. Interfacial free energy[J]. The Journal of Chemical Physics,1958,28(2):258-267.

[17] Cahn J W, Hilliard J E. Free Energy of a nonuniform system. Ⅲ. Nucleation in a two-component incompressible fluid[J]. The Journal of Chemical Physics,1959,31(3):688-699.

[18] Evans,R. The nature of the liquid-vapour interface and other topics in the statistical mechanics of non-uniform,classical fluids[J]. Advances in Physics,1979,28(2):143-200.

[19] Siebert D N, Philippi P C, Mattila K K. Consistent lattice Boltzmann equations for phase transitions [J]. Physical Review E,2014,90(5):053310.

[20] Anderson D M,McFadden G B,Wheeler A A. Diffuse-interface methods in fluid mechanics[J]. Annual Review of Fluid Mechanics,1998,30(1):139-165.

[21] Huang H B, Sukop M C, Lu X Y. Multiphase Lattice Boltzmann Methods: Theory and Application [M]. John Wiley & Sons,2015.

[22] Inamuro T,Konishi N,Ogino F. A Galilean invariant model of the lattice Boltzmann method for multiphase fluid flows using free-energy approach[J]. Computer Physics Communications,2000,129(1):32-45.

[23] Kalarakis A N,Burganos V N,Payatakes A C. Galilean-invariant lattice-Boltzmann simulation of liquid-vapor interface dynamics[J]. Physical Review E,2002,65(5):056702.

[24] Wagner A J, Li Q. Investigation of Galilean invariance of multi-phase lattice Boltzmann methods [J]. Physica A,2006,362(1):105-110.

[25] Mazloomi M A, Chikatamarla S S, Karlin I V. Entropic lattice Boltzmann method for multiphase flows [J]. Physical Review Letters,2015,114(17):174502.

[26] Lee T,Fischer P F. Eliminating parasitic currents in the Lattice Boltzmann equation method for nonideal gases[J]. Physical Review E,2006,74(4):046709.

[27] Wen B H,Zhao L,Qiu W,et al. Chemical-potential multiphase lattice Boltzmann method with superlarge density ratios[J]. Physical Review E,2020,102(1):013303.

[28] Qiao Z H, Yang X G, Zhang Y Z. Thermodynamic-consistent multiple-relaxation-time lattice Boltzmann equation model for two-phase hydrocarbon fluids with Peng-Robinson equation of state[J]. International Journal of Heat and Mass Transfer,2019,141:1216-1226.

[29] Guo Z L, Zheng C G, Shi B C. Force imbalance in lattice Boltzmann equation for two-phase flows [J]. Physical Review E,2011,83(3):036707.

[30] Lou Q,Guo Z L. Interface-capturing lattice Boltzmann equation model for two-phase flows[J]. Physical Review E,2015,91(1):013302.

[31] Huang R Z,Wu H Y. Third-order analysis of pseudopotential lattice Boltzmann model for multiphase flow [J]. Journal of Computational Physics,2016,327:121-139.

[32] Li Q,Yu Y,Huang R Z. Achieving thermodynamic consistency in a class of free-energy multiphase lattice Boltzmann models[J]. Physical Review E,2021,103(1):013304.

[33] Briant A J,Wagner A J,Yeomans J M. Lattice Boltzmann simulations of contact line motion. I. Liquid-gas systems[J]. Physical Review E,2004,69(3):031602.

[34] Cahn J W. Critical point wetting[J]. The Journal of Chemical Physics,1977,66(8):3667-3672.

[35] Yu Y,Li Q,Huang R Z. Alternative wetting boundary condition for the chemical-potential-based free-

energy lattice Boltzmann model[J]. Physical Review E,2021,104(1):015303.

[36] Krüger T, Kusumaatmaja H, Kuzmin A, et al. The Lattice Boltzmann Method-Principles and Practice [M]. Springer Nature,2017.

[37] Shan X W, Chen H D. Lattice Boltzmann model for simulating flows with multiple phases and components [J]. Physical Review E,1993,47:1815-1820.

[38] Shan X W, Chen H D. Simulation of nonideal gases and liquid-gas phase transitions by the lattice Boltzmann equation[J]. Physical Review E,1994,49:2941-2948.

[39] Shan X W. Analysis and reduction of the spurious current in a class of multiphase lattice Boltzmann models[J]. Physical Review E,2006,73(4):047701.

[40] Shan X W, Doolen G. Multicomponent lattice-Boltzmann model with interparticle interaction[J]. Journal of Statistical Physics,1995,81(1-2):379-393.

[41] Huang H B, Krafczyk M, Lu X Y. Forcing term in single-phase and Shan-Chen-type multiphase lattice Boltzmann models[J]. Physical Review E,2011,84(4):046710.

[42] Yu Z, Fan L S. An interaction potential based lattice Boltzmann method with adaptive mesh refinement (AMR) for two-phase flow simulation [J]. Journal of Computational Physics, 2009, 228 (17): 6456-6478.

[43] Li Q, Zhou P, Yan H J. Revised Chapman-Enskog analysis for a class of forcing schemes in the lattice Boltzmann method[J]. Physical Review E,2016,94(4):043313.

[44] Guo Z L, Zheng C G, Shi B C. Discrete lattice effects on the forcing term in the lattice Boltzmann method [J]. Physical Review E,2002,65(4):046308.

[45] Shan X W. Pressure tensor calculation in a class of nonideal gas lattice Boltzmann models[J]. Physical Review E,2008,77(6):066702.

[46] Li Q, Luo K H, Li X J. Lattice Boltzmann modeling of multiphase flows at large density ratio with an improved pseudopotential model[J]. Physical Review E,2013,87(5):053301.

[47] Succi S. Lattice Boltzmann 2038[J]. Europhysics Letters,2015,109:50001-50007.

[48] Li Q, Luo K H, Kang Q J, et al. Lattice Boltzmann methods for multiphase flow and phase-change heat transfer[J]. Progress in Energy and Combustion Science,2016,52:62-105.

[49] Chen L, Kang Q J, Mu Y T, et al. A critical review of the pseudopotential multiphase lattice Boltzmann model: Methods and applications [J]. International Journal of Heat and Mass Transfer, 2014, 76: 210-236.

[50] Li Q, Luo K H, Li X J. Forcing scheme in pseudopotential lattice Boltzmann model for multiphase flows [J]. Physical Review E,2012,86(1):016709.

[51] Zhang D, Papadikis K, Gu S. Three-dimensional multi-relaxation time lattice-Boltzmann model for the drop impact on a dry surface at large density ratio[J]. International Journal of Multiphase Flow, 2014, 64: 11-18.

[52] Xu A, Zhao T S, An L, et al. A three-dimensional pseudo-potential-based lattice Boltzmann model for multiphase flows with large density ratio and variable surface tension[J]. International Journal of Heat and Fluid Flow,2015,56:261-271.

[53] Li Q, Du D H, Fei L L, et al. Three-dimensional non-orthogonal MRT pseudopotential lattice Boltzmann model for multiphase flows[J]. Computers & Fluids,2019,186:128-140.

[54] Lycett-Brown D, Luo K H. Cascaded lattice Boltzmann method with improved forcing scheme for large-density-ratio multiphase flow at high Reynolds and Weber numbers[J]. Physical Review E, 2016, 94 (5):053313.

[55] Hou Y Z, Deng H, Du Q, et al. Multi-component multi-phase lattice Boltzmann modeling of droplet

coalescence in flow channel of fuel cell[J]. Journal of Power Sources,2018,393:83-91.

[56] Deng H,Jiao K,Hou Y Z,et al. A lattice Boltzmann model for multi-component two-phase gas-liquid flow with realistic fluid properties[J]. International Journal of Heat and Mass Transfer,2019,128:536-549.

[57] Zheng S F,Eimann F,Philipp C,et al. Single droplet condensation in presence of non-condensable gas by a multi-component multi-phase thermal lattice Boltzmann model[J]. International Journal of Heat and Mass Transfer,2019,139:254-268.

[58] Li Q,Kang Q J,Francois M M,et al. Lattice Boltzmann modeling of boiling heat transfer:The boiling curve and the effects of wettability[J]. International Journal of Heat and Mass Transfer,2015,85:787-796.

[59] Fang W Z,Chen L,Kang Q J,et al. Lattice Boltzmann modeling of pool boiling with large liquid-gas density ratio[J]. International Journal of Thermal Sciences,2017,114:172-183.

[60] Li Q,Yu Y,Zhou P,et al. Enhancement of boiling heat transfer using hydrophilic-hydrophobic mixed surfaces:A lattice Boltzmann study[J]. Applied Thermal Engineering,2018,132:490-499.

[61] Feng Y,Li H X,Guo K K,et al. Numerical investigation on bubble dynamics during pool nucleate boiling in presence of a non-uniform electric field by LBM[J]. Applied Thermal Engineering,2019,155:637-649.

[62] Sun T Z,Gui N,Yang X T,et al. Pseudo-potential MRT-thermal LB simulation of flow boiling in vertical tubes[J]. Heat and Mass Transfer,2018,54(10):3035-3045.

[63] Shen L Y,Tang G H,Li Q,et al. Hybrid wettability-induced heat transfer enhancement for condensation with noncondensable gas[J]. Langmuir,2019,35(29):9430-9440.

[64] Zheng S F,Eimann F,Fieback T,et al. Numerical investigation of convective dropwise condensation flow by a hybrid thermal lattice Boltzmann method[J]. Applied Thermal Engineering,2018,145:590-602.

[65] Hou Y Z,Deng H,Zamel N,et al. 3D lattice Boltzmann modeling of droplet motion in PEM fuel cell channel with realistic GDL microstructure and fluid properties[J]. International Journal of Hydrogen Energy,2020,45(22):12476-12488.

[66] Chen H Y,Zhang J Y,Zhang Y X,et al. Simulation on a gravity-driven dripping of droplet into micro-channels using the lattice Boltzmann method[J]. International Journal of Heat and Mass Transfer,2018,126:61-71.

[67] Sayyari M J,Kharmiani S F,Esfahani J A. A lattice Boltzmann study on dripping process during 2D droplet impact onto a wetted rotating cylinder[J]. Journal of Molecular Liquids,2018,275:409-420.

[68] Zhang L,Ku T,Cheng X D,et al. Inkjet droplet deposition dynamics into square microcavities for OLEDs manufacturing[J]. Microfluidics and Nanofluidics,2018,22(4):47.

[69] Shan M L,Zhu Y P,Yao C,et al. Modeling for collapsing cavitation bubble near rough solid wall by Mulit-Relaxation-Time pseudopotential lattice Boltzmann model[J]. Journal of Applied Mathematics and Physics,2017,05(6):1243-1256.

[70] He X L,Zhang J M,Xu W L. Study of cavitation bubble collapse near a rigid boundary with a multi-relaxation-time pseudo-potential lattice Boltzmann method[J]. AIP Advances,2020,10(3):035315.

[71] Yang Y,Shan M L,Kan X F,et al. Thermodynamic of collapsing cavitation bubble investigated by pseudopotential and thermal MRT-LBM[J]. Ultrasonics Sonochemistry,2020,62:104873.

[72] Asadi M B,Zendehboudi S. Hybridized method of pseudopotential lattice Boltzmann and cubic-plus-association equation of state assesses thermodynamic characteristics of associating fluids[J]. Physical Review E,2019,100(4):043302.

[73] Sbragaglia M,Benzi R,Biferale L,et al. Generalized lattice Boltzmann method with multirange pseudopotential[J]. Physical Review E,2007,75(2):026702.

[74] Li Q,Luo K H. Achieving tunable surface tension in the pseudopotential lattice Boltzmann modeling of

multiphase flows[J]. Physical Review E,2013,88(5):053307.

[75] Ammar S, Pernaudat G, Trépanier J Y. A multiphase three-dimensional multi-relaxation time (MRT) lattice Boltzmann model with surface tension adjustment[J]. Journal of Computational Physics,2017,343: 73-91.

[76] Fei L L, Du J Y, Luo K H, et al. Modeling realistic multiphase flows using a non-orthogonal multiple-relaxation-time lattice Boltzmann method[J]. Physics of Fluids,2019,31(4):042105.

[77] Huang R Z,Wu H Y,Adams N A. Eliminating cubic terms in the pseudopotential lattice Boltzmann model for multiphase flow[J]. Physical Review E,2018,97(5):053308.

[78] Wu Y Y,Gui N,Yang X T,et al. Fourth-order analysis of force terms in multiphase pseudopotential lattice Boltzmann model[J]. Computers and Mathematics with Applications,2018,76(7):1699-1712.

[79] Martys N S,Chen H D. Simulation of multicomponent fluids in complex three-dimensional geometries by the lattice Boltzmann method[J]. Physical Review E,1996,53(1):743-750.

[80] Raiskinmäki P,Koponen A,Merikoski J,et al. Spreading dynamics of three-dimensional droplets by the lattice-Boltzmann method[J]. Computational Materials Science,2000,18(1):7-12.

[81] Raiskinmäki P,Shakib-Manesh A,Jäsberg A,et al. Lattice-Boltzmann simulation of capillary rise dynamics [J]. Journal of Statistical Physics,2002,107(1-2):143-158.

[82] Li Q,Luo K H,Kang Q J,et al. Contact angles in the pseudopotential lattice Boltzmann modeling of wetting[J]. Physical Review E,2014,90(5):053301.

[83] Benzi R,Biferale L,Sbragaglia M,et al. Mesoscopic modeling of a two-phase flow in the presence of boundaries:the contact angle[J]. Physical Review E,2006,74(2):021509.

[84] Li Q,Yu Y,Luo K H. Implementation of contact angles in pseudopotential lattice Boltzmann simulations with curved boundaries[J]. Physical Review E,2019,100(5):053313.

[85] Ding H,Spelt P D M. Wetting condition in diffuse interface simulations of contact line motion[J]. Physical Review E,2007,75(4):046708.

[86] Li Q,Zhou P,Yan H J. Pinning-depinning mechanism of the contact line during evaporation on chemically patterned surfaces:A lattice Boltzmann study[J]. Langmuir,2016,32:9389-9396.

[87] Yu Y,Li Q,Zhou C Q,et al. Investigation of droplet evaporation on heterogeneous surfaces using a three-dimensional thermal multiphase lattice Boltzmann model[J]. Applied Thermal Engineering,2017,127: 1346-1354.

[88] Liu H R,Ding H. A diffuse-interface immersed-boundary method for two-dimensional simulation of flows with moving contact lines on curved substrates[J]. Journal of Computational Physics,2015,294: 484-502.

[89] Bao J,Schaefer L. Lattice Boltzmann equation model for multi-component multi-phase flow with high density ratios[J]. Applied Mathematical Modelling,2013,37(4):1860-1871.

[90] Porter M L,Coon E T,Kang Q,et al. Multicomponent interparticle-potential lattice Boltzmann model for fluids with large viscosity ratios[J]. Physical Review E,2012,86(3):036701.

[91] He X Y,Shan X W,Doolen G D. Discrete Boltzmann equation model for nonideal gases[J]. Physical Review E,1998,57(1):R13-R16.

[92] Sbragaglia M,Belardinelli D. Interaction pressure tensor for a class of multicomponent lattice Boltzmann models[J]. Physical Review E,2013,88(1):013306.

[93] Chai Z H,Zhao T S. A pseudopotential-based multiple-relaxation-time lattice Boltzmann model for multicomponent/multiphase flows[J]. Acta Mechanica Sinica,2012,28(4):983-992.

[94] Gunstensen A K,Rothman D H,Zaleski S,et al. Lattice Boltzmann model of immiscible fluids[J]. Physical Review A,1991,43(8):4320-4327.

[95] Rothman D H, Keller J M. Immiscible cellular-automaton fluids[J]. Journal of Statistical Physics, 1988, 52: 1119-1127.

[96] Tölke J, Krafczyk M, Schulz M, et al. Lattice Boltzmann simulations of binary fluid flow through porous media[J]. Philosophical Transactions of the Royal Society A, 2002, 360(1792): 535-545.

[97] Reis T, Phillips T N. Lattice Boltzmann model for simulating immiscible two-phase flows[J]. Journal of Physics A: Mathematical and Theoretical, 2007, 40(14): 4033.

[98] Grunau D, Chen S Y, Eggert K. A lattice Boltzmann model for multiphase fluid flows[J]. Physics of Fluids A, 1993, 5(10): 2557-2562.

[99] Ba Y, Liu H H, Li Q, et al. Multiple-relaxation-time color-gradient lattice Boltzmann model for simulating two-phase flows with high density ratio[J]. Physical Review E, 2016, 94(2): 023310.

[100] d'Ortona U, Salin D, Cieplak M, et al. Two-color nonlinear Boltzmann cellular automata: Surface tension and wetting[J]. Physical Review E, 1995, 51(4): 3718-3728.

[101] Latva-Kokko M, Rothman D H. Diffusion properties of gradient-based lattice Boltzmann models of immiscible fluids[J]. Physical Review E, 2005, 71(5): 056702.

[102] Halliday I, Hollis A P, Care C M. Lattice Boltzmann algorithm for continuum multicomponent flow [J]. Physical Review E, 2007, 76(2): 026708.

[103] Leclaire S, Reggio M, Trépanier J Y. Numerical evaluation of two recoloring operators for an immiscible two-phase flow lattice Boltzmann model [J]. Applied Mathematical Modelling, 2012, 36(5): 2237-2252.

[104] Liu H H, Valocchi A J, Kang Q J. Three-dimensional lattice Boltzmann model for immiscible two-phase flow simulations[J]. Physical Review E, 2012, 85(4): 046309.

[105] Wen Z X, Li Q, Yu Y, et al. Improved three-dimensional color-gradient lattice Boltzmann model for immiscible multiphase flows[J]. Physical Review E, 2019, 100(2): 023301.

[106] Lishchuk S V, Care C M, Halliday I. Lattice Boltzmann algorithm for surface tension with greatly reduced microcurrents[J]. Physical Review E, 2003, 67(3): 036701.

[107] Gupta A, Kumar R. Effect of geometry on droplet formation in the squeezing regime in a microfluidic T-junction[J]. Microfluidics and Nanofluidics, 2010, 8(6): 799-812.

[108] Liu H H, Zhang Y H. Droplet formation in microfluidic cross-junctions[J]. Physics of Fluids, 2011, 23(8): 082101.

[109] Xu Z Y, Liu H H, Valocchi A J. Lattice Boltzmann simulation of immiscible two-phase flow with capillary valve effect in porous media[J]. Water Resources Research, 2017, 53(5): 3770-3790.

[110] Akai T, Bijeljic B, Blunt M J. Wetting boundary condition for the color-gradient Lattice Boltzmann Method: Validation with analytical and experimental data[J]. Advances in Water Resources, 2018, 116: 56-66.

[111] Hollis A P, Halliday I, Law R. Kinematic condition for multicomponent lattice Boltzmann simulation [J]. Physical Review E, 2007, 76(2): 026709.

[112] Huang H B, Huang J J, Lu X Y, et al. On simulations of high-density ratio flows using color-gradient multiphase lattice Boltzmann models [J]. International Journal of Modern Physics C, 2013, 24(4): 1350021.

[113] Li Q, He Y L, Wang Y, et al. Coupled double-distribution-function lattice Boltzmann method for the compressible Navier-Stokes equations[J]. Physical Review E, 2007, 76(5): 056705.

[114] Li Q, Luo K H, He Y L, et al. Coupling lattice Boltzmann model for simulation of thermal flows on standard lattices[J]. Physical Review E, 2012, 85(1): 016710.

[115] Latva-Kokko M, Rothman D H. Static contact angle in lattice Boltzmann models of immiscible fluids

[J]. Physical Review E,2005,72(4):046701.

[116] Liu H H,Valocchi A J,Werth C,et al. Pore-scale simulation of liquid CO_2 displacement of water using a two-phase lattice Boltzmann model[J]. Advances in Water Resources,2014,73:144-158.

[117] Huang H B,Huang J J,Lu X Y. Study of immiscible displacements in porous media using a color-gradient-based multiphase lattice Boltzmann method[J]. Computers & Fluids,2014,93:164-172.

[118] Leclaire S,Abahri K,Belarbi R,et al. Modeling of static contact angles with curved boundaries using a multiphase lattice Boltzmann method with variable density and viscosity ratios[J]. International Journal for Numerical Methods in Fluids,2016,82:451-470.

[119] Leclaire S,Parmigiani A,Malaspinas O,et al. Generalized three-dimensional lattice Boltzmann color-gradient method for immiscible two-phase pore-scale imbibition and drainage in porous media[J]. Physical Review E,2017,95(3):033306.

[120] 汤涛,乔中华. 相场方程的高效数值算法[J]. 中国科学:数学,2020,50:775-794.

[121] Antanovskii L K. A phase field model of capillarity[J]. Physics of Fluids,1995,7(4):747-753.

[122] Jacqmin D. Calculation of two-phase Navier-Stokes flows using phase-field modeling[J]. Journal of Computational Physics,1999,155(1):96-127.

[123] Rowlinson J S. Translation of J. D. van der Waals "The thermodynamik theory of capillarity under the hypothesis of a continuous variation of density"[J]. Journal of Statistical Physics, 1979, 20(2): 197-200.

[124] Novick-Cohen A,Segel L A. Nonlinear aspects of the Cahn-Hilliard equation[J]. Physica D:Nonlinear Phenomena,1984,10(3):277-298.

[125] Badalassi V E,Ceniceros H D,Banerjee S. Computation of multiphase systems with phase field models [J]. Journal of Computational Physics,2003,190(2):371-397.

[126] Li Q,Luo K H,Gao Y J,et al. Additional interfacial force in lattice Boltzmann models for incompressible multiphase flows[J]. Physical Review E,2012,85(2):026704.

[127] Yang K, Guo Z L. Lattice Boltzmann method for binary fluids based on mass-conserving quasi-incompressible phase-field theory[J]. Physical Review E,2016,93(4):043303.

[128] Wang H L,Chai Z H,Shi B C,et al. Comparative study of the lattice Boltzmann models for Allen-Cahn and Cahn-Hilliard equations[J]. Physical Review E,2016,94(3):033304.

[129] Ren F,Song B W,Sukop M C,et al. Improved lattice Boltzmann modeling of binary flow based on the conservative Allen-Cahn equation[J]. Physical Review E,2016,94(2):023311.

[130] Hu Y,Li D C,Jin L C,et al. Hybrid Allen-Cahn-based lattice Boltzmann model for incompressible two-phase flows:The reduction of numerical dispersion[J]. Physical Review E,2019,99(2):023302.

[131] Chai Z H,Sun D K,Wang H L,et al. A comparative study of local and nonlocal Allen-Cahn equations with mass conservation[J]. International Journal of Heat and Mass Transfer,2018,122:631-642.

[132] Zheng H W,Shu C,Chew Y T. A lattice Boltzmann model for multiphase flows with large density ratio [J]. Journal of Computational Physics,2006,218(1):353-371.

[133] He X Y,Chen S Y,Zhang R Y. A lattice Boltzmann scheme for incompressible multiphase flow and its application in simulation of Rayleigh-Taylor instability[J]. Journal of Computational Physics,1999,152(2):642-663.

[134] Inamuro T,Ogata T,Tajima S,et al. A lattice Boltzmann method for incompressible two-phase flows with large density differences[J]. Journal of Computational Physics,2004,198(2):628-644.

[135] Lou Q,Guo Z L,Shi B C. Effects of force discretization on mass conservation in lattice Boltzmann equation for two-phase flows[J]. Europhysics Letters,2012,99(6):64005.

[136] Fakhari A,Rahimian M H. Phase-field modeling by the method of lattice Boltzmann equations[J].

Physical Review E,2010,81(3):036707.

[137] Lee T,Liu L. Lattice Boltzmann simulations of micron-scale drop impact on dry surfaces[J]. Journal of Computational Physics,2010,229(20):8045-8063.

[138] Wang Y,Shu C,Huang H B,et al. Multiphase lattice Boltzmann flux solver for incompressible multiphase flows with large density ratio[J]. Journal of Computational Physics,2015,280:404-423.

[139] Wang Y,Shu C,Yang L M. An improved multiphase lattice Boltzmann flux solver for three-dimensional flows with large density ratio and high Reynolds number[J]. Journal of Computational Physics,2015,302:41-58.

[140] 赵宁,王东红. 多介质流体界面问题的数值模拟[M]. 北京:科学出版社,2016.

[141] Fakhari A, Bolster D. Diffuse interface modeling of three-phase contact line dynamics on curved boundaries:A lattice Boltzmann model for large density and viscosity ratios[J]. Journal of Computational Physics,2017,334:620-638.

[142] Liang H,Xu J R,Chen J X,et al. Phase-field-based lattice Boltzmann modeling of large-density-ratio two-phase flows[J]. Physical Review E,2018,97(3):033309.

[143] 景思睿,张鸣远. 流体力学[M]. 西安:西安交通大学出版社,2001.

[144] Liu L,Lee T. Wall free energy based polynomial boundary conditions for non-ideal gas lattice Boltzmann equation[J]. International Journal of Modern Physics C,2009,20(11):1749-1768.

[145] Huang J J,Huang H B,Wang X Z. Wetting boundary conditions in numerical simulation of binary fluids by using phase-field method:some comparative studies and new development[J]. International Journal for Numerical Methods in Fluids,2015,77(3):123-158.

[146] Yue P T,Zhou C F,Feng J J. Sharp-interface limit of the Cahn-Hilliard model for moving contact lines[J]. Journal of Fluid Mechanics,2010,645:279-294.

[147] Yan Y Y,Zu Y Q. A lattice Boltzmann method for incompressible two-phase flows on partial wetting surface with large density ratio[J]. Journal of Computational Physics,2007,227(1):763-775.

第 6 章　流固两相流模型及固液相变格子 Boltzmann模型

流固两相流是由固体颗粒和气相或液相流体组成的一种混合流体流动，它广泛存在于热能动力、化工、冶金以及环保等多个领域[1]，通常可分为气固两相流和液固两相流。在这种流动中，固体颗粒与气相或液相流体之间有着密切的联系并在运动中相互影响。除了两相之间的相互作用，还存在着固相与固相之间、流体相与流体相之间的作用[2]。按照固体颗粒的处理方法，流固两相流的描述方法可分为点源颗粒方法和有限体积颗粒方法[3]。其中，点源颗粒方法将固体颗粒看作质点，不考虑颗粒的大小与形状，流固之间的相互作用通过经验表达式来描述。有限体积颗粒方法则考虑颗粒的大小和形状，且流固之间的相互作用通过流体在固体颗粒表面的无滑移边界条件来实现。这类方法的优点在于可以考察固体颗粒周边的流动细节，适合流固两相流的机理研究。在本章，我们将基于有限体积颗粒方法来介绍相应的流固两相流格子 Boltzmann 模型。

流固两相流的拉格朗日和欧拉方法

本章要介绍的另一部分内容是固液相变格子 Boltzmann 模型。固液相变材料（phase change materials）由于相变过程具有相变潜热大、温度变化范围小、体积变化小等特点，在能源存储与利用、建筑节能、制冷空调、航空航天等诸多领域[4-7]都有着广泛的应用，例如太阳能热利用中的相变材料、被动式太阳房、工业余热相变储热式回收利用技术、电力“移峰填谷”的冰蓄冷空调技术、航天器相变温控技术等。固液相变问题属于存在移动相界面的非稳态问题，具有强非线性、多尺度及多物理场耦合的特征，这给相关研究带来了挑战。对于实际应用中的复杂固液相变问题，传统数值模拟方法在处理移动相界面边界、捕获相变糊状区详细的流动与传热信息、实现固相自由运动等方面存在着一定的困难。从本质上讲，固液相变行为是固液两相之间微观作用的宏观体现，格子 Boltzmann 方法的介观物理背景和粒子特性使其在处理固液相变问题时具有相应的优势和特色。

方腔内固液相变问题的 LBM 模拟

6.1　流固两相流格子 Boltzmann 模型

6.1.1　基本原理

在格子 Boltzmann 方法领域，Ladd[8-9]率先提出了基于有限体积颗粒的流固两相流模型。在 Ladd 的模型中，每个颗粒被视为流体中的运动物体，其大小和形状用格子来描述，固体颗粒的边界用格线的中点来表示，如图 6.1 所示。这种阶梯逼近的近似方法会引入一定的误差，逼近精度为 $O(\delta_x/r)$，其中 r 是颗粒特征尺寸[3]。对刚性颗粒，其运动控制方程为[10]

$$M_p \frac{d\boldsymbol{U}_p}{dt}=\boldsymbol{F} \tag{6.1}$$

$$\boldsymbol{I}_p \cdot \frac{d\boldsymbol{\omega}_p}{dt}+\boldsymbol{\omega}_p\times(\boldsymbol{I}_p\cdot\boldsymbol{\omega}_p)=\boldsymbol{T} \tag{6.2}$$

式中：M_p 为颗粒的有效质量；$\boldsymbol{U}_p$ 和 $\boldsymbol{\omega}_p$ 分别为颗粒的平动速度和角速度；$\boldsymbol{F}$ 和 $\boldsymbol{T}$ 分别为作用在颗粒上的总力和总力矩；$\boldsymbol{I}_p$ 为颗粒的转动惯量张量；"×"为叉乘运算符号。对二维问题，转动惯量张量退化为标量 I_p。相应地，式(6.2)变为

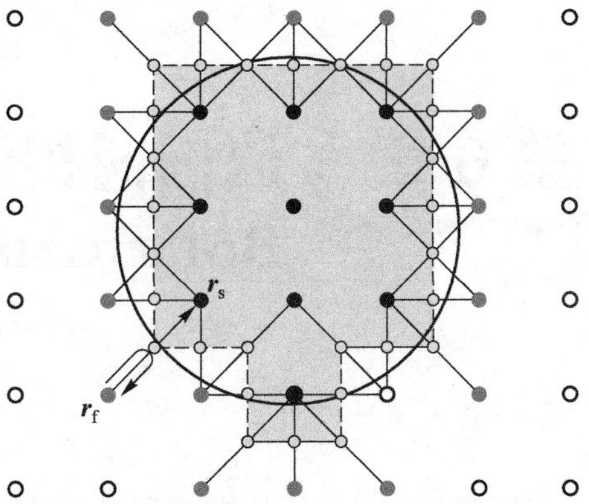

图 6.1 固体颗粒的边界近似

$$I_p \frac{d\boldsymbol{\omega}_p}{dt}=\boldsymbol{T} \tag{6.3}$$

对二维圆形颗粒，M_p 和 I_p 分别为 $M_p=\pi r^2\rho_p$ 和 $I_p=0.5r^2M_p$，其中 ρ_p 为固体颗粒的密度。针对一些相对简单的问题，可采用梯形法求解式(6.1)和式(6.3)。例如，由式(6.1)可求得固体颗粒的平动速度为

$$\boldsymbol{U}_p(t+\delta_t)=\boldsymbol{U}_p(t)+\frac{\delta_t}{2M_p}[\boldsymbol{F}(t)+\boldsymbol{F}(t+\delta_t)]\approx\boldsymbol{U}_p(t)+\frac{\delta_t}{M_p}\boldsymbol{F}\left(t+\frac{\delta_t}{2}\right) \tag{6.4}$$

颗粒位置可通过如下式子更新：

$$\boldsymbol{r}_p(t+\delta_t)=\boldsymbol{r}_p(t)+\frac{\delta_t}{2}[\boldsymbol{U}_p(t)+\boldsymbol{U}_p(t+\delta_t)] \tag{6.5}$$

式(6.3)可通过类似方式进行求解。

计算固体颗粒所受到的总力 $\boldsymbol{F}$ 和总力矩 $\boldsymbol{T}$ 是流固两相流格子 Boltzmann 模拟中最关键的步骤。在 Ladd[8] 的模型中，颗粒内部的格子也被流体粒子占据，并且颗粒内部的流体和外部流体具有相同的密度，但黏性不同。在这种处理方式下，颗粒内部流体和外部流体都会对颗粒施加作用力，该模型的优点是无须对穿越固体颗粒边界的流体做特别处理[3]，但缺点是固体颗粒和流体的密度比 ρ_p/ρ_f 受到了限制。为了解决这一问题，Aidun 等[11]提出了不使用颗粒内部流体的流固两相流格子 Boltzmann 模型，在他们的模型中，只计算外部流体施加在固体颗粒上的力和力矩，力的计算方式采用**动量交换法**(momentum-exchange method)。

以图 6.2 为例，图中 $\boldsymbol{r}_f$ 为流体节点，$\boldsymbol{r}_s$ 为固体节点，曲线为物理边界，$\boldsymbol{r}_b=\boldsymbol{r}_f+0.5\boldsymbol{e}_\alpha\delta_t$ 为数值计算中的边界位置。t 时刻流体节点 $\boldsymbol{r}_f$ 处发生粒子碰撞，碰撞后 $\boldsymbol{e}_\alpha$ 方向的分布函数为 $f_\alpha^*(\boldsymbol{r}_f,t)$，该分布函数将在 $t+0.5\delta_t$ 时刻达到 $\boldsymbol{r}_b$ 处，并发生反弹。根据半步长反弹格式[8]，反弹后的分布函数为

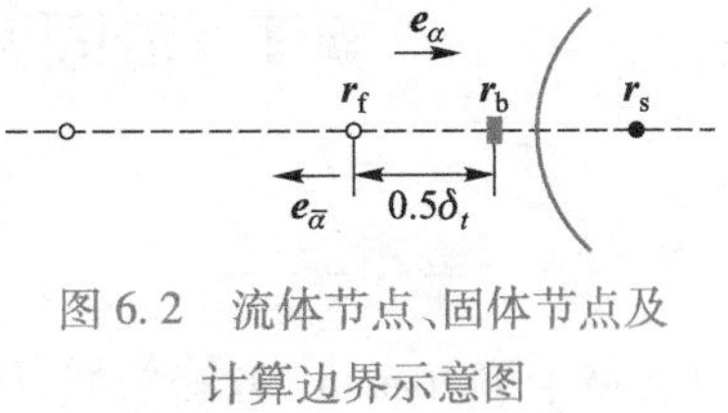

图 6.2 流体节点、固体节点及计算边界示意图

$$f_{\bar{\alpha}}\left(\boldsymbol{r}_b,t+\frac{\delta_t}{2}\right)=f_\alpha^*(\boldsymbol{r}_f,t)+2\omega_{\bar{\alpha}}\rho\frac{\boldsymbol{e}_{\bar{\alpha}}\cdot\boldsymbol{u}_b}{c_s^2} \tag{6.6}$$

式中：$\bar{\alpha}$ 为 α 的反方向，即 $\boldsymbol{e}_{\bar{\alpha}}=-\boldsymbol{e}_\alpha$；速度 $\boldsymbol{u}_b=\boldsymbol{U}_p+\boldsymbol{\omega}_p\times(\boldsymbol{r}_b-\boldsymbol{R})$，其中 $\boldsymbol{R}$ 为固体颗粒的质心位置。粒子分布函数 $f_{\bar{\alpha}}(\boldsymbol{r}_b,t+0.5\delta_t)$ 将在 $t+\delta_t$ 时刻达到 $\boldsymbol{r}_f$ 处，即

$$f_{\bar{\alpha}}(\boldsymbol{r}_{\mathrm{f}},t+\delta_t)=f_{\bar{\alpha}}\left(\boldsymbol{r}_{\mathrm{b}},t+\frac{\delta_t}{2}\right) \tag{6.7}$$

由图 6.2 可知，在 $t+0.5\delta_t$ 时刻，通过计算边界 $\boldsymbol{r}_{\mathrm{b}}$ 进入固体颗粒的动量为 $f_{\alpha}^{*}(\boldsymbol{r}_{\mathrm{f}},t)\boldsymbol{e}_{\alpha}$，而离开固体颗粒的动量则为 $f_{\bar{\alpha}}(\boldsymbol{r}_{\mathrm{b}},t+0.5\delta_t)\boldsymbol{e}_{\bar{\alpha}}$，也即 $f_{\bar{\alpha}}(\boldsymbol{r}_{\mathrm{f}},t+\delta_t)\boldsymbol{e}_{\bar{\alpha}}$。根据动量交换法，流体粒子施加在固体颗粒上的力为

$$\delta\boldsymbol{F}_{\mathrm{p}}\left(\boldsymbol{r}_{\mathrm{b}},\boldsymbol{e}_{\alpha},t+\frac{\delta_t}{2}\right)=f_{\alpha}^{*}(\boldsymbol{r}_{\mathrm{f}},t)\boldsymbol{e}_{\alpha}-f_{\bar{\alpha}}(\boldsymbol{r}_{\mathrm{f}},t+\delta_t)\boldsymbol{e}_{\bar{\alpha}} \tag{6.8}$$

根据式(6.6)和式(6.7)以及 $\boldsymbol{e}_{\bar{\alpha}}=-\boldsymbol{e}_{\alpha}$，式(6.8)可写成

$$\delta\boldsymbol{F}_{\mathrm{p}}\left(\boldsymbol{r}_{\mathrm{b}},\boldsymbol{e}_{\alpha},t+\frac{\delta_t}{2}\right)=2\left[f_{\alpha}^{*}(\boldsymbol{r}_{\mathrm{f}},t)-\omega_{\alpha}\rho\frac{\boldsymbol{e}_{\alpha}\cdot\boldsymbol{u}_{\mathrm{b}}}{c_{\mathrm{s}}^{2}}\right]\boldsymbol{e}_{\alpha} \tag{6.9}$$

因此，固体颗粒所受到的外部流体施加的力和力矩分别为

$$\boldsymbol{F}_{\mathrm{p}}\left(t+\frac{\delta_t}{2}\right)=\sum_{\boldsymbol{r}_{\mathrm{b}}}\sum_{\alpha}\delta\boldsymbol{F}_{\mathrm{p}}\left(\boldsymbol{r}_{\mathrm{b}},\boldsymbol{e}_{\alpha},t+\frac{\delta_t}{2}\right) \tag{6.10}$$

$$\boldsymbol{T}_{\mathrm{p}}\left(t+\frac{\delta_t}{2}\right)=\sum_{\boldsymbol{r}_{\mathrm{b}}}\sum_{\alpha}(\boldsymbol{r}_{\mathrm{b}}-\boldsymbol{R})\times\delta\boldsymbol{F}_{\mathrm{p}}\left(\boldsymbol{r}_{\mathrm{b}},\boldsymbol{e}_{\alpha},t+\frac{\delta_t}{2}\right) \tag{6.11}$$

式中：$\boldsymbol{r}_{\mathrm{b}}=\boldsymbol{r}_{\mathrm{f}}+0.5\boldsymbol{e}_{\alpha}\delta_t$；每个式子中的第一个求和符号针对所有边界位置 $\boldsymbol{r}_{\mathrm{b}}$，第二个求和符号针对边界位置 $\boldsymbol{r}_{\mathrm{b}}$ 相邻的流体节点，即 $\boldsymbol{r}_{\mathrm{f}}=\boldsymbol{r}_{\mathrm{b}}-0.5\boldsymbol{e}_{\alpha}\delta_t$ 为流体节点的所有 α 方向。在固体颗粒的运动过程中，一些原先被颗粒覆盖的固体节点会在新的时刻变成流体节点（见图 6.3 中的正方形节点），它们的分布函数、密度和速度需要进行处理；相应地，一些之前被流体覆盖的节点会在新的时刻被固体颗粒覆盖（见图 6.3 中的菱形节点），因此需要考虑进出固体颗粒的节点所携带的动量对颗粒的影响。

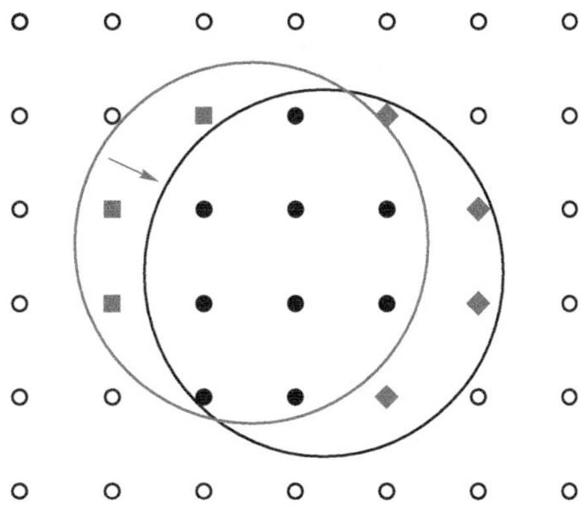

图 6.3　固体颗粒运动过程中进出颗粒的节点示意图

离开固体颗粒的节点对颗粒的作用力和力矩为[11]

$$\boldsymbol{F}_{\mathrm{un}}=-\sum_{\mathrm{un}}\rho(\boldsymbol{r}_{\mathrm{un}})\boldsymbol{u}(\boldsymbol{r}_{\mathrm{un}}),\quad \boldsymbol{T}_{\mathrm{un}}=-\sum_{\mathrm{un}}(\boldsymbol{r}_{\mathrm{un}}-\boldsymbol{R})\times\rho(\boldsymbol{r}_{\mathrm{un}})\boldsymbol{u}(\boldsymbol{r}_{\mathrm{un}}) \tag{6.12}$$

式中，下标 un 表示“uncover nodes”，即对所有离开颗粒的节点求和。类似地，对进入固体颗粒的节点，它们所携带的动量进入了固体颗粒，因此它们对固体颗粒的作用力和力矩为[11]

$$\boldsymbol{F}_{\mathrm{cn}}=\sum_{\mathrm{cn}}\rho(\boldsymbol{r}_{\mathrm{cn}})\boldsymbol{u}(\boldsymbol{r}_{\mathrm{cn}}),\quad \boldsymbol{T}_{\mathrm{cn}}=\sum_{\mathrm{cn}}(\boldsymbol{r}_{\mathrm{cn}}-\boldsymbol{R})\times\rho(\boldsymbol{r}_{\mathrm{cn}})\boldsymbol{u}(\boldsymbol{r}_{\mathrm{cn}}) \tag{6.13}$$

式中，下标 cn 表示“cover nodes”，即对所有进入颗粒的节点求和。结合式(6.10)～式(6.13)及重力 $\boldsymbol{F}_{\mathrm{g}}$，式(6.1)和式(6.2)中的总力 $\boldsymbol{F}$ 和总力矩 $\boldsymbol{T}$ 可表示为

$$\boldsymbol{F}=\boldsymbol{F}_{\mathrm{p}}+\boldsymbol{F}_{\mathrm{un}}+\boldsymbol{F}_{\mathrm{cn}}+\boldsymbol{F}_{\mathrm{g}},\quad \boldsymbol{T}=\boldsymbol{T}_{\mathrm{p}}+\boldsymbol{T}_{\mathrm{un}}+\boldsymbol{T}_{\mathrm{cn}} \tag{6.14}$$

式中，重力 $\boldsymbol{F}_{\mathrm{g}}$ 为

$$\boldsymbol{F}_{\mathrm{g}}=M_{\mathrm{p}}\frac{\rho_{\mathrm{p}}-\rho_{\mathrm{f}}}{\rho_{\mathrm{p}}}\boldsymbol{g} \tag{6.15}$$

式中,$\boldsymbol{g}$ 为重力加速度。在得到总力和总力矩之后,即可通过数值求解式(6. 1)和式(6. 2)更新固体颗粒的平动速度、角速度及其空间位置。

6. 1. 2 动量交换法的修正

2014 年,Wen 等[12]对 Aidun 等[11]的模型进行了修正。前面提到,分布函数 $f_\alpha^*(\boldsymbol{r}_\mathrm{f},t)$ 在到达图 6. 2 中的 $\boldsymbol{r}_\mathrm{b}$ 处时发生反弹,并在该处发生动量交换。Wen 等[12]指出,由于计算边界 $\boldsymbol{r}_\mathrm{b}$ 处本身具有速度 $\boldsymbol{u}_\mathrm{b}$,因此在采用动量交换法计算流体对固体颗粒的作用力时应当采用相对速度,即式(6. 8)应修改为

$$\delta \boldsymbol{F}_\mathrm{p}\left(\boldsymbol{r}_\mathrm{b},\boldsymbol{e}_\alpha,t+\frac{\delta_t}{2}\right)=f_\alpha^*(\boldsymbol{r}_\mathrm{f},t)(\boldsymbol{e}_\alpha-\boldsymbol{u}_\mathrm{b})-f_{\bar{\alpha}}(\boldsymbol{r}_\mathrm{f},t+\delta_t)(\boldsymbol{e}_{\bar{\alpha}}-\boldsymbol{u}_\mathrm{b}) \tag{6. 16}$$

根据式(6.6)和式(6.7),式(6.16)可整理为

$$\begin{aligned}\delta \boldsymbol{F}_\mathrm{p}\left(\boldsymbol{r}_\mathrm{b},\boldsymbol{e}_\alpha,t+\frac{\delta_t}{2}\right)&=f_\alpha^*(\boldsymbol{r}_\mathrm{f},t)\boldsymbol{e}_\alpha-f_{\bar{\alpha}}(\boldsymbol{r}_\mathrm{f},t+\delta_t)\boldsymbol{e}_{\bar{\alpha}}-\boldsymbol{u}_\mathrm{b}[f_\alpha^*(\boldsymbol{r}_\mathrm{f},t)-f_{\bar{\alpha}}(\boldsymbol{r}_\mathrm{f},t+\delta_t)]\\&=f_\alpha^*(\boldsymbol{r}_\mathrm{f},t)\boldsymbol{e}_\alpha-f_{\bar{\alpha}}(\boldsymbol{r}_\mathrm{f},t+\delta_t)\boldsymbol{e}_{\bar{\alpha}}-2\omega_\alpha\rho\frac{\boldsymbol{e}_\alpha\cdot\boldsymbol{u}_\mathrm{b}}{c_\mathrm{s}^2}\boldsymbol{u}_\mathrm{b}\end{aligned} \tag{6. 17}$$

对比式(6. 17)和式(6. 8)可以发现,式(6. 17)右端多了一项,该项与 Chen 等[13]针对式(6. 8)所提出的修正项是一致的[14]。需要注意的是,在采用基于相对速度的动量交换法时,无须再计算式(6. 12)和式(6. 13),即式(6. 14)应修改为

$$\boldsymbol{F}=\boldsymbol{F}_\mathrm{p}+\boldsymbol{F}_\mathrm{g},\quad \boldsymbol{T}=\boldsymbol{T}_\mathrm{p} \tag{6. 18}$$

式中,$\boldsymbol{F}_\mathrm{p}$ 和 $\boldsymbol{T}_\mathrm{p}$ 分别依据式(6. 10)和式(6. 11)计算,不过此时 $\delta\boldsymbol{F}_\mathrm{p}(\boldsymbol{r}_\mathrm{b},\boldsymbol{e}_\alpha,t+0.5\delta_t)$ 由式(6. 16)或式(6. 17)给出。

6. 1. 3 再填充格式

前面曾提到,对于不使用颗粒内部流体的模型,在颗粒的运动过程中,一些原先被固体颗粒覆盖的节点会在新的时刻变成流体节点,例如图 6. 3 所示的正方形节点。在数值模拟中,需要对这些节点的分布函数、密度和速度进行处理,这一过程在文献中通常被称为**再填充**(refilling)过程[14],相应的格式为再填充格式。文献中已有的再填充格式主要可分为平衡态格式、外推格式以及迭代格式。

在 Aidun 等[11]的模型中,采用的是平衡态格式。首先,新流体节点的密度取其周围原有流体节点的平均密度

$$\rho(\boldsymbol{r}_\mathrm{new})=\frac{1}{N_\mathrm{b}}\sum_{\alpha=0}^{m-1}S_\mathrm{w}\rho(\boldsymbol{r}_\mathrm{new}+\boldsymbol{e}_\alpha\delta_t) \tag{6. 19}$$

式中:$\boldsymbol{r}_\mathrm{new}$ 为新流体节点的空间位置;m 为离散速度的个数;S_w 为开关函数,当 $\boldsymbol{r}_\mathrm{new}+\boldsymbol{e}_\alpha\delta_t$ 为原有流体节点时,$S_\mathrm{w}=1$,其他情况下取零;N_b 为新流体节点周围原有流体节点的个数。新流体节点的速度则通过如下公式进行计算[11]:

$$\boldsymbol{u}(\boldsymbol{r}_\mathrm{new})=\boldsymbol{U}_\mathrm{p}+\boldsymbol{\omega}_\mathrm{p}\times(\boldsymbol{r}_\mathrm{new}-\boldsymbol{R}) \tag{6. 20}$$

确定新流体节点的密度和速度之后,分布函数直接赋其平衡态分布[11],即 $f_\alpha(\boldsymbol{r}_\mathrm{new})=f_\alpha^\mathrm{eq}(\boldsymbol{r}_\mathrm{new})$。

随后,Lallemand 和 Luo[15]提出了一种基于法向外推的再填充格式,其基本思路是,根

据新流体节点附近的固体边界法线方向(如图 6.4 所示的 $\boldsymbol{n}_w$),找到与之最接近的离散速度方向 $\boldsymbol{e}_c$,继而确定该方向上的三个流体节点 $\boldsymbol{r}'$、$\boldsymbol{r}''$和 $\boldsymbol{r}'''$,并采用二阶外推可得[15]

$$f_\alpha(\boldsymbol{r}_{\text{new}})=3f_\alpha(\boldsymbol{r}')-3f_\alpha(\boldsymbol{r}'')+f_\alpha(\boldsymbol{r}''') \tag{6.21}$$

相应的宏观密度和速度更新为

$$\rho(\boldsymbol{r}_{\text{new}})=\sum_\alpha f_\alpha(\boldsymbol{r}_{\text{new}}),$$
$$\rho(\boldsymbol{r}_{\text{new}})\boldsymbol{u}(\boldsymbol{r}_{\text{new}})=\sum_\alpha \boldsymbol{e}_\alpha f_\alpha(\boldsymbol{r}_{\text{new}}) \tag{6.22}$$

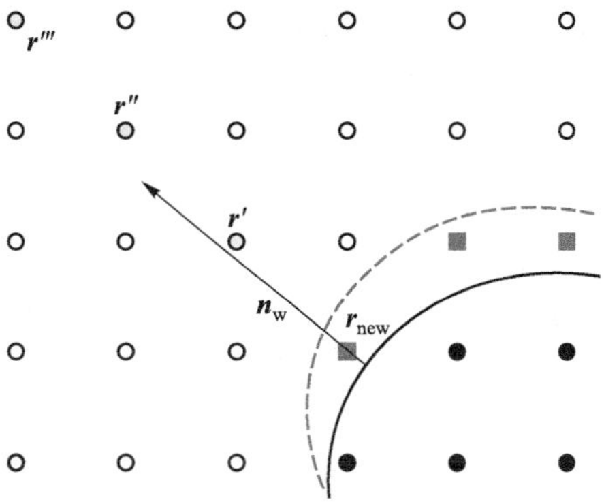

图 6.4 基于法向外推的再填充格式示意图

结合 Lallemand 和 Luo 的法向外推及 Guo 等的非平衡态外推[16],Chen 等[17]提出了基于法向非平衡态外推的再填充格式,即

$$f_\alpha(\boldsymbol{r}_{\text{new}})=f_\alpha^{\text{eq}}(\boldsymbol{r}_{\text{new}})+f_\alpha^{\text{neq}}(\boldsymbol{r}_{\text{new}}) \tag{6.23}$$

式中,新流体节点分布函数的非平衡态部分为

$$f_\alpha^{\text{neq}}(\boldsymbol{r}_{\text{new}})=3f_\alpha^{\text{neq}}(\boldsymbol{r}')-3f_\alpha^{\text{neq}}(\boldsymbol{r}'')+f_\alpha^{\text{neq}}(\boldsymbol{r}''') \tag{6.24}$$

平衡态分布函数 $f_\alpha^{\text{eq}}(\boldsymbol{r}_{\text{new}})$ 中的密度可通过法向外推确定[17]:$\rho(\boldsymbol{r}_{\text{new}})=3\rho(\boldsymbol{r}')-3\rho(\boldsymbol{r}'')+\rho(\boldsymbol{r}''')$。

采用法向外推时,每一时层每个新流体节点,都需要计算其对应的固体边界法向矢量 $\boldsymbol{n}_w$,在程序编写上具有一定的复杂性。为此,Wen 等[18]提出采用平均外推格式,即

$$f_\alpha(\boldsymbol{r}_{\text{new}})=\frac{1}{N_b}\sum_{\alpha=1}^{m-1}S_w\left[3f_\alpha(\boldsymbol{r}')-3f_\alpha(\boldsymbol{r}'')+f_\alpha(\boldsymbol{r}''')\right] \tag{6.25}$$

与式(6.19)一样,N_b 为新流体节点周围原有流体节点的个数。

此外,He 等[19]提出了另一种形式的非平衡态外推格式,新流体节点的分布函数仍由式(6.23)定义,但非平衡态部分计算如下:

$$f_\alpha^{\text{neq}}(\boldsymbol{r}_{\text{new}})=\frac{1}{\sum_{\gamma=1}^{m-1}S_w\omega_\gamma}\sum_{\gamma=1}^{m-1}S_w\omega_\gamma f_\alpha^{\text{neq}}(\boldsymbol{r}_{\text{new}}+\boldsymbol{e}_\gamma\delta_t) \tag{6.26}$$

式中,S_w 为开关函数,当 $\boldsymbol{r}_{\text{new}}+\boldsymbol{e}_\gamma\delta_t$ 为原有流体节点时,$S_w=1$,其他情况下取零。新流体节点的密度计算与式(6.26)类似。

除上述格式外,一些学者[17]提出了基于迭代的再填充格式,基本思路是通过迭代的方式使得新流体节点的分布函数、密度及速度与周围流场的信息相容,具体可参阅文献[17]的介绍。Tao 等[14]对三类再填充格式进行了数值模拟研究,他们发现,迭代格式的表现最好,其次是非平衡态外推格式,平衡态格式则表现较差。需要指出的是,在固体颗粒的运动过程中,通常难以保证每个时刻新流体节点的数量与新固体节点的数量是一致的,因此在一定程度上会造成质量不守恒。对流体密度近似为常数的不可压缩流固两相流,这一问题的影响可能不大,但涉及气液固三相流时,应当予以注意。

6.1.4 颗粒间相互作用力

当大量固体颗粒共存时,可能会存在颗粒间的间距非常小的情形,如图 6.5 所示。图

中两个颗粒表面的最小距离小于一个格子的长度，颗粒间没有足够的流体节点来支撑前面所提到的力和力矩的计算。此时，必须考虑颗粒间的排斥力(repellent force)。目前文献中的排斥力模型主要包括润滑力(lubrication force)模型、弹簧力(spring force)模型、碰撞力(collision force)模型等。对二维问题，Kromkamp等[20]提出了如下形式的润滑力

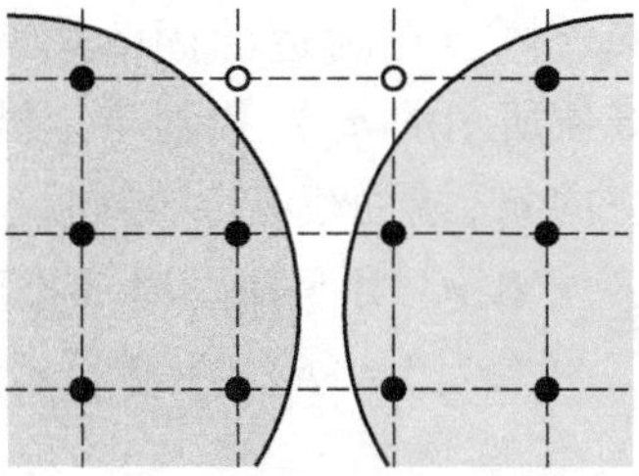

图 6.5 固体颗粒近距接触示意图

$$\boldsymbol{F}_{\text{lub}}=-\frac{1}{2}\mu\boldsymbol{U}_{12}\cdot\hat{\boldsymbol{R}}_{12}\left[\left(\frac{h}{a}\right)^{-3/2}\left(F_0+\frac{h}{a}F_1\right)-\left(\frac{h_c}{a}\right)^{-3/2}\left(F_0+\frac{h_c}{a}F_1\right)\right],\quad h<h_c \tag{6.27}$$

式中：μ 为动力黏度；$\boldsymbol{U}_{12}$ 为颗粒 1 相对于颗粒 2 的速度，即 $\boldsymbol{U}_{12}=\boldsymbol{U}_p(1)-\boldsymbol{U}_p(2)$；$\hat{\boldsymbol{R}}_{12}=\boldsymbol{R}_{12}/|\boldsymbol{R}_{12}|$ 是颗粒 1 的质心指向颗粒 2 的质心的单位矢量；h 为两个固体颗粒表面之间的最小距离；a 是两个颗粒的半径之和，即 $a=r_1+r_2$；h_c 是设定的临界距离，即当 $h\geqslant h_c$ 时润滑力为零，当 $h<h_c$ 时润滑力由式(6.27)给出，通常可设置 h_c 为 2 个格子的长度[21]；F_0 和 F_1 为常数，且分别为 $F_0=3\pi\sqrt{2}/4$、$F_1=207\pi\sqrt{2}/80$。式(6.27)及相关参数的推导参见文献[20,21]的附录。

对三维问题，可采用 Nguyen 和 Ladd[22] 提出的润滑力

$$\boldsymbol{F}_{\text{lub}}=-6\pi\mu\boldsymbol{U}_{12}\cdot\hat{\boldsymbol{R}}_{12}\frac{r_1^2r_2^2}{(r_1+r_2)^2}\left(\frac{1}{h}-\frac{1}{h_c}\right),\quad h<h_c \tag{6.28}$$

式中，$\boldsymbol{U}_{12}$、$\hat{\boldsymbol{R}}_{12}$等的定义与式(6.27)一致。此外，Wan 和 Turek[23] 提出了如下形式的碰撞力来处理颗粒间的相互作用力：

$$\boldsymbol{F}_{\text{cf}}=\begin{cases}\dfrac{1}{\varepsilon_p}\boldsymbol{R}_{12}(r_1+r_2+h_c-d_{12})^2, & r_1+r_2\leqslant d_{12}\leqslant r_1+r_2+h_c\\[2ex] \dfrac{1}{\varepsilon_p'}\boldsymbol{R}_{12}(r_1+r_2-d_{12}), & d_{12}\leqslant r_1+r_2\end{cases} \tag{6.29}$$

式中：d_{12}为两个颗粒质心间的距离，即 $d_{12}=|\boldsymbol{R}_{12}|$；$h_c$ 为临界距离，当 $d_{12}>r_1+r_2+h_c$ 时，碰撞力 $\boldsymbol{F}_{\text{cf}}$为零；$\varepsilon_p$ 和 ε_p'为常数。这一碰撞力亦可推广至固体颗粒和壁面之间的作用力，具体可参阅文献[23]。

6.2 固液相变格子 Boltzmann 模型

6.2.1 固液相变简介

固液相变储热技术是一种利用相变材料熔化或凝固过程中吸收或释放潜热(latent heat)进行热能利用的技术，以解决能源利用系统中能量供需之间在时间、空间及强度上不匹配的矛盾，从而提高能源利用率。固液相变的主要特点有：(1)相变潜热大；(2)潜热的吸收或释放可以在某一恒定温度下或较小的温度区间内进行；(3)相变材料在相变前后的体积变化较小。目前，固液相变的研究大致可分为两类：一类是固液相变过程流动与传热问题的研究，包括相变过程的传热机理、提高相变材料导热及储热性能的措施等；另一类是相变材料的研究，包括相变材料的制备、稳定性、使用寿命以及材料与设备的相容

性等。固液相变的应用可以追溯到很早以前。在古代,人们就从冬天河面或湖面冻结的厚冰层中获取冰块,储存于利用锯末隔热的“冰屋”中,冰块可以存放至夏末。自 20 世纪以来,国内外学者对固液相变的机理和应用进行了大量研究,相关研究成果已得到了广泛的应用[4-7],如太阳能热利用中的相变材料、被动式太阳房、工业余热相变储热式回收利用技术、建筑空调系统中起到电力“移峰填谷”作用的冰蓄冷空调技术等。在航空航天领域,以固液相变储热技术为基础发展而来的相变温控技术是保障航天任务安全实施的关键技术,例如,“阿波罗 15 号”登月任务的月球车采用了三套不同的固液相变温控系统[24],而我国发射的“嫦娥一号”绕月卫星则采用了基于相变材料的复合热管来控制温度对光学器件的影响[25]。

在固液相变过程中,固液两相之间存在着移动相界面,在相界面附近,相态转变伴随着大量潜热的释放或吸收。此外,液相及相变糊状区流动所引起的对流作用使得流动与传热之间的耦合更加复杂。只有极少数热传导熔化/凝固相变问题存在理论分析解,因此实验和数值模拟是目前研究固液相变问题的主要手段。实验能够直观地再现实际应用中的固液相变过程,有利于揭示一般性的规律和发现新现象,其研究结果可用于判断模型和数值模拟结果的可靠性。然而,实验所测得的数据通常是多种因素综合作用下的物理量,对于相变过程中局部流动与传热信息的精确测量较为困难。随着科学技术的发展,各种新材料不断涌现。具有高导热率、高渗透性、高比表面积等特性的多孔材料(如泡沫金属)的出现为固液相变的传热强化提供了有效手段。将多孔材料填充到相变材料中形成复合相变材料能够有效地提高相变材料的导热性能,从而达到提高储/放热速率的目的。对于多孔介质复合相变材料的固液相变过程,由于实际应用中多孔载体的结构复杂多样、尺寸不一,实验研究往往难以揭示孔隙内部的流动特性与传热规律,而格子 Boltzmann 方法的介观物理背景和粒子特性使其在处理这一问题时具有显著的优势[26]。

泡沫金属

1998 年,de Fabritiis 等[27]首次将格子 Boltzmann 方法应用于固液相变问题的研究。在他们提出的固液相变格子 Boltzmann 模型中,液相和固相被当作两种不同组分,分别采用两个不同的格子 Boltzmann 方程来描述,而相变则通过在格子 Boltzmann 方程中添加反应项来实现。自 de Fabritiis 等的研究之后,格子 Boltzmann 方法被广泛应用于各类固液相变问题。相界面的追踪或捕捉是固液相变模拟的关键,根据相界面追踪或捕捉方式的不同,固液相变格子 Boltzmann 模型主要可分为**相场模型**(phase-field model)、**焓法模型**(enthalpy-based model)和**界面追踪模型**(interface-tracking model)。下面分别介绍相场模型和焓法模型,界面追踪模型[28-29]将作为移动相界面的处理方法在 6.2.4 节进行介绍。此外,有关多孔介质的固液相变模型将在“第 7 章 多孔介质流动与传热的格子 Boltzmann 模型”中进行介绍。

6.2.2 相场模型概述

在基于相场的固液相变格子 Boltzmann 模型中,固液相界面采用相场法捕捉。相场法使用序参数 ϕ 来区分固相和液相,例如 $\phi=-1$ 代表固相、$\phi=1$ 代表液相;在固液两相之间,序参数光滑连续地变化。这一点与气液两相流的相场模型是一致的。对于相场固液相变格子 Boltzmann 模型来说,关键是建立序参数的演化方程,即相场方程,并通过求解该方程来捕捉固液相界面,同时得到相应的潜热源项用于温度场的求解。2001 年,Miller 等[30]率先建立了基于相场的固液相变格子 Boltzmann 模型,相应的序参数的演化方程为

$$\phi(\boldsymbol{r},t)=\phi(\boldsymbol{r},t-\delta_t)+R(\boldsymbol{r},t) \tag{6.30}$$

反应项 $R(\boldsymbol{r},t)$ 为

$$R(\boldsymbol{r},t)=f_+K^+(\boldsymbol{r},t)[1-\phi(\boldsymbol{r},t-\delta_t)]-f_-K^-(\boldsymbol{r},t)[1+\phi(\boldsymbol{r},t-\delta_t)] \tag{6.31}$$

式中：f_+ 和 f_- 为频率系数；$K^+(\boldsymbol{r},t)$ 和 $K^-(\boldsymbol{r},t)$ 为开关函数，控制着熔化或凝固的启动。Miller 等[30]通过一阶尺度分析导出了序参数演化方程所对应的偏微分方程，并将其与经典的相场方程进行了对比从而确定了各参数的物理意义。在 Miller 等的模型中，温度被视为跟随流体运动的被动标量，速度场和温度场由一组分布函数在四维面心超立方体格子上求解（第四维分量代表温度）。

应用该模型，Miller 和 Succi[31-32]模拟了侧壁加热腔内镓的熔化过程，证明了 Miller 等[30]的模型适用于固液相变问题，并将其应用于过冷液相中枝晶生长过程的数值模拟。随后，Miller 等[33]进一步将该模型用于模拟二元合金体系的枝晶生长过程。此外，Rasin 等[34]提出可以直接采用有限差分法求解如下形式的宏观相场方程：

$$\tau_\phi\partial_t\phi=\zeta^2\nabla^2\phi-g(\phi)-\lambda_\phi p(\phi)(T-T_{\mathrm{m}}) \tag{6.32}$$

式中：τ_ϕ 是相场特征松弛时间；ζ 是界面宽度；T_{m} 是熔点温度；$g(\phi)$ 和 $p(\phi)$ 分别为

$$g(\phi)=\phi^2(1-\phi)^2 \tag{6.33}$$

$$p(\phi)=\phi^3(6\phi^2-15\phi+10) \tag{6.34}$$

温度场则采用对流扩散方程的多松弛格子 Boltzmann 模型求解。基于该模型，Rasin 等[34]研究了各向异性导热、Stefan 数、毛细数等对导热作用下纯物质枝晶生长过程的影响。

2005 年，Medvedev 和 Kassner[35-36]将 Karma 和 Rappel[37]所提出的相场模型和被动标量热格子 Boltzmann 模型相结合，构建了一个相场固液相变格子 Boltzmann 模型，并应用该模型模拟了剪切流作用下的纯物质枝晶生长过程。2015 年，Rojas 等[38]在 Ohno 和 Matsuura[39]所提出的低浓度二元合金系统相场模型的基础上，考虑了相场的对流输运过程，建立了可实现枝晶运动和生长过程直接数值模拟的相场固液相变格子 Boltzmann 模型。在该模型中，相场方程和溶质浓度场方程均采用有限差分法求解，温度场则由于热扩散系数远大于溶质扩散系数而假设为恒定温度，因此仅流场采用格子 Boltzmann 方法求解。在实际应用中，相场固液相变格子 Boltzmann 模型主要用于材料科学领域凝固过程微观组织形成的模拟研究，而较少用于热科学领域固液相变过程流动与传热问题的研究。

6.2.3 焓法模型

与相场模型不同，在焓法模型中，固液相界面通过“潜焓”$\Delta H=Lf_l$ 进行捕捉，其中 L 为相变潜热，f_l 为液相分数，且 $f_l=0$ 表示固相、$f_l=1$ 表示液相、$0<f_l<1$ 表示固液相界面（也称糊状区）。显然，当相变材料为固相时，对应的潜焓为零；当相变材料为液相时，对应的潜焓为固相完全熔化时所吸收的相变潜热。在固液相界面处，潜焓光滑连续地变化。在焓法模型中，考虑对流作用时，能量守恒方程为

$$\frac{\partial(\rho c_p T)}{\partial t}+\nabla\cdot(\rho c_p T\boldsymbol{u})=\nabla\cdot(\lambda\nabla T)+q \tag{6.35}$$

式中：c_p 为比定压热容；λ 为导热系数；q 为固液相变的潜热源项[40-41]，其表达式为

$$q=-\left[\frac{\partial(\rho\Delta H)}{\partial t}+\nabla\cdot(\rho\Delta H\boldsymbol{u})\right] \tag{6.36}$$

对纯导热固液相变问题，式(6.35)和式(6.36)退化为

$$\frac{\partial(\rho c_p T)}{\partial t}=\nabla\cdot(\lambda\nabla T)-L\frac{\partial(\rho f_l)}{\partial t} \tag{6.37}$$

2001 年，Jiaung 等[42]针对纯导热固液相变问题，构建了基于焓法的固液相变格子 Boltzmann 模型。在该模型中，式(6.37)右端第二项以离散源项的形式添加到热格子 Boltzmann 方程中，导致温度分布函数的演化方程成为隐式方程。为了求解隐式的热格子 Boltzmann 方程，Jiaung 等[42]在每一个时间步内采用迭代计算，从而有

$$g_\alpha^k(\boldsymbol{r}+\boldsymbol{e}_\alpha\delta_t,t+\delta_t)-g_\alpha(\boldsymbol{r},t)=-\frac{1}{\tau_g}(g_\alpha-g_\alpha^{\mathrm{eq}})-\omega_\alpha\frac{L}{c_p}[f_l^{k-1}(t+\delta_t)-f_l(t)] \tag{6.38}$$

式中：g_α 为温度分布函数；上标 k 和 $k-1$ 分别表示迭代的当前步和上一步，收敛判据为

$$\left|\frac{f_l^k-f_l^{k-1}}{f_l^{k-1}}\right|<10^{-8} \tag{6.39}$$

为了加强迭代计算的收敛特性，Chatterjee 和 Chakraborty[43]引入了计算流体力学方法中的欠松弛焓值更新格式。

此外，Chatterjee 和 Chakraborty[44]结合不可压缩格子 Boltzmann 模型，构建了考虑液相流动的焓法固液相变格子 Boltzmann 模型。随后，Chatterjee 和 Chakraborty[40]建立了焓法固液相变混合格子 Boltzmann 模型，其中速度场采用格子 Boltzmann 方法求解，而温度场则采用有限差分法求解。该模型中采用了伪势模型来调节状态方程。为了考虑黏性热耗散和可压缩功，Chatterjee[45-46]结合双分布函数格子 Boltzmann 方法，构造了相应的固液相变模型。基于 Jiaung 等[42]的迭代焓法模型，Huber 等[47]亦构建了一个耦合自然对流的焓法固液相变格子 Boltzmann 模型，为了减小迭代过程的计算量，他们在模拟中强制在每一个时间步内只进行一次内迭代。虽然上述模型的出发点不尽相同，但在处理固液相变过程的潜热源项时，采用了相似的方式，即在热格子 Boltzmann 方程中引入了相应的离散潜热源项，并在每一个时间步内进行迭代求解，从而丧失了格子 Boltzmann 方法显式时间推进的计算优点。

如果固液相变潜热源项中的时间偏导数项$\partial_t f_l$ 采用如下的向后差分格式求解：

$$\frac{\partial f_l}{\partial t}=\frac{f_l(\boldsymbol{r},t)-f_l(\boldsymbol{r},t-\delta_t)}{\delta_t} \tag{6.40}$$

则可将热格子 Boltzmann 方程(6.38)变为显式，在数值模拟中只需将上一时层的液相分数值$f_l(\boldsymbol{r},t-\delta_t)$存储起来即可，从而避免了迭代求解过程。Luo 等[48]对这种潜热源项的处理方式进行了研究，通过对低 Stefan 数下的方腔自然对流熔化问题进行模拟，他们发现，在固液相界面区域附近出现了液相温度低于初始固相温度的现象，这与纯相变材料的实际熔化过程不相符。式(6.40)所给出的潜热源项处理方式会带来 $O(\delta_x)$ 量级的误差，在固液相界面区域附近，这种数值误差可能会被放大，在应用中应当注意。

2012 年，Eshraghi 和 Felicelli[49]构造了一个局部隐式的焓法固液相变格子 Boltzmann 模型，新时层的温度分布函数通过隐式方程进行求解：

$$\begin{aligned}&g_\alpha(\boldsymbol{r},t+\delta_t)+\frac{\omega_\alpha L}{c_p(T_l-T_s)}\sum_{\beta=0}^{8}g_\beta(\boldsymbol{r},t+\delta_t)\\&=g_\alpha(\boldsymbol{r},t)-\frac{1}{\tau_g}[g_\alpha(\boldsymbol{r},t)-g_\alpha^{\mathrm{eq}}(\boldsymbol{r},t)]+\frac{\omega_\alpha L}{c_p(T_l-T_s)}T(\boldsymbol{r},t)\end{aligned} \tag{6.41}$$

式中,T_l 和 T_s 分别为液相线和固相线温度。该模型通过在每个格点处求解小规模的线性方程组,避免了全局的焓值迭代计算过程。基于 Eshraghi 和 Felicelli 的研究工作,Feng 等[50]考虑了液相的流动过程,并研究了纳米颗粒增强相变材料的底部加热熔化过程。虽然 Eshraghi 和 Felicelli[49]提出的处理方式避免了全局迭代计算,但在每个格点处需要求解一个线性方程组,计算量仍较大,且该处理方式无法直接处理等温固液相变问题,即式(6.41)需满足 $T_l \neq T_s$。

为了解决上述问题,2013 年,Huang 和 Wu[51]提出了一个显式的焓法固液相变格子 Boltzmann 模型。对于纯物质的固液相变,式(6.36)右端第二项可以忽略,于是式(6.35)可写成

$$\frac{\partial(\rho H)}{\partial t}+\nabla \cdot (\rho c_p T \boldsymbol{u})=\nabla \cdot (\lambda \nabla T) \tag{6.42}$$

式中,$H=c_pT+Lf_l$ 为相变材料的焓。基于此,Huang 和 Wu[51]提出了以焓为基本变量的演化方程,形式如下:

$$h_\alpha(\boldsymbol{r}+\boldsymbol{e}_\alpha\delta_t, t+\delta_t)-h_\alpha(\boldsymbol{r}, t)=-\frac{1}{\tau_h}(h_\alpha-h_\alpha^{\mathrm{eq}}) \tag{6.43}$$

式中,h_α 为焓分布函数。不考虑固液相变过程的密度变化,则平衡态焓分布函数可选取为[51]

$$h_\alpha^{\mathrm{eq}}=\begin{cases} H-c_pT+\omega_\alpha c_pT\left(1-\dfrac{|\boldsymbol{u}|^2}{2c_s^2}\right), & \alpha=0 \\ \omega_\alpha c_pT\left[1+\dfrac{\boldsymbol{e}_\alpha \cdot \boldsymbol{u}}{c_s^2}+\dfrac{(\boldsymbol{e}_\alpha \cdot \boldsymbol{u})^2}{2c_s^4}-\dfrac{|\boldsymbol{u}|^2}{2c_s^2}\right], & \alpha \neq 0 \end{cases} \tag{6.44}$$

相变材料的焓 H 计算如下:

$$H=\sum_{\alpha=0}^{m-1} h_\alpha \tag{6.45}$$

相变材料的温度 T 通过下式计算:

$$T=\begin{cases} \dfrac{H}{c_p}, & H<H_s \\ T_s+\dfrac{H-H_s}{H_l-H_s}(T_l-T_s), & H_s \leqslant H \leqslant H_l \\ T_l+\dfrac{H-H_l}{c_p}, & H>H_l \end{cases} \tag{6.46}$$

式中,H_s 和 H_l 分别为 T_s 和 T_l 对应的焓值。液相分数 f_l 计算如下:

$$f_l=\begin{cases} 0, & H<H_s \\ \dfrac{H-H_s}{H_l-H_s}, & H_s \leqslant H \leqslant H_l \\ 1, & H>H_l \end{cases} \tag{6.47}$$

与之前的焓法模型不同,Huang 和 Wu[51]所提出的焓法模型不需要迭代计算。为了使导热系数与比热容解耦合,Huang 和 Wu[52]随后对该模型进行了修正,修正的平衡态焓分布函数为

$$h_\alpha^{\mathrm{eq}}=\begin{cases}H-c_{p,\mathrm{ref}}T+\omega_\alpha c_p T\left(\dfrac{c_{p,\mathrm{ref}}}{c_p}-\dfrac{|\boldsymbol{u}|^2}{2c_{\mathrm{s}}^2}\right), & \alpha=0\\ \omega_\alpha c_p T\left[\dfrac{c_{p,\mathrm{ref}}}{c_p}+\dfrac{\boldsymbol{e}_\alpha\cdot\boldsymbol{u}}{c_{\mathrm{s}}^2}+\dfrac{(\boldsymbol{e}_\alpha\cdot\boldsymbol{u})^2}{2c_{\mathrm{s}}^4}-\dfrac{|\boldsymbol{u}|^2}{2c_{\mathrm{s}}^2}\right], & \alpha\neq 0\end{cases}\tag{6.48}$$

式中，$c_{p,\mathrm{ref}}$为参考比定压热容，它的引入可以实现导热系数与比热容的独立自由调节，从而可以处理相界面处变热物性以及非稳态耦合传热。为了获得较优的数值稳定性，参考比定压热容通常取为固相和液相比定压热容的调和平均值，即 $c_{p,\mathrm{ref}}=2c_{p,\mathrm{s}}c_{p,l}/(c_{p,\mathrm{s}}+c_{p,l})$。在此基础上，结合多松弛格子 Boltzmann 方法，Huang 和 Wu[52] 构造了基于焓法的固液相变多松弛格子 Boltzmann 模型，并对松弛因子的选取进行了分析，通过选取适当的松弛因子可以减少固液相界面附近由于潜热吸收或释放所引起的数值扩散（numerical diffusion）。

此外，Li 等[53]结合轴对称格子 Boltzmann 模型，构建了轴对称固液相变问题的无迭代焓法多松弛格子 Boltzmann 模型。随后，Li 等[54]进一步构造了基于 D3Q7 格子的三维固液相变问题的焓法多松弛格子 Boltzmann 模型。基于 Huang 等[51]的模型，Hu 等[55]提出了三维固液相变问题的焓法单松弛格子 Boltzmann 模型，他们将**平滑轮廓法**（smoothed profile method）引入模型中用以处理移动相界面的无滑移速度边界。Hu 等[55]的模型能够在固定网格上对固相可自由运动的固液相变问题进行模拟。

近年来，一些学者[56-57]还提出了基于双松弛（two-relaxation-time）碰撞算子的焓法固液相变格子 Boltzmann 模型，这些模型的基本思路与焓法固液相变多松弛格子 Boltzmann 模型一致。焓法固液相变双松弛格子 Boltzmann 模型也可以通过调节松弛因子来降低固液相界面附近数值扩散效应的影响，详见文献[56,57]。为了模拟相变材料密度变化的固液相变问题，2020 年，Zhao 等[58]提出了一个焓法固液相变双松弛格子 Boltzmann 模型，该模型采用如下形式的平衡态焓分布函数（不考虑对流作用）：

$$g_\alpha^{\mathrm{eq}}=\begin{cases}\rho_{\mathrm{f}}H-\rho_0 c_{p,\mathrm{ref}}T+\omega_0\,\rho_0 c_{p,\mathrm{ref}}T, & \alpha=0\\ \omega_\alpha\,\rho_0 c_{p,\mathrm{ref}}T, & \alpha\neq 0\end{cases}\tag{6.49}$$

式中：$\rho_{\mathrm{f}}=f_l\,\rho_l+(1-f_l)\rho_{\mathrm{s}}$为固液两相混合物的密度；$\rho_0$ 为参考密度。相变材料的焓由 $\rho_{\mathrm{f}}H=\sum_{\alpha=0}^{m-1}g_\alpha$ 确定，相变材料的温度 T 和液相分数 f_l 可利用 $\rho_{\mathrm{f}}H$ 计算得到，计算公式为

$$T=\begin{cases}\dfrac{\rho_{\mathrm{f}}H}{\rho_{\mathrm{s}}c_{p,\mathrm{s}}}, & \rho_{\mathrm{f}}H<\rho_{\mathrm{s}}H_{\mathrm{s}}\\ T_{\mathrm{s}}+\dfrac{\rho_{\mathrm{f}}H-\rho_{\mathrm{s}}H_{\mathrm{s}}}{\rho_l H_l-\rho_{\mathrm{s}}H_{\mathrm{s}}}(T_l-T_{\mathrm{s}}), & \rho_{\mathrm{s}}H_{\mathrm{s}}\leqslant\rho_{\mathrm{f}}H\leqslant\rho_l H_l\\ T_l+\dfrac{\rho_{\mathrm{f}}H-\rho_l H_l}{\rho_l c_{p,l}}, & \rho_{\mathrm{f}}H>\rho_l H_l\end{cases}\tag{6.50}$$

$$f_l=\begin{cases}0, & \rho_{\mathrm{f}}H<\rho_{\mathrm{s}}H_{\mathrm{s}}\\ \dfrac{\rho_{\mathrm{f}}H-\rho_{\mathrm{s}}H_{\mathrm{s}}}{\rho_l H_l-\rho_{\mathrm{s}}H_{\mathrm{s}}}, & \rho_{\mathrm{s}}H_{\mathrm{s}}\leqslant\rho_{\mathrm{f}}H\leqslant\rho_l H_l\\ 1, & \rho_{\mathrm{f}}H>\rho_l H_l\end{cases}\tag{6.51}$$

式中,下标“l”和“s”分别表示液相和固相相变材料的物性参数。与式(6.46)和式(6.47)相比,Zhao 等的模型将相变材料的密度耦合到温度和液相分数的计算式中,从而纳入了相变材料密度变化的影响。

6.2.4 固液相界面的边界处理

对于耦合自然对流作用的固液相变问题来说,移动相界面的边界处理非常关键。为了实现移动相界面的无滑移速度边界,国内外学者从不同角度提出了各种各样的方法。在相场固液相变格子 Boltzmann 模型中[30-32],液固两相之间的作用通过在流场的演化方程中添加一个经验的阻力项来实现,其表达式为

$$\boldsymbol{F}_{\mathrm{fs}} = -\omega \boldsymbol{u}\,\frac{1-\phi}{2} \tag{6.52}$$

式中,ω 为松弛参数。$\boldsymbol{F}_{\mathrm{fs}}$ 的作用是强制固相区域无流动,从而在移动相界面处实现无滑移速度边界,这种处理与宏观数值模拟方法[41]中的处理是类似的。

Medvedev 和 Kassner[35-36]也提出了移动相界面无滑移速度边界的处理方法。具体地,他们在流场的演化方程中添加了一个耗散力(dissipative force)$\boldsymbol{F}_{\mathrm{d}}$ 来实现移动相界面的无滑移速度边界,形式如下:

$$\boldsymbol{F}_{\mathrm{d}} = -\frac{2h\rho\nu}{W_0^2}\left(\frac{1+\phi}{2}\right)^2 \boldsymbol{u} \tag{6.53}$$

式中:$h=2.757$;W_0 为与相界面宽度有关的参数(文献[35,36]取 $W_0=1$);ν 为运动黏性系数;$(1+\phi)/2$ 为固相分数(序参数 ϕ 取值:$\phi=-1$ 代表液相,$\phi=1$ 代表固相)。Rojas 等[38]的模型采用了与 Medvedev 和 Kassner[35-36]相同的处理方法。此外,Miller 等[33]和 Rasin 等[34]从其他角度提出了相场模型中移动相界面无滑移速度边界的处理方法,感兴趣的读者可参考相关文献。

针对焓法固液相变格子 Boltzmann 模型,国内外学者也提出了多种实现移动相界面无滑移速度边界的方法。在 Huber 等[47]的模型中,半步长反弹格式被用于处理移动相界面的无滑移速度边界。具体地,数值计算中的固液相界面由 $f_l=0.5$ 确定,流场演化方程的碰撞过程在 $f_l>0.5$ 对应的区域执行,在其他区域则采用半步长反弹格式处理分布函数。需要注意的是,对于局部非热平衡效应明显的泡沫金属基复合相变材料的熔化过程,由于固液相界面是一个具有一定宽度的区域,采用反弹或半步长反弹格式处理移动相界面的无滑移速度边界可能会引起较大的误差[59]。

在文献[40,44,45]中,Chakraborty 和 Chatterjee 采用等效摩擦阻力格式来处理移动相界面的无滑移速度边界。在他们的模型中,相界面的无滑移速度边界通过在流场的演化方程中引入如下形式的等效摩擦阻力 $\boldsymbol{F}_{\mathrm{s}}$ 来实现:

$$\boldsymbol{F}_{\mathrm{s}} = -\frac{\kappa}{\rho}\,\frac{(1-f_l)^2}{f_l^3+b}\boldsymbol{u} \tag{6.54}$$

式中:κ 为形态常数;b 为防止分母为零的计算常数(在固相区域 $f_l=0$)。显然,Chakraborty 和 Chatterjee 的处理方法与 Miller 等[30]以及 Medvedev 和 Kassner[35-36]的处理方法在本质上是一致的。

在 Huang 等[51]的无迭代焓法固液相变格子 Boltzmann 模型中,浸入移动边界格式[60-61]被用来实现移动相界面的无滑移速度边界,相应的流场演化方程被修改为

$$f_\alpha(\boldsymbol{r}+\boldsymbol{e}_\alpha\delta_t,t+\delta_t)=f_\alpha(\boldsymbol{r},t)-\frac{1-B}{\tau_f}[f_\alpha(\boldsymbol{r},t)-f_\alpha^{\mathrm{eq}}(\boldsymbol{r},t)]+B\Omega_\alpha^s+\delta_t F_\alpha \tag{6.55}$$

式中:B 为权函数;Ω_α^s 为附加碰撞项;F_α 为外力项。权函数取决于液相分数 f_l 和松弛时间,具体表达式为

$$B=\frac{(1-f_l)(\tau_f-0.5)}{f_l+\tau_f-0.5} \tag{6.56}$$

式(6.55)中的附加碰撞项 Ω_α^s 为

$$\Omega_\alpha^s=f_{\bar{\alpha}}(\boldsymbol{r},t)-f_\alpha(\boldsymbol{r},t)+f_\alpha^{\mathrm{eq}}(\rho,\boldsymbol{u}_s)-f_{\bar{\alpha}}^{\mathrm{eq}}(\rho,\boldsymbol{u}) \tag{6.57}$$

式中:离散速度方向 $\bar{\alpha}$ 与 α 存在 $\boldsymbol{e}_{\bar{\alpha}}=-\boldsymbol{e}_\alpha$ 的关系;$\boldsymbol{u}_s$ 为固相速度。Ω_α^s 的作用是在固液相界面处将分布函数的非平衡态部分"弹回",从而实现移动相界面的无滑移速度边界。

2016 年,Huang 和 Wu[62] 指出,等效摩擦阻力格式和浸入移动边界格式均无法准确地实现移动相界面的无滑移速度边界(这两种格式往往导致固相区的速度非零,即流体"穿透"到固相区)。为此,Huang 和 Wu 提出了一种体积格子 Boltzmann 格式(volumetric lattice Boltzmann scheme)来处理移动相界面的无滑移速度边界。在该格式中,考虑固相作用后的密度分布函数通过线性插值来确定:

$$f_\alpha=f_l f_\alpha^*+(1-f_l)f_\alpha^{\mathrm{eq}}(\rho,\boldsymbol{u}_s) \tag{6.58}$$

式中,f_α^* 为碰撞后的分布函数,固相无运动时 $\boldsymbol{u}_s=0$。当 $f_l=1$ 时,式(6.58)就是标准形式的演化方程所确定的密度分布函数;当 $f_l=0$ 时,式(6.58)变为 $f_\alpha=f_\alpha^{\mathrm{eq}}(\rho,\boldsymbol{u}_s)$,即无滑移速度边界能够得到严格满足。

2017 年,Hu 等[55]将平滑轮廓法用于处理移动相界面的无滑移速度边界。平滑轮廓法是一种类似于浸入边界法(immersed boundary method)的计算方法,该方法能够用一套简单、固定的网格处理复杂几何结构的计算,同时也可用于具有移动界面问题的模拟。在 Hu 等的模型中,流场的演化方程为

$$f_\alpha(\boldsymbol{r}+\boldsymbol{e}_\alpha\delta_t,t+\delta_t)=f_\alpha(\boldsymbol{r},t)-\frac{1}{\tau_f}[f_\alpha(\boldsymbol{r},t)-f_\alpha^{\mathrm{eq}}(\boldsymbol{r},t)]+\frac{\omega_\alpha\delta_t}{c_s^2}[(\boldsymbol{f}_s+\boldsymbol{f}_e)\cdot\boldsymbol{e}_\alpha] \tag{6.59}$$

式中:$\boldsymbol{f}_s$ 为液固作用力;$\boldsymbol{f}_e$ 为其他体积力。在平滑轮廓法中,液固作用力的表达式为

$$\boldsymbol{f}_s=(1-f_l)\frac{\boldsymbol{u}_s-\boldsymbol{u}}{\delta_t} \tag{6.60}$$

模拟结果表明,平滑轮廓法能够简单高效地处理移动相界面的无滑移速度边界。结合适当的固相运动方程,Hu 等[55]的模型能够模拟固相可自由运动的固液相变问题。

在文献[28]中,Huang 和 Wu 提出了用于固液相变问题的浸入边界格子 Boltzmann 模型,该模型采用浸入边界法处理移动相界面上的速度边界和温度边界。在浸入边界格子 Boltzmann 模型中,相界面的移动速度 $\boldsymbol{u}_{sn}$定义如下:

$$\boldsymbol{u}_{sn}=\frac{c_{p,s}Q_s(s_n,t+\delta_t)}{L}\boldsymbol{n}_n \tag{6.61}$$

式中,L 为相变潜热;s_n 为拉格朗日节点(对应相界面节点);$\boldsymbol{n}_n$ 为相界面节点处的法向矢量(从固相指向液相);$Q_s(s_n,t+\delta_t)$为能量力。由式(6.61)可知,Huang 和 Wu 的模型利用相变潜热的吸收或释放来确定相界面的移动速度,该模型也能够对固相可自由运动的固液相变问题进行模拟。

Li 等[29]提出了基于界面追踪方法的固液相变格子 Boltzmann 模型，该模型的液相分数定义为

$$f_l=\frac{S-r_{n-1}-0.5\delta r}{\delta r} \tag{6.62}$$

式中：S 为相界面位置；r_{n-1} 为液相区与相界面相邻的计算节点。当 $f_l=1$ 时，相界面位置移到下一个节点 r_{n+1}。相界面位置 S 根据相变潜热的吸收或释放来确定，而相界面节点 r_n 的运动速度 $\boldsymbol{u}(r_n,t)$ 和温度 $T(r_n,t)$ 则通过插值公式进行计算，即

$$\boldsymbol{u}(r_n,t)=\frac{f_l-0.5}{f_l+0.5}\boldsymbol{u}(r_{n-1},t) \tag{6.63}$$

$$T(r_n,t)=\frac{f_l-0.5}{f_l+0.5}T(r_{n-1},t) \tag{6.64}$$

Li 等[29]的界面追踪模型采用了固定网格追踪固液相界面，而 Huang 和 Wu[28]的浸入边界格子 Boltzmann 模型则采用了拉格朗日节点对固液相界面进行标记。需要指出的是，这两类模型都利用了相变潜热的吸收或释放来追踪固液相界面，这在本质上与焓法模型是一致的。

6.3 小　结

格子 Boltzmann 方法清晰的粒子运动图像使其在处理流体与固体之间的相互作用时具有较大的优势。本章介绍了基于有限体积颗粒的流固两相流格子 Boltzmann 模型以及固液相变格子 Boltzmann 模型。考虑颗粒的大小和形状时，流固之间的相互作用通过流体在固体颗粒表面的无滑移边界条件来实现。在早期的流固两相流格子 Boltzmann 模型中，颗粒内部也存在着流体，从而导致固体颗粒与流体的密度比受到了限制。随后，一些学者提出了不使用颗粒内部流体的流固两相流格子 Boltzmann 模型，只计算外部流体施加在固体颗粒上的力和力矩，力的计算方式采用动量交换法。在颗粒的运动过程中，一些原先被固体颗粒覆盖的节点会在新的时刻变成流体节点，从而需要对这些节点的分布函数、密度和速度进行处理，这一过程通常被称为再填充过程。此外，当大量固体颗粒共存时，可能会存在颗粒间的间距非常小的情形，此时需要处理颗粒间的相互作用。

固液相变格子 Boltzmann 模型通常可分为相场模型、焓法模型和界面追踪模型。在文献中，相场固液相变格子 Boltzmann 模型主要用于材料科学领域凝固过程微观组织形成的模拟研究，而焓法模型则主要应用于热科学领域固液相变过程流动与传热问题的研究。早期的焓法模型需要进行迭代求解，丧失了格子 Boltzmann 方法显式时间推进的优点。为了解决这一问题，一些学者提出了无迭代的焓法固液相变格子 Boltzmann 模型。在不同的模型中，移动相界面的边界处理也不尽相同。总体而言，与传统数值模拟方法相比，固液相变格子 Boltzmann 模型在处理移动相界面边界、实现固相自由运动等方面具有一定的优势和特色。

习　题

6-1　根据 6.1 节所介绍的颗粒运动的流固两相流模型，利用伪代码或流程图的形式

写出模型的计算步骤，包括颗粒边界节点速度计算、流体节点分布函数的碰撞迁移、颗粒边界处的半步长反弹、颗粒受力和力矩计算、颗粒刚体运动方程计算、新流体节点再填充和进出固体颗粒节点与颗粒的动量交换等步骤。

6-2　考虑图 6.6 所示的情况，固体边界为竖直边界，沿 x 方向以速度 U 匀速运动。以流体边界节点 $\boldsymbol{r}_{\mathrm{f}}$ 为例，采用半步长反弹格式即式(6.6)进行边界处理，试分析流体区域的质量变化。

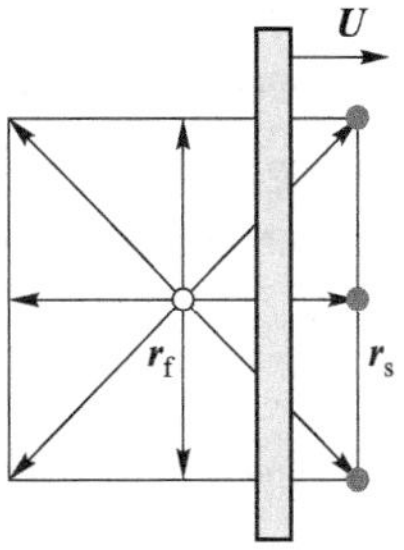

图 6.6　习题 6-2 附图

6-3　针对修正的平衡态焓分布函数(6.48)，采用 D2Q9 模型，试列出分布函数的零阶～二阶矩，结合 4.2 节的内容，对该模型进行 Chapman-Enskog 多尺度展开分析，推导其对应的宏观焓方程及导热系数的表达式。对照平衡态分布函数(6.44)，分析式(6.48)如何实现导热系数与比热容的独立自由调节。

6-4　针对式(6.42)～式(6.47)所介绍的显式焓法固液相变格子 Boltzmann 模型，利用伪代码或流程图的形式写出模型的计算步骤(注意焓值与温度的转换)。

6-5　试采用 D3Q7 模型构建三维多松弛焓法固液相变格子 Boltzmann 模型，给出其平衡态矩函数及松弛参数与导热系数的关系，可参考式(4.41)所采用的平衡态分布函数。

参考文献

[1] 林宗虎. 变幻流动的科学:多相流体力学[M]. 北京:清华大学出版社,2000.

[2] 倪晋仁,王光谦,张红武. 固液两相流基本理论及其最新应用[M]. 北京:科学出版社,1991.

[3] 郭照立,郑楚光. 格子 Boltzmann 方法的原理及应用[M]. 北京:科学出版社,2009.

[4] 张寅平,胡汉平,孔祥冬,等. 相变贮能:理论和应用[M]. 合肥:中国科学技术大学出版社,1996.

[5] 闵桂荣,郭舜. 航天器热控制[M]. 北京:科学出版社,1998.

[6] Sharma A, Tyagi V V, Chen C R, et al. Review on thermal energy storage with phase change materials and applications[J]. Renewable and Sustainable Energy Reviews, 2009, 13(2): 318-345.

[7] Tao Y B, He Y L. A review of phase change material and performance enhancement method for latent heat storage system[J]. Renewable and Sustainable Energy Reviews, 2018, 93: 245-259.

[8] Ladd A J C. Numerical simulations of particulate suspensions via a discretized Boltzmann equation. Part 1. Theoretical foundation[J]. Journal of Fluid Mechanics, 1994, 271: 285-309.

[9] Ladd A J C. Numerical simulations of particulate suspensions via a discretized Boltzmann equation. Part Ⅱ. Numerical results[J]. Journal of Fluid Mechanics, 1994, 271: 311-339.

[10] Aidun C K, Clausen J R. Lattice-Boltzmann method for complex flows[J]. Annual Review of Fluid Mechanics, 2010, 42: 439-472.

[11] Aidun C K, Lu Y N, Ding E J. Direct analysis of particulate suspensions with inertia using the discrete Boltzmann equation[J]. Journal of Fluid Mechanics, 1998, 373: 287-311.

[12] Wen B H, Zhang C Y, Tu Y S, et al. Galilean invariant fluid-solid interfacial dynamics in lattice Boltzmann simulations[J]. Journal of Computational Physics, 2014, 266: 161-170.

[13] Chen Y, Cai Q D, Xia Z H, et al. Momentum-exchange method in lattice Boltzmann simulations of particle-fluid interactions[J]. Physical Review E, 2013, 88(1): 013303.

[14] Tao S, Hu J J, Guo Z L. An investigation on momentum exchange methods and refilling algorithms for lattice Boltzmann simulation of particulate flows[J]. Computers & Fluids, 2016, 133: 1-14.

[15] Lallemand P, Luo L S. Lattice Boltzmann method for moving boundaries[J]. Journal of Computational Physics, 2003, 184(2): 406-421.

[16] Guo Z L, Zheng C G, Shi B C. An extrapolation method for boundary conditions in lattice Boltzmann method[J]. Physics of Fluids, 2002, 14(6): 2007-2010.

[17] Chen L, Yu Y, Lu J H, et al. A comparative study of lattice Boltzmann methods using bounce-back schemes and immersed boundary ones for flow acoustic problems[J]. International Journal for Numerical Methods in Fluids, 2014, 74(6): 439-467.

[18] Wen B H, Zhang C Y, Fang H P. Hydrodynamic force evaluation by momentum exchange method in lattice Boltzmann simulations[J]. Entropy, 2015, 17(12): 8240-8266.

[19] He Q, Li Y J, Huang W F, et al. Lattice Boltzmann model for dense suspended particles based on improved bounce-back method[J]. Computers & Mathematics with Applications, 2020, 80(3): 552-567.

[20] Kromkamp J, Van Den Ende D T M, Kandhai D, et al. Shear-induced self-diffusion and microstructure in non-Brownian suspensions at non-zero Reynolds numbers[J]. Journal of Fluid Mechanics, 2005, 529: 253-278.

[21] Kromkamp J, van den Ende D, Kandhai D, et al. Lattice Boltzmann simulation of 2D and 3D non-Brownian suspensions in Couette flow[J]. Chemical Engineering Science, 2006, 61(2): 858-873.

[22] Nguyen N Q, Ladd A J C. Lubrication corrections for lattice-Boltzmann simulations of particle suspensions [J]. Physical Review E, 2002, 66(4): 046708.

[23] Wan D C, Turek S. Direct numerical simulation of particulate flow via multigrid FEM techniques and the fictitious boundary method[J]. International Journal for Numerical Methods in Fluids, 2010, 51(5): 531-566.

[24] 陈江平,黄家荣,范宇峰,等. "阿波罗"登月飞行器热控系统方案概述[J]. 载人航天,2012,18(1): 40-47.

[25] 邵兴国,向艳超,谭沧海. 嫦娥一号卫星热控设计中热管的应用及验证[J]. 航天器工程,2008,17(1): 63-67.

[26] He Y L, Liu Q, Li Q, et al. Lattice Boltzmann methods for single-phase and solid-liquid phase-change heat transfer in porous media: A review[J]. International Journal of Heat and Mass Transfer, 2019, 129: 160-197.

[27] de Fabritiis G, Mancini A, Mansutti D, et al. Mesoscopic models of liquid/solid phase transitions[J]. International Journal of Modern Physics C, 1998, 9(8): 1405-1415.

[28] Huang R Z, Wu H Y. An immersed boundary-thermal lattice Boltzmann method for solid-liquid phase change[J]. Journal of Computational Physics, 2014, 277: 305-319.

[29] Li Z, Yang M, Zhang Y W. Numerical simulation of melting problems using the lattice Boltzmann method with the interfacial tracking method[J]. Numerical Heat Transfer, Part A: Applications, 2015, 68(11): 1175-1197.

[30] Miller W, Succi S, Mansutti D. Lattice Boltzmann model for anisotropic liquid-solid phase transition [J]. Physical Review Letters, 2001, 86(16): 3578-3581.

[31] Miller W, Succi S. A lattice Boltzmann model for anisotropic crystal growth from melt[J]. Journal of Statistical Physics, 2002, 107(1-2): 173-186.

[32] Miller W. The lattice Boltzmann method: A new tool for numerical simulation of the interaction of growth kinetics and melt flow[J]. Journal of Crystal Growth, 2001, 230(1): 263-269.

[33] Miller W, Rasin I, Succi S. Lattice Boltzmann phase-field modelling of binary-alloy solidification[J]. Physica A: Statistical Mechanics and its Applications, 2006, 362(1): 78-83.

[34] Rasin I, Miller W, Succi S. Phase-field lattice kinetic scheme for the numerical simulation of dendritic

growth[J]. Physical Review E,2005,72(6):066705.

[35] Medvedev D, Kassner K. Lattice Boltzmann scheme for crystal growth in external flows [J]. Physical Review E,2005,72(5):056703.

[36] Medvedev D, Kassner K. Lattice-Boltzmann scheme for dendritic growth in presence of convection [J]. Journal of Crystal Growth,2005,275(1-2):e1495-e1500.

[37] Karma A,Rappel W-J. Quantitative phase-field modeling of dendritic growth in two and three dimensions [J]. Physical Review E,1998,57(4):4323-4349.

[38] Rojas R,Takaki T,Ohno M. A phase-field-lattice Boltzmann method for modeling motion and growth of a dendrite for binary alloy solidification in the presence of melt convection[J]. Journal of Computational Physics,2015,298:29-40.

[39] Ohno M,Matsuura K. Quantitative phase-field modeling for dilute alloy solidification involving diffusion in the solid[J]. Physical Review E,2009,79(3):031603.

[40] Chakraborty S,Chatterjee D. An enthalpy-based hybrid lattice-Boltzmann method for modelling solid-liquid phase transition in the presence of convective transport [J]. Journal of Fluid Mechanics, 2007, 592: 155-175.

[41] Brent A D, Voller V R, Reid K J. Enthalpy-porosity technique for modeling convection-diffusion phase change:Application to the melting of a pure metal[J]. Numerical Heat Transfer, Part A: Applications, 1988,13(3):297-318.

[42] Jiaung W S, Ho J R, Kuo C P. Lattice Boltzmann method for the heat conduction problem with phase change[J]. Numerical Heat Transfer:Part B:Fundamentals,2001,39(2):167-187.

[43] Chatterjee D,Chakraborty S. An enthalpy-source based lattice Boltzmann model for conduction dominated phase change of pure substances[J]. International Journal of Thermal Sciences,2008,47(5):552-559.

[44] Chatterjee D,Chakraborty S. A hybrid lattice Boltzmann model for solid-liquid phase transition in presence of fluid flow[J]. Physics Letters A,2006,351(4):359-367.

[45] Chatterjee D. An enthalpy-based thermal lattice Boltzmann model for non-isothermal systems [J]. Europhysics Letters,2009,86(1):14004.

[46] Chatterjee D. Lattice Boltzmann smulation of incompressible transport phenomena in macroscopic solidification processes[J]. Numerical Heat Transfer,Part B:Fundamentals,2010,58(1):55-72.

[47] Huber C, Parmigiani A, Chopard B, et al. Lattice Boltzmann model for melting with natural convection [J]. International Journal of Heat and Fluid Flow,2008,29(5):1469-1480.

[48] Luo K,Yao F J,Yi H L,et al. Lattice Boltzmann simulation of convection melting in complex heat storage systems filled with phase change materials[J]. Applied Thermal Engineering,2015,86:238-250.

[49] Eshraghi M, Felicelli S D. An implicit lattice Boltzmann model for heat conduction with phase change [J]. International Journal of Heat and Mass Transfer,2012,55(9):2420-2428.

[50] Feng Y C,Li H X,Li L X,et al. Numerical investigation on the melting of nanoparticle-enhanced phase change materials (NEPCM) in a bottom-heated rectangular cavity using lattice Boltzmann method [J]. International Journal of Heat and Mass Transfer,2015,81:415-425.

[51] Huang R Z, Wu H Y, Cheng P. A new lattice Boltzmann model for solid-liquid phase change [J]. International Journal of Heat and Mass Transfer,2013,59:295-301.

[52] Huang R Z,Wu H Y. Phase interface effects in the total enthalpy-based lattice Boltzmann model for solid-liquid phase change[J]. Journal of Computational Physics,2015,294:346-362.

[53] Li D, Ren Q L, Tong Z X, et al. Lattice Boltzmann models for axisymmetric solid-liquid phase change [J]. International Journal of Heat and Mass Transfer,2017,112:795-804.

[54] Li D,Tong Z X,Ren Q L,et al. Three-dimensional lattice Boltzmann models for solid-liquid phase change

[J]. International Journal of Heat and Mass Transfer, 2017, 115: 1334–1347.

[55] Hu Y, Li D C, Shu S, et al. Lattice Boltzmann simulation for three-dimensional natural convection with solid-liquid phase change[J]. International Journal of Heat and Mass Transfer, 2017, 113: 1168–1178.

[56] Lu J H, Lei H Y, Dai C S. An optimal two-relaxation-time lattice Boltzmann equation for solid-liquid phase change: The elimination of unphysical numerical diffusion[J]. International Journal of Thermal Sciences, 2019, 135: 17–29.

[57] Zhao Y, Wang L, Chai Z H, et al. Comparative study of natural convection melting inside a cubic cavity using an improved two-relaxation-time lattice Boltzmann model[J]. International Journal of Heat and Mass Transfer, 2019, 143: 118449.

[58] Zhao Y, Pereira G G, Kuang S B, et al. A generalized lattice Boltzmann model for solid-liquid phase change with variable density and thermophysical properties [J]. Applied Mathematics Letters, 2020, 104: 106250.

[59] Ren Q L, Meng F L, Guo P H. A comparative study of PCM melting process in a heat pipe-assisted LHTES unit enhanced with nanoparticles and metal foams by immersed boundary-lattice Boltzmann method at pore-scale[J]. International Journal of Heat and Mass Transfer, 2018, 121: 1214–1228.

[60] Noble D R, Torczynski J R. A lattice-Boltzmann method for partially saturated computational cells [J]. International Journal of Modern Physics C, 1998, 9(8): 1189–1201.

[61] Cook B K, Noble D R, Williams J R. A direct simulation method for particle-fluid systems[J]. Engineering Computations, 2004, 21: 151–168.

[62] Huang R Z, Wu H Y. Total enthalpy-based lattice Boltzmann method with adaptive mesh refinement for solid-liquid phase change[J]. Journal of Computational Physics, 2016, 315: 65–83.

第 7 章　多孔介质流动与传热的格子 Boltzmann模型

多孔介质是指由固体物质组成的骨架以及由骨架分隔成大量密集成群的微小孔隙所构成的物质。微小孔隙中的介质可以是气体、液体或气液两相流体。相对于其中一相来说，其他相弥散在其中。多孔介质中的流动与传热是自然界常见的现象，同时也是应用科学和工程技术领域所广泛涉及的一类基础问题[1]。例如，油气资源开发和二氧化碳地质封存中的多孔介质渗流、利用多孔介质骨架强化相变储热[2]、泡沫陶瓷太阳能集热单元中的流动与传热[3]、航空航天隔热材料内的热量传递[4]等。数值模拟是多孔介质流动与传热问题的重要研究手段之一。格子 Boltzmann 方法具有清晰的粒子运动图像，能够非常方便地从底层刻画流体与固体之间的相互作用，从而在多孔介质流动与传热的数值模拟特别是孔隙尺度模拟方面具有显著的优势。

二氧化碳封存

多孔介质中的流动与传热问题是一个典型的多尺度问题，一般涉及三个尺度，即宏观区域尺度、**表征体元**（representative elementary volume，REV）尺度和**孔隙尺度**（pore scale），如图 7.1 所示。在孔隙尺度上，通常研究流体在单个或若干个微孔内的输运过程，固体骨架被视作流场的边界，流体与固体骨架之间的流固相互作用通过适当的边界条件予以反映。REV 尺度则比孔隙尺度大得多，它是指多孔介质中包含足够多微孔的一个控制体积，类似于流体力学中的流体微团。相对于宏观的多孔介质区域，REV 可以看作是一个点，且 REV 中的基本参数随空间变化的幅度很小，因此可以将 REV 作为基本单元来定义相应的（体积平均）流体物理量，如密度、速度、压力、温度等，并建立这些物理量的守恒方程。一般来说，孔隙尺度的模拟可用来探索微观渗流的机理和基本规律，而 REV 尺度的模拟则往往应用于大尺度的工程渗流问题。

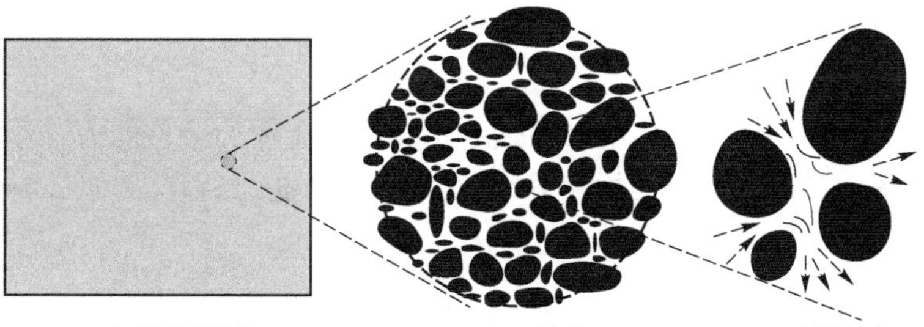

(a) 宏观区域尺度　　(b) REV尺度　　(c) 孔隙尺度

图 7.1　多孔介质流动与传热涉及的三个尺度[5]

相应地，格子 Boltzmann 方法在孔隙尺度和 REV 尺度也发展出了相应的模型。其中，孔隙尺度的格子 Boltzmann 模拟可以充分利用该方法粒子运动图像清晰和边界条件处理简单的优势，采用半步长反弹等边界处理格式，实现孔隙尺度复杂多孔结构中流动与传热问题的数值计算；REV 尺度的格子 Boltzmann 模型则通过修改平衡态分布函数和作用力格式等，在模型中引入孔隙率、热容比等物性参数以及反映多孔介质影响的 REV 尺度作用力项，其构建过程充分体现了格子 Boltzmann 方法的灵活性。接下来，将分别介绍 REV 尺度和孔隙尺度的格子 Boltzmann 模型以及多孔介质固液相变格子 Boltzmann 模型。

7.1 表征体元尺度格子 Boltzmann 模型

7.1.1 REV 尺度流动与传热的控制方程

宏观或表观平均物理量是多孔介质 REV 尺度模拟的基础。假设表征体元的体积为 V，而表征体元内流体区域的体积为 V_f，则与流体相关的物理量 φ 的宏观平均可定义为

$$\overline{\varphi}=\frac{1}{V}\int_{V_f}\varphi \mathrm{d}V \tag{7.1}$$

当 $\varphi=1$ 时，该平均值为多孔介质的孔隙率 ϕ，即多孔介质内的孔隙体积占多孔介质总体积的比例。当 φ 为流体速度时，可以得到流体的表观平均速度，即 Darcy 速度或渗流速度。此外，物理量的固有平均为该物理量在流体区域的平均值，即

$$\langle\varphi\rangle=\frac{1}{V_f}\int_{V_f}\varphi \mathrm{d}V \tag{7.2}$$

以此为基础，可通过对质量、动量及能量守恒方程进行平均，引入适当的封闭关系，建立 REV 尺度的宏观控制方程。例如，REV 尺度的质量守恒方程为

$$\frac{\partial\overline{\rho}}{\partial t}+\nabla\cdot(\overline{\rho}\,\overline{\boldsymbol{u}})=0 \tag{7.3}$$

在不可压缩条件下，该方程可以简化为

$$\nabla\cdot\overline{\boldsymbol{u}}=0 \tag{7.4}$$

式中，$\overline{\boldsymbol{u}}$ 为渗流速度。为了叙述方便，在后面的表述中将该速度简写为 $\boldsymbol{u}$。

基于一些适当的简化假设，可以建立起反映平均物理量变化规律的渗流模型，如 Darcy 模型、Brinkman-Darcy 模型、Forchheimer-Darcy 模型等。其中，最基本的渗流模型是 Darcy 模型，即

$$\boldsymbol{u}=-\frac{K}{\mu}\nabla p \tag{7.5}$$

式中：K 为多孔介质的渗透率，它用来描述多孔介质对流体的渗透能力；μ 为流体的动力黏度。渗透率 K 与多孔介质的结构相关，一般是孔隙率 ϕ 的函数，例如 Kozeny-Carman 关系式给出了堆积床结构的渗透率

$$K=a\,\frac{\phi^3 D_p^2}{(1-\phi)^2} \tag{7.6}$$

式中：D_p 为堆积颗粒的等效直径；a 为与多孔材料相关的常数。

Darcy 方程适用于流速较低的多孔介质流动（一般指基于孔尺度的雷诺数 $Re_p<1$）。当流速较高时，需要对 Darcy 方程进行修正，例如引入非线性阻力项建立 Forchheimer-Darcy 方程

$$\nabla p=-\frac{\mu}{K}\boldsymbol{u}-\frac{F_{\phi}}{\sqrt{K}}\rho|\boldsymbol{u}|\boldsymbol{u} \tag{7.7}$$

式中，F_{ϕ} 为阻力系数，其值与多孔介质的结构有关，且随着流动向湍流状态的转变（$Re_{\mathrm{p}}>100$），在一定速度范围内与速度呈线性关系。Darcy 方程的另一类修正是 Brinkman-Darcy 方程，即

$$\nabla p=-\frac{\mu}{K}\boldsymbol{u}+\mu_{\mathrm{e}}\nabla^{2}\boldsymbol{u} \tag{7.8}$$

式中，μ_{e} 为等效动力黏度。研究表明，该方程适用于多孔介质孔隙率较高的情况。

在 REV 尺度的格子 Boltzmann 模型研究中，目前广泛采用的是综合考虑惯性项、线性阻力（Darcy）项、Brinkman 黏性项和 Forchheimer 非线性阻力项的广义非达西方程，即

$$\frac{\partial\boldsymbol{u}}{\partial t}+(\boldsymbol{u}\cdot\nabla)\left(\frac{\boldsymbol{u}}{\phi}\right)=-\frac{1}{\rho}\nabla(\phi p)+\nu_{\mathrm{e}}\nabla^{2}\boldsymbol{u}+\boldsymbol{F} \tag{7.9}$$

式中：ρ 为密度；p 为体积平均的压力；ν_{e} 为等效运动黏度。作用力 $\boldsymbol{F}$ 包含 Forchheimer-Darcy 模型的介质阻力以及流体所受到的外力，其表达式为

$$\boldsymbol{F}=-\frac{\phi\nu}{K}\boldsymbol{u}-\frac{\phi F_{\phi}}{\sqrt{K}}|\boldsymbol{u}|\boldsymbol{u}+\phi\boldsymbol{G} \tag{7.10}$$

式中：第一项和第二项分别对应线性和非线性阻力；$\nu=\mu/\rho$ 为流体的运动黏度；$\boldsymbol{G}$ 为外力场施加的体积力。广义非达西方程可通过对 Navier-Stokes 方程在 REV 尺度进行体积平均，并引入适当的封闭关系获得。由于引入了体积平均，与式（7.7）相比，式（7.10）的线性和非线性阻力项均出现了孔隙率 ϕ。

对于多孔介质中的传热过程，可针对固体骨架和流体区域分别建立传热方程，即

$$(1-\phi)(\rho c)_{\mathrm{s}}\frac{\partial T_{\mathrm{s}}}{\partial t}=\nabla\cdot(k_{\mathrm{e,s}}\nabla T_{\mathrm{s}})+(1-\phi)\dot{\Phi}_{\mathrm{s}}+h_{v}(T_{\mathrm{f}}-T_{\mathrm{s}}) \tag{7.11}$$

$$\phi(\rho c_{p})_{\mathrm{f}}\frac{\partial T_{\mathrm{f}}}{\partial t}+(\rho c_{p})_{\mathrm{f}}\boldsymbol{u}\cdot\nabla T_{\mathrm{f}}=\nabla\cdot(k_{\mathrm{e,f}}\nabla T_{\mathrm{f}})+\phi\dot{\Phi}_{\mathrm{f}}+h_{v}(T_{\mathrm{s}}-T_{\mathrm{f}}) \tag{7.12}$$

式中：下标“s”和“f”分别表示固体和流体；c 为固体的比热容；c_{p} 为流体的比定压热容；k_{e} 为等效导热系数；$\dot{\Phi}$ 为内热源所产生的单位体积加热速率；h_{v} 为固体和流体间的容积换热系数。

根据多孔介质局部的骨架温度与流体温度是否相等，可将多孔介质传热模型分为局部热平衡模型和局部非热平衡模型。对于局部非热平衡模型，需分别求解式（7.11）和式（7.12），获得固体骨架和流体的温度分布 T_{s} 和 T_{f}；局部热平衡模型则假定固体骨架和流体的温度相等，即 $T_{\mathrm{s}}=T_{\mathrm{f}}=T$，因此可将式（7.11）和式（7.12）求和，得到多孔介质整体的传热方程

$$(\rho c)_{\mathrm{m}}\frac{\partial T}{\partial t}+(\rho c_{p})_{\mathrm{f}}\boldsymbol{u}\cdot\nabla T=\nabla\cdot(k_{\mathrm{e}}\nabla T)+\dot{\Phi} \tag{7.13}$$

式中，多孔介质的表观热容、表观导热系数和表观内热源分别为

$$(\rho c)_{\mathrm{m}}=(1-\phi)(\rho c)_{\mathrm{s}}+\phi(\rho c_{p})_{\mathrm{f}} \tag{7.14}$$

$$k_{\mathrm{e}}=k_{\mathrm{e,s}}+k_{\mathrm{e,f}} \tag{7.15}$$

$$\dot{\Phi}=(1-\phi)\dot{\Phi}_{\mathrm{s}}+\phi\dot{\Phi}_{\mathrm{f}} \tag{7.16}$$

式(7.14)和式(7.16)表明,多孔介质的表观热容和表观内热源分别是固体和流体变量的体积平均。但是,多孔介质中固体和流体的等效导热系数,以及多孔介质的表观导热系数,均与多孔介质的具体结构有关,不能简单地表示为体积平均。在简化的情况下,如果假设多孔介质由许多相互平行的通道组成,且传热方向与通道方向平行,则表观导热系数为固体和流体导热系数的体积平均,即

$$k_{\mathrm{A}}=(1-\phi)k_{\mathrm{s}}+\phi k_{\mathrm{f}} \tag{7.17}$$

如果假设多孔介质为分层结构且传热方向依次经过固体和流体,则表观导热系数为固体和流体导热系数的调和平均:

$$\frac{1}{k_{\mathrm{H}}}=\frac{(1-\phi)}{k_{\mathrm{s}}}+\frac{\phi}{k_{\mathrm{f}}} \tag{7.18}$$

多孔介质实际的表观导热系数通常在 k_{A} 和 k_{H} 之间。在应用中,k_{e} 的另一种近似估算方法为

$$k_{\mathrm{e}}=k_{\mathrm{s}}^{1-\phi}+k_{\mathrm{f}}^{\phi} \tag{7.19}$$

此外,利用孔隙尺度的传热计算,可以给出更精确的等效导热系数。

7.1.2 REV 尺度多孔介质流动的格子 Boltzmann 模型

1997 年,Spaid 和 Phelan[6] 发展了基于 Brinkman-Darcy 方程的 REV 尺度格子 Boltzmann 模型,相应的演化方程为

$$f_{\alpha}(\boldsymbol{r}+\boldsymbol{e}_{\alpha}\delta_t,t+\delta_t)-f_{\alpha}(\boldsymbol{r},t)=-\frac{1}{\tau}[f_{\alpha}-f_{\alpha}^{\mathrm{eq}}(\rho,\boldsymbol{u}^{\mathrm{eq}})] \tag{7.20}$$

式中,平衡态密度分布函数$f_{\alpha}^{\mathrm{eq}}(\rho,\boldsymbol{u}^{\mathrm{eq}})$中的速度定义如下:

$$\boldsymbol{u}^{\mathrm{eq}}=\boldsymbol{u}+s(\boldsymbol{r})\frac{\tau\boldsymbol{F}}{\rho} \tag{7.21}$$

式中:$\boldsymbol{F}=-\mu\boldsymbol{u}/K$ 为介质阻力;$s(\boldsymbol{r})$为一个指标函数,对多孔介质区取 1,对自由流动区取 0。因此,Spaid 和 Phelan 的模型是通过修正平衡态密度分布函数中的速度来体现流体与固体之间的相互影响。随后,Spaid 和 Phelan[7] 将他们的模型与多组分伪势模型[8] 相结合,构建了用于多孔介质多组分流动的格子 Boltzmann 模型。

2001 年,Martys[9] 在 Spaid 和 Phelan 研究的基础上提出了一个改进模型,其中介质阻力采用 Luo 所提出的作用力格式[10] 来描述。此外,Dardis 和 McCloskey[11-12] 提出了间接求解 Brinkman-Darcy 方程的格子 Boltzmann 模型,其演化方程为

$$f_{\alpha}(\boldsymbol{r}+\boldsymbol{e}_{\alpha}\delta_t,t+\delta_t)-f_{\alpha}(\boldsymbol{r},t)=-\frac{1}{\tau}(f_{\alpha}-f_{\alpha}^{\mathrm{eq}})+\Delta f_{\alpha}^{\mathrm{PM}}(\boldsymbol{r}+\boldsymbol{e}_{\alpha}\delta_t,t) \tag{7.22}$$

式中,$\Delta f_{\alpha}^{\mathrm{PM}}$为部分反弹项。

$$\Delta f_{\alpha}^{\mathrm{PM}}(\boldsymbol{r}+\boldsymbol{e}_{\alpha}\delta_t,t)=n_s(\boldsymbol{r}+\boldsymbol{e}_{\alpha}\delta_t)[f_{\bar{\alpha}}(\boldsymbol{r}+\boldsymbol{e}_{\alpha}\delta_t,t)-f_{\alpha}(\boldsymbol{r},t)] \tag{7.23}$$

式中:n_s 表示每一固体格点的平均固体散射密度(取值在 0 和 1 之间的连续变量);$\bar{\alpha}$ 表示与方向 α 相反的离散速度方向。部分反弹项 $\Delta f_{\alpha}^{\mathrm{PM}}$表示流体粒子与固体骨架之间的碰撞效应。当 $n_s=0$ 时,$\Delta f_{\alpha}^{\mathrm{PM}}=0$;当 $n_s=1$ 时,$\Delta f_{\alpha}^{\mathrm{PM}}$的作用类似于反弹格式。从本质上来说,$\Delta f_{\alpha}^{\mathrm{PM}}$反映了固体骨架对流动的阻力作用。

由于 Brinkman-Darcy 方程只适用于低速流动,并且要求多孔介质的孔隙率足够高,从而限制了它的应用范围。2002 年,Guo 和 Zhao[13] 基于广义非达西方程,即式(7.9),构

造了相应的 REV 尺度多孔介质流动的格子 Boltzmann 模型，其基本思路是将孔隙率引入到格子 Boltzmann 模型的平衡态密度分布函数中，并在外力项中引入线性和非线性阻力，相应的演化方程为

$$f_\alpha(\boldsymbol{r}+\boldsymbol{e}_\alpha\delta_t,t+\delta_t)-f_\alpha(\boldsymbol{r},t)=-\frac{1}{\tau}(f_\alpha-f_\alpha^{\mathrm{eq}})+\rho\omega_\alpha\left(1-\frac{1}{2\tau}\right)\delta_t\left[\frac{\boldsymbol{e}_\alpha\cdot\boldsymbol{F}}{c_s^2}+\frac{\boldsymbol{uF}:(\boldsymbol{e}_\alpha\boldsymbol{e}_\alpha-c_s^2\boldsymbol{I})}{\phi c_s^4}\right] \tag{7.24}$$

式中，作用力 $\boldsymbol{F}$ 由式(7.10)给出，平衡态密度分布函数为

$$f_\alpha^{\mathrm{eq}}=\rho\omega_\alpha\left[1+\frac{\boldsymbol{e}_\alpha\cdot\boldsymbol{u}}{c_s^2}+\frac{\boldsymbol{uu}:(\boldsymbol{e}_\alpha\boldsymbol{e}_\alpha-c_s^2\boldsymbol{I})}{2\phi c_s^4}\right] \tag{7.25}$$

对式(7.25)求矩可以发现，平衡态密度分布函数满足下列关系式：

$$\sum_{\alpha=0}^{8}f_\alpha^{\mathrm{eq}}=\rho,\quad \sum_{\alpha=0}^{8}\boldsymbol{e}_\alpha f_\alpha^{\mathrm{eq}}=\rho\boldsymbol{u},\quad \sum_{\alpha=0}^{8}\boldsymbol{e}_\alpha\boldsymbol{e}_\alpha f_\alpha^{\mathrm{eq}}=\rho c_s^2\boldsymbol{I}+\frac{\rho\boldsymbol{uu}}{\phi} \tag{7.26}$$

通过在 f_α^{eq} 的二阶矩中引入 ϕ，从而可以在惯性项中体现与广义非达西方程相对应的孔隙率。密度和速度分别定义如下：

$$\rho=\sum_{\alpha=0}^{8}f_\alpha,\quad \rho\boldsymbol{u}=\sum_{\alpha=0}^{8}\boldsymbol{e}_\alpha f_\alpha+\frac{\delta_t}{2}\rho\boldsymbol{F} \tag{7.27}$$

由于作用力 $\boldsymbol{F}$ 是速度 $\boldsymbol{u}$ 的二次函数，因此需要求解式(7.27)获得速度 $\boldsymbol{u}$ 的显式表达式。若令

$$\rho\boldsymbol{v}=\sum_{\alpha=0}^{8}\boldsymbol{e}_\alpha f_\alpha+\frac{\delta_t}{2}\rho\phi\boldsymbol{G} \tag{7.28}$$

$$c_0=\frac{1}{2}\left(1+\phi\frac{\delta_t}{2}\frac{\nu}{K}\right),\quad c_1=\phi\frac{\delta_t}{2}\frac{F_\phi}{\sqrt{K}} \tag{7.29}$$

则速度 $\boldsymbol{u}$ 的计算式为

$$\boldsymbol{u}=\frac{\boldsymbol{v}}{c_0+\sqrt{c_0^2+c_1|\boldsymbol{v}|}} \tag{7.30}$$

通过 Chapman-Enskog 多尺度展开分析，可得该模型所对应的宏观方程为

$$\frac{\partial\rho}{\partial t}+\nabla\cdot(\rho\boldsymbol{u})=0 \tag{7.31}$$

$$\frac{\partial(\rho\boldsymbol{u})}{\partial t}+\nabla\cdot\left(\frac{\rho\boldsymbol{uu}}{\phi}\right)=-\nabla(\phi p)+\nabla\cdot\{\rho\nu_e[\nabla\boldsymbol{u}+(\nabla\boldsymbol{u})^{\mathrm{T}}]\}+\rho\boldsymbol{F} \tag{7.32}$$

式中：$p=\rho c_s^2/\phi$；$\nu_e=c_s^2(\tau-0.5)\delta_t$。在不可压缩条件下（密度 ρ 近似为常数，$\nabla\cdot\boldsymbol{u}=0$），式(7.32)退化为广义非达西方程，即式(7.9)。该模型具有以下几个特点[5,13]：(1)消除了离散效应所引起的数值模拟误差，可以恢复准确的宏观方程；(2)孔隙率被引入平衡态密度分布函数中，使得该模型可用于均质及非均质多孔介质流动问题，并可用于瞬态流动；(3)该模型不但包含线性介质阻力，而且包含 Forchheimer 非线性介质阻力，可用于流速较大的多孔介质流动问题。

2014 年，Liu 等[14]基于广义非达西方程构造了 REV 尺度的多松弛格子 Boltzmann 模型，演化方程与式(3.252)一致，平衡态矩函数为

$$\boldsymbol{m}^{\mathrm{eq}}=\rho\left(1,-2+\frac{3u^2}{\phi},\beta_1+\beta_2\frac{u^2}{\phi},u_x,-u_x,u_y,-u_y,\frac{u_x^2-u_y^2}{\phi},\frac{u_xu_y}{\phi}\right)^{\mathrm{T}} \tag{7.33}$$

式中：$\beta_1=1$；$\beta_2=-3$。该模型采用的作用力项为

$$\boldsymbol{S}=\rho\left(0,\frac{6\boldsymbol{u}\cdot\boldsymbol{F}}{\phi},-\frac{6\boldsymbol{u}\cdot\boldsymbol{F}}{\phi},F_x,-F_x,F_y,-F_y,2\frac{u_xF_x-u_yF_y}{\phi},\frac{u_xF_y+u_yF_x}{\phi}\right)^{\mathrm{T}} \tag{7.34}$$

当不同矩函数的松弛因子相等时，上述多松弛模型退化为 Guo 和 Zhao 的单松弛模型。为了消除该模型的可压缩效应，Liu 和 He[15] 随后构建了改进的 REV 尺度多松弛格子 Boltzmann 模型，具体可参阅文献[14,15]。

7.1.3 REV 尺度多孔介质传热的格子 Boltzmann 模型

1）局部热平衡模型

如前所述，根据多孔介质局部的骨架温度与流体温度是否相等，可将 REV 尺度的多孔介质传热模型分为局部热平衡模型和局部非热平衡模型。其中，局部热平衡模型认为固体骨架与流体之间的温差可以忽略，因此它们之间没有热量交换，相应的控制方程由式(7.13)给出，也即

$$\sigma\frac{\partial T}{\partial t}+\boldsymbol{u}\cdot\nabla T=\nabla\cdot(a_{\mathrm{e}}\nabla T)+\dot{\varphi} \tag{7.35}$$

式中，$\sigma=(\rho c)_{\mathrm{m}}/(\rho c_p)_{\mathrm{f}}$ 为多孔介质（包括固体骨架与流体）与流体的热容比；$a_{\mathrm{e}}=k_{\mathrm{e}}/(\rho c_p)_{\mathrm{f}}$ 为等效热扩散系数；$\dot{\varphi}=\dot{\Phi}/(\rho c_p)_{\mathrm{f}}$。

2005 年，Guo 和 Zhao[16] 将 REV 尺度多孔介质流动的格子 Boltzmann 模型推广传热问题，根据被动标量的概念，他们采用如下形式的温度分布函数演化方程：

$$g_\alpha(\boldsymbol{r}+\boldsymbol{e}_\alpha\delta_t,t+\delta_t)-g_\alpha(\boldsymbol{r},t)=-\frac{1}{\tau_T}[g_\alpha(\boldsymbol{r},t)-g_\alpha^{\mathrm{eq}}(\boldsymbol{r},t)] \tag{7.36}$$

式中，平衡态温度分布函数为

$$g_\alpha^{\mathrm{eq}}=\omega_\alpha T\left(\sigma+\frac{\boldsymbol{e}_\alpha\cdot\boldsymbol{u}}{c_{\mathrm{s}}^2}\right) \tag{7.37}$$

该模型采用 D2Q9 格子，格子声速 $c_{\mathrm{s}}=c/\sqrt{3}$。对式(7.37)求矩可得

$$\sum_{\alpha=0}^{8}g_\alpha^{\mathrm{eq}}=\sigma T,\quad \sum_{\alpha=0}^{8}\boldsymbol{e}_\alpha g_\alpha^{\mathrm{eq}}=\boldsymbol{u}T,\quad \sum_{\alpha=0}^{8}\boldsymbol{e}_\alpha\boldsymbol{e}_\alpha g_\alpha^{\mathrm{eq}}=\sigma T\boldsymbol{I} \tag{7.38}$$

通过 Chapman-Enskog 多尺度展开分析，可得式(7.36)所对应的宏观方程为

$$\frac{\partial(\sigma T)}{\partial t}+\nabla\cdot(\boldsymbol{u}T)=\nabla\cdot[c_{\mathrm{s}}^2(\tau_T-0.5)\delta_t\nabla(\sigma T)+\delta_t(\tau_T-0.5)\varepsilon_K\partial_{t_1}(\boldsymbol{u}T)] \tag{7.39}$$

式(7.39)右端与 $\varepsilon_K\partial_{t_1}(\boldsymbol{u}T)$ 有关的项为偏差项，其中 ε_K 为 Chapman-Enskog 多尺度展开的展开系数。关于这一偏差项的产生可参阅本书 4.2 节的分析。对于不可压缩流动与传热问题，这一偏差项的影响往往可以忽略。

当热容比 $\sigma=(\rho c)_{\mathrm{m}}/(\rho c_p)_{\mathrm{f}}$ 为常数或其空间变化较小时，$\nabla(\sigma T)\approx\sigma\nabla T$，此时上述模型可以恢复相应的多孔介质传热方程，且等效热扩散系数 a_{e} 与松弛因子 τ_T 之间的关系为 $a_{\mathrm{e}}=\sigma c_{\mathrm{s}}^2(\tau_T-0.5)\delta_t$。当多孔介质不同区域的热容比 σ 有显著差异时，$\nabla(\sigma T)\neq\sigma\nabla T$，该模型将带来一定的数值模拟误差。为了解决这一问题，Chen 等[17] 通过引入参考热容比 σ_0 构建了如下形式的平衡态温度分布函数：

$$g_\alpha^{\mathrm{eq}}=\begin{cases}\sigma T-(1-\omega_0)\sigma_0 T, & \alpha=0\\ \omega_\alpha T\left(\sigma_0+\dfrac{\boldsymbol{e}_\alpha\cdot\boldsymbol{u}}{c_{\mathrm{s}}^2}\right), & \alpha\neq 0\end{cases}\tag{7.40}$$

上述平衡态温度分布函数所对应的零阶到二阶矩分别为

$$\sum_{\alpha=0}^{8}g_\alpha^{\mathrm{eq}}=\sigma T,\quad \sum_{\alpha=0}^{8}\boldsymbol{e}_\alpha g_\alpha^{\mathrm{eq}}=\boldsymbol{u}T,\quad \sum_{\alpha=0}^{8}\boldsymbol{e}_\alpha\boldsymbol{e}_\alpha g_\alpha^{\mathrm{eq}}=\sigma_0 T\boldsymbol{I}\tag{7.41}$$

对比式(7.41)与式(7.38)可以发现,式(7.40)的二阶矩已与热容比 σ 解耦,因此式(7.39)中的$\nabla(\sigma T)$将由$\nabla(\sigma_0 T)$替换。由于 σ_0 为常数,从而有$\nabla(\sigma_0 T)\equiv\sigma_0\nabla T$,因此该模型可以适用于热容比 σ 为变量的多孔介质传热问题。

与 Chen 等[17]的研究类似,刘清[5]基于 D2Q5 格子构造了如下形式的平衡态温度分布函数:

$$g_\alpha^{\mathrm{eq}}=\begin{cases}\sigma T-\varpi T, & \alpha=0\\ \dfrac{1}{4}\varpi T\left(1+\dfrac{\boldsymbol{e}_\alpha\cdot\boldsymbol{u}}{c_{\mathrm{s}T}^2}\right), & \alpha\neq 0\end{cases}\tag{7.42}$$

式中:ϖ 为常数,且 $0<\varpi<1$;格子声速 $c_{sT}=c\sqrt{\varpi/2}$。采用多松弛碰撞算子时,可得如下形式的平衡态矩函数[5]:

$$\boldsymbol{n}^{\mathrm{eq}}=T(\sigma,u_x,u_y,-4\sigma+5\varpi,0)^{\mathrm{T}}\tag{7.43}$$

上述建模思路可拓展至三维,具体可参阅 Liu 等[18]的研究工作。

2) 局部非热平衡模型

对于不可压缩流动($\nabla\cdot\boldsymbol{u}=0$),多孔介质局部非热平衡传热过程的控制方程式(7.11)和式(7.12)可转化为

$$\frac{\partial T_{\mathrm{s}}}{\partial t}=\nabla\cdot(a_{\mathrm{e,s}}\nabla T_{\mathrm{s}})+\frac{\dot{\Phi}_{\mathrm{s}}}{(\rho c)_{\mathrm{s}}}+\frac{h_v(T_{\mathrm{f}}-T_{\mathrm{s}})}{(1-\phi)(\rho c)_{\mathrm{s}}}\tag{7.44}$$

$$\frac{\partial T_{\mathrm{f}}}{\partial t}+\nabla\cdot\left(\frac{\boldsymbol{u}T_{\mathrm{f}}}{\phi}\right)=\nabla\cdot(a_{\mathrm{e,f}}\nabla T_{\mathrm{f}})+\frac{\dot{\Phi}_{\mathrm{f}}}{(\rho c_p)_{\mathrm{f}}}+\frac{h_v(T_{\mathrm{s}}-T_{\mathrm{f}})}{\phi(\rho c_p)_{\mathrm{f}}}\tag{7.45}$$

式中,固体骨架和流体的等效热扩散系数分别为

$$a_{\mathrm{e,s}}=\frac{k_{\mathrm{e,s}}}{(1-\phi)(\rho c)_{\mathrm{s}}},\quad a_{\mathrm{e,f}}=\frac{k_{\mathrm{e,f}}}{\phi(\rho c_p)_{\mathrm{f}}}\tag{7.46}$$

因此,多孔介质局部非热平衡传热过程的热格子 Boltzmann 模型通常需要采用两个温度分布函数来分别求解式(7.44)和式(7.45)。

根据热格子 Boltzmann 模型的多尺度展开分析,宏观温度方程的对流项产生于平衡态温度分布函数的一阶矩,因此针对流体的平衡态温度分布函数,需要修改其中与($\boldsymbol{e}_\alpha\cdot\boldsymbol{u}$)有关的项。此外,由于热源的存在,需要在温度分布函数的演化方程中加入相应的源项。基于这一思路,Gao 等[19]构建了多孔介质局部非热平衡传热的单松弛格子 Boltzmann 模型,该模型固体和流体区域温度分布函数的演化方程分别为

$$g_{\alpha,\mathrm{s}}(\boldsymbol{r}+\boldsymbol{e}_\alpha\delta_t,t+\delta_t)-g_{\alpha,\mathrm{s}}(\boldsymbol{r},t)=-\frac{1}{\tau_{T,\mathrm{s}}}(g_{\alpha,\mathrm{s}}-g_{\alpha,\mathrm{s}}^{\mathrm{eq}})+\delta_t Sr_{\alpha,\mathrm{s}}+\frac{1}{2}\delta_t^2\partial_t Sr_{\alpha,\mathrm{s}}\tag{7.47}$$

$$g_{\alpha,\mathrm{f}}(\boldsymbol{r}+\boldsymbol{e}_\alpha\delta_t,t+\delta_t)-g_{\alpha,\mathrm{f}}(\boldsymbol{r},t)=-\frac{1}{\tau_{T,\mathrm{f}}}(g_{\alpha,\mathrm{f}}-g_{\alpha,\mathrm{f}}^{\mathrm{eq}})+\delta_t Sr_{\alpha,\mathrm{f}}+\frac{1}{2}\delta_t^2\partial_t Sr_{\alpha,\mathrm{f}}+\delta_t Su_{\alpha,\mathrm{f}}\tag{7.48}$$

式中,固体和流体区域的源项 $Sr_{\alpha,s}$ 和 $Sr_{\alpha,f}$ 分别为

$$Sr_{\alpha,s}=\omega_\alpha\left[\frac{\dot{\Phi}_s}{(\rho c)_s}+\frac{h_v(T_f-T_s)}{(1-\phi)(\rho c)_s}\right],\quad Sr_{\alpha,f}=\omega_\alpha\left[\frac{\dot{\Phi}_f}{(\rho c_p)_f}+\frac{h_v(T_s-T_f)}{\phi(\rho c_p)_f}\right] \tag{7.49}$$

相应的平衡态分布函数 $g_{\alpha,s}^{eq}$ 和 $g_{\alpha,f}^{eq}$ 分别为

$$g_{\alpha,s}^{eq}=\omega_\alpha T_s,\quad g_{\alpha,f}^{eq}=\omega_\alpha T_f\left(1+\frac{\boldsymbol{e}_\alpha\cdot\boldsymbol{u}}{\phi c_s^2}\right) \tag{7.50}$$

式(7.47)和式(7.48)右端所引入的时间导数项用于消除源项的离散效应[20]。在数值计算中,可采用显式差分来计算源项的时间导数,即 $\partial_t Sr_\alpha=[Sr_\alpha(\boldsymbol{r},t)-Sr_\alpha(\boldsymbol{r},t-\delta_t)]/\delta_t$。

2020年,Liu 等[21]构建了多孔介质局部非热平衡传热的多松弛格子 Boltzmann 模型,其固体和流体区域温度分布函数的演化方程为

$$g_{\alpha,s}(\boldsymbol{r}+\boldsymbol{e}_\alpha\delta_t,t+\delta_t)-g_{\alpha,s}(\boldsymbol{r},t)=-\Lambda_{\alpha\beta,s}(g_{\beta,s}-g_{\beta,s}^{eq})\big|_{(\boldsymbol{r},t)}+\delta_t(\delta_{\alpha\beta}-0.5\Lambda_{\alpha\beta,s})\boldsymbol{\Psi}_{\beta,s}\big|_{(\boldsymbol{r},t)} \tag{7.51}$$

$$g_{\alpha,f}(\boldsymbol{r}+\boldsymbol{e}_\alpha\delta_t,t+\delta_t)-g_{\alpha,f}(\boldsymbol{r},t)=-\Lambda_{\alpha\beta,f}(g_{\beta,f}-g_{\beta,f}^{eq})\big|_{(\boldsymbol{r},t)}+\delta_t(\delta_{\alpha\beta}-0.5\Lambda_{\alpha\beta,f})\boldsymbol{\Psi}_{\beta,f}\big|_{(\boldsymbol{r},t)} \tag{7.52}$$

该模型采用 D2Q9 格子,相应的矩空间平衡态分布函数分别为

$$\boldsymbol{n}_s^{eq}=T_s(1,-2,2,0,0,0,0,0,0)^T \tag{7.53}$$

$$\boldsymbol{n}_f^{eq}=T_f\left(1,-2,2,\frac{u_x}{\phi},-\frac{u_x}{\phi},\frac{u_y}{\phi},-\frac{u_y}{\phi},0,0\right)^T \tag{7.54}$$

此外,矩空间作用力项 $\boldsymbol{\Psi}_k$(k 表示 s 或 f)可以表示为

$$\boldsymbol{\Psi}_k=Sr_k(1,2,-2,0,0,0,0,0,0)^T \tag{7.55}$$

式中,Sr_k 为

$$Sr_s=\left[\frac{\dot{\Phi}_s}{(\rho c)_s}+\frac{h_v(T_f-T_s)}{(1-\phi)(\rho c)_s}\right],\quad Sr_f=\left[\frac{\dot{\Phi}_f}{(\rho c_p)_f}+\frac{h_v(T_s-T_f)}{\phi(\rho c_p)_f}\right] \tag{7.56}$$

固体骨架和流体区域的温度 T_s 和 T_f 分别定义如下:

$$T_s=\sum_{\alpha=0}^{8}g_{\alpha,s}+\frac{\delta_t}{2}Sr_s \tag{7.57}$$

$$T_f=\sum_{\alpha=0}^{8}g_{\alpha,f}+\frac{\delta_t}{2}Sr_f \tag{7.58}$$

由于 Sr_s 和 Sr_f 中包含 T_s 和 T_f,因此需要联立式(7.57)和式(7.58)求解 T_s 和 T_f。通过分析这两个式子可以发现,它们是关于温度 T_s 和 T_f 的线性方程组,求解该线性方程组即可得到温度 T_s 和 T_f 的显式计算公式,具体表达式可参阅文献[21]。

7.2 孔隙尺度格子 Boltzmann 模拟

7.2.1 多孔介质孔隙结构的重构与生成

多孔介质孔隙尺度模拟的前提是获取多孔介质的孔隙结构数据[22],常用的获取方法包括**孔隙空间成像**(pore-space imaging)和人工构造方法。对于天然多孔介质,如岩石、土壤、木材等,可利用孔隙空间成像来获取微米级分辨率的三维图像。成像的原理和方法类

似于医学检查中的 CT 扫描,X 射线能够清晰地对比岩石和流体之间的差异,岩石对 X 射线吸收作用很强,而流体几乎不阻碍 X 射线穿过。在实际应用中,可通过电镜扫描、CT 扫描等手段获得多孔介质的二维或三维图像,并采用图像处理方法获取多孔介质的孔隙结构数据。例如,图 7.2 所示的颗粒堆积结构电镜扫描,可基于其灰度图像进行二值化处理,区分结构中的颗粒和孔隙。图 7.3 展示了泡沫陶瓷三维多孔结构的重构过程,其中采用 CT 扫描获得泡沫陶瓷逐个截面的图像,并采用增强对比度、降噪、锐化、二值化等图像处理手段,得到逐个截面的结构图,最终重构生成泡沫陶瓷的三维结构[3]。

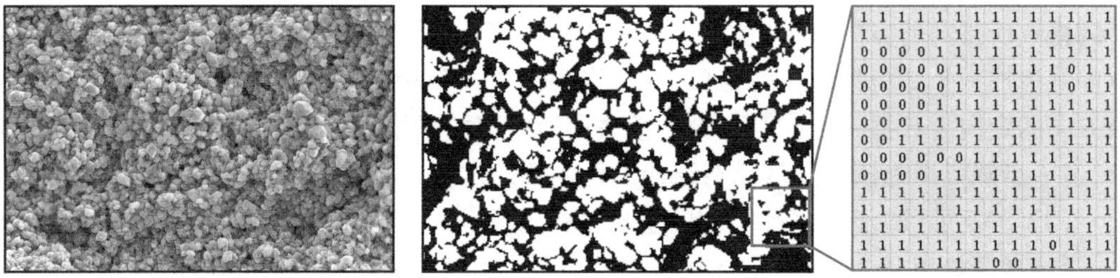

(a) 电镜扫描图片　(b) 二值化处理后图片　(c) 多孔介质结构矩阵

图 7.2 多孔介质的图像处理与结构存储示意图

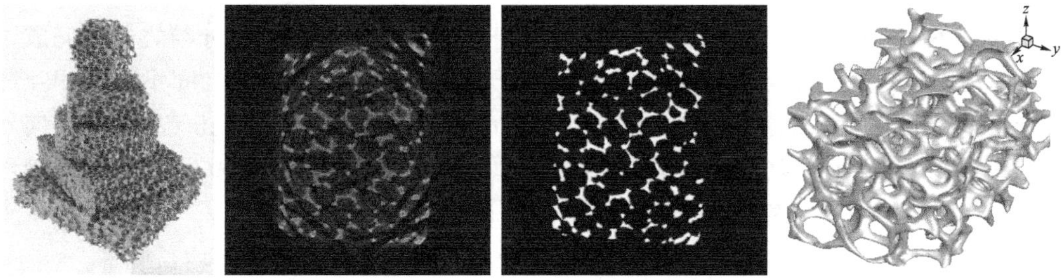

(a) 泡沫陶瓷　(b) CT扫描截面图　(c) 图像处理　(d) 三维重构

图 7.3 基于 CT 扫描的泡沫陶瓷三维多孔介质孔隙结构重构[3]

为了实现上述重构过程,在实际建模过程中一个比较容易实现的方法是借助 MATLAB 中的 imread 函数对多孔介质的二维或三维图像进行读取,获取图像中各个像素点的 RGB 值或灰度值,并将 RGB 值或灰度值及其对应的位置信息存入相关格式的信息文件中。然后利用自编程手段对信息文件进行读取,采用设定阈值的方式判别各个位置是固体骨架还是孔隙,也即将灰度高于该阈值的像素点标记为固体骨架,而低于该阈值的像素点标记为孔隙。最后将该信息存储于对应位置的网格中完成多孔介质孔隙结构的重构。

基于孔隙空间成像的重构方法能够得到绝大部分真实的孔隙结构信息,但一般成本较高,而且孔隙结构的精度取决于仪器的分辨率,一些起重要作用的微/纳米孔隙可能检测不到。另一种获取多孔介质孔隙结构数据的方法是人工构造方法,即根据实际多孔介质的特点和特征参数如孔隙率、迂曲率等,通过人工构造的方式生成具有相同特征参数的多孔介质孔隙结构。由于不同应用领域多孔介质的结构特征不同,往往需要根据具体问题提取关键的特征参数,发展相应的生成方法。目前,人工生成多孔介质结构常用的方法有球体沉降法、硬球 Monte-Carlo 方法、随机生成方法等。由 Wang 等[23]提出的**四参数随机生成方法**(quartet structure generation set,QSGS)是一种与格子 Boltzmann 方法紧密结合的生成方法。该方法生成的介质形貌特征可通过参数调整进行控制,孔隙的尺寸分布、连通性以及迂曲率等统计特性也能够得到较好的体现。这里我们以三维结构为例,简述其主要步骤。

(1) 确定需要生成的多孔结构的计算域 $N_x \times N_y \times N_z$,并为计算域中的每个网格点生成一个随机数 $s_{\text{int}}(i,j,k) \in (0,\ 1)$。当某一网格点所标记的随机数 $s_{\text{int}}(i,j,k)$ 小于或等于核心分布概率时,该网格点将成为多孔结构所对应的组分,并被称为多孔结构核心。

(2) 确定多孔结构在各个方向上的定向生长概率 D_q,其中 q 代表方向。以三维模拟中的一种多孔结构生长方式为例,q 的取值范围为 1~26,分别代表某一网格点邻域中 26 个网格点所指向的方向。随后以已成为多孔结构所对应组分的网格点为核心,在每一个方向上生成一个随机数 $s_{\text{sec}}(q) \in (0,\ 1)$,当某一方向所标记的随机数 $s_{\text{sec}}(q)$ 小于或等于该方向的定向生长概率 D_q 时,邻域中该方向所对应的网格点将成为多孔结构所对应的组分,这一步骤可视为该组分从已有结构向外生长的过程。

(3) 重复第(2)步中多孔结构的生长过程,直至该多孔结构所对应的组分达到指定的体积分数(若该组分为气相,则体积分数一般由孔隙率替代)。在实际数值模拟中,如果只需要对单一组分进行多孔结构的构建(如对木质多孔介质中单一流体流动过程的模拟,只需要对木质多孔结构进行构建),采用步骤(1)至步骤(3)即可。在其他情况下,还需继续实施以下步骤。

(4) 在对第 n 个组分($n>1$)进行多孔结构构建时,需要根据该组分与之前已完成了多孔结构构建的组分之间是否存在相互影响分为两种情况。若组分之间不存在相互影响,即各组分互为等效离散相,此时只需要对第 n 个组分生成合适的多孔结构核心并向外生长即可,该过程与步骤(1)至步骤(3)基本一致;若各组分之间存在相互影响,则在处理多孔结构的构建时更为复杂,需要通过设置组分间相互作用增长概率 $I_q^{n,m}$ 来实现第 n 个组分的多孔结构生长,$I_q^{n,m}$ 表示 n 组分在 m 组分多孔结构上沿 q 方向的增长概率。

(5) 重复步骤(4),直至第 n 个组分达到其指定的体积分数。

(6) 对每一个组分实施步骤(4)和步骤(5),直至所有需要进行多孔结构构建的组分均完成多孔结构生长过程。

图 7.4 中展示了四参数随机生成方法采用不同定向生长概率所构建的单组分多孔介质结构。

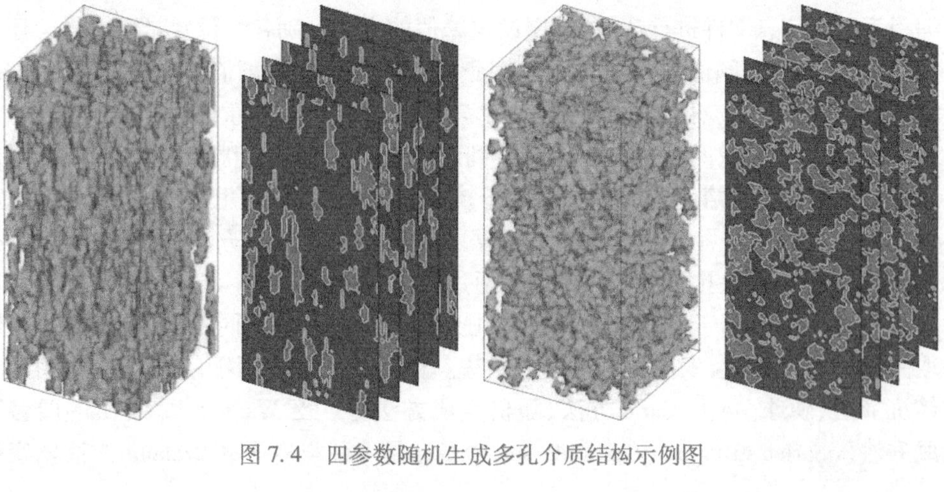

图 7.4 四参数随机生成多孔介质结构示例图

何超等[24]提出了纤维增强气凝胶复合材料中纤维结构的生成方法。该生成方法假设纤维为圆柱,并根据实际材料确定纤维的半径、长度等特征参数。生成过程中,首先令网格中所有节点为气凝胶节点,接着随机或按照一定规律确定每一条纤维的中心坐标和空间取向,利用几何关系将包含在该纤维圆柱体内的节点变为纤维节点,如此逐条添加纤维直至达到给定的体积分数。生成过程中可引入相应的约束条件,防止纤维之间的相互交叠。同样的方法还可用于在材料中增加颗粒状添加物,如图 7.5 所示。

一些复合材料和多孔结构的重构方法

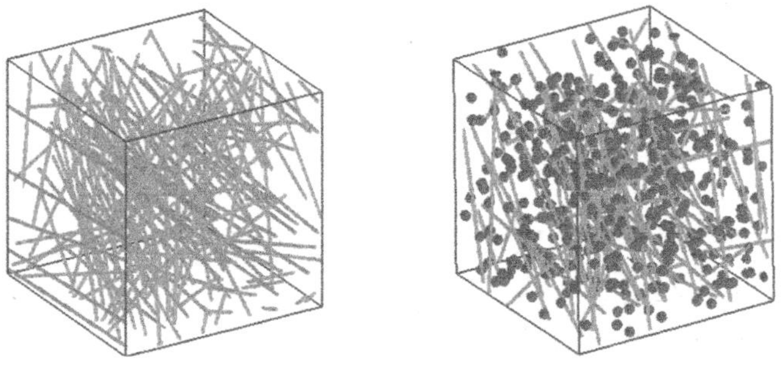

(a) 纤维结构　　(b) 含颗粒状添加物的纤维结构

图 7.5　纤维结构生成示例图[24]

此外,还可采用特定的孔隙结构模型,例如 Lu 等[25]利用三维立方体单元结构模型来表征开孔泡沫金属材料,Ashby 等[26]采用立方体单元结构模型对多孔金属材料进行微观结构及力学性能分析。一些学者还提出了其他的结构模型(图 7.6),如 Kelvin 十四面体泡沫金属结构模型[27]。从图中可以看到,与多孔泡沫金属材料的电镜扫描图像对比,Kelvin 十四面体泡沫金属结构模型能够近似描述多孔金属材料的实际孔隙状况。

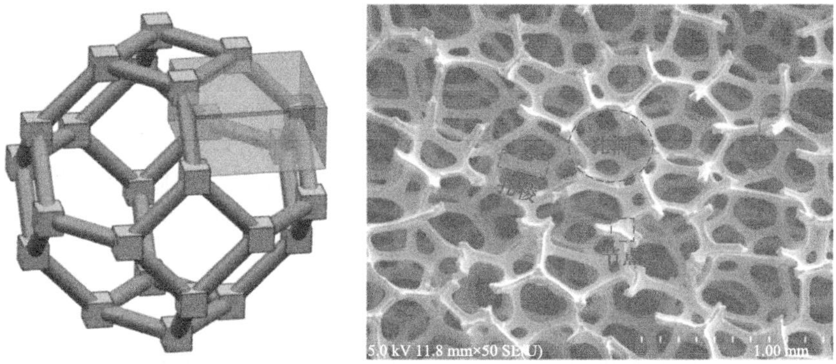

(a) Kelvin 十四面体泡沫金属结构模型　　(b) 电镜扫描图像

图 7.6　泡沫金属结构模型[27]

7.2.2　孔隙尺度模拟中的边界处理

在建立多孔介质孔隙结构的数值化存储和表示方式后,便可开展孔隙尺度多孔介质流动与传热问题的数值模拟。流场的模拟可以采用标准格子 Boltzmann 模型,在流体区域执行格子 Boltzmann 方程的碰撞步和迁移步,同时采用相应的边界格式处理流体与固体交界面的边界条件。

以图 7.7 所示的半步长反弹格式为例,其中实线为网格线,实心圆点表示固体骨架区

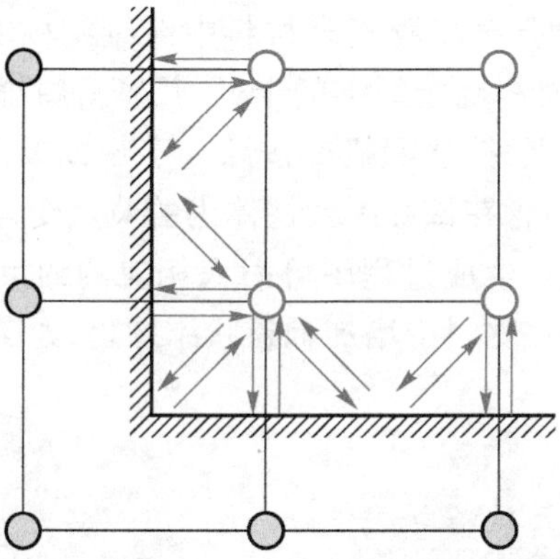

图 7.7 半步长反弹格式的计算示意图

域的网格节点,空心圆点表示流体区域的网格节点,阴影线为计算中固体与流体区域的分界线,处于网格连线的1/2处。该格式可在格子 Boltzmann 方程的迁移步骤中通过判断相邻节点的属性直接进行处理,其计算步骤如下:

遍历计算节点,判断节点是否为流体节点,若为流体节点:

{

执行碰撞步骤,得到每个节点的碰撞后分布函数 f_α^*;

执行迁移步骤,遍历对应于 $\boldsymbol{e}_\alpha$ 离散速度方向的相邻网格节点 $\boldsymbol{r}-\boldsymbol{e}_\alpha\delta_t$;

若相邻节点为流体节点

从该节点迁移分布函数:$f_\alpha(\boldsymbol{r},t+\delta_t)=f_\alpha^*(\boldsymbol{r}-\boldsymbol{e}_\alpha\delta_t,t)$;

否则(相邻节点为固体)

执行半步长反弹格式:$f_\alpha(\boldsymbol{r},t+\delta_t)=f_{\bar{\alpha}}^*(\boldsymbol{r},t)$,其中 $\boldsymbol{e}_{\bar{\alpha}}$ 为 $\boldsymbol{e}_\alpha$ 的反方向;

计算密度和速度;

}

多孔介质孔隙尺度的传热问题通常在固体和流体区域内统一计算(采用统一的热格子 Boltzmann 模型),其中可能同时包含流体区域以及不同组分的固体区域。考虑共轭传热时,流体与固体的交界面应当满足温度和热流连续的条件[28]:

$$T^{\text{int},+}=T^{\text{int},-} \tag{7.59}$$

$$\boldsymbol{n}\cdot(-k\,\nabla T+\rho c_p T\boldsymbol{u})^{\text{int},+}=\boldsymbol{n}\cdot(-k\,\nabla T+\rho c_p T\boldsymbol{u})^{\text{int},-} \tag{7.60}$$

式中:上标"int"表示流固交界面;上标"+"表示流体侧;"-"表示固体侧;$\boldsymbol{n}$ 为流固交界面的法向矢量。对存在温度跳跃和热流跳跃的交界面,边界条件可参阅文献[29,30]。对无滑移边界(流固交界面处 $\boldsymbol{u}=0$),式(7.60)退化为

$$\boldsymbol{n}\cdot(-k\,\nabla T)^{\text{int},+}=\boldsymbol{n}\cdot(-k\,\nabla T)^{\text{int},-} \tag{7.61}$$

针对上述边界条件,可采用 Li 等[31-32]提出的共轭传热边界处理格式:

$$g_{\bar{\alpha}}(\boldsymbol{r}_{\text{fn}},t+\delta_t)=\left(\frac{1-\sigma}{1+\sigma}\right)g_\alpha^*(\boldsymbol{r}_{\text{fn}},t)+\left(\frac{2\sigma}{1+\sigma}\right)g_{\bar{\alpha}}^*(\boldsymbol{r}_{\text{sn}},t) \tag{7.62}$$

$$g_\alpha(\boldsymbol{r}_{\text{sn}},t+\delta_t)=-\left(\frac{1-\sigma}{1+\sigma}\right)g_{\bar{\alpha}}^*(\boldsymbol{r}_{\text{sn}},t)+\left(\frac{2}{1+\sigma}\right)g_\alpha^*(\boldsymbol{r}_{\text{fn}},t) \tag{7.63}$$

为避免混淆,式(7.62)和式(7.63)采用正体下标"fn"和"sn"表示流固交界面两侧相邻的

流体节点和固体节点，且有 $\boldsymbol{r}_{sn}=\boldsymbol{r}_{fn}+\boldsymbol{e}_\alpha\delta_t$；$g_\alpha^*$ 为碰撞后、迁移前的分布函数；α 对应从流体节点指向固体节点的离散速度方向，$\bar{\alpha}$ 为其反方向；σ 为固体介质与流体的热容比，即 $\sigma=(\rho c)_s/(\rho c_p)_f$。当 $\sigma=1$ 时，可得

$$g_{\bar{\alpha}}(\boldsymbol{r}_{fn},t+\delta_t)=g_{\bar{\alpha}}^*(\boldsymbol{r}_{sn},t),g_\alpha(\boldsymbol{r}_{sn},t+\delta_t)=g_\alpha^*(\boldsymbol{r}_{fn},t) \tag{7.64}$$

式(7.64)表明，当 $(\rho c)_s=(\rho c_p)_{fn}$ 时，相邻流体节点与固体节点之间的温度分布函数可以直接执行碰撞-迁移步而无须特别处理。

7.3 多孔介质固液相变格子 Boltzmann 模型

传统相变材料的导热系数通常在 0.1~0.6 W/(m·K)的范围[33]，这极大地影响了相变储热系统的储/放热速率，从而限制了其应用范围。强化固液相变传热成为相变储热技术推广应用中亟待解决的重要课题之一。作为一种超轻多孔金属材料，泡沫金属(如泡沫铜、泡沫铝)具有高导热率、高渗透性、高比表面积等特性，将其填充到相变材料中形成**复合相变材料**(composite phase change materials)能够有效地提高相变材料的传热性能，从而达到提高储热系统储/放热速率的目的[33-35]。格子 Boltzmann 方法的介观物理背景和粒子特性使其在多孔介质固液相变问题的研究中也得到了成功的应用[36]。目前，格子 Boltzmann 方法模拟多孔介质固液相变问题主要在孔隙尺度和 REV 尺度进行，下面分别进行介绍。

7.3.1 孔隙尺度

多孔介质固液相变问题的孔隙尺度模拟是直接采用焓法格子 Boltzmann 方法模拟孔隙内的固液相变过程。多孔介质固液相变孔隙尺度模拟中一个需要注意的问题是相界面(固液相界面以及流体-固体骨架界面)的温度和热流边界的处理问题。以无迭代焓法固液相变格子 Boltzmann 模型[37]为例，该模型满足 $T^{\text{int},+}=T^{\text{int},-}$ 及

$$\boldsymbol{n}\cdot[-\alpha\nabla(\rho c_pT)+\rho c_pT\boldsymbol{u}]^{\text{int},+}=\boldsymbol{n}\cdot[-\alpha\nabla(\rho c_pT)+\rho c_pT\boldsymbol{u}]^{\text{int},-} \tag{7.65}$$

当 $(\rho c_p)^{\text{int},+}=(\rho c_p)^{\text{int},-}$，该模型满足式(7.60)，此时式(7.65)与式(7.60)是等价的。由文献[38,39]可知，在相变材料温度场演化方程的平衡态分布函数中引入参考比定压热容 $c_{p,\text{ref}}$ 和参考密度 ρ_0 可以使 $(\rho c_p)^{\text{int},+}=(\rho c_p)^{\text{int},-}$ 得到满足。2017 年，Gao 等[40]开展了多孔介质共轭传热格子 Boltzmann 模型的研究，他们通过引入一个类似于参考比定压热容的自由参数 γ 并取 $\rho_0=1$ 从而使 $(\rho c_p)^{\text{int},+}=(\rho c_p)^{\text{int},-}$ 得到了满足，于是无须对相界面的温度和热流边界做特别处理。除了上述处理方法，一些学者还从不同角度提出了共轭传热问题的相界面温度和热流边界处理格式，具体可参阅文献[36]。

从公开的文献资料来看，已有的多孔介质固液相变格子 Boltzmann 模型几乎都可以归入焓法模型这一类。2006 年，钱吉裕等[41]基于 Jiaung 等[42]的迭代焓法模型建立了热传导固液相变格子 Boltzmann 模型，并模拟了孔隙尺度泡沫铝内的固液相变导热问题，分析了多孔介质骨架对相变材料熔化过程的影响。2008 年，Huber 等[43]也开展了多孔介质固液相变问题的孔隙尺度模拟，他们利用迭代焓法模拟了孔隙率为 0.45 的随机生成多孔介质内的相变材料熔化过程。随后，杲东彦[34]构造了局部热平衡和局部非热平衡条件下泡沫金属内固液相变的焓法格子 Boltzmann 模型，并开展了一系列孔隙尺度泡沫金属固液相变问题的研究，揭示了对流作用、孔隙率、骨架纤维形状、结构缺陷等因素对固液相变传热

及流动过程的影响。近年来,多孔介质固液相变的孔隙尺度模拟引起了越来越广泛的关注,如文献[44,45]基于迭代焓法模型[42-43]模拟了随机生成多孔介质内的固液相变问题,文献[46,47]则基于无迭代焓法模型[37,48]开展了类似问题的研究。

多孔泡沫金属材料的固液相变过程是典型的多相多尺度耦合问题,其相变传热行为和内在规律远未得到真正揭示。特别是,固体骨架表面特性及各向异性孔空间拓扑结构对固液相变过程的影响机制尚不清晰。孔隙尺度的格子 Boltzmann 模拟可为揭示多孔泡沫金属材料的相变传热行为和内在规律提供新的思路和途径。

7.3.2 REV 尺度

REV 尺度的焓法固液相变格子 Boltzmann 方法不再关注孔隙内固液相变过程的具体细节,而是通过求解体积平均的控制方程来获得固液相变过程的宏观信息。在局部非热平衡条件下,REV 尺度多孔介质固液相变过程的能量守恒方程(双温度方程)为[34,49]

$$\frac{\partial}{\partial t}\{\phi[f_l\rho_l c_{p,l}+(1-f_l)\rho_s c_{p,s}]T_f\}+\nabla\cdot(\rho_l c_{p,l}T_f\boldsymbol{u})$$
$$=\nabla\cdot(k_{e,f}\nabla T_f)+h_v(T_m-T_f)-\frac{\partial}{\partial t}(\phi\rho_l L_a f_l) \tag{7.66}$$

$$\frac{\partial}{\partial t}[(1-\phi)\rho_m c_{p,m}T_m]=\nabla\cdot(k_{e,m}\nabla T_m)+h_v(T_f-T_m) \tag{7.67}$$

式中:ϕ 为多孔介质的孔隙率;h_v 为容积换热系数;$k_{e,f}$为相变材料的等效导热系数;$k_{e,m}$为多孔介质固体骨架的等效导热系数,相关的经验关联式参见文献[34,50]。为避免混淆,这里的下标“f”和“m”分别表示流体(相变材料)和多孔介质(固体骨架)。

液相、固相相变材料和多孔介质固体骨架的焓分别为

$$H_l=c_{p,l}T_f+L_a,\quad H_s=c_{p,s}T_f,\quad H_m=c_{p,m}T_m \tag{7.68}$$

利用式(7.68),可将式(7.66)改写为

$$\frac{\partial H_f}{\partial t}+\nabla\cdot\left(\frac{c_{p,l}T_f\boldsymbol{u}}{\phi}\right)=\nabla\cdot\left(\frac{k_{e,f}}{\phi\rho_l}\nabla T_f\right)+\frac{h_v(T_m-T_f)}{\phi\rho_l} \tag{7.69}$$

式中,相变材料的焓为 $H_f=\sigma_f c_{p,l}T_f+L_a f_l$,其中 σ_f 为热容比,定义如下

$$\sigma_f=\frac{\rho_f c_{p,f}}{\rho_l c_{p,l}}=\frac{f_l\rho_l c_{p,l}+(1-f_l)\rho_s c_{p,s}}{\rho_l c_{p,l}} \tag{7.70}$$

在局部热平衡条件($T_f=T_m=T$)下,式(7.66)和式(7.67)退化为如下单温度方程[36,51]:

$$\frac{\partial}{\partial t}(\overline{\rho c_p}T)+\nabla\cdot(\rho_l c_{p,l}T\boldsymbol{u})=\nabla\cdot(k_e\nabla T)-\frac{\partial}{\partial t}(\phi\rho_l L_a f_l) \tag{7.71}$$

式中:$\overline{\rho c_p}=\phi[f_l\rho_l c_{p,l}+(1-f_l)\rho_s c_{p,s}]+(1-\phi)\rho_m c_{p,m}$;$k_e$ 为等效导热系数(包括相变材料和固体骨架),详见文献[51]。式(7.71)可进一步简化为

$$\frac{\partial}{\partial t}(\sigma T)+\nabla\cdot(T\boldsymbol{u})=\nabla\cdot(a_e\nabla T)-\phi\frac{L_a}{c_{p,l}}\frac{\partial f_l}{\partial t} \tag{7.72}$$

式中:$\sigma=\overline{\rho c_p}/(\rho c_p)_l$ 为混合物(包括液相、固相相变材料和多孔介质固体骨架)与液相相变材料的热容比;$a_e=k_e/(\rho c_p)_l$ 为等效热扩散系数。

2011 年,Gao 和 Chen[52]将焓法格子 Boltzmann 模型应用于求解 REV 尺度多孔介质固液相变问题。他们的模型对应于单温度方程,并采用迭代焓法模型[42-43]处理固液相变的

潜热源项。2015 年,Liu 和 He[15]提出了一个耦合自然对流作用的多孔介质固液相变焓法多松弛格子 Boltzmann 模型,该模型以相变材料的温度作为基本演化变量,采用 Jiaung 等[42]的迭代焓法处理固液相变的潜热源项。为了避免焓值迭代计算过程,依据文献[37]的思路,应当基于格子 Boltzmann 方法的自身特性将相变材料的焓定义为演化方程的基本演化变量。根据这一思路,Wu 等[53]将式(7.71)右端的相变潜热源项与左端的瞬态项结合,得到了如下形式的能量方程:

$$\partial_t(\rho_l H_e)+\nabla\cdot(\rho_l c_{p,l} T\boldsymbol{u})=\nabla\cdot(k_e\nabla T) \tag{7.73}$$

式中,$H_e=\sigma c_{p,l}T+\phi L_a f_l$ 为等效焓。基于式(7.73),Wu 等[53]建立了多孔介质固液相变的焓法单松弛和多松弛格子 Boltzmann 模型,平衡态焓分布函数为

$$g_\alpha^{eq}=\begin{cases} H_e-\sigma_0 c_{p,l}T+\omega_0\sigma c_{p,l}T\left(\dfrac{\sigma_0}{\sigma}-\dfrac{|\boldsymbol{u}|^2}{2\sigma^2 c_s^2}\right), & \alpha=0 \\ \omega_\alpha\sigma c_{p,l}T\left[\dfrac{\sigma_0}{\sigma}+\dfrac{\boldsymbol{e}_\alpha\cdot\boldsymbol{u}}{c_s^2\sigma}+\dfrac{\boldsymbol{uu}:(\boldsymbol{e}_\alpha\boldsymbol{e}_\alpha-c_s^2\boldsymbol{I})}{2\sigma^2 c_s^4}\right], & \alpha\neq 0 \end{cases} \tag{7.74}$$

式中,σ_0 为参考热容比。多松弛模型的平衡态矩函数可通过 $\boldsymbol{m}_g^{eq}=\boldsymbol{M}\boldsymbol{g}^{eq}$ 得到。该模型无须在演化方程中添加相变潜热源项,从而避免了迭代计算过程。

2017 年,Gao 等[54]也发展了多孔介质固液相变的焓法单松弛和多松弛格子 Boltzmann 模型,相应的平衡态焓分布函数为

$$g_\alpha^{eq}=\begin{cases} H_{sys}-\gamma T+\omega_0\gamma T, & \alpha=0 \\ \omega_\alpha T\left(\gamma+\rho c_p\dfrac{\boldsymbol{e}_\alpha\cdot\boldsymbol{u}}{c_s^2}\right), & \alpha\neq 0 \end{cases} \tag{7.75}$$

式中:H_{sys}为混合系统(包括固相、液相相变材料和固体骨架)的焓;γ 为可调参数。同样地,多松弛模型的平衡态矩函数也可通过 $\boldsymbol{m}_g^{eq}=\boldsymbol{M}\boldsymbol{g}^{eq}$得到。与 Wu 等[53]的模型相比,Gao 等[54]的模型舍弃了平衡态焓分布函数中的速度平方项,形式上更为简洁。实际上,式(7.75)中的可调参数 γ 与 Huang 和 Wu[38]模型中的参考比定压热容 $c_{p,ref}$的作用是类似的,即 Gao 等[54]的模型也可以实现导热系数与比热容解耦。此外,Liu 等[18]也开展了多孔介质固液相变的无迭代焓法格子 Boltzmann 模型研究,并基于式(7.73)构建了一个三维多孔介质固液相变焓法多松弛格子 Boltzmann 模型,其平衡态矩函数 $\boldsymbol{m}_g^{eq}$ 为

$$\boldsymbol{m}_g^{eq}=(H_e, c_{p,l}Tu_x, c_{p,l}Tu_y, c_{p,l}Tu_z, \varpi c_{p,ref}T, 0, 0)^{T} \tag{7.76}$$

式中,$\varpi\in(0,1)$为模型参数。文献[18,53,54]中的焓法格子 Boltzmann 模型均避免了采用迭代焓法求解温度和液相分数,从而提高了模型的计算效率和计算精度。

上述 REV 尺度焓法固液相变格子 Boltzmann 模型是在局部热平衡条件下建立的,没有考虑相变材料与多孔介质固体骨架之间的热交换。对泡沫金属(如泡沫铜、泡沫铝)基复合相变材料来说,金属骨架的导热系数通常比相变材料的导热系数高 2~3 个数量级,此时需要采用局部非热平衡模型(也即双温度方程模型)来描述固体骨架与相变材料之间的内部热交换。基于局部非热平衡模型,2010 年,Gao 等[55]建立了考虑固体骨架与相变材料之间热交换的焓法固液相变格子 Boltzmann 模型,该模型采用迭代焓法模型[42]处理固液相变潜热源项。2014 年,Gao 等[19]在原有模型的基础上,进一步构建了一个局部非热平衡条件下的焓法固液相变格子 Boltzmann 模型。与原有模型相比,该模型在温度演化方程中考虑了离散效应的影响。

上述局部非热平衡格子 Boltzmann 模型由于采用迭代焓法求解温度和液相分数，导致计算效率较低。为了克服这一缺陷，Gao 等[56]提出了一个改进的焓法固液相变单松弛格子 Boltzmann 模型，该模型的演化方程采用如下形式的平衡态焓分布函数：

$$g_{\alpha,\mathrm{f}}^{\mathrm{eq}}=\begin{cases}\phi H_{\mathrm{f}}-\gamma_{\mathrm{f}}T_{\mathrm{f}}+\omega_0\gamma_{\mathrm{f}}T_{\mathrm{f}}, & \alpha=0\\ \omega_\alpha T_{\mathrm{f}}\left(\gamma_{\mathrm{f}}+\rho_l c_{p,l}\dfrac{\boldsymbol{e}_\alpha\cdot\boldsymbol{u}}{c_{\mathrm{s}}^2}\right), & \alpha\neq 0\end{cases}\tag{7.77}$$

式中，γ_{f} 为可调参数，可实现导热系数与比热容解耦。该模型避免了迭代计算过程，并且与迭代焓法模型[42]相比，可以有效降低固液相界面区域附近数值扩散效应的影响。

Liu 等[57]也开展了局部非热平衡条件下焓法固液相变格子 Boltzmann 模型的研究。基于式(7.69)，他们构造了一个二维焓法固液相变多松弛格子 Boltzmann 模型，该模型的演化方程采用如下形式的平衡态矩函数：

$$\boldsymbol{m}_g^{\mathrm{eq}}=\left(H_{\mathrm{f}},-4H_{\mathrm{f}}+2c_{p,\mathrm{ref}}T_{\mathrm{f}},4H_{\mathrm{f}}-3c_{p,\mathrm{ref}}T_{\mathrm{f}},\frac{c_{p,l}T_{\mathrm{f}}u_x}{\phi c},-\frac{c_{p,l}T_{\mathrm{f}}u_x}{\phi c},\frac{c_{p,l}T_{\mathrm{f}}u_y}{\phi c},-\frac{c_{p,l}T_{\mathrm{f}}u_y}{\phi c},0,0\right)^{\mathrm{T}}\tag{7.78}$$

该模型同样避免了迭代计算过程，且由于采用了多松弛碰撞算子，可以通过调节相应的松弛因子减少固液相界面区域附近的数值扩散。

7.4 小　　结

在本章，我们介绍了用于多孔介质流动与传热的格子 Boltzmann 模型。具体地，在 REV 尺度，通过修改格子 Boltzmann 模型的平衡态分布函数和作用力格式等，可以在模型中引入孔隙率、热容比等物性参数以及反映多孔介质影响的 REV 尺度作用力项。根据多孔介质局部的骨架温度与流体温度是否相等，REV 尺度多孔介质的传热模型可分为局部热平衡模型和局部非热平衡模型。其中，局部热平衡模型认为固体骨架与流体之间的温差可以忽略，相应的双温度方程将退化为单温度方程。在孔隙尺度，固体骨架通常被看作流场的边界，流体与固体骨架之间的流固相互作用通过适当的边界条件予以反映。格子 Boltzmann 方法可以充分利用其粒子运动图像清晰和边界条件处理简单的优势，采用半步长反弹等边界处理格式，实现孔隙尺度复杂多孔结构中流动与传热问题的数值计算。一般来说，孔隙尺度的模拟可用来探索微观渗流的机理和基本规律，而 REV 尺度的模拟则往往应用于大尺度的工程渗流问题。

习　　题

7-1　试推导 REV 尺度流动模型中速度 $\boldsymbol{u}$ 与临时速度 $\boldsymbol{v}$ 之间的关联式(7.30)。

7-2　参考第 4.2 节中温度分布函数热格子 Boltzmann 模型的多尺度分析，结合式(7.41)，分析为何采用式(7.40)的平衡态分布函数，可以实现热容比的可调及其与导热系数的解耦。

7-3　给出局部非热平衡模型(7.47)~(7.50)中通过分布函数计算温度的公式，是否需要在计算中考虑源项的影响，如式(7.57)和(7.58)所示？总结模型(7.47)~(7.50)

和模型(7.51)~(7.58)中两种源项添加方法的不同点。

7-4 结合参考文献[23],编写程序,在二维200×200的网格区域中,利用四参数随机生成方法生成孔隙率0.7的二维多孔介质结构,其核心的生成概率可设为0.001,各方向的生长概率可设为0.001,生成各向同性的多孔结构。调整核心生成概率,观察其对生成结构的影响。增大左右、上下或对角相邻节点的生长概率,生成各向异性多孔结构。

7-5 基于上题所生成的多孔介质结构,结合7.2.2节所给出的编程思路及附录D中的顶盖驱动流代码,编写孔隙尺度内流动过程的格子Boltzmann计算程序(注意附录D中代码的碰撞和迁移步骤融合在演化函数的一个计算中)。

参考文献

[1] 刘伟,范爱武,黄晓明. 多孔介质传热传质理论与应用[M]. 北京:科学出版社,2006.

[2] Xu Y, Ren Q L, Zheng Z J, et al. Evaluation and optimization of melting performance for a latent heat thermal energy storage unit partially filled with porous media[J]. Applied Energy, 2017, 193: 84-95.

[3] Du S, Li M J, Ren Q L, et al. Pore-scale numerical simulation of fully coupled heat transfer process in porous volumetric solar receiver[J]. Energy, 2017, 140: 1267-1275.

[4] He Y L, Xie T. Advances of thermal conductivity models of nanoscale silica aerogel insulation material [J]. Applied Thermal Engineering, 2015, 81: 28-50.

[5] 刘清. 多孔介质单相和固-液相变传热的多松弛格子Boltzmann方法研究[D]. 西安:西安交通大学,2018.

[6] Spaid M A A, Phelan F R. Lattice Boltzmann methods for modeling microscale flow in fibrous porous media [J]. Physics of Fluids, 1997, 9(9): 2468-2474.

[7] Spaid M A A, Phelan F R. Modeling void formation dynamics in fibrous porous media with the lattice Boltzmann method[J]. Composites Part A: Applied Science and Manufacturing, 1998, 29(7): 749-755.

[8] Shan X W, Chen H D. Lattice Boltzmann model for simulating flows with multiple phases and components [J]. Physical Review E, 1993, 47: 1815-1820.

[9] Martys N S. Improved approximation of the Brinkman equation using a lattice Boltzmann method[J]. Physics of Fluids, 2001, 13(6): 1807-1810.

[10] Luo L S. Unified theory of lattice Boltzmann models for nonideal gases[J]. Physical Review Letters, 1998, 81(8): 1618-1621.

[11] Dardis O, McCloskey J. Lattice Boltzmann scheme with real numbered solid density for the simulation of flow in porous media[J]. Physical Review E, 1998, 57(4): 4834-4837.

[12] Dardis O, McCloskey J. Permeability porosity relationships from numerical simulations of fluid flow [J]. Geophysical Research Letters, 1998, 25(9): 1471-1474.

[13] Guo Z L, Zhao T S. Lattice Boltzmann model for incompressible flows through porous media[J]. Physical Review E, 2002, 66(3): 036304.

[14] Liu Q, He Y L, Li Q, et al. A multiple-relaxation-time lattice Boltzmann model for convection heat transfer in porous media[J]. International Journal of Heat and Mass Transfer, 2014, 73: 761-775.

[15] Liu Q, He Y L. Double multiple-relaxation-time lattice Boltzmann model for solid-liquid phase change with natural convection in porous media[J]. Physica A: Statistical Mechanics and its Applications, 2015, 438: 94-106.

[16] Guo Z L,Zhao T S. A lattice Boltzmann model for convection heat transfer in porous media[J]. Numerical Heat Transfer,Part B:Fundamentals,2005,47(2):157-177.

[17] Chen S,Yang B,Zheng C G. A lattice Boltzmann model for heat transfer in porous media[J]. International Journal of Heat and Mass Transfer,2017,111:1019-1022.

[18] Liu Q,Feng X B,He Y L,et al. Three-dimensional multiple-relaxation-time lattice Boltzmann models for single-phase and solid-liquid phase-change heat transfer in porous media at the REV scale[J]. Applied Thermal Engineering,2019,152:319-337.

[19] Gao D Y,Chen Z Q,Chen L H. A thermal lattice Boltzmann model for natural convection in porous media under local thermal non-equilibrium conditions[J]. International Journal of Heat and Mass Transfer, 2014,70:979-989.

[20] Shi B C,Guo Z L. Lattice Boltzmann model for nonlinear convection-diffusion equations[J]. Physical Review E,2009,79(1):016701.

[21] Liu Q,Feng X B,Wang X L. Multiple-relaxation-time lattice Boltzmann model for convection heat transfer in porous media under local thermal non-equilibrium condition[J]. Physica A:Statistical Mechanics and its Applications,2020,545:123794.

[22] Blunt M J. Multiphase Flow in Permeable Media. A Pore-Scale Perspective[M]. Cambridge:Cambridge University Press,2017.

[23] Wang M R,Wang J K,Pan N,et al. Mesoscopic predictions of the effective thermal conductivity for microscale random porous media[J]. Physical Review E,2007,75(3):036702.

[24] 何超,何雅玲,谢涛,等. 基于格子 Boltzmann 方法的纤维增强气凝胶复合材料等效热导率求解[J]. 工程热物理学报,2013,33(4):742-745.

[25] Lu T J,Stone H A,Ashby M F. Heat transfer in open-cell metal foams[J]. Acta Materialia,1998,46(10):3619-3635.

[26] Ashby M F,Evans A,Fleck N A,et al. Metal Foams:A Design Guide[M]. Elsevier,2000.

[27] Yang H Z,Li Y Y,Yang Y,et al. Effective thermal conductivity of high porosity open-cell metal foams[J]. International Journal of Heat and Mass Transfer,2020,147:118974.

[28] Karani H,Huber C. Lattice Boltzmann formulation for conjugate heat transfer in heterogeneous media[J]. Physical Review E,2015,91(2):023304.

[29] Guo K K,Li L K,Xiao G,et al. Lattice Boltzmann method for conjugate heat and mass transfer with interfacial jump conditions[J]. International Journal of Heat and Mass Transfer,2015,88:306-322.

[30] Mu Y T,Gu Z L,He P,et al. Lattice Boltzmann method for conjugated heat and mass transfer with general interfacial conditions[J]. Physical Review E,2018,98(4):043309.

[31] Li L K,Chen C,Mei R W,et al. Conjugate heat and mass transfer in the lattice Boltzmann equation method[J]. Physical Review E,2014,89(4):043308.

[32] Korba D,Wang N Q,Li L K. Accuracy of interface schemes for conjugate heat and mass transfer in the lattice Boltzmann method[J]. International Journal of Heat and Mass Transfer,2020,156:119694.

[33] Pielichowska K,Pielichowski K. Phase change materials for thermal energy storage[J]. Progress in Materials Science,2014,65:67-123.

[34] 杲东彦. 基于格子 Boltzmann 方法的泡沫金属内固液相变传热过程的研究[D]. 南京:东南大学,2011.

[35] Wu S F,Yan T,Kuai Z H,et al. Thermal conductivity enhancement on phase change materials for thermal energy storage:A review[J]. Energy Storage Materials,2020,25:251-295.

[36] He Y L,Liu Q,Li Q,et al. Lattice Boltzmann methods for single-phase and solid-liquid phase-change heat

transfer in porous media: A review[J]. International Journal of Heat and Mass Transfer, 2019, 129: 160-197.

[37] Huang R Z, Wu H Y, Cheng P. A new lattice Boltzmann model for solid-liquid phase change[J]. International Journal of Heat and Mass Transfer, 2013, 59: 295-301.

[38] Huang R Z, Wu H Y. Phase interface effects in the total enthalpy-based lattice Boltzmann model for solid-liquid phase change[J]. Journal of Computational Physics, 2015, 294: 346-362.

[39] Zhao Y, Pereira G G, Kuang S B, et al. A generalized lattice Boltzmann model for solid-liquid phase change with variable density and thermophysical properties[J]. Applied Mathematics Letters, 2020, 104: 106250.

[40] Gao D Y, Chen Z Q, Chen L H, et al. A modified lattice Boltzmann model for conjugate heat transfer in porous media[J]. International Journal of Heat and Mass Transfer, 2017, 105: 673-683.

[41] 钱吉裕,李强,宣益民. Lattice-Boltzmann 方法计算多孔介质内固液相变问题[J]. 自然科学进展, 2006, 16(4): 504-507.

[42] Jiaung W S, Ho J R, Kuo C P. Lattice Boltzmann method for the heat conduction problem with phase change[J]. Numerical Heat Transfer: Part B: Fundamentals, 2001, 39(2): 167-187.

[43] Huber C, Parmigiani A, Chopard B, et al. Lattice Boltzmann model for melting with natural convection [J]. International Journal of Heat and Fluid Flow, 2008, 29(5): 1469-1480.

[44] Fang W Z, Tang Y Q, Yang C, et al. Pore scale investigations on melting of phase change materials considering the interfacial thermal resistance[J]. International Communications in Heat and Mass Transfer, 2020, 115: 104631.

[45] Han Q, Wang H, Yu C, et al. Lattice Boltzmann simulation of melting heat transfer in a composite phase change material[J]. Applied Thermal Engineering, 2020, 176: 115423.

[46] Ren Q L, He Y L, Su K Z, et al. Investigation of the effect of metal foam characteristics on the PCM melting performance in a latent heat thermal energy storage unit by pore-scale lattice Boltzmann modeling [J]. Numerical Heat Transfer, Part A: Applications, 2017, 72(10): 745-764.

[47] Hu Y R, Zuo D H, Zhang Y N, et al. Thermal performances of saturated porous soil during freezing process using lattice Boltzmann method[J]. Journal of Thermal Analysis and Calorimetry, 2020, 141(5): 1529-1541.

[48] Hu Y, Li D C, Shu S, et al. Lattice Boltzmann simulation for three-dimensional natural convection with solid-liquid phase change[J]. International Journal of Heat and Mass Transfer, 2017, 113: 1168-1178.

[49] Krishnan S, Murthy J Y, Garimella S V. A two-temperature model for solid/liquid phase change in metal foams[J]. Journal of Heat Transfer, 2005, 127(9): 995-1004.

[50] Calmidi V V, Mahajan R L. Forced convection in high porosity metal foams[J]. Journal of Heat Transfer, 2000, 122(3): 557-565.

[51] Beckermann C, Viskanta R. Natural convection solid/liquid phase change in porous media[J]. International Journal of Heat and Mass Transfer, 1988, 31(1): 35-46.

[52] Gao D Y, Chen Z Q. Lattice Boltzmann simulation of natural convection dominated melting in a rectangular cavity filled with porous media[J]. International Journal of Thermal Sciences, 2011, 50(4): 493-501.

[53] Wu W, Zhang S L, Wang S F. A novel lattice Boltzmann model for the solid-liquid phase change with the convection heat transfer in the porous media[J]. International Journal of Heat and Mass Transfer, 2017, 104: 675-687.

[54] Gao D Y, Chen Z Q, Zhang D L, et al. Lattice Boltzmann modeling of melting of phase change materials in porous media with conducting fins[J]. Applied Thermal Engineering, 2017, 118: 315-327.

[55] Gao D Y, Chen Z Q, Shi M H, et al. Study on the melting process of phase change materials in metal foams using lattice Boltzmann method[J]. Science China Technological Sciences, 2010, 53(11): 3079-3087.

[56] Gao D Y, Tian F B, Chen Z Q, et al. An improved lattice Boltzmann method for solid-liquid phase change in porous media under local thermal non-equilibrium conditions[J]. International Journal of Heat and Mass Transfer, 2017, 110: 58-62.

[57] Liu Q, He Y L, Li Q. Enthalpy-based multiple-relaxation-time lattice Boltzmann method for solid-liquid phase-change heat transfer in metal foams[J]. Physical Review E, 2017, 96(2): 023303.

第 8 章　格子Boltzmann方法的边界处理

在前面几章,我们介绍了格子 Boltzmann 方法的基础理论以及常用的单相模型、多相模型、多孔介质流动与传热模型等。对于流动与传热问题,边界条件起着至关重要的作用。以稳态问题为例,当系统达到稳定后,流场和温度场与初始条件无关,模拟结果主要由边界条件决定。对宏观数值模拟方法,如基于 Euler 方程或 Navier-Stokes 方程的有限差分法、有限容积法、有限元法等,描述流体运动的变量为速度、温度等宏观变量,因而可在数值模拟中根据流体的受力、温度、边界情况等设置相应的边界条件,并选择合适的处理方法。

在本章,将介绍格子 Boltzmann 方法的边界处理,它是格子 Boltzmann 方法实施过程中非常关键的步骤,对数值模拟的精度、稳定性以及计算效率都有着重要的影响。在格子 Boltzmann 方法中,描述流场或温度场演化的是密度分布函数或其他类型的分布函数(如温度分布函数、总能分布函数等)。考虑无外力项的情况,格子 Boltzmann-BGK 方程的碰撞步和迁移步分别为

$$\text{碰撞:}\quad f_\alpha^*(\boldsymbol{r},t)=f_\alpha(\boldsymbol{r},t)-\frac{1}{\tau}\left[f_\alpha(\boldsymbol{r},t)-f_\alpha^{\mathrm{eq}}(\boldsymbol{r},t)\right] \tag{8.1}$$

$$\text{迁移:}\quad f_\alpha(\boldsymbol{r}+\boldsymbol{e}_\alpha\delta_t,t+\delta_t)=f_\alpha^*(\boldsymbol{r},t) \tag{8.2}$$

式中:f_α为密度分布函数;f_α^{eq}为平衡态密度分布函数;$\boldsymbol{e}_\alpha$ 为离散速度;τ 为量纲为 1 的松弛时间;$\boldsymbol{r}$ 为空间位置;δ_t 为时间步长。

在采用格子 Boltzmann 方法进行数值模拟时,无论是采用有限差分法或有限容积法求解微分形式的离散 Boltzmann-BGK 方程,还是采用碰撞迁移形式的式(8.1)和式(8.2),每个时步之后,流场内部节点上的分布函数均可获得,但边界节点上部分方向的分布函数是未知的。只有确定边界节点上的分布函数之后,才能进行下一时层的计算。因此,需要根据已知的边界条件确定边界节点上的未知分布函数,此过程中使用的方法被称为格子 Boltzmann 方法的**边界处理方法**,相应的数值计算格式则称为**边界处理格式**。

由于一些边界处理格式可以同时适用于多种边界条件,例如外推格式可适用于固体边界、开口边界等,因此本章将根据边界处理格式的特性对其进行分类,而不是按照应用范围进行划分。具体地,格子 Boltzmann 方法的边界处理格式可分为启发式格式、动力学格式、外推格式、复杂边界处理格式等。本章将分别介绍各类边界处理格式的理论基础和实施过程。为了简化,我们将基于 D2Q9 模型以及格子 Boltzmann 方程的碰撞迁移规则对各类边界处理格式进行阐述。

8.1 节点设置方式

与有限容积法类似，格子 Boltzmann 方法中也有两种典型的节点设置方式，分别为**湿节点式**（wet-node）和**链接式**（link-wise），如图 8.1 所示。采用湿节点式设置方式时，边界节点的连线即为计算边界，而采用链接式设置方式时，计算边界往往位于边界节点和固体节点连线的中间位置。针对这两种节点设置方式，需注意以下区别。从图 8.1 可以看到，当这两种方式描述相同大小的计算区域时（即计算边界所围成的区域），链接式设置方式所使用的有效节点数少于湿节点式。这里的有效节点是指流体内节点和流体边界节点，不包括固体节点（固体节点为虚拟节点）。相应地，当这两种设置方式采用相同数量的内节点和边界节点时，链接式设置方式所描述的计算区域将大于湿节点式所对应的计算区域。

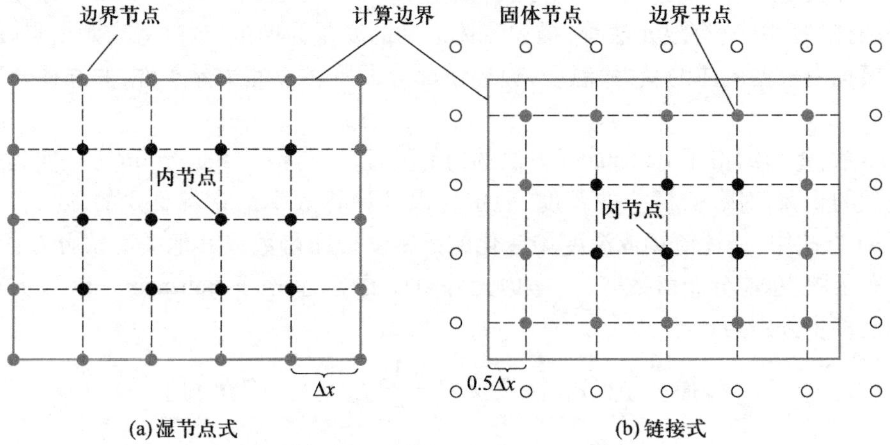

图 8.1 两种节点设置方式

8.2 启发式格式

启发式格式（heuristic scheme）主要根据边界上诸如周期性、对称性、充分发展等宏观物理特性，通过微观粒子的运动规则直接确定边界节点上的未知分布函数。与动力学格式、复杂边界处理格式等相比，启发式格式不需要复杂的数学推导或公式求解，形式简单易于实施，主要包括周期性边界处理格式、对称边界处理格式、充分发展边界处理格式以及用于固体壁面处理的反弹格式、镜面反射格式等。

8.2.1 周期性边界处理格式

在数值计算中，如果流场在空间呈现出周期性变化或在某个方向无穷大，则常常将周期性单元取出作为计算区域，并在相应的边界应用**周期性边界条件**（periodic boundary）：当流体粒子从一侧边界离开流场时，在下一时步会从流场的另一侧边界进入流场。周期性边界能够严格保证整个系统的质量和动量守恒[1]，其具体实施如下。

以图 8.2 所示的两无限大平行平板间的 Couette 流动为例，图中实心圆点表示边界节点，空心圆点表示格式所需要增加的虚拟节点。流场被均匀网格$\{i=1,\cdots,N_x;j=1,\cdots,N_y\}$覆盖，$x$ 方向采用周期性边界条件，y 方向采用无滑移边界条件（实施方法参见反弹格式、动力学格式、外推格式等相关内容）。由此，需要在 x 方向上增设两层虚拟节点，即$\{i=0;j=1,\cdots,N_y\}$与$\{i=N_x+1;j=1,\cdots,N_y\}$。周期性边界处理格式可以表示为

$$f_\alpha^*(0,j,t)=f_\alpha^*(N_x,j,t) \tag{8.3}$$

$$f_\alpha^*(N_x+1,j,t)=f_\alpha^*(1,j,t) \tag{8.4}$$

式中，f_α^* 为碰撞后（post-collision）的分布函数，由式（8.1）给出。

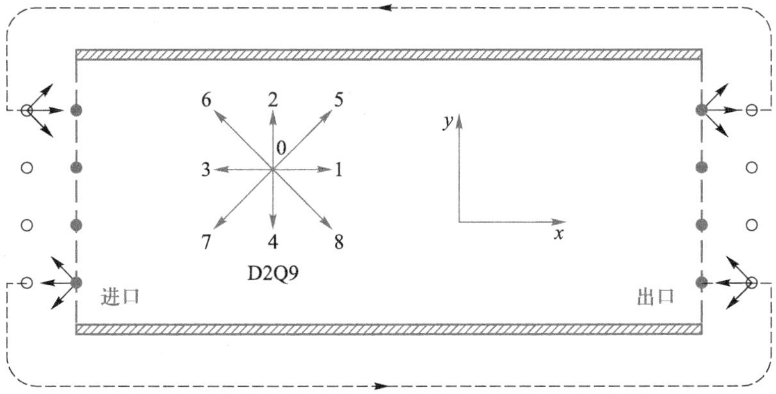

图 8.2 周期性边界处理示意图

式（8.3）表示左侧虚拟节点上 t 时刻的碰撞后分布函数 f_α^* 等于右侧边界节点上的碰撞后分布函数；相应地，式（8.4）表示右侧虚拟节点上 t 时刻的碰撞后分布函数 f_α^* 等于左侧边界节点上的碰撞后分布函数。在确定左右两侧虚拟节点的碰撞后分布函数 f_α^* 之后，即可执行迁移步式（8.2）：左侧边界节点上 $t+\delta_t$ 时刻的未知分布函数 f_1、f_5 和 f_8 是由 t 时刻左侧虚拟节点处的碰撞后分布函数迁移而来；右侧边界节点上 $t+\delta_t$ 时刻的未知分布函数 f_3、f_6 和 f_7 是由 t 时刻右侧虚拟节点处的碰撞后分布函数迁移而来。

需要注意的是，通常情况下，边界角点处的未知分布函数个数多于非角点的边界节点，因此需要特殊处理。例如，图 8.2 边界角点$(0,N_y)$上的未知分布函数为 f_1、f_4、f_5、f_7 和 f_8，其具体处理方法可参考文献[2]。

8.2.2 对称边界处理格式

对于对称性问题，为了节省计算资源，可以取物理模型的一半作为计算区域，并在对称轴上采用**对称边界**（symmetric boundary）处理，相应的处理格式实施如下。

参见图 8.3，图中实心圆点表示流体节点，空心圆点表示格式所需要增加的虚拟节点。流场被均匀网格$\{i=1,\cdots,N_x;j=1,\cdots,N_y\}$覆盖，且$\{i=1,\cdots,N_x;j=1\}$为对称轴。模拟时，需要在 y 方向上增设一层虚拟网格$\{i=1,\cdots,N_x;j=0\}$。对称边界处理格式可以表示为

$$f_{2,5,6}^*(i,0,t)=f_{4,8,7}^*(i,2,t) \tag{8.5}$$

即以 $j=1$ 为对称边界，$j=0$ 虚拟层上的碰撞后分布函数依据 $j=2$ 流体层对应节点的碰撞后分布函数进行镜面对称。在确定虚拟节点的碰撞后分布函数后，可执行迁移步：对称边界上 $t+\delta_t$ 时刻的未知分布函数 f_2、f_5 和 f_6 是由 t 时刻虚拟节点处的碰撞后分布函数迁移

而来，于是有

$$f_2(i,1,t+\delta_t)=f_2^*(i,0,t) \tag{8.6}$$

$$f_5(i,1,t+\delta_t)=f_5^*(i-1,0,t) \tag{8.7}$$

$$f_6(i,1,t+\delta_t)=f_6^*(i+1,0,t) \tag{8.8}$$

采用链接式节点设置方式时，对称边界位于边界节点和虚拟节点的中间位置，可进行类似的处理。

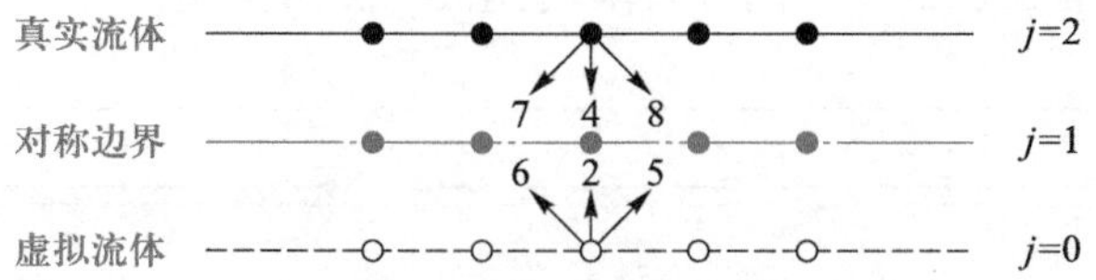

图 8.3 对称边界处理示意图

8.2.3 充分发展边界处理格式

当流体在通道内流动达到**充分发展**(fully developed)时，密度、速度等物理量在主流方向上不再发生变化，即它们的空间导数为零，相应的边界处理格式实施如下。以二维平板通道内的充分发展 Poiseuille 流动为例，出口边界处的三个未知分布函数 f_3、f_6 和 f_7 可近似认为等于内层流体节点相应方向上的分布函数，其数学表示为

$$f_{3,6,7}(N_x,j)=f_{3,6,7}(N_x-1,j) \tag{8.9}$$

这是格子 Boltzmann 方法中处理充分发展边界最简单的方法。同时，速度更新法也常用于充分发展边界的处理。它首先采用速度边界更新方法，获得出口边界上的速度，即

$$\boldsymbol{u}(N_x,j)=\boldsymbol{u}(N_x-1,j) \tag{8.10}$$

然后，假设边界节点处的未知分布函数 f_3、f_6 和 f_7 满足相应的平衡态分布，即

$$f_{3,6,7}(N_x,j)=f_{3,6,7}^{\mathrm{eq}}[\rho(N_x,j),\boldsymbol{u}(N_x,j)] \tag{8.11}$$

该方法在处理充分发展边界条件时表现出良好的数值稳定性。同时，对于已知入口各物理参数的流动，也可以采用类似式(8.11)的方法确定入口边界的未知分布函数。此外，后面将提到的非平衡反弹格式、外推格式等也可用于出口边界的处理。

8.2.4 半步长反弹格式

半步长反弹格式(halfway bounce-back scheme)是 1993 年由 Ziegler 提出的[3]。该格式适用于链接式节点设置方式(参见图 8.1)，它具有清晰的粒子运动图像，能够严格保证质量守恒，故而在格子 Boltzmann 方法领域得到了广泛应用。如图 8.4 所示，图中实心圆点为边界节点，空心圆点为固体节点，固体壁面位于边界节点和固体节点连线的中间位置。t 时刻，流体粒子在边界节点发生碰撞，碰撞后 $\boldsymbol{e}_\alpha$ 方向的粒子沿着节点连线向固体壁面迁移，在半个时间步长后，即 $t+0.5\delta_t$ 时刻流体粒子达到固体壁面并发生反弹(粒子速度大小不变，方向变为 $\boldsymbol{e}_{\bar{\alpha}}$ 且有 $\boldsymbol{e}_{\bar{\alpha}}=-\boldsymbol{e}_\alpha$)，随后流体粒子沿原路返回，再经过半个时间步长，在 $t+\delta_t$ 时刻达到其出发的位置。

根据半步长反弹格式的粒子运动图像，其数学表达式为

$$f_{\bar{\alpha}}(\boldsymbol{r}_\mathrm{b},t+\delta_t)=f_\alpha^*(\boldsymbol{r}_\mathrm{b},t) \tag{8.12}$$

式中：$\boldsymbol{r}_\mathrm{b}$ 为流体边界节点；f_α^* 为碰撞后分布函数，由式(8.1)给出；离散速度方向 $\bar{\alpha}$ 与 α 存

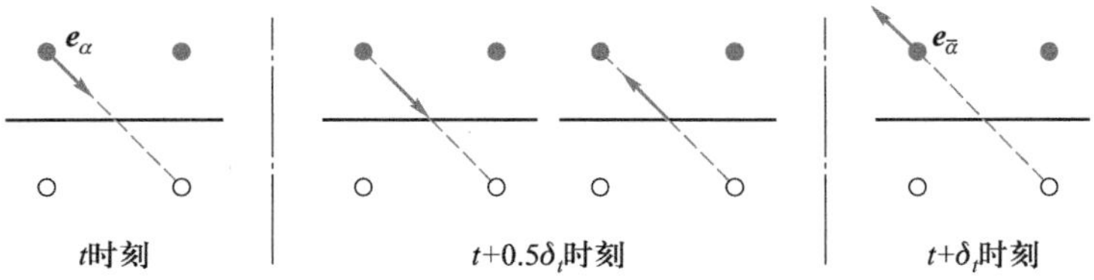

图 8.4 半步长反弹示意图

半步长反弹

在 $\boldsymbol{e}_{\bar{\alpha}}=-\boldsymbol{e}_\alpha$ 的关系。具体地,对 D2Q9 模型,依据图 8.4 和式(8.12)有

$$f_2(\boldsymbol{r}_\mathrm{b},t+\delta_t)=f_4^*(\boldsymbol{r}_\mathrm{b},t) \tag{8.13}$$

$$f_5(\boldsymbol{r}_\mathrm{b},t+\delta_t)=f_7^*(\boldsymbol{r}_\mathrm{b},t) \tag{8.14}$$

$$f_6(\boldsymbol{r}_\mathrm{b},t+\delta_t)=f_8^*(\boldsymbol{r}_\mathrm{b},t) \tag{8.15}$$

从上述式子可以看到,半步长反弹格式实施简便,且具有完全局部特性,特别适合复杂结构中的流动问题如多孔介质流动。需要注意的是,当半步长反弹格式与单松弛格子 Boltzmann 方程结合使用时,可能会在固体壁面产生虚假滑移速度,这一问题可通过采用双松弛或多松弛碰撞算子并选取适当的松弛参数来解决[4-6]。

8.2.5 镜面反射格式

反弹格式主要应用于无滑移壁面,而对于光滑壁面,如果壁面对流体没有摩擦作用,即壁面与流体之间没有动量交换,则通常采用**镜面反射格式**(specular reflection scheme)来实现自由滑移边界。在图 8.4 中,流体粒子撞击到壁面时,沿原路弹回,为反弹处理;如果以壁面法线的对称方向进行反射,则为镜面反射处理,如图 8.5 所示。

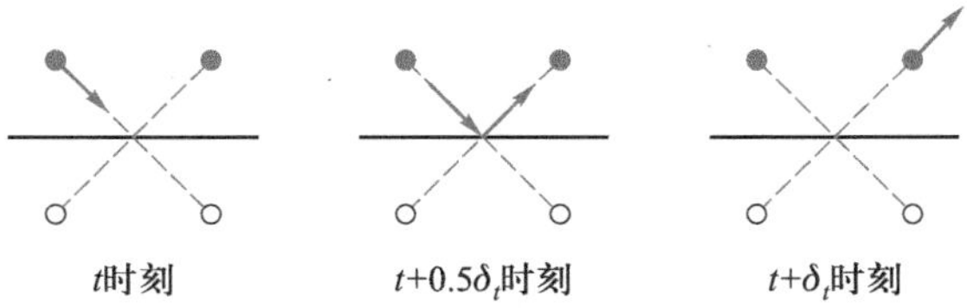

图 8.5 镜面反射示意图

镜面反射

实际上,对于平直表面,镜面反射格式与对称边界格式的结果是一样的。以固体壁面为对称轴,将 t 时刻流体边界节点的碰撞后分布函数 f_α^* 按照镜面对称方式(参见图 8.3)赋给相应的固体节点,之后可按照迁移规则式(8.2)确定边界节点 $t+\delta_t$ 时刻的未知分布函数。

在一些场合,如微通道气体流动,既不能简单地用反弹格式处理,也不能用镜面反射格式来描述气体与固体壁面之间的相互作用及动量交换。为此,一些学者提出可将反弹格式和镜面反射格式结合起来,用于描述真实的气体与固体之间的相互作用[7-8]。这样处理的一个理由源自稀薄气体动力学研究先驱 Knudsen 在分子层面所做的一个实验[9]。在实验中,他让所有分子以一个固定的入射角向墙面运动,可是分子在碰到墙面后,却任意地朝四面八方散开。如果把这个现象应用于格子 Boltzmann 方法中,则意味着反弹和镜面反射各占一半。因此,可以定义一个弹回比例系数 $r_\mathrm{b}(0\leqslant r_\mathrm{b}\leqslant 1)$,来表示粒子在与壁面作用时沿原路弹回所占的比例,具体处理可参考相关文献[7-8]。

8.3 动力学格式

动力学格式(hydrodynamic scheme)主要利用边界上宏观物理量的定义,直接求解边界节点上未知分布函数的方程组

$$\sum_{\alpha=0}^{m-1} f_\alpha(\boldsymbol{r}) = \rho, \quad \sum_{\alpha=0}^{m-1} \boldsymbol{e}_\alpha f_\alpha(\boldsymbol{r}) = \rho\boldsymbol{u} \tag{8.16}$$

以获得边界节点上待定的分布函数。动力学格式主要包括非平衡反弹格式、反滑移格式、质量修正格式等。

8.3.1 非平衡态反弹格式

1997 年,Zou 和 He[10]对二维及三维的格子 Boltzmann-BGK 模型进行了研究,提出了一种动力学边界处理格式,本书称之为**非平衡态反弹格式**(non-equilibrium bounce-back scheme)。其相应的边界处理方法如下。

如图 8.6 所示,"3-0-1"处在边界上,"6-2-5"和"7-4-8"分别位于流体区域内和非流体区域内。当每一个迁移步完成后,在边界节点 O 处,分布函数 f_0、f_1、f_3、f_4、f_7 和 f_8 已知,而 f_2、f_5、f_6 和密度 ρ 未知。以速度边界(如流固边界、已知速度的进出口边界等)为例,流体宏观速度的水平分量 u_x 和竖直分量 u_y 已知,根据式(8.16)并考虑 $\delta_x=\delta_t=1$ 均取单位量,可得如下方程组:

$$f_2+f_5+f_6=\rho-(f_0+f_1+f_3+f_4+f_7+f_8) \tag{8.17}$$

$$f_5-f_6=\rho u_x-(f_1-f_3+f_8-f_7) \tag{8.18}$$

$$f_2+f_5+f_6=\rho u_y+(f_4+f_7+f_8) \tag{8.19}$$

联立式(8.17)和式(8.19)可得边界节点 O 处的密度为

$$\rho=\frac{1}{1-u_y}[f_0+f_1+f_3+2(f_4+f_7+f_8)] \tag{8.20}$$

边界节点处有 4 个未知量,却只有 3 个独立方程。为了使方程组封闭,Zou 和 He 假设边界节点的法向方向上,反弹格式对分布函数的非平衡态部分成立,即

$$f_2-f_2^{\mathrm{eq}}=f_4-f_4^{\mathrm{eq}} \tag{8.21}$$

联立式(8.17)~式(8.21),即可求得边界节点 O 处的未知分布函数。以无滑移边界为例($u_x=u_y=0$, $f_2^{\mathrm{eq}}=f_4^{\mathrm{eq}}$),联立上述方程可得

$$f_2=f_4 \tag{8.22}$$

$$f_5=f_7-0.5(f_1-f_3) \tag{8.23}$$

$$f_6=f_8+0.5(f_1-f_3) \tag{8.24}$$

需要指出的是,式(8.22)~式(8.24)是建立在式(8.16)的基础上。当系统存在外力时,速度的定义式通常需修改为 $\rho\boldsymbol{u}=\sum_{\alpha=0}^{8}\boldsymbol{e}_\alpha f_\alpha+0.5\delta_t\boldsymbol{F}$(参见第 3 章),此时应当根据新的速度定义式推导相应的分布函数[11]。此外,边界角点处的未知分布函数需进行特殊处理,具体处理方式可参阅文献[10]。

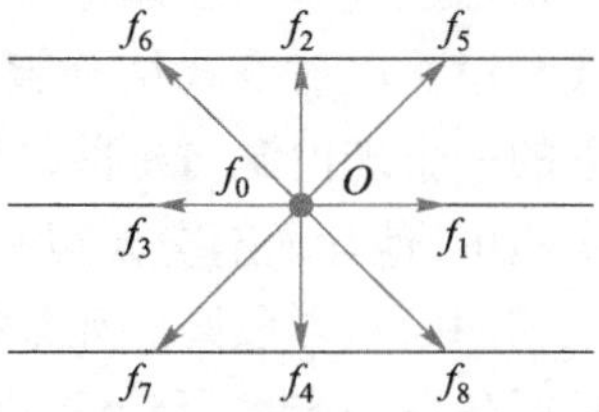

图 8.6 非平衡态反弹格式

按照类似的思路,可设计出压力和密度边界条件[10]。Zou 和 He 采用这种非平衡态反弹格式模拟了二维 Poiseuille 流动和三维管道流动,并与半步长反弹格式等边界处理格式进行了对比,研究结果表明这种格式可取得近似二阶精度,并具有较好的数值稳定性。

8.3.2 反滑移格式

1995 年,Inamuro 等[12]提出了**反滑移速度**(counter-slip velocity)处理格式。其基本思想是假设边界节点上的未知分布函数可以直接用一个新的平衡态分布函数来计算,但为了克服壁面滑移,需要在这个平衡态分布函数中加入一个反滑移速度来进行修正,从而使得壁面上的流体速度与壁面速度相等。

参照图 8.6,以 D2Q9 模型为例,边界节点 O 处的未知分布函数 f_2、f_5 和 f_6 可以表示为

$$f_2=\frac{\rho'}{9}\left\{1+3u_{y,\mathrm{w}}+\frac{9}{2}u_{y,\mathrm{w}}^2-\frac{3}{2}\left[(u_{x,\mathrm{w}}+u')^2+u_{y,\mathrm{w}}^2\right]\right\} \tag{8.25}$$

$$f_5=\frac{\rho'}{36}\left\{1+3(u_{x,\mathrm{w}}+u'+u_{y,\mathrm{w}})+\frac{9}{2}(u_{x,\mathrm{w}}+u'+u_{y,\mathrm{w}})^2-\frac{3}{2}\left[(u_{x,\mathrm{w}}+u')^2+u_{y,\mathrm{w}}^2\right]\right\} \tag{8.26}$$

$$f_6=\frac{\rho'}{36}\left\{1+3(-u_{x,\mathrm{w}}-u'+u_{y,\mathrm{w}})+\frac{9}{2}(-u_{x,\mathrm{w}}-u'+u_{y,\mathrm{w}})^2-\frac{3}{2}\left[(u_{x,\mathrm{w}}+u')^2+u_{y,\mathrm{w}}^2\right]\right\} \tag{8.27}$$

式中:$u_{x,\mathrm{w}}$ 和 $u_{y,\mathrm{w}}$ 为壁面速度分量;ρ' 和 u' 待定,且 u' 为前面提到的反滑移速度。此外,另有一待定参数,即壁面处流体的密度 ρ_w。根据密度和速度的定义,可得 ρ_w、ρ' 和 u' 的计算式为

$$\rho_\mathrm{w}=\frac{1}{1-u_{y,\mathrm{w}}}\left[f_0+f_1+f_3+2(f_4+f_7+f_8)\right] \tag{8.28}$$

$$\rho'=6\left(\frac{\rho_\mathrm{w}u_{y,\mathrm{w}}+f_4+f_7+f_8}{1+3u_{y,\mathrm{w}}+3u_{y,\mathrm{w}}^2}\right) \tag{8.29}$$

$$u'=\frac{1}{1+3u_{y,\mathrm{w}}}\left[6\frac{\rho_\mathrm{w}u_{x,\mathrm{w}}-(f_1-f_3-f_7+f_8)}{\rho'}-u_{x,\mathrm{w}}-3u_{x,\mathrm{w}}u_{y,\mathrm{w}}\right] \tag{8.30}$$

进而,可通过式(8.25)~式(8.27)求得未知分布函数。数值模拟表明,该格式能够严格满足无滑移速度条件,且具有二阶精度。随后,基于 He 等[13]的双分布函数不可压缩热模型,D'Orazio 等[14-15]提出了**反滑移热能**(counter-slip thermal energy)的概念,并构造了相应的温度边界处理格式。

总体说来,反滑移格式能够严格满足边界条件,且可以获得较好的数值稳定性和计算精度。但在实施过程中,需要确定壁面上的未知宏观物理量,并且需要用到平衡态分布函数的定义式,而对于不同的格子 Boltzmann 模型,分布函数及宏观物理量的定义也不尽相同,因而相关表达式往往需要重新推导。此外,在处理拐角边界或复杂边界时,需要根据具体情况,确定边界节点上未知分布函数的个数及方向,再计算边界节点上的宏观物理量和未知分布函数,实施比较麻烦,通用性较差。

8.3.3 质量修正格式

对充分发展流动来说,如果边界条件不能使流体在整个流动区域内满足质量守恒,则数值模拟的精度就会降低,而且数值计算的收敛速度及稳定性都会受到影响。因而,如果

在处理充分发展边界条件时,采取一定的方法,使得流体在整个流动区域内的总体质量守恒得到满足,那么数值模拟的精度以及计算过程的收敛状况都会有较大程度的提升。基于这一思路,Tong 等[16]提出了一种处理充分发展流动的**质量修正格式**。

所谓质量修正,就是通过计算进口和出口的质量流率,确定一个质量修正系数,以此系数来修正出口速度,并用修正后的出口速度更新边界节点上的未知粒子分布函数,具体做法如下。

首先,假设出口截面上各点速度的 x 方向分量的相对变化率为常数,即

$$\frac{u_x(N_x,j)-u_x(N_x-1,j)}{u_x(N_x-1,j)}=\phi=\text{const} \tag{8.31}$$

式中:(N_x,j)为出口边界上的节点;(N_x-1,j)为其相邻的流体节点。于是

$$u_x(N_x,j)=(\phi+1)u_x(N_x-1,j)=\varphi u_x(N_x-1,j) \tag{8.32}$$

式中,φ 为质量修正系数。考虑整体质量守恒,也即进出口的质量流率相等,则

$$\sum_{j=1}^{N_y}\rho(1,j)u_x(1,j)\Delta y=\sum_{j=1}^{N_y}\rho(N_x,j)u_x(N_x,j)\Delta y=\sum_{j=1}^{N_y}\varphi\rho(N_x,j)u_x(N_x-1,j)\Delta y \tag{8.33}$$

式中,N_y为进出口边界的节点数。于是可得质量修正系数为

$$\varphi=\frac{\sum_{j=1}^{N_y}\rho(1,j)u_x(1,j)\Delta y}{\sum_{j=1}^{N_y}\rho(N_x,j)u_x(N_x-1,j)\Delta y} \tag{8.34}$$

确定质量修正系数 φ 之后就可根据式(8.32)确定出口速度 $u_x(N_x,j)$,然后基于 $u_x(N_x,j)$ 计算出口边界上的未知粒子分布函数。

综上,质量修正处理格式包括以下步骤:

(1) 计算进、出口截面的质量流率,$\sum_{j=1}^{N_y}\rho(1,j)u_x(1,j)\Delta y$ 和 $\sum_{j=1}^{N_y}\rho(N_x,j)u_x(N_x-1,j)\Delta y$;

(2) 依据式(8.34)计算质量修正系数 φ;

(3) 按照式(8.32)修正流体在出口边界上的速度 $u_x(N_x,j)$;

(4) 根据新求得的 $u_x(N_x,j)$ 按所采用的边界处理格式,重新计算出口边界上的未知粒子分布函数。

Tong 等[16]考核了二维平板通道、后掠台阶充分发展等流动,数值模拟结果表明,该格式数值稳定性良好,计算结果受松弛时间的影响很小,在松弛时间较大或较小的情况下,仍能获得比较稳定和精确的数值模拟结果;即使在稀疏网格下,也能获得精度较高的结果,且计算时间较短,操作简单。

8.4 外推格式

动力学格式可以处理一些特定的边界条件,但局限性较大。一方面,这些格式依赖于所使用的格子 Boltzmann 模型;另一方面,实施过程中往往需要对边界节点处的一些物理性质(如速度、密度、压力等)进行假设,从而很难推广到一般的边界处理。鉴于格子

Boltzmann 方程是连续 Boltzmann 方程的一种特殊离散形式，因此，可以借鉴传统计算流体力学方法（如有限差分法）中的边界处理方法来构造格子 Boltzmann 方法的边界处理格式。基于这一思路，国内外学者先后提出了多种**外推格式**（extrapolation scheme）。

8.4.1 Chen 格式

1996 年，Chen 等[17]提出了一种**外推格式**。如图 8.7 所示，假设物理边界外还有一层虚拟节点，且虚拟节点 $(i,0)$ 上的分布函数参与迁移过程。因此，在每次迁移之前，需要知道虚拟节点 $(i,0)$ 上的碰撞后分布函数 f_2^*、f_5^* 和 f_6^*（这些分布函数将迁移至物理边界上）。借鉴有限差分的概念，可对边界节点 $(i,1)$ 采取中心差分处理，由此可确定虚拟节点 $(i,0)$ 上的 f_2^*、f_5^* 和 f_6^* 为

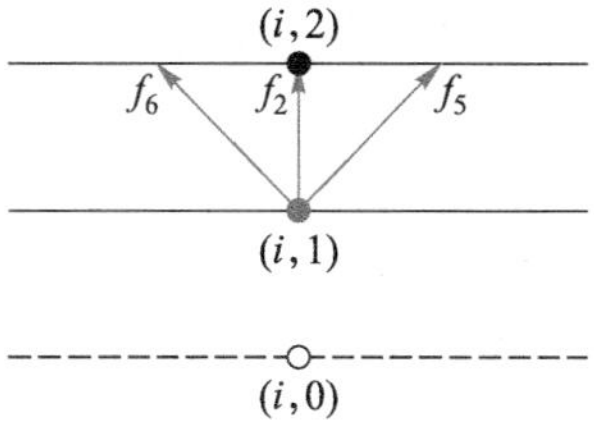

图 8.7 外推格式

$$f_{2,5,6}^*(i,0)=2f_{2,5,6}^*(i,1)-f_{2,5,6}^*(i,2) \tag{8.35}$$

然后，即可执行迁移步骤确定物理边界节点上新时刻的未知分布函数。

研究表明，这种外推格式具有二阶精度，因此能够保证格子 Boltzmann 方法的整体精度。同时，这种格式适用范围广，计算简单，容易实现，并可做相应的推广。例如，文献[18]曾使用一种**修正外推方法**，该方法假设出口边界附近的分布函数的变化率相等，即

$$\frac{f_{1,5,8}(N_x,j)-f_{1,5,8}(N_x-1,j)}{\Delta x_{N_x-1}}=\frac{f_{1,5,8}(N_x-1,j)-f_{1,5,8}(N_x-2,j)}{\Delta x_{N_x-2}}=\text{const} \tag{8.36}$$

式中：节点 (N_x,j) 为出口边界节点；(N_x-1,j) 与 (N_x-2,j) 为相邻的流体节点；Δx_{N_x-1} 与 Δx_{N_x-2} 代表不同网格步长。对均匀网格来说，$\Delta x_{N_x-1}=\Delta x_{N_x-2}$，此时式(8.36)与式(8.35)是一致的。需要指出的是，采用外推格式时，可能计算出负的分布函数，因此要特别注意其数值稳定性。

8.4.2 非平衡态外推格式

受 Chen 等的外推格式以及 Zou 和 He 的非平衡反弹格式的启发，Guo 等[19]提出了一种**非平衡态外推格式**（non-equilibrium extrapolation scheme）。其基本思想是将边界节点上的分布函数分为平衡态和非平衡态两部分，其中，平衡态部分根据边界条件的定义近似获得，而非平衡态部分则通过外推的方式确定。

如图 8.8 所示，COA 位于边界上，EBD 位于流场中。每次碰撞之前，需要知道边界节点 O 处的分布函数 $f_\alpha(O,t)$，Guo 等[19]将其分解成平衡态和非平衡态两部分，即

$$f_\alpha(O,t)=f_\alpha^{\text{eq}}(O,t)+f_\alpha^{\text{neq}}(O,t) \tag{8.37}$$

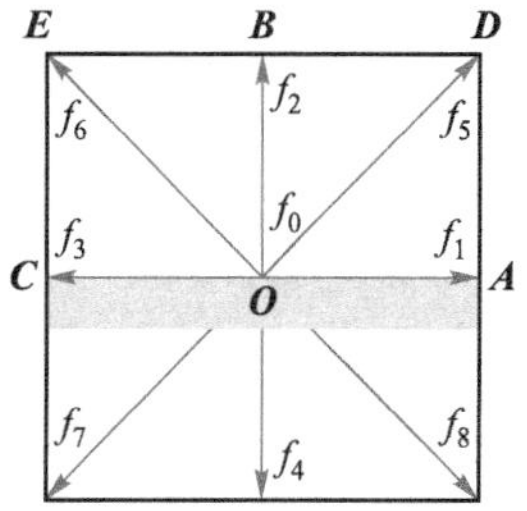

图 8.8 非平衡态外推格式

对平衡态部分 $f_\alpha^{\text{eq}}(O,t)$，可根据边界节点上的宏观物理量求得，如果节点 O 存在未知宏观物理量，则由点 B 的相应值代替。以速度边界条件为例，已知点 O 的速度 $\boldsymbol{u}(O,t)$，而密度 $\rho(O,t)$ 未知，则点 O 的平衡态分布函数可表示为

$$f_\alpha^{\text{eq}}(O,t)=f_\alpha^{\text{eq}}[\rho(A,t),\boldsymbol{u}(O,t)] \tag{8.38}$$

对非平衡态部分 $f_\alpha^{\text{neq}}(O,t)$，由于流体节点 B 处的分布函数 $f_\alpha(B,t)$、宏观速度 $\boldsymbol{u}(B,t)$

和密度$\rho(B,t)$均已知,因此可计算出点B的非平衡态分布函数为

$$f_\alpha^{\text{neq}}(B,t)=f_\alpha(B,t)-f_\alpha^{\text{eq}}[\rho(B,t),\boldsymbol{u}(B,t)] \tag{8.39}$$

同时,考虑到O、B两点的非平衡态分布函数具有如下关系:

$$f_\alpha^{\text{neq}}(B,t)=f_\alpha^{\text{neq}}(O,t)+O(\delta_x^2) \tag{8.40}$$

式中,δ_x为空间步长。由此,可用点B的非平衡态部分代替点O的非平衡态部分。

综上所述,边界节点O的分布函数可表示为

$$f_\alpha(O,t)=f_\alpha^{\text{eq}}(O,t)+[f_\alpha(B,t)-f_\alpha^{\text{eq}}(B,t)] \tag{8.41}$$

若考虑碰撞过程,则点O碰撞后的分布函数可表示为

$$f_\alpha^*(O,t)=f_\alpha^{\text{eq}}(O,t)+\left(1-\frac{1}{\tau}\right)[f_\alpha(B,t)-f_\alpha^{\text{eq}}(B,t)] \tag{8.42}$$

式中:$f_\alpha^*(O,t)$为碰撞后分布函数;τ为量纲为1的松弛时间。

Guo 等所提出的非平衡态外推格式最初主要用于速度边界(密度分布函数)的处理。随后,这一格式被推广应用于温度边界(热力学能分布函数、温度分布函数、总能分布函数等)的处理[20-22]。以温度分布函数为例,根据式(8.41)类推可得

$$T_\alpha(O,t)=T_\alpha^{\text{eq}}(O,t)+[T_\alpha(B,t)-T_\alpha^{\text{eq}}(B,t)] \tag{8.43}$$

研究表明,非平衡态外推格式具有二阶精度。与此同时,这一格式兼具了非平衡态反弹格式及外推格式的优点,具有操作简单、数值稳定性好、应用范围广等特点。根据我们的经验,在有限差分格子 Boltzmann 方法中,对多速度模型以及耦合双分布函数模型,与其他边界处理格式相比,非平衡态外推格式具有较大的优势。

8.5 复杂边界处理格式

当所要处理的物理区域不规则时,一种处理方式是采用适体网格、非结构化网格等网格处理技术,这部分内容将在下一章进行介绍;另一种处理方式则是仍然基于结构化的直角正交网格,但在边界处采用阶梯逼近或插值处理,以满足物理边界上的条件。阶梯逼近可与前面提到的半步长反弹格式结合使用。以下主要介绍基于直角坐标系正交网格下的插值处理。

为了说明具体情况,需要先分清网格节点的类型。如图 8.9 所示,图中曲线代表实际的物理边界,实心圆点为流体节点(包括内节点和边界节点),空心圆点为固体节点,椭圆点是物理边界和格线的交点。具体的,图中$\boldsymbol{r}_\text{b}$为流体边界节点,$\boldsymbol{r}_\text{s}$是其在$\boldsymbol{e}_\alpha$方向相邻的固体节点(虚拟节点),而$\boldsymbol{r}_\text{f}$则是其在$\boldsymbol{e}_{\bar\alpha}$方向($\boldsymbol{e}_{\bar\alpha}=-\boldsymbol{e}_\alpha$)相邻的流体内节点。流体边界节点和内节点的区别在于边界节点至少有一个相邻节点是固体节点。

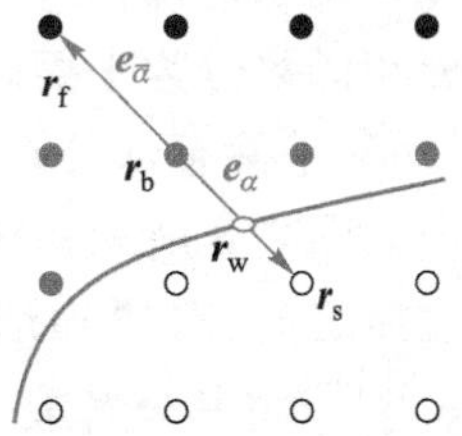

图 8.9 复杂边界处理的节点类型示意图

对边界节点和固体节点应用碰撞迁移规则可知,$t+\delta_t$时刻边界节点$\boldsymbol{r}_\text{b}$处的分布函数$f_{\bar\alpha}(\boldsymbol{r}_\text{b},t+\delta_t)$是由$t$时刻固体节点$\boldsymbol{r}_\text{s}$处的碰撞后分布函数$f_{\bar\alpha}^*(\boldsymbol{r}_\text{s},t)$迁移而来,即

$$f_{\bar\alpha}(\boldsymbol{r}_\text{b},t+\delta_t)=f_{\bar\alpha}^*(\boldsymbol{r}_\text{s},t) \tag{8.44}$$

式中,$f_{\bar\alpha}^*(\boldsymbol{r}_\text{s},t)$是未知的,需要根据给定的边界条件确定。实际上,各种复杂边界处理格

式的区别就在于如何计算$f_{\bar{\alpha}}^{*}(\boldsymbol{r}_{\mathrm{s}},t)$或$f_{\bar{\alpha}}(\boldsymbol{r}_{\mathrm{b}},t+\delta_t)$。

以下,介绍较常用的 Filippova-Hänel 格式及其改进形式、Bouzidi 格式、Guo 格式等。

8.5.1 Filippova-Hänel 格式及其改进形式

1998 年,Filippova 与 Hänel[23]提出构建固体节点的**虚拟平衡态分布函数**,并采用线性插值的方式来计算由壁面弹回流体区域的分布函数值。如图 8.9 所示,引入变量 q 来表征边界节点和固体节点连线上边界节点到物理边界的相对距离,即

$$q=\frac{|\boldsymbol{r}_{\mathrm{b}}-\boldsymbol{r}_{\mathrm{w}}|}{|\boldsymbol{r}_{\mathrm{b}}-\boldsymbol{r}_{\mathrm{s}}|},\quad 0<q\leqslant 1 \tag{8.45}$$

式中,$\boldsymbol{r}_{\mathrm{w}}$ 为边界节点和固体节点连线与实际物理边界的交点。

为了重构$f_{\bar{\alpha}}^{*}(\boldsymbol{r}_{\mathrm{s}},t)$,Filippova 与 Hänel 针对固体节点构造了如下形式的虚拟平衡态分布函数:

$$f_{\alpha}^{\mathrm{eq}}(\boldsymbol{r}_{\mathrm{s}},t)=\omega_{\alpha}\rho(\boldsymbol{r}_{\mathrm{b}},t)\left[1+\frac{\boldsymbol{e}_{\alpha}\cdot\boldsymbol{u}_{\mathrm{s}}}{c_{\mathrm{s}}^{2}}+\frac{(\boldsymbol{e}_{\alpha}\cdot\boldsymbol{u}_{\mathrm{b}})^{2}}{2c_{\mathrm{s}}^{4}}-\frac{(\boldsymbol{u}_{\mathrm{b}}\cdot\boldsymbol{u}_{\mathrm{b}})}{2c_{\mathrm{s}}^{2}}\right] \tag{8.46}$$

式中:$\boldsymbol{u}_{\mathrm{b}}=\boldsymbol{u}(\boldsymbol{r}_{\mathrm{b}},t)$为边界节点处的流体速度;$\boldsymbol{u}_{\mathrm{s}}$ 为待定的固体节点虚拟速度。

首先,考虑一种特殊情形,即 $q=0.5$,此时 $\boldsymbol{r}_{\mathrm{w}}$ 处在边界节点和固体节点连线的中间位置,可以采用半步长反弹格式,即$f_{\bar{\alpha}}(\boldsymbol{r}_{\mathrm{b}},t+\delta_t)=f_{\alpha}^{*}(\boldsymbol{r}_{\mathrm{b}},t)$,故有$f_{\bar{\alpha}}^{*}(\boldsymbol{r}_{\mathrm{s}},t)=f_{\alpha}^{*}(\boldsymbol{r}_{\mathrm{b}},t)$。结合这一特殊情形及固体节点的虚拟平衡态分布函数,Filippova 与 Hänel 针对一般情况提出了如下形式的线性组合:

$$f_{\bar{\alpha}}^{*}(\boldsymbol{r}_{\mathrm{s}},t)=(1-\chi)f_{\alpha}^{*}(\boldsymbol{r}_{\mathrm{b}},t)+\chi f_{\alpha}^{\mathrm{eq}}(\boldsymbol{r}_{\mathrm{s}},t)-2\omega_{\alpha}\rho(\boldsymbol{r}_{\mathrm{b}},t)\frac{\boldsymbol{e}_{\alpha}\cdot\boldsymbol{u}_{\mathrm{w}}}{c_{\mathrm{s}}^{2}} \tag{8.47}$$

式中:$\boldsymbol{u}_{\mathrm{w}}$ 为物理边界处的速度(对无滑移固体壁面,$\boldsymbol{u}_{\mathrm{w}}=\boldsymbol{0}$);$\chi$ 为线性插值因子,其取值与 $\boldsymbol{u}_{\mathrm{s}}$ 的选择有关。式(8.47)右端第三项代表物理边界运动时所引起的动量增加。显然,当线性插值因子$\chi=0$,式(8.47)退化为半步长反弹格式。

Filippova 与 Hanel 提出由以下关系式来确定固体节点虚拟速度 $\boldsymbol{u}_{\mathrm{s}}$ 和线性插值因子χ。

当 $q<1/2$ 时,

$$\boldsymbol{u}_{\mathrm{s}}=\boldsymbol{u}_{\mathrm{b}},\quad \chi=\frac{2q-1}{\tau-1} \tag{8.48}$$

当 $q\geqslant 1/2$ 时,

$$\boldsymbol{u}_{\mathrm{s}}=\frac{q-1}{q}\boldsymbol{u}_{\mathrm{b}}+\frac{1}{q}\boldsymbol{u}_{\mathrm{w}},\quad \chi=\frac{2q-1}{\tau} \tag{8.49}$$

考察式(8.48)可知,对于 $q<1/2$ 的情况,若松弛因子 τ 趋近于 1,χ 会变得无穷大,从而导致数值不稳定。Mei 等[24]基于 Poiseuille 流动考察了上述格式的稳定性,并指出通过构造恰当的$f_{\alpha}^{\mathrm{eq}}(\boldsymbol{r}_{\mathrm{s}},t)$可以提高数值稳定性和计算精度。他们在文献[24]中提出了如下改进格式。

当 $q<1/2$ 时,将上述格式修改为

$$\boldsymbol{u}_{\mathrm{s}}=\boldsymbol{u}_{\mathrm{f}},\quad \chi=\frac{2q-1}{\tau-2} \tag{8.50}$$

式中,$\boldsymbol{u}_{\mathrm{f}}=\boldsymbol{u}(\boldsymbol{r}_{\mathrm{f}},t)$为内节点 $\boldsymbol{r}_{\mathrm{f}}=\boldsymbol{r}_{\mathrm{b}}+\boldsymbol{e}_{\bar{\alpha}}\delta_t$ 处的流体速度(参见图 8.9)。

当 $q \geq 1/2$ 时,依然采用 Filippova 与 Hänel 的插值方式,即式(8.49)。

当松弛因子较小时,Mei 等的改进格式与 Filippova-Hänel 格式相比可以明显提高数值稳定性,但当 $\tau>1.25$ 时仍然可能导致数值发散[25]。随后,Mei 等[26]将其改进格式拓展到了三维问题。

8.5.2 Bouzidi 格式

2001 年,Bouzidi 等[27]提出了一种结合反弹格式和空间插值的边界处理方法,用以处理任意曲率的复杂几何边界。如图 8.10 所示,A 为流体边界节点,B 为固体节点,C 为物理边界与格线的交点,且 $q=|AC|/|AB|$,E 和 F 为流体内节点。Bouzidi 等的基本思路是利用空间插值的方式来确定边界节点 A 在 $t+\delta_t$ 时刻的未知分布函数 $f_{\bar{\alpha}}(A,t+\delta_t)$,他们依据 q 值的大小分了两种情况。

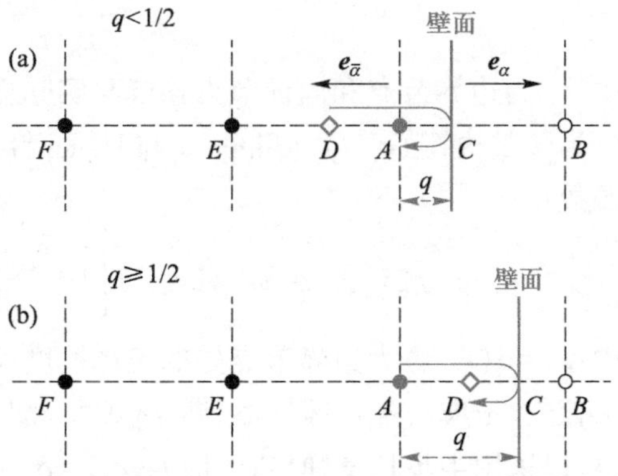

图 8.10 Bouzidi 格式插值示意图

当 $q<1/2$ 时,边界节点 A 的分布函数 $f_{\bar{\alpha}}(A,t+\delta_t)$ 可通过以下方式得到:点 D 是流体节点 AE 连线上的一个点,它与边界节点 A 的相对距离为 $|AD|/|AE|=(1-2q)$,根据碰撞迁移及反弹规则,t 时刻点 D 的碰撞后分布函数 $f_\alpha^*(D,t)$ 在经历 $(1-q)\delta_t$ 时间的运动后将达到固体壁面,并发生反弹,然后再经历 $q\delta_t$ 时间的运动,在 $t+\delta_t$ 时刻将达到点 A,于是有 $f_{\bar{\alpha}}(A,t+\delta_t)=f_\alpha^*(D,t)$。因此,只要确定 $f_\alpha^*(D,t)$,即可获得边界节点 A 的待定分布函数 $f_{\bar{\alpha}}(A,t+\delta_t)$。由于点 D 位于点 A 和点 E 之间,且它与点 A 的相对距离为 $(1-2q)$,与点 E 的相对距离为 $2q$,故可采用线性插值,即

$$f_{\bar{\alpha}}(A,t+\delta_t)=f_\alpha^*(D,t)=2qf_\alpha^*(A,t)+(1-2q)f_\alpha^*(E,t) \tag{8.51}$$

当 $q\geq 1/2$ 时,插值点 D 位于边界节点 A 和固体节点 B 的连线上,且它与点 A 的相对距离为 $|AD|/|AB|=(2q-1)$,点 A 的碰撞后分布函数 $f_\alpha^*(A,t)$ 在经历 $q\delta_t$ 时间的运动后将达到固体壁面,并在壁面处发生反弹,随后再经历 $(1-q)\delta_t$ 时间的运动,在 $t+\delta_t$ 时刻达到点 D,因此有 $f_{\bar{\alpha}}(D,t+\delta_t)=f_\alpha^*(A,t)$。由于 $t+\delta_t$ 时刻点 E 的分布函数 $f_{\bar{\alpha}}(E,t+\delta_t)$ 是已知的[根据碰撞迁移规则有 $f_{\bar{\alpha}}(E,t+\delta_t)=f_{\bar{\alpha}}^*(A,t)$],因此可根据线性插值利用点 E 和点 D 新时刻的分布函数确定边界节点 A 的待定分布函数 $f_{\bar{\alpha}}(A,t+\delta_t)$,即

$$\begin{aligned} f_{\bar{\alpha}}(A,t+\delta_t) &= \frac{1}{2q}f_{\bar{\alpha}}(D,t+\delta_t)+\frac{2q-1}{2q}f_{\bar{\alpha}}(E,t+\delta_t) \\ &= \frac{1}{2q}f_\alpha^*(A,t)+\frac{2q-1}{2q}f_{\bar{\alpha}}^*(A,t) \end{aligned} \tag{8.52}$$

由式(8.52)可知,对于 $q\geq 1/2$ 的情况,$f_{\bar{\alpha}}(A,t+\delta_t)$ 的计算是完全局部的。结合图 8.9 和图 8.10,可将式(8.51)和式(8.52)写成下列一般形式。

当 $q<1/2$ 时,

$$f_{\bar{\alpha}}(\boldsymbol{r}_\text{b},t+\delta_t)=2q\,f_\alpha^*(\boldsymbol{r}_\text{b},t)+(1-2q)f_\alpha^*(\boldsymbol{r}_\text{b}+\boldsymbol{e}_{\bar{\alpha}}\delta_t,t) \tag{8.53}$$

当 $q\geq 1/2$ 时,

$$f_{\bar{\alpha}}(\boldsymbol{r}_{\mathrm{b}},t+\delta_t)=\frac{1}{2q}f_{\alpha}^{*}(\boldsymbol{r}_{\mathrm{b}},t)+\frac{2q-1}{2q}f_{\bar{\alpha}}^{*}(\boldsymbol{r}_{\mathrm{b}},t) \tag{8.54}$$

式中，$\boldsymbol{e}_{\bar{\alpha}}=-\boldsymbol{e}_{\alpha}$。为了获得高阶精度，还可采用以下二次插值：

当 $q<1/2$ 时，

$$f_{\bar{\alpha}}(\boldsymbol{r}_{\mathrm{b}},t+\delta_t)=q(1+2q)f_{\alpha}^{*}(\boldsymbol{r}_{\mathrm{b}},t)+(1-4q^2)f_{\alpha}^{*}(\boldsymbol{r}_{\mathrm{b}}+\boldsymbol{e}_{\bar{\alpha}}\delta_t,t)-q(1-2q)f_{\alpha}^{*}(\boldsymbol{r}_{\mathrm{b}}+2\boldsymbol{e}_{\bar{\alpha}}\delta_t,t) \tag{8.55}$$

当 $q\geqslant 1/2$ 时，

$$f_{\bar{\alpha}}(\boldsymbol{r}_{\mathrm{b}},t+\delta_t)=\frac{1}{q(2q+1)}f_{\alpha}^{*}(\boldsymbol{r}_{\mathrm{b}},t)+\frac{2q-1}{q}f_{\bar{\alpha}}^{*}(\boldsymbol{r}_{\mathrm{b}},t)-\frac{2q-1}{2q+1}f_{\bar{\alpha}}^{*}(\boldsymbol{r}_{\mathrm{b}}+\boldsymbol{e}_{\bar{\alpha}}\delta_t,t) \tag{8.56}$$

此外，对运动边界，由于流体与边界的相互作用，因此需要在式(8.53)~式(8.56)等式右边添加适当的源项[27]。Bouzidi 格式需要按照参数 q 使用不同的计算格式，因此当 q 的变化跨越 1/2 时，计算格式的不同会引起分布函数的跳变。为此，Yu 等提出了一个统一的插值格式，感兴趣的读者可参阅文献[28]。

8.5.3 Guo 格式

2002 年，Guo 等[29]提出了一种结合非平衡态外推格式和空间插值的曲线边界处理方法。参见图 8.9，边界节点 $\boldsymbol{r}_{\mathrm{b}}$ 在 $t+\delta_t$ 时刻的未知分布函数 $f_{\bar{\alpha}}(\boldsymbol{r}_{\mathrm{b}},t+\delta_t)$ 可认为是从 t 时刻固体节点的碰撞后分布函数 $f_{\bar{\alpha}}^{*}(\boldsymbol{r}_{\mathrm{s}},t)$ 迁移而来，为了确定 $f_{\bar{\alpha}}^{*}(\boldsymbol{r}_{\mathrm{s}},t)$，Guo 等提出将固体节点的分布函数分成平衡态和非平衡态两部分，即

$$f_{\bar{\alpha}}(\boldsymbol{r}_{\mathrm{s}},t)=f_{\bar{\alpha}}^{\mathrm{eq}}(\boldsymbol{r}_{\mathrm{s}},t)+f_{\bar{\alpha}}^{\mathrm{neq}}(\boldsymbol{r}_{\mathrm{s}},t) \tag{8.57}$$

式中，平衡态部分 $f_{\bar{\alpha}}^{\mathrm{eq}}(\boldsymbol{r}_{\mathrm{s}},t)$ 由一个虚拟的平衡态分布函数代替

$$f_{\bar{\alpha}}^{\mathrm{eq}}(\boldsymbol{r}_{\mathrm{s}},t)=\omega_{\bar{\alpha}}\rho(\boldsymbol{r}_{\mathrm{b}},t)\left[1+\frac{\boldsymbol{e}_{\bar{\alpha}}\cdot\boldsymbol{u}_{\mathrm{s}}}{c_{\mathrm{s}}^2}+\frac{(\boldsymbol{e}_{\bar{\alpha}}\cdot\boldsymbol{u}_{\mathrm{s}})^2}{2c_{\mathrm{s}}^4}-\frac{(\boldsymbol{u}_{\mathrm{s}}\cdot\boldsymbol{u}_{\mathrm{s}})}{2c_{\mathrm{s}}^2}\right] \tag{8.58}$$

式中，$\boldsymbol{u}_{\mathrm{s}}$ 为固体节点虚拟速度，其表达式为

$$\boldsymbol{u}_{\mathrm{s}}=\begin{cases} q\boldsymbol{u}^{(1)}+(1-q)\boldsymbol{u}^{(2)}, & q<0.75 \\ \boldsymbol{u}^{(1)}, & q\geqslant 0.75 \end{cases} \tag{8.59}$$

式中，$\boldsymbol{u}^{(1)}$ 和 $\boldsymbol{u}^{(2)}$ 分别为

$$\boldsymbol{u}^{(1)}=\frac{(q-1)\boldsymbol{u}(\boldsymbol{r}_{\mathrm{b}},t)+\boldsymbol{u}(\boldsymbol{r}_{\mathrm{w}},t)}{q},\quad \boldsymbol{u}^{(2)}=\frac{(q-1)\boldsymbol{u}(\boldsymbol{r}_{\mathrm{f}},t)+2\boldsymbol{u}(\boldsymbol{r}_{\mathrm{w}},t)}{1+q} \tag{8.60}$$

相应地，式(8.57)中的非平衡态部分也类似给出

$$f_{\bar{\alpha}}^{\mathrm{neq}}(\boldsymbol{r}_{\mathrm{s}},t)=\begin{cases} qf_{\bar{\alpha}}^{\mathrm{neq}}(\boldsymbol{r}_{\mathrm{b}},t)+(1-q)f_{\bar{\alpha}}^{\mathrm{neq}}(\boldsymbol{r}_{\mathrm{f}},t), & q<0.75 \\ f_{\bar{\alpha}}^{\mathrm{neq}}(\boldsymbol{r}_{\mathrm{b}},t), & q\geqslant 0.75 \end{cases} \tag{8.61}$$

式中，$f_{\bar{\alpha}}^{\mathrm{neq}}(\boldsymbol{r}_{\mathrm{b}},t)=f_{\bar{\alpha}}(\boldsymbol{r}_{\mathrm{b}},t)-f_{\bar{\alpha}}^{\mathrm{eq}}(\boldsymbol{r}_{\mathrm{b}},t)$。在确定 $f_{\bar{\alpha}}(\boldsymbol{r}_{\mathrm{s}},t)$ 之后，固体节点的碰撞后分布函数可根据式(8.1)获得，即

$$f_{\bar{\alpha}}^{*}(\boldsymbol{r}_{\mathrm{s}},t)=f_{\bar{\alpha}}^{\mathrm{eq}}(\boldsymbol{r}_{\mathrm{s}},t)+\frac{\tau-1}{\tau}f_{\bar{\alpha}}^{\mathrm{neq}}(\boldsymbol{r}_{\mathrm{s}},t) \tag{8.62}$$

算例表明，Guo 格式具有二阶数值精度，且具有良好的数值稳定性。

8.5.4 质量损失及修正格式

由于存在插值步骤，上述格式破坏了质量守恒。如图 8.11 所示，图中实心圆点为流体边界节点，空心圆点为固体节点（虚拟节点），在迁移过程中，边界节点 t 时刻的三个碰撞后分布函数 $f_4^*(t)$、$f_7^*(t)$ 和 $f_8^*(t)$ 将迁移至固体节点，因此对该边界节点而言，流出流体区域的质量为 $m_{out}=f_4^*(t)+f_7^*(t)+f_8^*(t)$；相应地，边界节点 $t+\delta_t$ 时刻的 3 个分布函数 $f_2(t+\delta_t)$、$f_5(t+\delta_t)$ 和 $f_6(t+\delta_t)$ 可以看作是从固体节点迁移而来，因此流入该边界节点的质量为 $m_{in}=f_2(t+\delta_t)+f_5(t+\delta_t)+f_6(t+\delta_t)$，于是该边界节点处的质量损失为

$$\Delta m(t+\delta_t)=f_4^*(t)+f_7^*(t)+f_8^*(t)-[f_2(t+\delta_t)+f_5(t+\delta_t)+f_6(t+\delta_t)] \tag{8.63}$$

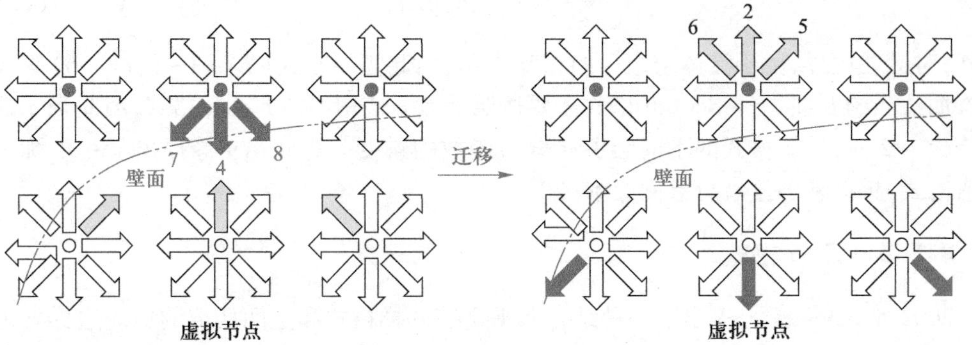

图 8.11 边界节点上的质量流出、流入示意图

显然，采用半步长反弹格式时，由式(8.13)～式(8.15)可知，$\Delta m=0$，即半步长反弹格式严格满足质量守恒。式(8.63)写成通用形式为[30]

$$\Delta m(\boldsymbol{r}_b,t+\delta_t)=\sum_{\text{outgoing}} f_\alpha^*(\boldsymbol{r}_b,t)-\sum_{\text{incoming}} f_{\bar{\alpha}}(\boldsymbol{r}_b,t+\delta_t) \tag{8.64}$$

式中：下标"α"代表流出流体区域的离散速度方向；"outgoing"代表对所有流出流体区域的分布函数求和；"$\bar{\alpha}$"代表进入流体区域的离散速度方向，且有 $\boldsymbol{e}_\alpha=-\boldsymbol{e}_{\bar{\alpha}}$；"incoming"代表对所有进入流体区域的分布函数求和。根据式(8.64)可对前面提及的复杂边界处理格式进行质量修正，一种较简单的处理方法是将系统通过边界节点损失的质量加回到边界节点上，例如 Yu 等[30]提出对边界节点 $t+\delta_t$ 时刻的静止分布函数做如下修正：

$$f_0(\boldsymbol{r}_b,t+\delta_t)=f_0^*(\boldsymbol{r}_b,t)+\sum_{\text{outgoing}} f_\alpha^*(\boldsymbol{r}_b,t)-\sum_{\text{incoming}} f_{\bar{\alpha}}(\boldsymbol{r}_b,t+\delta_t) \tag{8.65}$$

数值模拟研究表明，修正后的复杂边界处理格式能够严格保证系统的质量守恒[30]。

8.6 温度边界处理格式

在研究传热问题时，通常有两类温度边界条件，即定壁温（Dirichlet 边界条件）和定热流（Neumann 边界条件）。针对这两类温度边界条件，文献中已有多种相应的边界处理格式。实际上，许多温度边界处理格式就是从前面所介绍的速度边界处理格式推广或衍生而来的，例如，在 8.4.1 节我们就曾提到非平衡态外推格式可直接推广至温度边界条件。下面，对文献中几类典型的温度边界处理格式做一个简单介绍。

Dirichlet 边界条件和 Neumann 边界条件

2005 年，Ginzburg[31]提出了一种处理温度场 Dirichlet 边界条件的镜像半步长反弹格式(anti-bounce-back，ABB)格式。如图 8.12 所示，图中 $\boldsymbol{r}_{\mathrm{b}}$ 为流体边界节点，$\boldsymbol{r}_{\mathrm{s}}$ 为固体节点，固体壁面位于边界节点和固体节点连线的中间位置(与半步长反弹格式一致)。对 Dirichlet 边界条件，它给定了固体壁面的温度，即 $T(\boldsymbol{r}_{\mathrm{w}})=T_{\mathrm{w}}$，相应的 ABB 格式为[31]

$$g_{\bar{\alpha}}(\boldsymbol{r}_{\mathrm{b}},t+\delta_t)=-g_{\alpha}^{*}(\boldsymbol{r}_{\mathrm{b}},t)+2g_{\alpha}^{\mathrm{eq}}(\boldsymbol{r}_{\mathrm{w}},t+\delta_t) \tag{8.66}$$

式中：$g_{\alpha}^{*}(\boldsymbol{r}_{\mathrm{b}},t)$ 为边界节点 $\boldsymbol{e}_{\alpha}$ 方向的碰撞后温度分布函数；g_{α}^{eq} 为平衡态温度分布函数，对无滑移固体壁面有 $g_{\alpha}^{\mathrm{eq}}(\boldsymbol{r}_{\mathrm{w}},t+\delta_t)=\omega_{\alpha}T_{\mathrm{w}}$。该格式因为概念清晰，实施简便，在文献中得到了广泛的应用。

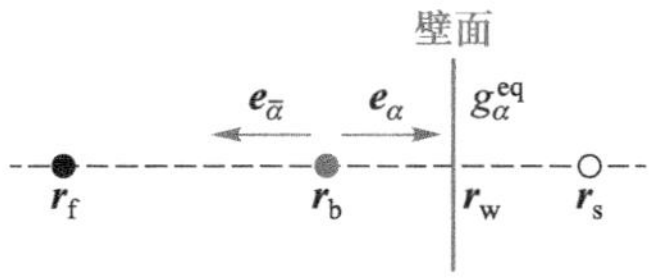

图 8.12 定壁温边界 ABB 格式示意图

针对采用湿节点布置方式的网格系统，Inamuro 等[32]将反滑移格式的基本思想拓展到了温度场的 Dirichlet 边界条件，即假定边界节点上的未知温度分布函数可以用一个新的平衡态温度分布函数来计算。以 D2Q9 模型为例，参见图 8.6，图中边界节点 O 位于固体壁面上，在 $t+\delta_t$ 时刻边界节点的未知温度分布函数 g_2、g_5 和 g_6 可表示为

$$g_{2,5,6}=g_{2,5,6}^{\mathrm{eq}}(T',\boldsymbol{u}_{\mathrm{w}}) \tag{8.67}$$

式中：T' 为待定参数，没有实际物理意义；$\boldsymbol{u}_{\mathrm{w}}$ 为固体壁面的速度，对无滑移速度边界有 $\boldsymbol{u}_{\mathrm{w}}=\boldsymbol{0}$。结合边界处宏观温度的计算式 $T_{\mathrm{w}}=\sum_{\alpha=0}^{8}g_{\alpha}$ 可得

$$T_{\mathrm{w}}=g_0+g_1+g_3+g_4+g_7+g_8+g_2^{\mathrm{eq}}(T',\boldsymbol{u}_{\mathrm{w}})+g_5^{\mathrm{eq}}(T',\boldsymbol{u}_{\mathrm{w}})+g_6^{\mathrm{eq}}(T',\boldsymbol{u}_{\mathrm{w}}) \tag{8.68}$$

根据式(8.68)即可确定待定参数 T'，继而可通过式(8.67)计算边界节点处新时刻的未知温度分布函数 g_2、g_5 和 g_6。

2013 年，Li 等[33]结合反弹思想和空间插值构造了适用于曲线边界的温度边界格式。与 Bouzidi 等的线性插值格式类似(参见图 8.10)，对温度场的 Dirichlet 边界条件可采用如下格式：

当 $q<1/2$ 时，

$$g_{\bar{\alpha}}(\boldsymbol{r}_{\mathrm{b}},t+\delta_t)=-2qg_{\alpha}^{*}(\boldsymbol{r}_{\mathrm{b}},t)+(1-2q)g_{\alpha}^{*}(\boldsymbol{r}_{\mathrm{b}}+\boldsymbol{e}_{\bar{\alpha}}\delta_t,t)+\Delta g_{\mathrm{b}} \tag{8.69}$$

当 $q\geqslant 1/2$ 时，

$$g_{\bar{\alpha}}(\boldsymbol{r}_{\mathrm{b}},t+\delta_t)=-\frac{1}{2q}g_{\alpha}^{*}(\boldsymbol{r}_{\mathrm{b}},t)+\frac{2q-1}{2q}g_{\bar{\alpha}}^{*}(\boldsymbol{r}_{\mathrm{b}},t)+\frac{1}{2q}\Delta g_{\mathrm{b}} \tag{8.70}$$

式中，$\Delta g_{\mathrm{b}}=2\omega_{\alpha}T_{\mathrm{w}}$。显然，当 $q=0.5$ 时，该格式退化为 ABB 格式，也即式(8.66)。为了获得高阶精度，对温度边界处理亦可采用二次插值[34]。此外，上述提及的几类格式均可推广至 Neumann 边界条件(定热流)，具体实施过程可分别参阅文献[35,36,33]。

8.7 小 结

在使用格子 Boltzmann 方法进行数值模拟计算时，必须设法根据已知的宏观边界条件确定边界节点的未知分布函数，因而需要设计适用于各种边界条件下的边界处理格式。不同的边界处理格式对数值模拟的精度、稳定性以及计算效率会产生较大的影响。

边界处理格式大体上可分为启发式格式、动力学格式、外推格式、复杂边界处理格式等四类。其中,启发式格式主要根据边界条件的物理本质,通过微观粒子的运动规则来确定边界节点的未知分布函数;动力学格式则利用边界上的宏观物理量或宏观物理量梯度值的定义,直接求解未知分布函数的方程组;外推格式是将格子 Boltzmann 方法看作连续 Boltzmann 方程的一种特殊差分形式,并借鉴传统计算流体力学中的边界处理方法来构造相应的边界处理格式;复杂边界处理格式主要用于处理曲线边界和运动边界。由于各种边界处理格式具有不同的适用范围,其精度和数值稳定性也不尽相同,因此需要根据具体物理问题选择合适的边界处理格式。

除了本章所介绍的边界处理格式外,文献中还有一些边界处理格式,例如单点格式(single-node scheme)[37-38]、体积法(volumetric method)[39-40]、浸入边界法(immersed boundary method)[41-47]等,感兴趣的读者可以参阅相关文献。

浸入边界法简介

习 题

8-1 结合 8.3.1 节的内容,推导密度 ρ(或压力)和宏观速度水平分量 u_x 已知时的非平衡态反弹格式。

8-2 在第 8.4.2 节的非平衡态外推格式中,式(8.41)和式(8.42)有何区别?对照附录 D 中顶盖驱动流的格子 Boltzmann 程序,分析其采用的是哪一种处理。

8-3 结合 8.5.2 节中一次插值方法的介绍,给出二次插值边界条件式(8.55)和式(8.56)的推导过程。

8-4 编写一个计算圆柱绕流的二维格子 Boltzmann 程序,其中左侧入口为速度边界、右侧为密度边界,可采用非平衡态反弹格式(8.3.1 节)或非平衡态外推格式(8.4.2 节),上下边界采用周期性边界(8.2.1 节)或对称边界(8.2.2 节),圆柱采用阶梯近似及半步长反弹格式(8.2.4 节),可参考作为多孔介质固体区域处理(7.2 节)。

8-5 结合所编写的圆柱绕流程序,思考并回答下列问题:1)在入口边界的非平衡外推格式中速度已知,而密度采用了相邻节点的密度,是否可以在入口边界同时给定速度和密度,会带来怎样的问题?2)周期性边界或对称边界中的虚拟节点是否必需,如何在程序的迁移步骤中通过判断语句直接实现边界处未知分布函数的迁移?3)出口边界的切向速度如何确定?

参 考 文 献

[1] 郭照立,郑楚光,李青,等. 流体动力学的格子 Boltzmann 方法[M]. 武汉:湖北科学技术出版社,2002.

[2] Succi S. Lattice Boltzmann Equation for Fluid Dynamics and Beyond[M]. Oxford:Clarendon press,2001.

[3] Ziegler D P. Boundary conditions for lattice Boltzmann simulations[J]. Journal of Statistical Physics,1993,71(5/6):1171-1177.

[4] Ginzburg I,d'Humières D. Multireflection boundary conditions for lattice Boltzmann models[J]. Physical Review E,2003,68(6):066614.

[5] Guo Z L,Zheng C G. Analysis of lattice Boltzmann equation for microscale gas flows:Relaxation times,boundary conditions and the Knudsen layer[J]. International Journal of Computational Fluid Dynamics,2008,22(7):465-473.

[6] Fei L L, Luo K H, Li Q. Three-dimensional cascaded lattice Boltzmann method: Improved implementation and consistent forcing scheme[J]. Physical Review E, 2018, 97(5): 053309.
[7] Cornubert R. A Knudsen layer theory lattice gases[J]. Physica D, 1991, 47: 241-259.
[8] Tang G H, Tao W Q, He Y L. Lattice Boltzmann method for simulating gas flow in microchannels[J]. International Journal of Modern Physics C, 2004, 15(2): 335-347.
[9] Knudsen M. The Kinetic Theory of Gases[M]. London: Methuen Monographs, 1934.
[10] Zou Q S, He X Y. On pressure and velocity boundary conditions for the lattice Boltzmann BGK model [J]. Physics of Fluids, 1997, 9(6): 1591-1598.
[11] Li Q, Luo K H, Kang Q J, et al. Contact angles in the pseudopotential lattice Boltzmann modeling of wetting[J]. Physical Review E, 2014, 90(5): 053301.
[12] Inamuro T, Yoshino M, Ogino F. A non-slip boundary condition for lattice Boltzmann simulations[J]. Physics of Fluids, 1995, 7(12): 2928-2930.
[13] He X Y, Chen S Y, Doolen G D. A novel thermal model for the lattice Boltzmann method in incompressible limit[J]. Journal of Computational Physics, 1998, 146(1): 282-300.
[14] D'Orazio A, Succi S, Arrighetti C. Lattice Boltzmann simulation of open flows with heat transfer[J]. Physics of Fluids, 2003, 15(9): 2778-2781.
[15] D'Orazio A, Succi S. Simulating two-dimensional thermal channel flows by means of a lattice Boltzmann method with new boundary conditions[J]. Future Generation Computer Systems, 2004, 20: 935-944.
[16] Tong C Q, He Y L, Tang G H, et al. Mass modified outlet boundary for a fully developed flow in the lattice Boltzmann equation[J]. International Journal of Modern Physics C, 2007, 18(7): 1209-1221.
[17] Chen S Y, Martínez D, Mei R W. On boundary conditions in lattice Boltzmann methods[J]. Physics of Fluids, 1996, 8(9): 2527-2536.
[18] Yu D Z, Mei R W, Shyy W. Improved treatment of the open boundary in the method of Lattice Boltzmann equation: general description of the method[J]. Progress in Computational Fluid Dynamics, 2005, 5(1): 3-12.
[19] Guo Z L, Zheng C G, Shi B C. Non-equilibrium extrapolation method for velocity and boundary conditions in the lattice Boltzmann method[J]. Chinese Physics, 2002, 11(4): 0366-0374.
[20] Tang G H, Tao W Q, He Y L. Thermal boundary condition for the thermal lattice Boltzmann equation [J]. Physics Review E, 2005, 72(1): 016703.
[21] Wang Y, He Y L, Li Q, et al. Numerical simulations of gas resonant oscillations in a closed tube using lattice Boltzmann method[J]. International Journal of Heat and Mass Transfer, 2007, 51: 3082-3090.
[22] Guo Z L, Zheng C G, Shi B C, et al. Thermal lattice Boltzmann equation for low Mach number flows: Decoupling model[J]. Physical Review E, 2007, 75(3): 036704.
[23] Filippova O, Hänel D. Grid refinement for lattice-BGK models[J]. Journal of Computational Physics, 1998, 147: 219-228.
[24] Mei R W, Luo L S, Shyy W. An accurate curved boundary treatment in the lattice Boltzmann method [J]. Journal of Computational Physics, 1999, 155(2): 307-330.
[25] 李华兵. 晶格玻尔兹曼方法对血液流的初步研究[D]. 上海:复旦大学, 2004.
[26] Mei R W, Shyy W, Yu D Z, et al. Lattice Boltzmann method for 3-D flows with curved boundary [J]. Journal of Computational Physics, 2000, 161: 680-699.
[27] Bouzidi M H, Firdaouss M, Lallemand P. Momentum transfer of a Boltzmann-lattice fluid with boundaries [J]. Physics of Fluids, 2001, 13(11): 3452-3459.
[28] Yu D Z, Mei R W, Shyy W. A unified boundary treatment in lattice Boltzmann method[C]. 41st Aerospace Sciences Meeting and Exhibit, Aerospace Sciences Meetings. Nevada. 2003: AIAA Paper No. 2003-2953.

[29] Guo Z L, Zheng C G, Shi B C. An extrapolation method for boundary conditions in lattice Boltzmann method[J]. Physics of Fluids, 2002, 14(6): 2007–2010.

[30] Yu Y, Li Q, Wen Z X. Modified curved boundary scheme for two-phase lattice Boltzmann simulations [J]. Computers & Fluids, 2020, 208: 104638.

[31] Ginzburg I. Generic boundary conditions for lattice Boltzmann models and their application to advection and anisotropic dispersion equations[J]. Advances in Water Resources, 2005, 28(11): 1196–1216.

[32] Inamuro T, Yoshino M, Inoue H, et al. A lattice Boltzmann method for a binary miscible fluid mixture and its application to a heat-transfer problem[J]. Journal of Computational Physics, 2002, 179(1): 201–215.

[33] Li L K, Mei R W, Klausner J F. Boundary conditions for thermal lattice Boltzmann equation method [J]. Journal of Computational Physics, 2013, 237: 366–395.

[34] Rosemann T, Kravets B, Reinecke S R, et al. Comparison of numerical schemes for 3D lattice Boltzmann simulations of moving rigid particles in thermal fluid flows[J]. Powder Technology, 2019, 356: 528–546.

[35] Zhang T, Shi B C, Guo Z L, et al. General bounce-back scheme for concentration boundary condition in the lattice-Boltzmann method[J]. Physical Review E, 2012, 85(1): 016701.

[36] Yoshino M, Inamuro T. Lattice Boltzmann simulations for flow and heat/mass transfer problems in a three-dimensional porous structure[J]. International Journal for Numerical Methods in Fluids, 2003, 43(2): 183–198.

[37] Zhao W F, Yong W A. Single-node second-order boundary schemes for the lattice Boltzmann method [J]. Journal of Computational Physics, 2017, 329: 1–15.

[38] Tao S, He Q, Chen B M, et al. One-point second-order curved boundary condition for lattice Boltzmann simulation of suspended particles[J]. Computers & Mathematics with Applications, 2018, 76(7): 1593–1607.

[39] Rohde M, Derksen J J, Van den Akker H E A. Volumetric method for calculating the flow around moving objects in lattice-Boltzmann schemes[J]. Physical Review E, 2002, 65(5): 056701.

[40] Rohde M, Kandhai D, Derksen J J, et al. Improved bounce-back methods for no-slip walls in lattice-Boltzmann schemes: Theory and simulations[J]. Physical Review E, 2003, 67(6): 066703.

[41] Feng Z G, Michaelides E E. The immersed boundary-lattice Boltzmann method for solving fluid-particles interaction problems[J]. Journal of Computational Physics, 2004, 195(2): 602–628.

[42] Feng Z G, Michaelides E E. Proteus: a direct forcing method in the simulations of particulate flows [J]. Journal of Computational Physics, 2005, 202(1): 20–51.

[43] Wu J, Shu C. Implicit velocity correction-based immersed boundary-lattice Boltzmann method and its applications[J]. Journal of Computational Physics, 2009, 228(6): 1963–1979.

[44] Wu J, Wu J, Zhan J P, et al. A robust immersed boundary-lattice Boltzmann method for simulation of fluid-structure interaction problems[J]. Communications in Computational Physics, 2016, 20(1): 156–178.

[45] Zhang C Z, Cheng Y G, Zhu L D, et al. Accuracy improvement of the immersed boundary-lattice Boltzmann coupling scheme by iterative force correction[J]. Computers & Fluids, 2016, 124: 246–260.

[46] Mittal R, Iaccarino G. Immersed boundary methods[J]. Annual Review of Fluid Mechanics, 2005, 37(1): 239–261.

[47] Griffith B E, Patankar N A. Immersed methods for fluid-structure interaction[J]. Annual Review of Fluid Mechanics, 2020, 52(1): 421–448.

第 9 章 格子Boltzmann方法的网格技术

格子 Boltzmann 方法在带来种种优点的同时,也带来了它的一些固有缺点,例如网格划分必须对称均匀(正方形、正六边形等)、计算精度不可调等,从而使得格子 Boltzmann 方法在某些方面的应用受到了限制。为了拓展格子 Boltzmann 方法的应用范围,国内外学者围绕格子 Boltzmann 方法的网格技术开展了大量的研究。

如第 3 章所述,流体粒子的密度分布函数 f 依赖于时间、空间和粒子速度,即 $f(\boldsymbol{r},\boldsymbol{\xi},t)$。在不考虑外力的情况下,密度分布函数满足如下形式的连续 Boltzmann 方程:

$$\frac{\partial f}{\partial t}+\boldsymbol{\xi}\cdot\nabla f=\Omega_f \tag{9.1}$$

式中:$\boldsymbol{\xi}$ 为粒子速度;Ω_f 为碰撞算子。在格子 Boltzmann 方法中,需要对式(9.1)在粒子速度空间进行离散,在此基础上采用 BGK 碰撞算子,可得相应的离散 Boltzmann-BGK 方程

$$\frac{\partial f_\alpha}{\partial t}+\boldsymbol{e}_\alpha\cdot\nabla f_\alpha=-\frac{1}{\tau_0}(f_\alpha-f_\alpha^{\mathrm{eq}}) \tag{9.2}$$

式中:$\boldsymbol{e}_\alpha$ 为离散速度;τ_0 为松弛时间。对式(9.2)进行相应的时间和空间离散,可得不含外力项的格子 Boltzmann-BGK 方程

$$f_\alpha(\boldsymbol{r}+\boldsymbol{e}_\alpha\delta_t,t+\delta_t)-f_\alpha(\boldsymbol{r},t)=-\frac{1}{\tau}[f_\alpha(\boldsymbol{r},t)-f_\alpha^{\mathrm{eq}}(\boldsymbol{r},t)] \tag{9.3}$$

式中:$\boldsymbol{r}$ 为空间位置;τ 为量纲为 1 的松弛时间;δ_t 为时间步长。

本章所要介绍的各种格子 Boltzmann 方法网格技术,就是对式(9.2)或式(9.3)的不同处理形式。为了方便对比,本章将先归纳标准格子 Boltzmann 方法的局限性;随后,介绍插值格子 Boltzmann 方法,以及泰勒展开和最小二乘格子 Boltzmann 方法;作为格子 Boltzmann 方法与传统数值模拟方法的结合,有限差分格子 Boltzmann 方法、有限容积格子 Boltzmann 方法和有限元格子 Boltzmann 方法受到了广泛的关注,本章也将一一进行介绍;最后,将介绍多块网格和多重网格技术在格子 Boltzmann 方法中的实施。

在这些网格技术中,插值格子 Boltzmann 方法以及泰勒展开和最小二乘格子 Boltzmann 方法与标准格子 Boltzmann 方法一样都是基于式(9.3),而有限差分格子 Boltzmann 方法、有限容积格子 Boltzmann 方法和有限元格子 Boltzmann 方法则是基于式(9.2);多块网格以及多重网格技术可应用于上述两个方程。在未做特别说明的情况下,本章将以密度分布函数为例对相应的网格技术进行介绍。对于双分布函数模型,温度场的演化方程与流场的演化方程相似,因而这些网格技术同样适用于双分布函数模型。

9.1 标准格子 Boltzmann 方法

对于式(9.3),如果要求格子离散速度等于格子间距与时间步长的比值,则该方程为**标准格子 Boltzmann 方法**的控制方程。以二维问题为例,即要求 $e_{\alpha x}=\delta_x/\delta_t$ 和 $e_{\alpha y}=\delta_y/\delta_t$。其中,$\delta_x$ 和 δ_y 分别是 x 和 y 方向上的网格步长,均为常数。在标准格子 Boltzmann 方法中,通常取 $e_{\alpha x}=e_{\alpha y}=1$,因而有 $\delta_x=\delta_y=e_{\alpha x}\delta_t=e_{\alpha y}\delta_t=1$,此时物理量所采用的单位是**格子单位**(lattice unit)。格子单位与**物理单位**(physical unit)之间的换算,可参考附录 C。

前面曾多次提到,流体粒子的演化可分为碰撞和迁移两个过程。不考虑外力项时,格子 Boltzmann-BGK 方程的碰撞步和迁移步参见式(8.1)和式(8.2)。根据式(8.1),位于同一网格节点 $\boldsymbol{r}$ 的流体粒子相互碰撞,对不同离散速度方向上的分布函数进行重新分配;随后,不同离散速度方向上的粒子按照式(8.2)进行迁移,一个时间步长后在 $t+\delta_t$ 时刻运动至 $\boldsymbol{r}+\boldsymbol{e}_\alpha\delta_t$。由于网格设置满足 $e_{\alpha x}=\delta_x/\delta_t$ 和 $e_{\alpha y}=\delta_y/\delta_t$,因此可以严格保证 $\boldsymbol{r}+\boldsymbol{e}_\alpha\delta_t$ 在网格节点上,从而在每个时步,粒子从一个节点迁移至其相邻节点。需要再次强调的是,虽然式(9.3)是对式(9.2)在空间上进行一阶离散得到的,但由于离散误差可以包含于黏性项中,因此式(9.3)具有二阶精度[1]。

标准格子 Boltzmann 方法最初是由格子气自动机演变而来。由于格子气自动机只做布尔运算,因此它的网格划分必须与离散速度一致,以保证下一个时刻粒子运动到其相邻节点。由于在时间上采取显式递进,空间迁移仅涉及相邻节点,且只需对代数方程进行求解,因而标准格子 Boltzmann 方法具有原理简单、实施方便、并行性能好等优点。然而,随着格子 Boltzmann 方法的发展,标准的碰撞迁移过程也逐渐凸显出不足。一方面,由于网格划分必须对称均匀(正六边形、正方形等),使得格子 Boltzmann 方法在某些方面的应用受到了限制。例如,在靠近流场边界的区域,物理量一般变化比较剧烈,对这些区域需要使用较密的网格,而在远离边界的区域,物理量变化相对平缓,可以使用较稀疏的网格。如果对整个计算区域采用均匀的网格离散,要么会降低算法的效率,造成资源浪费,要么不能细致地描述流体的流动[2]。另一方面,对一些多速度模型[3],很难保证在所有的离散速度方向上都满足 $e_{\alpha x}=\delta_x/\delta_t$ 和 $e_{\alpha y}=\delta_y/\delta_t$,使得碰撞迁移过程不能满足计算的需要。再者,由于标准格子 Boltzmann 方法的空间和时间精度已定,因而不能根据实际物理问题的需要来选择计算精度。例如,对于强间断、强激波等声学问题的数值模拟,往往需要采用高阶的时间和空间计算格式[4],标准格子 Boltzmann 方法难以胜任。为了克服标准格子 Boltzmann 方法的不足,国内外学者开展了大量的相关研究,并提出了多种网格技术。

9.2 插值格子 Boltzmann 方法

由于格子 Boltzmann 方法中的密度分布函数为实数,并且在流场的整个计算域上是连续的[1],因而 $e_{\alpha x}=\delta_x/\delta_t$ 和 $e_{\alpha y}=\delta_y/\delta_t$ 的限制完全可以通过插值来消除。He 等[5-7]最早将插值的概念引入格子 Boltzmann 方法中,提出了**插值格子 Boltzmann 方法**(interpolation-supplemented lattice Boltzmann method)。在插值格子 Boltzmann 方法中,碰撞迁移过程仍然保留,但由于网格不再是均匀分布的,迁移之后粒子所处的位置并不一定是固定的网格节点,因而需要增加一个插值过程来确定网格节点上的分布函数。以下就直角坐标系和曲线坐标系两种情况分别进行介绍。

9.2.1 直角坐标系中的插值

考虑一直角坐标系下的非均匀的网格，如图 9.1 所示，网格节点用实心圆点表示，$\boldsymbol{r}=(x_i,y_j)$ 为节点坐标，$\mathrm{d}x_i$ 和 $\mathrm{d}y_j$ 表示网格宽度，且有

$$\mathrm{d}x_i=x_{i+1}-x_i,\quad \mathrm{d}y_i=y_{i+1}-y_i \tag{9.4}$$

定义如下比值：

$$r_x^i=\frac{\mathrm{d}x_i}{\delta_x},\quad r_y^i=\frac{\mathrm{d}y_i}{\delta_y} \tag{9.5}$$

式中：$\delta_x=e_{\alpha x}\delta_t$；$\delta_y=e_{\alpha y}\delta_t$；对 D2Q9 模型有 $\delta_x=\delta_y$。显然，在模拟中，可以根据所求解的实际物理问题，在不同区域选择不同的 r_x^i 和 r_y^i，即不同区域的网格疏密程度不同。

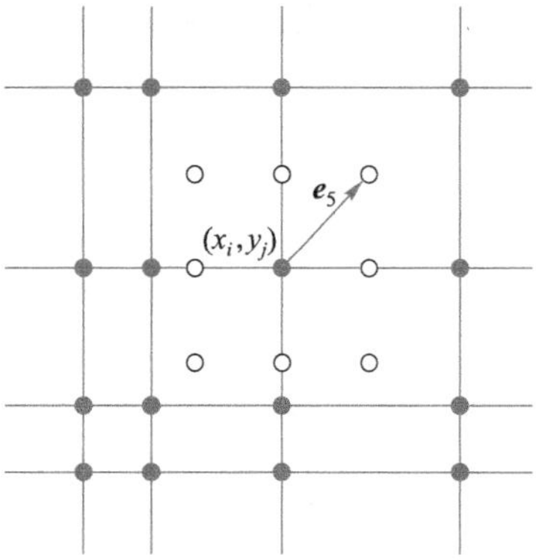

图 9.1 直角坐标系下的非均匀网格示意图

与标准格子 Boltzmann 方法相比，非均匀网格的计算需要在迁移过程之后、碰撞过程之前再增加一个插值过程，具体步骤如下[5]：

（1）根据给定的初场计算各网格节点上的初始平衡态分布函数；

（2）根据式(8.1)进行碰撞，计算出碰撞后的分布函数；

（3）接着进行迁移，网格节点 $\boldsymbol{r}$ 处的粒子运动到新的位置 $\boldsymbol{r}+\boldsymbol{e}_\alpha\delta_t$，如图 9.1 中空心圆点所示，并计算出 $f_\alpha(\boldsymbol{r}+\boldsymbol{e}_\alpha\delta_t,t+\delta_t)$ 的值；

（4）节点 $\boldsymbol{r}$ 处的分布函数 $f_\alpha(\boldsymbol{r},t+\delta_t)$ 通过点 $\boldsymbol{r}+\boldsymbol{e}_\alpha\delta_t$ 的分布函数 $f_\alpha(\boldsymbol{r}+\boldsymbol{e}_\alpha\delta_t,t+\delta_t)$ 插值重构得到；

（5）重复上述步骤(2)~步骤(4)，直至程序收敛。

其中，第(4)步的插值可以选择不同的插值格式。如文献[5]分别采用线性和二次插值模拟了二维突扩流，结果表明：在相同的网格下，二次插值的精度高于线性插值。由于插值过程不可避免地会引入数值误差，而格子 Boltzmann 方法本身具有二阶精度，因而推荐使用二阶及更高阶的插值格式，且为了计算稳定，应以迎风插值为主。在物理量变化较大的区域采用较密的网格，在物理量变化平缓的区域采用较稀疏的网格，在降低计算量的同时，可以提高计算精度。与此同时，采用这种方法还可对高雷诺数、高瑞利数下的宏观问题进行模拟[6,8-9]。

9.2.2 曲线坐标系中的插值

上述直角坐标系中的插值思想，同样可以应用于曲线坐标系。在标准格子 Boltzmann 方法的基础上引入适体网格的概念，粒子的碰撞迁移过程在计算平面（直角坐标系）完成，插值过程则依据物理平面（曲线坐标系）和计算平面的信息完成。下面以二维问题为例，简述相应的实施步骤[7,9-10]。

（1）在物理平面上选取适当的曲线坐标系 (ξ,η)，使得 (ξ,η) 上的网格为正方形。选取正方形网格是为了插值方便，并非必须。尽管 ξ 和 η 在前面的章节代表别的物理量，但为了遵循计算流体力学相关书籍的书写习惯，此处仍用它们来表示曲线坐标。取坐标变换 $x=x(\xi,\eta)$，$y=y(\xi,\eta)$ 得到网格节点的直角坐标 $\boldsymbol{r}(\xi,\eta)$，$\boldsymbol{r}=(x,y)$ 代表计算平面。

(2) 给定网格节点上的宏观量,由平衡态分布函数计算 $f_\alpha(\boldsymbol{r},t)$。

(3) 按照式(8.1)和式(8.2)在计算平面上进行演化,得到 $f_\alpha(\boldsymbol{r}+\boldsymbol{e}_\alpha\delta_t,t+\delta_t)$。

(4) 坐标变换 $x=x(\xi,\eta)$、$y=y(\xi,\eta)$ 对应逆变换 $\xi^x=\xi(\xi,\eta)$、$\eta^y=\eta(\xi,\eta)$。在 (ξ,η) 平面上寻找 $\boldsymbol{r}+\boldsymbol{e}_\alpha\delta_t$ 相对应的坐标 $(\xi_\alpha^x,\eta_\alpha^y)$,以及最接近的网格节点 (ξ^x,η^y)。取 $\mathrm{d}\xi=\xi^x-\xi_\alpha^x$,$\mathrm{d}\eta=\eta^y-\eta_\alpha^y$,按如下关系式进行插值:

$$f_\alpha(\boldsymbol{r},t+\delta_t)=\sum_{k=0}^{2}\sum_{l=0}^{2}a_{m,k}b_{n,l}f_\alpha(\xi_{m+k\cdot md}^x,\eta_{n+l\cdot nd}^y,t+\delta_t) \tag{9.6}$$

式中:$md=\mathrm{sgn}(1,\mathrm{d}\xi)$;$nd=\mathrm{sgn}(1,\mathrm{d}\eta)$,由它们确定插值节点,其中 sgn 为符号函数;$a_{m,k}$ 和 $b_{n,l}$ 为插值系数。以二次抛物线插值为例,相应的插值系数为

$$a_{m,0}=\frac{(|\mathrm{d}\xi|-\Delta\xi)(|\mathrm{d}\xi|-2\Delta\xi)}{2\Delta\xi^2},\quad b_{n,0}=\frac{(|\mathrm{d}\eta|-\Delta\eta)(|\mathrm{d}\eta|-2\Delta\eta)}{2\Delta\eta^2}$$

$$a_{m,1}=-\frac{|\mathrm{d}\xi|(|\mathrm{d}\xi|-2\Delta\xi)}{\Delta\xi^2},\quad b_{n,1}=-\frac{|\mathrm{d}\eta|(|\mathrm{d}\eta|-2\Delta\eta)}{\Delta\eta^2}$$

$$a_{m,2}=\frac{|\mathrm{d}\xi|(|\mathrm{d}\xi|-\Delta\xi)}{2\Delta\xi^2},\quad b_{n,1}=\frac{|\mathrm{d}\eta|(|\mathrm{d}\eta|-\Delta\eta)}{2\Delta\eta^2} \tag{9.7}$$

(5) 第(3)(4)步反复循环,直至程序收敛,并输出物理平面上的结果。

He 与 Doolen 在文献[7]中采用这一方法模拟了二维圆柱绕流。模拟中采用的物理平面 (ξ,η) 和计算平面 (x,y) 如图 9.2 所示,生成的曲线网格如图 9.3 所示。模拟结果表明,该方法具有良好的数值稳定性。由于插值系数可以预先计算好,每个时层仅需要对分布函数做插值处理,插值系数不需要再确定。与标准格子 Boltzmann 方法相比,虽然额外的插值步骤增加了计算量,但由于网格数减少,总体而言,计算效率反而可以提高。此外,为了降低数值误差,在权衡计算量和计算精度的情况下,应尽量采用高阶插值格式。

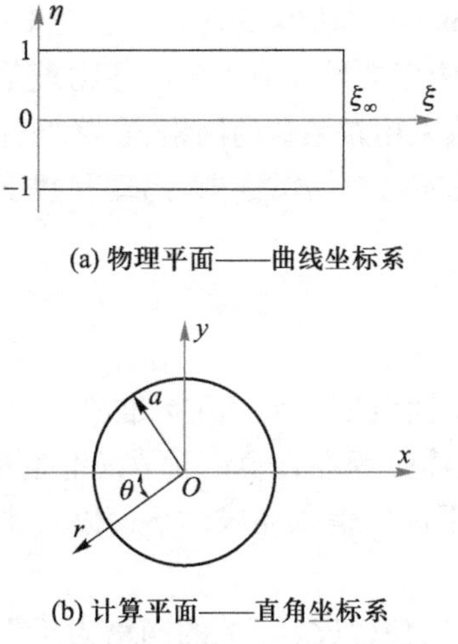

(a) 物理平面——曲线坐标系

(b) 计算平面——直角坐标系

图 9.2 曲线坐标系下的坐标变换图[7]

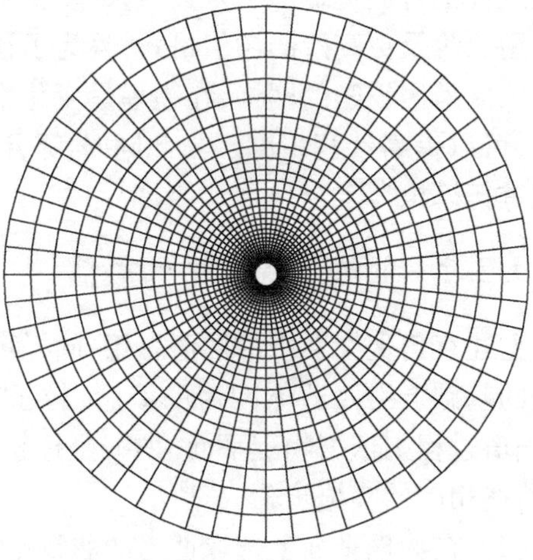

图 9.3 采用插值格子 Boltzmann 方法生成的圆柱绕流计算网格[7]

9.3 泰勒展开和最小二乘格子 Boltzmann 方法

如前所述,对于非均匀网格,粒子迁移之后所处的位置并不一定是网格节点。此时,除了采用插值格子 Boltzmann 方法对分布函数进行再分配以外,还可采用 Shu 等[11-17]提出的**泰勒展开和最小二乘格子 Boltzmann 方法**(Taylor series expansion and least square-based lattice Boltzmann method)。这种方法的基本思路是在标准格子 Boltzmann 方法的基础上引入空间上的泰勒级数展开和最小二乘法优化。下面以二维问题为例,对这一方法做简要介绍,更详尽的描述可参考文献[11-17]。

9.3.1 泰勒级数展开

考虑如图 9.4 所示的非均匀网格,实心圆点为网格节点,点 A 表示(x_A,y_A),点 P 表示(x_P,y_P);空心圆点为流体粒子沿 α 方向迁移后所处的位置,点 A'表示$(x_A+e_{\alpha x}\delta_t,y_A+e_{\alpha y}\delta_t)$。基于式(9.3),可得点 A'的密度分布函数为

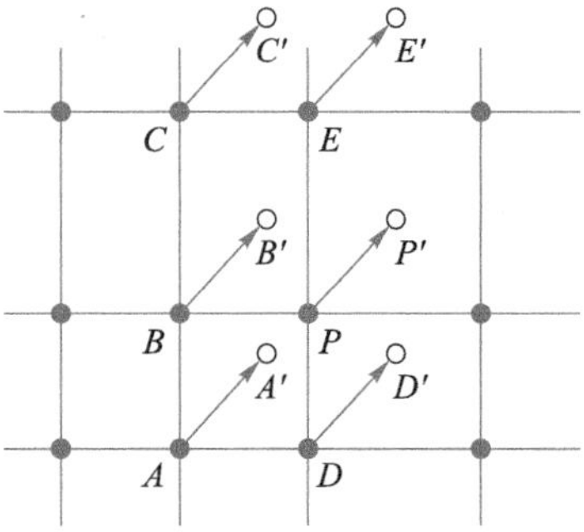

图 9.4 流体粒子沿 α 方向的迁移

$$f_\alpha(A',t+\delta_t)=f_\alpha(A,t)+\frac{f_\alpha^{\mathrm{eq}}(A,t)-f_\alpha(A,t)}{\tau} \tag{9.8}$$

由于通常情况下点 A'与点 P 并不重合,为了求得点 P 的分布函数,可对 $f_\alpha(A',t+\delta_t)$做泰勒级数展开,得

$$\begin{aligned}f_\alpha(A',t+\delta_t)=&f_\alpha(P,t+\delta_t)+\Delta x_A\frac{\partial f_\alpha(P,t+\delta_t)}{\partial x}+\Delta y_A\frac{\partial f_\alpha(P,t+\delta_t)}{\partial y}+\\&\frac{1}{2}(\Delta x_A)^2\frac{\partial^2 f_\alpha(P,t+\delta_t)}{\partial x^2}+\frac{1}{2}(\Delta y_A)^2\frac{\partial^2 f_\alpha(P,t+\delta_t)}{\partial y^2}+\\&\Delta x_A\Delta y_A\frac{\partial^2 f_\alpha(P,t+\delta_t)}{\partial x\partial y}+O[(\Delta y_A)^3,(\Delta y_A)^3]\end{aligned} \tag{9.9}$$

式中:$\Delta x_A=x_A+e_{\alpha x}\delta_t-x_P$;$\Delta y_A=y_A+e_{\alpha y}\delta_t-y_P$。显然,式(9.9)的截断误差为三阶。将式(9.9)代入式(9.8)可得

$$\begin{aligned}&f_\alpha(A,t)+\frac{f_\alpha^{\mathrm{eq}}(A,t)-f_\alpha(A,t)}{\tau}\\&=f_\alpha(P,t+\delta_t)+\Delta x_A\frac{\partial f_\alpha(P,t+\delta_t)}{\partial x}+\Delta y_A\frac{\partial f_\alpha(P,t+\delta_t)}{\partial y}+\\&\frac{1}{2}(\Delta x_A)^2\frac{\partial^2 f_\alpha(P,t+\delta_t)}{\partial x^2}+\frac{1}{2}(\Delta y_A)^2\frac{\partial^2 f_\alpha(P,t+\delta_t)}{\partial y^2}+\\&\Delta x_A\Delta y_A\frac{\partial^2 f_\alpha(P,t+\delta_t)}{\partial x\partial y}\end{aligned} \tag{9.10}$$

特别地，当 $\Delta x_A = \Delta y_A = 0$ 时，式(9.10)就退化为标准格子 Boltzmann 方程。在式(9.10)中，点 P 的密度分布函数及其一阶、二阶导数都是待定的，共计6个未知量。为了求解这6个未知量，还需再增加5个方程。考虑点 A 附近的点 P、B、C、D 和 E，这些点上的流体粒子在经过一个时间步长后，分别运动至点 P'、B'、C'、D' 和 E'。这些点的分布函数分别为

$$f_\alpha(P',t+\delta_t)=f_\alpha(P,t)+\frac{f_\alpha^{\mathrm{eq}}(P,t)-f_\alpha(P,t)}{\tau} \tag{9.11}$$

$$f_\alpha(B',t+\delta_t)=f_\alpha(B,t)+\frac{f_\alpha^{\mathrm{eq}}(B,t)-f_\alpha(B,t)}{\tau} \tag{9.12}$$

$$f_\alpha(C',t+\delta_t)=f_\alpha(C,t)+\frac{f_\alpha^{\mathrm{eq}}(C,t)-f_\alpha(C,t)}{\tau} \tag{9.13}$$

$$f_\alpha(D',t+\delta_t)=f_\alpha(D,t)+\frac{f_\alpha^{\mathrm{eq}}(D,t)-f_\alpha(D,t)}{\tau} \tag{9.14}$$

$$f_\alpha(E',t+\delta_t)=f_\alpha(E,t)+\frac{f_\alpha^{\mathrm{eq}}(E,t)-f_\alpha(E,t)}{\tau} \tag{9.15}$$

类似地，将 $f_\alpha(P',t+\delta_t)$、$f_\alpha(B',t+\delta_t)$、$f_\alpha(C',t+\delta_t)$、$f_\alpha(D',t+\delta_t)$ 和 $f_\alpha(E',t+\delta_t)$ 在点 P 做泰勒级数展开，并将结果分别代入式(9.11)~式(9.15)，可得

$$\begin{aligned}&f_\alpha(P,t)+\frac{f_\alpha^{\mathrm{eq}}(P,t)-f_\alpha(P,t)}{\tau}\\&=f_\alpha(P,t+\delta_t)+\Delta x_P\frac{\partial f_\alpha(P,t+\delta_t)}{\partial x}+\Delta y_P\frac{\partial f_\alpha(P,t+\delta_t)}{\partial y}+\\&\frac{1}{2}(\Delta x_P)^2\frac{\partial^2 f_\alpha(P,t+\delta_t)}{\partial x^2}+\frac{1}{2}(\Delta y_P)^2\frac{\partial^2 f_\alpha(P,t+\delta_t)}{\partial y^2}+\\&\Delta x_P\Delta y_P\frac{\partial^2 f_\alpha(P,t+\delta_t)}{\partial x\partial y}\end{aligned} \tag{9.16}$$

$$\begin{aligned}&f_\alpha(B,t)+\frac{f_\alpha^{\mathrm{eq}}(B,t)-f_\alpha(B,t)}{\tau}\\&=f_\alpha(P,t+\delta_t)+\Delta x_B\frac{\partial f_\alpha(P,t+\delta_t)}{\partial x}+\Delta y_B\frac{\partial f_\alpha(P,t+\delta_t)}{\partial y}+\\&\frac{1}{2}(\Delta x_B)^2\frac{\partial^2 f_\alpha(P,t+\delta_t)}{\partial x^2}+\frac{1}{2}(\Delta y_B)^2\frac{\partial^2 f_\alpha(P,t+\delta_t)}{\partial y^2}+\\&\Delta x_B\Delta y_B\frac{\partial^2 f_\alpha(P,t+\delta_t)}{\partial x\partial y}\end{aligned} \tag{9.17}$$

$$\begin{aligned}&f_\alpha(C,t)+\frac{f_\alpha^{\mathrm{eq}}(C,t)-f_\alpha(C,t)}{\tau}\\&=f_\alpha(P,t+\delta_t)+\Delta x_C\frac{\partial f_\alpha(P,t+\delta_t)}{\partial x}+\Delta y_C\frac{\partial f_\alpha(P,t+\delta_t)}{\partial y}+\\&\frac{1}{2}(\Delta x_C)^2\frac{\partial^2 f_\alpha(P,t+\delta_t)}{\partial x^2}+\frac{1}{2}(\Delta y_C)^2\frac{\partial^2 f_\alpha(P,t+\delta_t)}{\partial y^2}+\end{aligned}$$

$$\Delta x_C\Delta y_C\frac{\partial^2 f_\alpha(P,t+\delta_t)}{\partial x\partial y} \tag{9.18}$$

$$\begin{aligned}&f_\alpha(D,t)+\frac{f_\alpha^{\rm eq}(D,t)-f_\alpha(D,t)}{\tau}\\&=f_\alpha(P,t+\delta_t)+\Delta x_D\frac{\partial f_\alpha(P,t+\delta_t)}{\partial x}+\Delta y_D\frac{\partial f_\alpha(P,t+\delta_t)}{\partial y}+\\&\frac{1}{2}(\Delta x_D)^2\frac{\partial^2 f_\alpha(P,t+\delta_t)}{\partial x^2}+\frac{1}{2}(\Delta y_D)^2\frac{\partial^2 f_\alpha(P,t+\delta_t)}{\partial y^2}+\\&\Delta x_D\Delta y_D\frac{\partial^2 f_\alpha(P,t+\delta_t)}{\partial x\partial y}\end{aligned} \tag{9.19}$$

$$\begin{aligned}&f_\alpha(E,t)+\frac{f_\alpha^{\rm eq}(E,t)-f_\alpha(E,t)}{\tau}\\&=f_\alpha(P,t+\delta_t)+\Delta x_E\frac{\partial f_\alpha(P,t+\delta_t)}{\partial x}+\Delta y_E\frac{\partial f_\alpha(P,t+\delta_t)}{\partial y}+\\&\frac{1}{2}(\Delta x_E)^2\frac{\partial^2 f_\alpha(P,t+\delta_t)}{\partial x^2}+\frac{1}{2}(\Delta y_E)^2\frac{\partial^2 f_\alpha(P,t+\delta_t)}{\partial y^2}+\\&\Delta x_E\Delta y_E\frac{\partial^2 f_\alpha(P,t+\delta_t)}{\partial x\partial y}\end{aligned} \tag{9.20}$$

式中：$\Delta x_p=e_{\alpha x}\delta_t$，$\Delta y_p=e_{\alpha y}\delta_t$；$\Delta x_B=x_B+e_{\alpha x}\delta_t-x_P$，$\Delta y_B=y_B+e_{\alpha y}\delta_t-y_P$；$\Delta x_C=x_C+e_{\alpha x}\delta_t-x_P$，$\Delta y_C=y_C+e_{\alpha y}\delta_t-y_P$；$\Delta x_D=x_D+e_{\alpha x}\delta_t-x_P$，$\Delta y_D=y_D+e_{\alpha y}\delta_t-y_P$；$\Delta x_E=x_E+e_{\alpha x}\delta_t-x_P$，$\Delta y_E=y_E+e_{\alpha y}\delta_t-y_P$。

式(9.10)、式(9.16)~式(9.20)可组成一个封闭的方程组，从而可以求解 6 个未知量。该方程组还可表示为

$$g_{\alpha:i}=\{s_{\alpha:i}\}^{\rm T}\{V_\alpha\}=\sum_{j=1}^{6}s_{\alpha:i,j}V_{\alpha,j},\quad i=P,A,B,C,D,E \tag{9.21}$$

式中，

$$g_{\alpha:i}=f_\alpha(x_i,y_i,t)+\frac{f_\alpha^{\rm eq}(x_i,y_i,t)-f_\alpha(x_i,y_i,t)}{\tau} \tag{9.22}$$

$$\{s_{\alpha:i}\}^{\rm T}=\left\{1,\Delta x_i,\Delta y_i,\frac{(\Delta x_i)^2}{2},\frac{(\Delta y_i)^2}{2},\Delta x_i\Delta y_i\right\} \tag{9.23}$$

$$\{V_\alpha\}=\left\{f_\alpha,\frac{\partial f_\alpha}{\partial x},\frac{\partial f_\alpha}{\partial y},\frac{\partial^2 f_\alpha}{\partial x^2},\frac{\partial^2 f_\alpha}{\partial y^2},\frac{\partial^2 f_\alpha}{\partial x\partial y}\right\} \tag{9.24}$$

式中：$g_{\alpha:i}$为网格节点 i 在 t 时刻 α 方向上碰撞后的分布函数；$\{s_{\alpha:i}\}^{\rm T}$ 为由网格节点决定的向量；$\{V_\alpha\}$为网格节点 P 点在 α 方向上的未知量；$s_{\alpha:i,j}$和 $V_{\alpha:j}$分别为向量$\{s_{\alpha:i}\}^{\rm T}$ 和$\{V_\alpha\}$的第 j 个元素。显然，$V_{\alpha:1}=f_\alpha(P,t+\delta_t)$即为求解的最终目标。此外，式(9.21)还可写成如下矩阵形式：

$$[\boldsymbol{S}_\alpha]\{V_\alpha\}=\{g_\alpha\} \tag{9.25}$$

式中，

$$\{g_\alpha\}=\{g_{\alpha:P},g_{\alpha:A},g_{\alpha:B},g_{\alpha:C},g_{\alpha:D},g_{\alpha:E}\}^{\rm T} \tag{9.26}$$

$$[\boldsymbol{S}_\alpha]=[s_{\alpha:i}]^{\mathrm{T}}=\begin{bmatrix}\{s_{\alpha:P}\}^{\mathrm{T}}\\ \{s_{\alpha:A}\}^{\mathrm{T}}\\ \{s_{\alpha:B}\}^{\mathrm{T}}\\ \{s_{\alpha:C}\}^{\mathrm{T}}\\ \{s_{\alpha:D}\}^{\mathrm{T}}\\ \{s_{\alpha:E}\}^{\mathrm{T}}\end{bmatrix}=\begin{bmatrix}1 & \Delta x_P & \Delta y_P & \dfrac{(\Delta x_P)^2}{2} & \dfrac{(\Delta y_P)^2}{2} & \Delta x_P\Delta y_P\\ 1 & \Delta x_A & \Delta y_A & \dfrac{(\Delta x_A)^2}{2} & \dfrac{(\Delta y_A)^2}{2} & \Delta x_A\Delta y_A\\ 1 & \Delta x_B & \Delta y_B & \dfrac{(\Delta x_B)^2}{2} & \dfrac{(\Delta y_B)^2}{2} & \Delta x_B\Delta y_B\\ 1 & \Delta x_C & \Delta y_C & \dfrac{(\Delta x_C)^2}{2} & \dfrac{(\Delta y_C)^2}{2} & \Delta x_C\Delta y_C\\ 1 & \Delta x_D & \Delta y_D & \dfrac{(\Delta x_D)^2}{2} & \dfrac{(\Delta y_D)^2}{2} & \Delta x_D\Delta y_D\\ 1 & \Delta x_E & \Delta y_E & \dfrac{(\Delta x_E)^2}{2} & \dfrac{(\Delta y_E)^2}{2} & \Delta x_E\Delta y_E\end{bmatrix} \tag{9.27}$$

由于$[\boldsymbol{S}_\alpha]$是一个 6×6 的矩阵，对式(9.25)进行理论求解非常困难，一种可行的办法就是数值求解。此外，$[\boldsymbol{S}_\alpha]$仅与网格节点和离散速度有关，因而可以在初始时刻就计算好，并在以后的所有时层中使用。

9.3.2 最小二乘法优化

在实际应用中，$[\boldsymbol{S}_\alpha]$可能是奇异的或病态的，为了克服这一缺陷，使得该方法应用更广泛，Shu 等又提出采用最小二乘法来优化式(9.21)。前面提到，式(9.21)有 6 个未知量，如果将它应用于更多的网格节点，则方程组是超定的。在这种情况下，未知量可以采用最小二乘法来确定。以下用 $i=0$ 来表示网格节点 P，它的邻近网格节点用 $i=1,2,\cdots,N$ 表示，N 表示点 P 附近的网格点个数，它必须大于 5。在每个网格节点，定义一个误差项

$$err_{\alpha:i}=g_{\alpha:i}-\sum_{j=1}^{6}s_{\alpha:i,j}V_{\alpha:j},\quad i=0,1,2,\cdots,N \tag{9.28}$$

以及误差项的平方和

$$E=\sum_{i=0}^{N}err_{\alpha:i}=\sum_{i=0}^{N}\left(g_{\alpha:i}-\sum_{j=1}^{6}s_{\alpha:i,j}V_{\alpha:j}\right)^2 \tag{9.29}$$

为了使得 E 最小化，可以设$\partial E_\alpha/\partial V_{\alpha:j}=0(j=1,2,\cdots,6)$，并得到

$$[\boldsymbol{S}_\alpha]^{\mathrm{T}}[\boldsymbol{S}_\alpha]\{V_\alpha\}=[\boldsymbol{S}_\alpha]^{\mathrm{T}}\{g_\alpha\} \tag{9.30}$$

式中，$[\boldsymbol{S}_\alpha]$为一个$(N+1)\times 6$ 的矩阵，具体表达式为

$$[\boldsymbol{S}_\alpha]=\begin{bmatrix}1 & \Delta x_0 & \Delta y_0 & \dfrac{(\Delta x_0)^2}{2} & \dfrac{(\Delta y_0)^2}{2} & \Delta x_0\Delta y_0\\ 1 & \Delta x_1 & \Delta y_1 & \dfrac{(\Delta x_1)^2}{2} & \dfrac{(\Delta y_1)^2}{2} & \Delta x_1\Delta y_1\\ - & - & - & - & - & -\\ - & - & - & - & - & -\\ - & - & - & - & - & -\\ 1 & \Delta x_N & \Delta y_N & \dfrac{(\Delta x_N)^2}{2} & \dfrac{(\Delta y_N)^2}{2} & \Delta x_N\Delta y_N\end{bmatrix} \tag{9.31}$$

式中：$\Delta x_0=e_{\alpha x}\delta_t$；$\Delta y_0=e_{\alpha y}\delta_t$；$\Delta x_i=x_i+e_{\alpha x}\delta_t-x_0,i=1,2,\cdots,N$；$\Delta y_i=y_i+e_{\alpha y}\delta_t-y_0,i=1,2,\cdots,N$。相应地，$\{g_\alpha\}$也变为$\{g_\alpha\}=\{g_{\alpha:0},g_{\alpha:1},\cdots,g_{\alpha:N}\}^{\mathrm{T}}$。从式(9.30)可以导出

$$\{V_\alpha\}=([\boldsymbol{S}_\alpha]^{\mathrm{T}}[\boldsymbol{S}_\alpha])^{-1}[\boldsymbol{S}_\alpha]^{\mathrm{T}}\{g_\alpha\}=[\boldsymbol{A}_\alpha]\{g_\alpha\} \tag{9.32}$$

式中，$[\boldsymbol{A}_\alpha]$为一个 $6\times(N+1)$的矩阵。

最终，基于式(9.32)可得

$$f_\alpha(x_0,y_0,t+\delta_t)=V_{\alpha:1}=\sum_{j=1}^{N+1}a_{\alpha:1,j}g_{\alpha:j-1} \tag{9.33}$$

式中，$a_{\alpha:1,j}$为矩阵$[\boldsymbol{A}_\alpha]$的第一列元素。式(9.33)即为基于泰勒级数展开和最小二乘的格子 Boltzmann 方法的控制方程。

9.3.3　讨论

式(9.33)为一个代数方程，系数 $a_{\alpha:1,j}$仅由网格节点坐标和格子离散速度来确定，可以在初始时刻便计算出来，因而与标准格子 Boltzmann 方法相比，初始增加了一定的计算量。式(9.33)自右向左将分布函数从 t 时层更新到 $t+\delta_t$，因而仍然保留了标准格子 Boltzmann 方法的显式特征。此外，以上推导过程表明，该方法对网格结构没有任何要求，具有无网格特性，因而可以应用于均匀网格、非均匀网格等各种网格。且推导过程并未对方向 α 做限制，因而可以应用于任何离散速度模型。理论分析表明，该方法可以达到二阶精度[12]。

以上的推导过程仅以二维问题为例，三维问题的推导与此类似，感兴趣的读者可参考文献[15]。Shu 等[11-17]采用这一方法对多种流动与传热问题进行了数值模拟，模拟结果表明，在相同的网格数下，该方法计算精度高于标准格子 Boltzmann 方法，但由于需要存储及使用的系数较多(尤其是三维问题)，因而计算时间稍长。

9.4　有限差分格子 Boltzmann 方法

前面所介绍的插值格子 Boltzmann 方法以及泰勒展开和最小二乘格子 Boltzmann 方法都是基于式(9.3)。在这些方法中，流体粒子的演化通过碰撞迁移及必要的插值过程来完成。另一方面，式(9.3)本身可以看作是连续 Boltzmann-BGK 方程的一种特殊离散形式，因而也可以直接采用传统的有限差分法对式(9.2)进行数值求解。

有限差分法简介

本节将介绍**有限差分格子 Boltzmann 方法**(finite-difference lattice Boltzmann method)，其基本思想是采用有限差分法对式(9.2)进行求解。式(9.2)等号左边第一项为非稳态项，第二项为对流项，等号右边为碰撞项。在有限差分格子 Boltzmann 方法中，计算域被划分为离散的网格，用有限的网格节点来代替连续的计算域，这一点与标准格子 Boltzmann 方法相似。同时，非稳态项和对流项分别采用时间和空间离散格式来离散，即时间和空间导数分别用相应的差分表达式来近似代替，从而将式(9.2)转化为代数方程进行求解。在有限差分格子 Boltzmann 方法中，可以比较方便地使用非均匀网格，时间步长的选取也更加自由，因而可以应用传统有限差分法中的相关技术来提高计算精度和计算效率。

在已有的文献中，Reider 和 Sterling[18]首次将有限差分法应用于格子 Boltzmann 方法中，相应的空间和时间离散分别采用四阶中心差分格式和四阶 Runge-Kutta 格式，他们对

有限差分格子 Boltzmann 方法、标准格子 Boltzmann 方法和有限差分法进行了对比，证明了有限差分格子 Boltzmann 方法的可行性和有效性。Reider 和 Sterling 的方法仍然是基于均匀网格，随后，Cao 等[1]在有限差分格子 Boltzmann 方法中实施了非均匀网格。他们指出，由于格子 Boltzmann 方法的分布函数为实数，在流场的整个物理空间上是连续的，且格子 Boltzmann 方法的物理对称和格子对称可以分离开来，因而可以像有限差分法、有限元法等方法一样，在格子 Boltzmann 方法中实施非均匀网格。而且，数值模拟的稳定性同样可以通过 Courant-Friedrichs-Lewy 条件来保证。

为了处理曲线网格，Mei 和 Shyy[19]将适体网格引入有限差分格子 Boltzmann 方法中。他们指出，由于时间步长的选取不再与空间步长相关，因而可以采用有限差分格子 Boltzmann 方法模拟物理黏性较小的流动，即高雷诺数流动。在实际模拟中，为了保证数值稳定性，需要对碰撞项采用半隐式处理，其他项则采用显式处理。由于碰撞项的非线性特性，Mei 和 Shyy 对平衡态分布函数在时层上做了线性外推处理，实现了控制方程的显式求解。与此同时，却带来了数值不稳定性。基于 Mei 和 Shyy 的工作，Guo 和 Zhao[20]通过引入一个新的分布函数，消除了原有方法中碰撞项的隐式离散，从而构建了一个类似于标准格子 Boltzmann 方法的全显格式。

以上所述的多种有限差分格子 Boltzmann 方法在时间上均采用显式递进，保留了标准格子 Boltzmann 方法的演化特性，从而可以很方便地处理非稳态问题。另外，由于时间步长不能选取过大，导致计算效率较低。特别是对于**刚性问题**（stiff problem，即松弛时间 τ_0 远小于系统的特征时间），数值模拟将变得非常困难。因此，采用隐式处理来提高计算效率就显得非常必要。Tölke 等[21]讨论了隐式离散和网格优化技术在有限差分格子 Boltzmann 方法中的应用，并采用隐式迭代求解了高雷诺数下的边界层问题，但其方法仅能应用于稳态问题的模拟。

为了克服上述不足，我们提出了一种**隐式-显式有限差分格子 Boltzmann 方法**（implicit-explicit finite-difference lattice Boltzmann method）[22]。在这种方法中，时间离散采用隐式-显式 Runge-Kutta 格式，也即对碰撞项和对流项分别采用隐式和显式处理，而空间离散则可以任意选择差分格式。进一步，结合格子 Boltzmann 方法的碰撞不变量特性，隐式处理可以巧妙地消除掉，无须迭代。从而，整个方法同时保持了隐式求解快速，以及显式适合非稳态问题求解的双重优点。以下对隐式-显式有限差分格子 Boltzmann 方法进行介绍，读者也可借此了解有限差分格子 Boltzmann 方法的具体实施细节。

9.4.1 时间离散

龙格-库塔法简介

考察式(9.2)可知该方程为双曲型方程，因而可以采用诸如隐式-显式 Runge-Kutta 格式等对其进行求解，以获得较快的数值计算速率。与传统方法中的 Runge-Kutta 格式相同，为了将 t 时层的分布函数 f_α^t 递进到 $t+\delta_t$ 时层 $f_\alpha^{t+\delta_t}$，需要分多步来实现，即

$$f_\alpha^{(J)}=f_\alpha^t-\delta_t\sum_{k=1}^{J-1}\widetilde{m}_{Jk}(\boldsymbol{e}_\alpha\cdot\nabla f_\alpha^{(k)})+\delta_t\sum_{k=1}^{J}m_{Jk}\frac{f_\alpha^{\mathrm{eq}(k)}-f_\alpha^{(k)}}{\tau_0^{(k)}} \tag{9.34}$$

$$f_\alpha^{t+\delta_t}=f_\alpha^t-\delta_t\sum_{J=1}^{r}\widetilde{n}_J(\boldsymbol{e}_\alpha\cdot\nabla f_\alpha^{(J)})+\delta_t\sum_{J=1}^{r}n_J\frac{f_\alpha^{\mathrm{eq}(J)}-f_\alpha^{(J)}}{\tau_0^{(J)}} \tag{9.35}$$

式中：$f_\alpha^{(k)}$、$f_\alpha^{\mathrm{eq}(k)}$ 和 $\tau_0^{(k)}$ 分别为 t 时层第 k 步的密度分布函数、平衡态密度分布函数以及松

弛时间；$r\times r$ 矩阵 $\widetilde{\boldsymbol{M}}=[\widetilde{m}_{Jk}]$（当 $k\geqslant J$ 时，$\widetilde{m}_{Jk}=0$）和 $\boldsymbol{M}=[m_{Jk}]$（当 $k>J$ 时，$m_{Jk}=0$），以及向量 $\widetilde{\boldsymbol{n}}=\{\widetilde{n}_1,\cdots,\widetilde{n}_r\}^{\mathrm{T}}$ 和 $\boldsymbol{n}=\{n_1,\cdots,n_r\}^{\mathrm{T}}$ 则随着所采用的隐式-显式 Runge-Kutta 格式不同而不同，具体形式将在后面给出。可以看到，式(9.34)为隐式，而式(9.35)则为显式。

观察式(9.34)可以发现，除了等号左边有 $f_\alpha^{(J)}$ 之外，等号右边第三项在 $k=J$ 时也会出现 $f_\alpha^{(J)}$ 以及 $f_\alpha^{\mathrm{eq}(J)}$。此时，对式(9.34)的求解就会非常困难。这并不仅仅因为等号两边同时出现了 $f_\alpha^{(J)}$ 需要迭代，而且 $f_\alpha^{\mathrm{eq}(J)}$ 的求解还与该步的宏观物理量 $\sum\limits_{\alpha=0}^{m-1} f_\alpha^{(J)}\phi$ 有关，也即与 $f_\alpha^{(J)}$ 有关。此处，ϕ 为碰撞不变量，m 为离散速度的个数。

对式(9.34)求和，可得

$$\sum_{\alpha=0}^{m-1} f_\alpha^{(J)}\phi=\sum_{\alpha=0}^{m-1} f_\alpha^t\phi-\delta_t\sum_{k=1}^{J-1}\widetilde{m}_{Jk}\left[\sum_{\alpha=0}^{m-1}(\boldsymbol{e}_\alpha\cdot\nabla f_\alpha^{(k)})\phi\right]+\delta_t\sum_{k=1}^{J}\frac{m_{Jk}}{\tau_0^{(k)}}\left[\sum_{\alpha=0}^{m-1}(f_\alpha^{\mathrm{eq}(k)}-f_\alpha^{(k)})\phi\right] \tag{9.36}$$

考虑到 ϕ 为碰撞不变量，有

$$\sum_{\alpha=0}^{m-1}(f_\alpha^{\mathrm{eq}(k)}-f_\alpha^{(k)})\phi=0 \tag{9.37}$$

于是式(9.36)等号右边第三项为零，故式(9.36)可简化为

$$\sum_{\alpha=0}^{m-1} f_\alpha^{(J)}\phi=\sum_{\alpha=0}^{m-1} f_\alpha^t\phi-\delta_t\sum_{k=1}^{J-1}\widetilde{m}_{Jk}\left[\sum_{\alpha=0}^{m-1}(\boldsymbol{e}_\alpha\cdot\nabla f_\alpha^{(k)})\phi\right] \tag{9.38}$$

式(9.38)表明，t 时层第 J 步的宏观物理量 $\rho^{(J)}$ 和 $\boldsymbol{u}^{(J)}$ 与 $f_\alpha^{(J)}$、$f_\alpha^{\mathrm{eq}(J)}$ 和 $\tau_0^{(J)}$ 无关，仅需要前面第 $J-1$ 步的相关信息便可计算出。进而，利用 $\rho^{(J)}$ 和 $\boldsymbol{u}^{(J)}$，又可求出 $f_\alpha^{\mathrm{eq}(J)}$ 和 $\tau_0^{(J)}$。此处提到 $\tau_0^{(J)}$ 是因为在一些模型中，$\tau_0^{(J)}$ 与压力等宏观量有关，因而也需要实时更新。于是，$f_\alpha^{(J)}$ 可计算如下：

$$f_\alpha^{(J)}=\frac{f_\alpha^t-\delta_t\sum\limits_{k=1}^{J-1}\widetilde{m}_{Jk}(\boldsymbol{e}_\alpha\cdot\nabla f_\alpha^{(k)})+\delta_t\sum\limits_{k=1}^{J-1}m_{Jk}\dfrac{f_\alpha^{\mathrm{eq}(k)}-f_\alpha^{(k)}}{\tau^{(k)}}+\dfrac{\delta_t}{\tau^{(J)}}m_{JJ}f_\alpha^{\mathrm{eq}(J)}}{1+\dfrac{\delta_t}{\tau_0^{(J)}}m_{JJ}} \tag{9.39}$$

综上，式(9.39)和式(9.35)可以看作隐式-显式有限差分格子 Boltzmann 方法的时间离散控制方程。由于利用了碰撞不变量的固有特性，该方法虽然采用了隐式-显式 Runge-Kutta 格式，在保留了隐式和显式处理优点的同时，消除了隐式过程，从而避免了复杂的三角迭代求解。

隐式-显式 Runge-Kutta 格式的系数由如下形式表示：

$\widetilde{M}$	M
$\widetilde{n}$	n

其中，左边和右边分别为对流项和碰撞项系数，也即显式和隐式部分的系数。在实际应用中，可以根据需要选择不同精度的隐式-显式 Runge-Kutta 格式[23]，对应地，系数取值也不尽相同。例如，其三阶形式的系数见表 9.1。

表 9.1 三阶隐式-显式 Runge-Kutta 格式系数

0	0	0	0	m_{11}	0	0	0
0	0	0	0	$-m_{11}$	m_{11}	0	0
0	1	0	0	0	$1-m_{11}$	m_{11}	0
0	1/4	1/4	0	m_{41}	m_{42}	$1/2-m_{11}-m_{41}-m_{41}$	m_{11}
0	1/6	1/6	2/3	0	1/6	1/6	2/3

$m_{11}=0.241\ 694\ 260\ 788\ 21$, $m_{41}=0.060\ 423\ 565\ 197\ 05$, $m_{42}=0.129\ 152\ 869\ 605\ 90$

以上方法虽然在每个时层增加了额外的步骤，但思路简单，且使得时间步长 δ_t 的选取非常灵活，例如我们在文献[22]中曾采用了 $\delta_t=45\ 000\times\tau_0$，显著地提高了计算效率。如果为了计算简便，还可直接采用时间导数简单离散法[24]对式(9.2)进行逼近，如采用一阶向前格式对时间进行离散：

$$f_\alpha^{t+\delta_t}=f_\alpha^t-\delta_t\boldsymbol{e}_\alpha\cdot\nabla f_\alpha^t+\delta_t\frac{f_\alpha^{eq,t}-f_\alpha^t}{\tau_0^t} \tag{9.40}$$

9.4.2 空间离散

空间离散主要用于处理式(9.35)、式(9.39)和式(9.40)中的对流项 $\boldsymbol{e}_\alpha\cdot\nabla f_\alpha$。空间离散格式的选取对计算精度有着非常重要的影响。已有学者指出，采用有限差分格子 Boltzmann 方法对诸如存在激波等复杂物理问题的求解，需要采用高精度的空间离散格式[4,25]。以下，分别介绍一阶迎风格式、二阶 **NND**(non-oscillatory and non-free-parameter dissipation)格式[26]和五阶 **WENO**(weighted essentially non oscillatory)格式[23,27]。

1. 一阶迎风格式

考虑对流项 $\boldsymbol{e}_\alpha\cdot\nabla f_\alpha$ 在 x 方向上的分量 $\partial(e_{\alpha x}f_\alpha)/\partial x$，采用一阶迎风格式对其进行离散可得

$$\frac{\partial(e_{\alpha x}f_\alpha)}{\partial x}=\frac{e_{\alpha x}}{\delta_x}(f_{\alpha,i,j}-f_{\alpha,i-1,j}) \tag{9.41}$$

式中，$f_{\alpha,i,j}$ 和 $f_{\alpha,i-1,j}$ 分别为网格节点 (x_i,y_j) 和 (x_{i-1},y_j) 处的分布函数，i 和 j 为网格节点标号。

2. 二阶 NND 格式

在二阶 NND 格式中，分量 $\partial(e_{\alpha x}f_\alpha)/\partial x$ 可表示为

$$\frac{\partial(e_{\alpha x}f_\alpha)}{\partial x}=\frac{1}{\delta_x}(\hat{F}_{\alpha,i+1/2,j}-\hat{F}_{\alpha,i-1/2,j}) \tag{9.42}$$

式中，$\hat{F}_{\alpha,i+1/2,j}$ 为流过界面 $x_{i+1/2}$ 的数值通量，并计算如下：

$$\hat{F}_{\alpha,i+1/2,j}=\hat{F}^L_{\alpha,i+1/2,j}+\hat{F}^R_{\alpha,i+1/2,j} \tag{9.43}$$

$$\hat{F}^L_{\alpha,i+1/2,j}=F^+_{\alpha,i,j}+\frac{1}{2}\min\bmod(\Delta F^+_{\alpha,i+1/2,j},\Delta F^+_{\alpha,i-1/2,j}) \tag{9.44}$$

$$\hat{F}^R_{\alpha,i+1/2,j}=F^-_{\alpha,i+1,j}-\frac{1}{2}\min\bmod(\Delta F^-_{\alpha,i+1/2,j},\Delta F^-_{\alpha,i+3/2,j}) \tag{9.45}$$

$$F^{+}_{\alpha,i,j}=\frac{1}{2}(e_{\alpha x}+|e_{\alpha x}|)f_{\alpha,i,j} \tag{9.46}$$

$$F^{-}_{\alpha,i,j}=\frac{1}{2}(e_{\alpha x}-|e_{\alpha x}|)f_{\alpha,i,j} \tag{9.47}$$

$$\Delta F^{\pm}_{\alpha,i+1/2,j}=F^{\pm}_{\alpha,i+1,j}-F^{\pm}_{\alpha,i,j} \tag{9.48}$$

式中，min mod 函数为**通量限制器**(flux limiter)，其取值方式为当所含变量异号时取零，当所含变量同号时取这些变量中绝对值最小的那个变量的值，其数学表达式为

$$\min\bmod(a,b)=\begin{cases}0, & ab\leqslant 0\\ \mathrm{sgn}(a)\cdot\min(|a|,|b|), & ab>0\end{cases} \tag{9.49}$$

式中，sgn 函数表示变量的正负号，即

$$\mathrm{sgn}(a)=\begin{cases}1, & a>0\\ 0, & a=0\\ -1, & a<0\end{cases} \tag{9.50}$$

3. 五阶 WENO 格式

在五阶 WENO 格式中，分量$\partial(e_{\alpha x}f_\alpha)/\partial x$ 同样由式(9.42)表示，而 $\hat{F}_{\alpha,i+1/2,j}$则表示为

$$\hat{F}_{\alpha,i+1/2,j}=w_1\hat{F}^1_{\alpha,i+1/2,j}+w_2\hat{F}^2_{\alpha,i+1/2,j}+w_3\hat{F}^3_{\alpha,i+1/2,j} \tag{9.51}$$

在 $e_{\alpha x}\geqslant 0$ 的情况下，式(9.51)等号右边的三个位于不同插值片段上的三阶通量可由下面的式子计算获得

$$\hat{F}^1_{\alpha,i+1/2,j}=\frac{1}{3}F_{\alpha,i-2,j}-\frac{7}{6}F_{\alpha,i-1,j}+\frac{11}{6}F_{\alpha,i,j} \tag{9.52}$$

$$\hat{F}^2_{\alpha,i+1/2,j}=-\frac{1}{6}F_{\alpha,i-1,j}+\frac{5}{6}F_{\alpha,i,j}+\frac{1}{3}F_{\alpha,i+1,j} \tag{9.53}$$

$$\hat{F}^3_{\alpha,i+1/2,j}=\frac{1}{3}F_{\alpha,i,j}+\frac{5}{6}F_{\alpha,i+1,j}-\frac{1}{6}F_{\alpha,i+2,j} \tag{9.54}$$

式中，$F_{\alpha,i,j}=e_{\alpha x}f_{\alpha,i,j}$。

式(9.51)中的系数 $w_s(s=1,2,3)$定义为

$$w_1=\frac{\tilde{w}_1}{\tilde{w}_1+\tilde{w}_2+\tilde{w}_3},\quad w_2=\frac{\tilde{w}_2}{\tilde{w}_1+\tilde{w}_2+\tilde{w}_3},\quad w_3=\frac{\tilde{w}_3}{\tilde{w}_1+\tilde{w}_2+\tilde{w}_3} \tag{9.55}$$

$$\tilde{w}_1=\frac{1}{10\,(10^{-6}+\sigma_1)^2},\quad \tilde{w}_2=\frac{3}{5\,(10^{-6}+\sigma_2)^2},\quad \tilde{w}_3=\frac{3}{10\,(10^{-6}+\sigma_3)^2} \tag{9.56}$$

为了避免除零，在式(9.56)的分母中添加了一个小量10^{-6}。式(9.56)中的光滑因子 σ 计算如下：

$$\sigma_1=\frac{13}{12}(F_{\alpha,i-2,j}-2F_{\alpha,i-1,j}+F_{\alpha,i,j})^2+\frac{1}{4}(F_{\alpha,i-2,j}-4F_{\alpha,i-1,j}+3F_{\alpha,i,j})^2 \tag{9.57}$$

$$\sigma_2=\frac{13}{12}(F_{\alpha,i-1,j}-2F_{\alpha,i,j}+F_{\alpha,i+1,j})^2+\frac{1}{4}(F_{\alpha,i-1,j}-F_{\alpha,i+1,j})^2 \tag{9.58}$$

$$\sigma_3=\frac{13}{12}(F_{\alpha,i,j}-2F_{\alpha,i+1,j}+F_{\alpha,i+2,j})^2+\frac{1}{4}(3F_{\alpha,i,j}-4F_{\alpha,i+1,j}+F_{\alpha,i+2,j})^2 \tag{9.59}$$

当 $e_{\alpha x}<0$ 时，计算过程与上述过程类似，只需将式(9.51)~式(9.59)中的相应物理量关于 $i+1/2$ 做镜像对称即可。虽然以上三种空间离散格式都是在 x 方向做的论述，对二

维问题,只需要将不同格式同时应用于 x 和 y 方向即可。

9.4.3 讨论

在 9.4.1 节和 9.4.2 节中,我们以隐式-显式有限差分格子 Boltzmann 方法为例,介绍了有限差分格子 Boltzmann 方法的时间离散和空间离散。在有限差分格子 Boltzmann 方法中,可以很方便地使用非均匀网格,也可以应用传统有限差分法中的相关技术来改善计算精度和效率,此外,数值稳定性可以通过 Courant-Friedrichs-Lewy 条件来保证。基于以上优点,有限差分格子 Boltzmann 方法已广泛应用于多种物理问题的模拟求解,例如高雷诺数流动[22]、声波传播[28]、微流动[29]、气液两相流[30-31]、带有激波的可压缩流动[4,22,32-35]、二元流体及多组分流动[36-37]、血液流及非牛顿流体[38-39]等。有限差分格子 Boltzmann 方法的不足之处在于放弃了标准格子 Boltzmann 方法的碰撞迁移过程,使得程序操作较为复杂;且对于不同的空间离散格式,会在不同程度上引入人工黏性[22,40],增加数值误差。因此,离散格式的选择就显得尤为重要。此外,有限差分格子 Boltzmann 方法与有限差分法一样,对于复杂区域的适应性较差。

9.5 有限容积格子 Boltzmann 方法

有限容积法简介

有限容积法是以守恒型宏观方程为出发点,通过对计算域的离散来构造相应的离散方程,它可以保证守恒特性,且易于应用不规则网格来模拟具有复杂边界的物理问题。基于此,不少学者将有限容积法引入格子 Boltzmann 方法中,提出了**有限容积格子 Boltzmann 方法**(finite-volume lattice Boltzmann method)。

9.5.1 结构化网格

早在格子 Boltzmann 方法发展伊始,Succi 及其合作者就在这一方向做出了初步探索[41]。以二维问题为例,考虑计算域被离散成如图 9.5 所示的一系列控制容积 C_P(粗网格,coarse-graining mesh),每个 C_P 里包含有数个类似标准格子 Boltzmann 方法的细网格(fine-graining mesh)节点。图中,点 P 为控制容积 C_P 的中心,点 sw、nw、ne 和 se 分别代表控制容积 C_P 的西南、西北、东北和东南角点,并定义点 P 的分布函数为控制容积 C_P 内分布函数的平均值,即

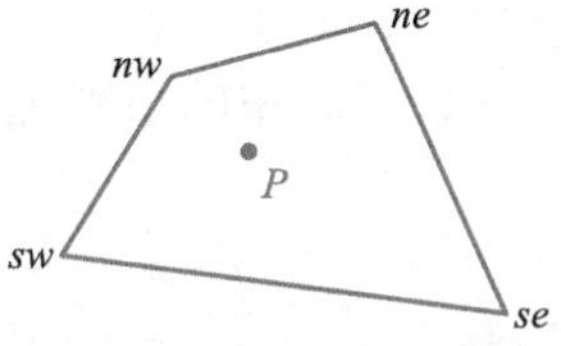

图 9.5 控制容积示意图(结构化)

$$F_\alpha(P)=\frac{1}{S_P}\int_{C_P}f_\alpha(x,y)\,\mathrm{d}S \tag{9.60}$$

式中:S_P 为控制容积 C_P 的面积;$f_\alpha(x,y)$ 为网格节点上的分布函数。将式(9.60)代入离散 Boltzmann-BGK 方程,并采用一阶迎风格式对时间项进行离散,可得[42]

$$F_\alpha(P,t+\delta_t)-F_\alpha(P,t)+\delta_t\sum_{s=1}^{4}\Phi_{\alpha,s}=\delta_t\,\frac{F_\alpha^{\mathrm{eq}}(P,t)-F_\alpha(P,t)}{\tau_0} \tag{9.61}$$

式中:

$$\Phi_{\alpha,s}=\frac{1}{S_P}\int_{\partial C_P}(\boldsymbol{e}_\alpha\cdot\boldsymbol{n}_s)f_\alpha(x,y)\,\mathrm{d}S \tag{9.62}$$

$$F_\alpha^{\mathrm{eq}}(P,t)=\frac{1}{S_P}\int_{C_P} f_\alpha^{\mathrm{eq}}(x,y)\,\mathrm{d}S \tag{9.63}$$

式(9.62)为流经控制容积 C_P 的 4 个边界界面(记为界面 $s,s=1,2,3,4$)上的通量;$\boldsymbol{n}_s$ 为相应界面的法线方向;∂C_P 为控制容积 C_P 的所有边界界面;而式(9.63)则对应 P 点的平衡态分布函数。

自此,式(9.61)可以看作是关于未知量 F_α 的方程组,该方程组共有 N_c 个方程(N_c 是计算域所包含的控制容积个数)。为了对该方程组进行求解,需要将通量 $\Phi_{\alpha,s}$ 用 F_α 表示出来,也即对应一种插值过程将 f_α 转变为 F_α。相应插值方法有片段常数、片段线性、片段抛物线插值等,具体可参阅文献[42]。此外,Amati 等[43]将这一方法发展到了三维,并应用于湍流的数值模拟研究。上述有限容积格子 Boltzmann 方法将物理量存储于控制容积中心,属于**结构化网格**。

9.5.2 非结构化网格

为了在格子 Boltzmann 方法中实现**非结构化网格**(unstructured grid),Peng 等[44-46]提出采用格点法,即将物理量存储在网格节点上。对二维问题,可以将计算域离散成一系列非结构化的三角元[44-45]。如图 9.6 所示,选取点 P 相邻的多边形 $P_1P_2P_3P_4P_5P_6$ 作为控制容积。其中,点 E 为界面 PP_2 的中点;点 C 为三角元 PP_1P_2 的中心,类似地,点 D 为三角元 PP_2P_3 的中心。为了得到积分型离散方程,需要对式(9.2)在控制容积上积分,也就是在三角形 PCE、三角形 PED 等之上做积分,并将积分值相加。下面以三角形 PCE 为例,介绍具体的积分过程。

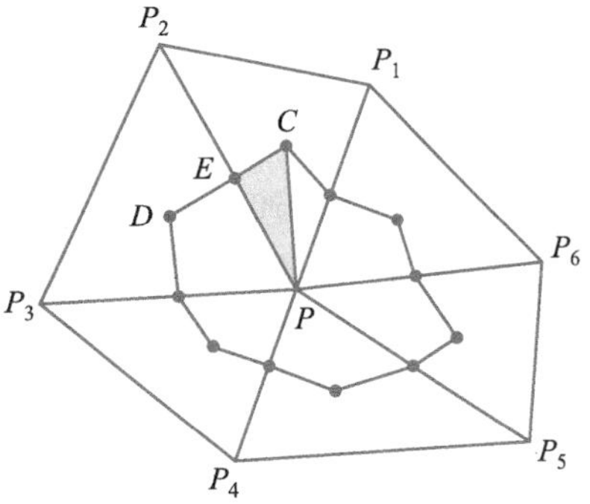

图 9.6 控制容积示意图(非结构化)

首先考虑式(9.2)等号左边第一项。为了简便,假设三角形 PCE 内的分布函数都为常数 $f_\alpha(P)$,且 $f_\alpha(P)$ 为点 P 的分布函数,故有

$$\int_{PCE}\frac{\partial f_\alpha}{\partial t}\mathrm{d}A=\frac{\partial f_\alpha(P)}{\partial t}A_{PCE} \tag{9.64}$$

式中,A_{PCE} 为三角形 PCE 的面积。

接着考虑式(9.2)等号左边第二项。对该项的积分会出现通过三个界面 PC、CE 和 EP 的通量,但由于后面需要对三角形 PCE、三角形 PED 等求和,通过 PC、CE 等内部界面的通量都会抵消掉,所以可以忽略内部界面对积分的影响,因而第二项的积分可以表示为

$$\int_{PCE}\nabla f_\alpha\mathrm{d}A=\boldsymbol{e}_\alpha\cdot\int_{CE}f_\alpha\mathrm{d}\boldsymbol{l}+I_s \tag{9.65}$$

式中:$\mathrm{d}\boldsymbol{l}$ 表示积分界面 CE 的法线方向;I_s 为通过内部界面的通量。假设每个三角元上的 f_α 是线性变化的,则式(9.65)可以转换为

$$\int_{PCE}\nabla f_\alpha\mathrm{d}A=\boldsymbol{e}_\alpha\cdot\boldsymbol{n}_{CE}l_{CE}\frac{f_\alpha(C)+f_\alpha(E)}{2}+I_s \tag{9.66}$$

式中:$\boldsymbol{n}_{CE}$ 为沿着界面 CE 法线方向的单位矢量;l_{CE} 为 CE 的长度;$f_\alpha(C)$ 和 $f_\alpha(E)$ 可通过线性插值得到

$$f_\alpha(C)=\frac{f_\alpha(P)+f_\alpha(P_1)+f_\alpha(P_2)}{3} \tag{9.67}$$

$$f_\alpha(E)=\frac{f_\alpha(P)+f_\alpha(P_2)}{2} \tag{9.68}$$

对式(9.2)等号右边的碰撞项进行积分,假设f_α 和f_α^{eq}在三角形 *PCE* 中线性变化,则可得

$$\begin{aligned}&\int_{PCE}-\frac{f_\alpha-f_\alpha^{\mathrm{eq}}}{\tau_0}\mathrm{d}A\\&=-\frac{A_{PCE}}{\tau_0}\frac{f_\alpha(P)-f_\alpha^{\mathrm{eq}}(P)+f_\alpha(C)-f_\alpha^{\mathrm{eq}}(C)+f_\alpha(E)-f_\alpha^{\mathrm{eq}}(E)}{3}\end{aligned} \tag{9.69}$$

式中,

$$f_\alpha^{\mathrm{eq}}(C)=\frac{f_\alpha^{\mathrm{eq}}(P)+f_\alpha^{\mathrm{eq}}(P_1)+f_\alpha^{\mathrm{eq}}(P_2)}{3} \tag{9.70}$$

$$f_\alpha^{\mathrm{eq}}(E)=\frac{f_\alpha^{\mathrm{eq}}(P)+f_\alpha^{\mathrm{eq}}(P_2)}{2} \tag{9.71}$$

至此,完成了对式(9.2)在三角形 *PCE* 上的积分。类似步骤可以在 *PED* 等三角形上进行,并最后求和。对求和后的式子在时间上采用一阶迎风离散,从而可得点 *P* 在 $t+\delta_t$ 时刻的分布函数

$$f_\alpha(P,t+\delta_t)=f_\alpha(P,t)+\frac{\delta_t}{A_P}\left(\sum_{i=1}^{N}\text{collisions}-\sum_{i=1}^{N}\text{fluxes}\right) \tag{9.72}$$

式中:A_P 为点 *P* 附近的控制容积面积;“collisions”和“fluxes”分别表示碰撞项和对流项在整个控制容积上积分后的值;*N* 为点 *P* 周围划分的三角形 *PCE*、三角形 *PED* 等的总数。

需要指出的是,对于边界节点分布函数的更新,处理方法与上面相同,但是控制容积只有一半。如果将计算域离散成一系列的四边形元,则上述线性插值过程可变为双线性插值[46]。

Peng 等[44-46]采用格点法在格子 Boltzmann 方法中成功实现了非结构化网格,如图 9.7 所示,并使用有限容积格子 Botlzmann 方法模拟了共轴旋转筒内的流动、Poiseuille 流动、变形的 Couette 流动以及 Taylor 涡流动,证明了这一方法的有效性。随后,Ubertini 等对这一方法做了进一步的发展,参见文献[47-50]。

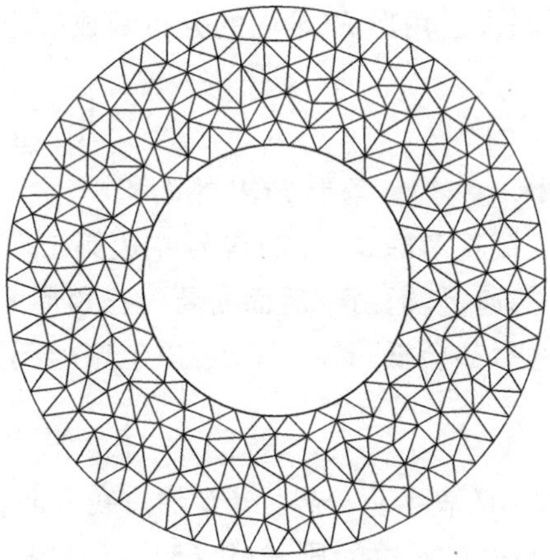

图 9.7 采用有限容积格子 Boltzmann 方法生成的共轴旋转圆筒流动计算网格

9.6 有限元格子 Boltzmann 方法

有限元方法是将计算域划分为一系列元体(在二维情况下,元体多为三角形和四边形),在每个元体上取数个点作为节点,然后通过对控制方程积分来获得离散方程,并通过近似解来逼近微分方程的准确解,通常采用的逼近方法为分段(块)逼近。它最大的优点是对不规则区域的适应性好,但计算量一般比较大。基于有限元方法,对离散 Boltzmann-BGK 方程进行求解的方法,就称为**有限元格子 Boltzmann 方法**(finite-element lattice Boltzmann method)。

Lee 和 Lin[51-52] 在这方面做了开拓性的工作。他们将特征线伽辽金有限元方法(characteristic Galerkin finite-element method)引入格子 Boltzmann 方法中,用于求解离散 Boltzmann-BGK 方程。在计算中,时间离散采用二阶预测修正方法(predictor-corrector method),生成的非结构化网格如图 9.8 所示。随后,Li 等[53-54] 提出采用更稳定的最小二乘有限元方法来代替前面提到的特征线伽辽金有限元方法。此外,还可采用间断伽辽金谱元方法(discontinuous Galerkin spectral element method)对离散 Boltzmann-BGK 方程进行求解[55-56]。

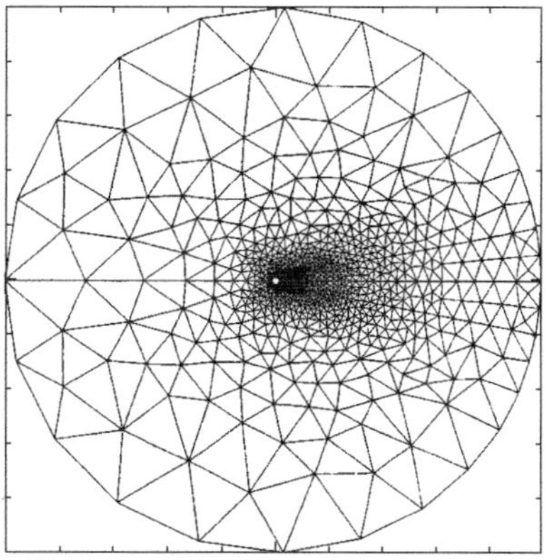

图 9.8 采用有限元格子 Boltzmann 方法生成的圆柱绕流计算网格

除了应用于流体流动的数值模拟,有限元格子 Boltzmann 方法还被用于研究固体中激波的传播,感兴趣的读者可参阅文献[57]。

9.7 多块网格

均分网格的一个挑战就是为了获得局部区域的精确解,必须把整个计算区域的网格加密,因而会大大提高所需的硬件资源和计算时间。前面提到的各种非均匀网格技术可以解决这一问题。与此同时,在物理量变化剧烈的区域采用比较密的网格,而在变化缓慢的区域采用比较稀疏的网格,把流动区域划分为多块结构化网格,也不失为一种有效的解决方法。虽然多块网格技术可以与插值格子 Boltzmann 方法、有限差分格子 Boltzmann 方法等配合使用,但为了简便,在许多文献中,每块结构化网格上,仍然采用的是标准 Boltzmann 方法。Filippova 和 Hänel 在文献[58]中首次提出了在格子 Boltzmann 方法中应用多块网格的基本思想,Yu 等[59]对该方法做了进一步的发展。但是,采用多块网格带来了一些问题,例如如何保证网格界面物理量的连续性、不同块之间的时层不一致等,都需要特别考虑。

多块网格如图 9.9 所示,MN 为细网格边界,AB 为粗网格边界,细网格边界在粗网格内部,粗网格边界在细网格内部,这样的安排便于界面上的数据交换。例如,点 Q 是细网格边界上的节点,同时又是粗网格的内部节点,其分布函数 f_2^* 、f_5^* 和 f_6^* 分别由点 E、D 和 F 的值得到("$*$"表示碰撞后的分布函数),其他方向的分布函数则由粗网格内部节点上的相应值得到。同理,细网格的边界也在粗网格的内部,但是细网格中有些节点的分布函数不能直接从粗网格中得到,例如图 9.9 所示的实心圆点,这些点的分布函数需要通过空间上的插值求得。此外,由于粗细网格的时间步长不等,不同网格之间还需要进行时层上的插值。

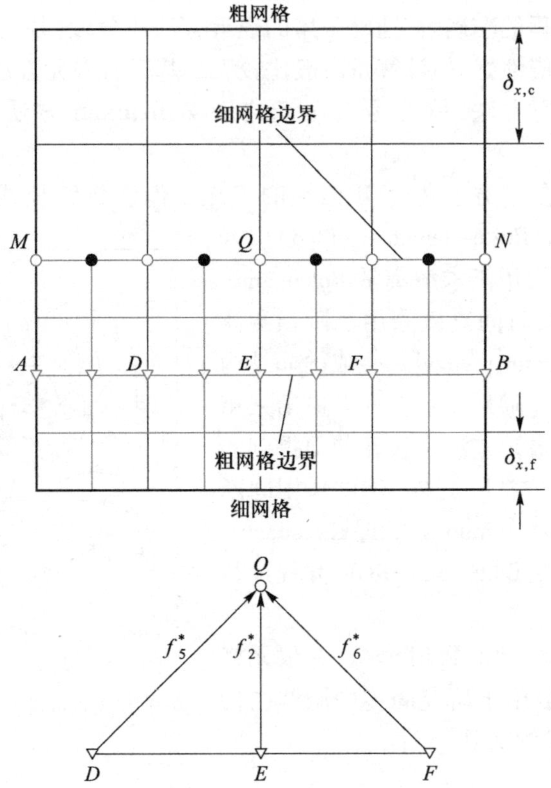

图 9.9 多块网格示意图

考虑粗细网格都均匀分布,满足 $\delta_{x,c}=\delta_{y,c}=\delta_{t,c}$ 和 $\delta_{x,f}=\delta_{y,f}=\delta_{t,f}$,且粗细网格间距比值为 $m=\delta_{x,c}/\delta_{x,f}$,可设 m 为整数。详细的计算流程如图 9.10 所示,图中下标“f”表示细网格、下标“c”表示粗网格,且 $m=2$。需要注意的是,文献[59]将时间递进放在了碰撞过程中,即

$$f_\alpha^*(\boldsymbol{r},t+\delta_t)-f_\alpha(\boldsymbol{r},t)=-\frac{1}{\tau}\left[f_\alpha(\boldsymbol{r},t)-f_\alpha^{\mathrm{eq}}(\boldsymbol{r},t)\right] \tag{9.73}$$

在本书其他章节,时间递进都是放在迁移过程中。为了与原文献保持一致,此处的表述基于式(9.73)。

下面介绍如何保证物理量的连续性。为了保证流体黏度和雷诺数在整场的一致性,粗网格中的松弛时间 τ_c 和细网格中的松弛时间 τ_f 需满足如下关系:

$$\tau_f=\frac{1}{2}+m\left(\tau_c-\frac{1}{2}\right) \tag{9.74}$$

为了保证不同网格中计算的物理量在交接面处的连续性,分布函数需进行特殊处理。将分布函数分解成平衡态和非平衡态两部分,即

$$f_\alpha(\boldsymbol{r},t)=f_\alpha^{\mathrm{eq}}(\boldsymbol{r},t)+f_\alpha^{\mathrm{neq}}(\boldsymbol{r},t) \tag{9.75}$$

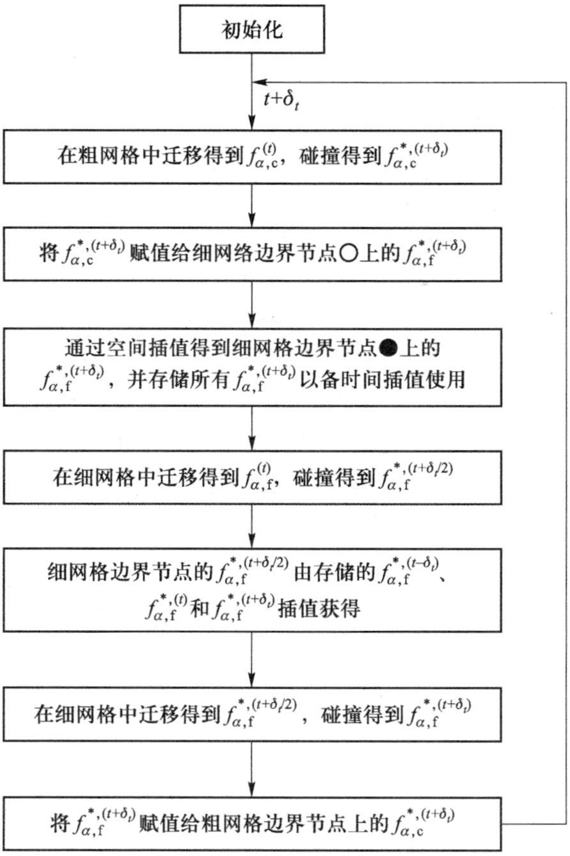

图 9.10　多块网格系统计算流程图

并将式(9.75)代入式(9.73)可得

$$f_\alpha^*(\boldsymbol{r},t+\delta_t)=f_\alpha^{\mathrm{eq}}(\boldsymbol{r},t)+\frac{\tau-1}{\tau}f_\alpha^{\mathrm{neq}}(\boldsymbol{r},t) \tag{9.76}$$

将式(9.76)分别应用到粗网格和细网格中,可得

$$f_{\alpha,\mathrm{c}}^*=f_{\alpha,\mathrm{c}}^{\mathrm{eq}}+\frac{\tau_\mathrm{c}-1}{\tau_\mathrm{c}}f_{\alpha,\mathrm{c}}^{\mathrm{neq}} \tag{9.77}$$

$$f_{\alpha,\mathrm{f}}^*=f_{\alpha,\mathrm{f}}^{\mathrm{eq}}+\frac{\tau_\mathrm{f}-1}{\tau_\mathrm{f}}f_{\alpha,\mathrm{f}}^{\mathrm{neq}} \tag{9.78}$$

为了保证粗细网格交界面处的密度和速度连续性,需满足

$$f_{\alpha,\mathrm{c}}^{\mathrm{eq}}=f_{\alpha,\mathrm{f}}^{\mathrm{eq}} \tag{9.79}$$

为了保证应力张量的连续性,还需满足

$$\left(1-\frac{1}{2\tau_\mathrm{c}}\right)f_{\alpha,\mathrm{c}}^{\mathrm{neq}}=\left(1-\frac{1}{2\tau_\mathrm{f}}\right)f_{\alpha,\mathrm{f}}^{\mathrm{neq}} \tag{9.80}$$

即

$$f_{\alpha,\mathrm{c}}^{\mathrm{neq}}=m\frac{\tau_\mathrm{c}}{\tau_\mathrm{f}}f_{\alpha,\mathrm{f}}^{\mathrm{neq}} \tag{9.81}$$

联立式(9.77)~式(9.79)及式(9.81)可得

$$f_{\alpha,c}^{*}=f_{\alpha,f}^{eq}+m\frac{\tau_c-1}{\tau_f-1}(f_{\alpha,f}-f_{\alpha,f}^{eq}) \tag{9.82}$$

同理可得

$$f_{\alpha,f}^{*}=f_{\alpha,c}^{eq}+\frac{\tau_f-1}{m(\tau_c-1)}(f_{\alpha,c}-f_{\alpha,c}^{eq}) \tag{9.83}$$

式(9.82)和式(9.83)由文献[58]首次给出。此外,由图9.9可知,在粗细网格的交界面上,每个$f_{\alpha,c}^{eq}$和$f_{\alpha,c}^{*}$对应m个$f_{\alpha,f}^{*}$。因此,需要对$f_{\alpha,c}^{eq}$和$f_{\alpha,c}^{*}$进行空间和时间插值,用来确定另外$m-1$个$f_{\alpha,f}^{*}$。文献[59]采用三次样条函数进行空间插值,时间上则采用三次 Lagrangian 插值,并用多块网格成功模拟了顶盖驱动流(图9.11)、圆柱绕流和 NACA0012 翼型绕流(图9.12)。且与标准格子 Boltzmann 相比,计算精度和计算效率均得到了提高。

多块网格和自适应网格

此外,郭照立等[2]提出了基于区域分裂的非均匀格子 Boltzmann 方法,王广超等[60-61]提出了一种发展的嵌套边界的非均匀格子 Boltzmann 方法。这些方法都是基于多块网格,只是在粗细网格界面采用了不同的插值方式,感兴趣的读者可参考相关文献。

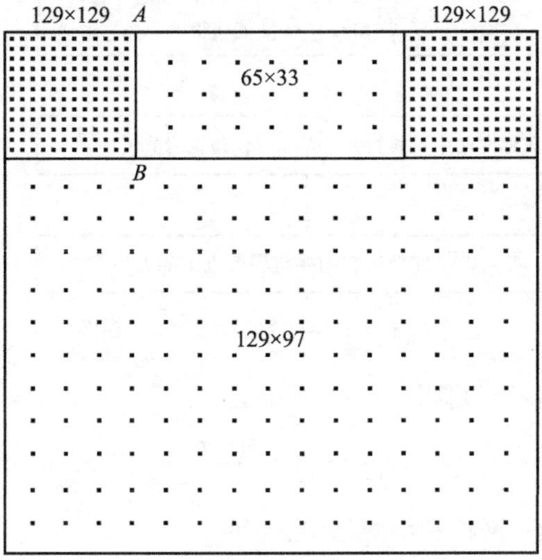

图9.11 顶盖驱动流计算网格

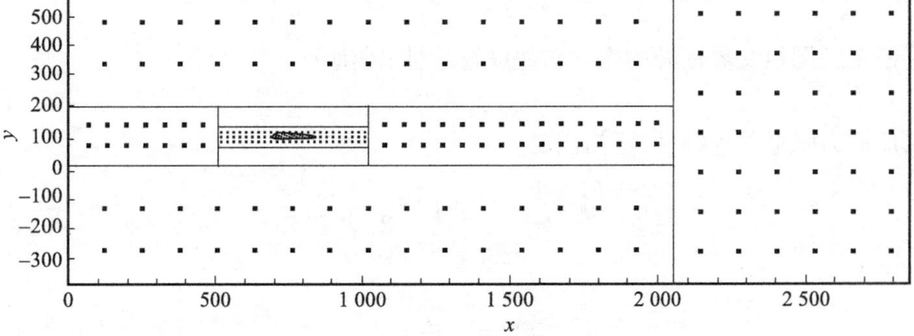

图9.12 NACA0012 翼型绕流计算网格

9.8 多重网格

多重网格(multi-grid)方法是促进迭代求解收敛速度的有效方法,并被广泛应用于流动与传热问题的数值计算[62]。Tölke 等[63]率先采用多重网格对离散 Boltzmann-BGK 方程进行了求解,计算中采用了有限差分法来代替碰撞迁移过程。随后,Mavriplis[64]发展了这一方法,在算法中保留了碰撞迁移过程,并将其应用于稳态流动。在国内,王兴勇等[65-66]提出了**双重网格**(double meshes)格子 Boltzmann 方法。这种算法将格子 Boltzmann 方法在两套网格上迭代演化,并利用粗网格和细网格各自不同的收敛特性,加快收敛速度,提高计算效率。本文以双重网格为例,介绍多重网格在格子 Boltzmann 方法中的实施。

双重网格的布置如图 9.13 所示,下标"c"和"f"分别代表粗网格和细网格,粗细网格间距比一般取为 2。计算中,先在细网格上迭代得到一个初步解答,然后转移到粗网格上,这个过程称为限定;在粗网格上迭代求出比较精确的值后,再转移到细网格上计算,称为延拓。如此循环,从而使误差不断减小,最后在细网格上获得所需的解。粗细网格的数据传递,也即限定与延拓过程,通过插值来实现。从理论上讲,在细网格上的松弛过程,使得误差中的短波(高频)分量迅速衰减,对长波(低频)分量的作用有限;而在粗网格上,由于网格间距的增大,松弛过程把长波分量衰减掉。由此,采用双重网格方法,不同波长分量的误差可以得到比较均匀的衰减,从而加快计算收敛速度,以达到提高计算效率的目的。

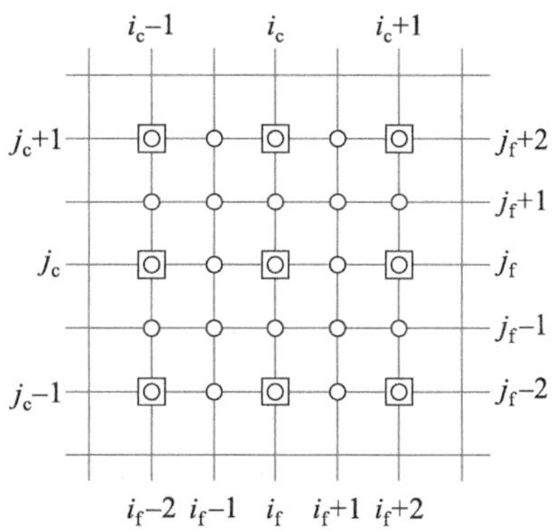

图 9.13 双重网格示意图

在文献[63,64]中,参与不同网格上迭代以及不同网格间传输的物理量都是分布函数 f_α,而在王兴勇等的双重网格格子 Boltzmann 方法中,不同网格上参与迭代的物理量为 f_α,但在粗细网格之间传输的则是密度 ρ 和速度 $\boldsymbol{u}$ 等宏观物理量。双重网格实施的具体步骤如下:

(1) 全场初始化,给定初始的 $\rho^{(0)}$、$\boldsymbol{u}^{(0)}$ 等,并计算细网格上的 $f_\alpha^{(0)}$。

(2) 根据式(9.3)在细网格上迭代 ι 次,得到细网格上的 $\rho_f^{(\iota)}$、$\boldsymbol{u}_f^{(\iota)}$ 等。

(3) 将 $\rho_f^{(\iota)}$、$\boldsymbol{u}_f^{(\iota)}$ 等由细网格向粗网格转移

$$\rho_c = I_f^c \rho_f^{(\iota)}, \quad \boldsymbol{u}_c = I_f^c \boldsymbol{u}_f^{(\iota)} \tag{9.84}$$

式中,I_f^c 为限定算子(restriction operator,即文献[65]中的完全加权转移算子)。常用的限定算子有直接引入(direct inject)和就近平均(averaging nearby values)两种,前者把细网格上的值直接引入粗网格的对应位置,后者则按照粗网格上的被插值点的位置,对该点附近细网格的值取平均。文献[65]所采用的是就近平均。

(4) 将 ρ_c、$\boldsymbol{u}_c$ 等作为初值,在粗网格上迭代 κ 次,得到粗网格上的 $\rho_c^{(\kappa)}$、$\boldsymbol{u}_c^{(\kappa)}$ 等。

(5) 将 $\rho_c^{(\kappa)}$、$\boldsymbol{u}_c^{(\kappa)}$ 等由粗网格向细网格转移

$$\rho_f = I_c^f \rho_c^{(\kappa)}, \quad \boldsymbol{u}_f = I_c^f \boldsymbol{u}_c^{(\kappa)} \tag{9.85}$$

式中,I_c^f 为延拓算子(prolongation operator,即文献[65]中的插值算子)。对二维问题,常用

的延拓算子有双线性插值(bilinear interpolation)和双二次插值(bi-quadratic interpolation),即分别按两个坐标方向进行线性或二次插值。文献[65]所采用的是双线性插值。

(6) 判断细网格上收敛与否,若不收敛,则以 ρ_f、$\boldsymbol{u}_f$ 等为初值,循环第(2)~第(5)步。

文献[65,66]采用以上步骤,用双重网格成功模拟了多种稳态流动,并与标准格子 Boltzmann 方法的结果进行了对比。模拟结果表明,采用双重网格可以成倍提高计算效率:在部分网格下,节约计算时间高达 86.49%;但随着计算网格的加大,粗细网格间数据传递随之增多,效率提高的幅度会有所下降。如果增设更多层网格,并改进循环方式[62-63],计算效率可进一步提高。此外,上述多重网格格子 Boltzmann 方法仅能应用于稳态问题的求解。对于非稳态问题,需要重新设计计算步骤。多重网格技术可以与前面所提及的各种非均匀网格技术结合在一起使用。

9.9 小 结

本章介绍了格子 Boltzmann 方法中的多种网格处理技术。由于采用均匀网格的标准格子 Boltzmann 方法的不足,从而促使了各种非均匀网格方法的出现,如插值格子 Boltzmann 方法、泰勒展开和最小二乘格子 Boltzmann 方法、有限差分格子 Boltzmann 方法、有限容积格子 Boltzmann 方法、有限元格子 Boltzmann 方法等,此外还有多块以及多重网格技术。毫无疑问,这些方法对于 D2Q9、D3Q15、D3Q19 等基本格子模型是完全通用的,且对于多速度模型以及双分布函数模型,大多数网格技术仍然适用。但是,多速度模型下的多块网格技术,粗细网格的交界面会变得更加复杂,需要重新设计计算步骤。总的说来,这些方法提高了格子 Boltzmann 方法的计算精度、计算效率以及对不规则区域的适应性,拓展了格子 Boltzmann 方法的应用范围,但也在一定程度上使得相关理论以及求解过程变得更复杂。为了扬长避短,在实际应用中应当以研究背景为依托,针对不同的物理问题及研究需要,选取适当的网格处理方法。

习 题

9-1 针对图 9.14 所示非均匀网格中的分布函数 f_5,通过迁移后位置(空心圆圈)上的分布函数插值出 (i,j) 节点的分布函数,给出以下插值关系式:1)通过节点 (i,j)、$(i-1,j)$、$(i,j-1)$、$(i-1,j-1)$ 建立一次插值;2)通过节点 $(i,j+1)$、$(i-1,j+1)$、$(i+1,j+1)$、(i,j)、$(i-1,j)$、$(i+1,j)$、$(i,j-1)$、$(i-1,j-1)$、$(i+1,j-1)$ 建立二次插值;3)通过节点 (i,j)、$(i-1,j)$、$(i-2,j)$、$(i,j-1)$、$(i-1,j-1)$、$(i-2,j-1)$、$(i,j-2)$、$(i-1,j-2)$、$(i-2,j-2)$ 建立二次插值。

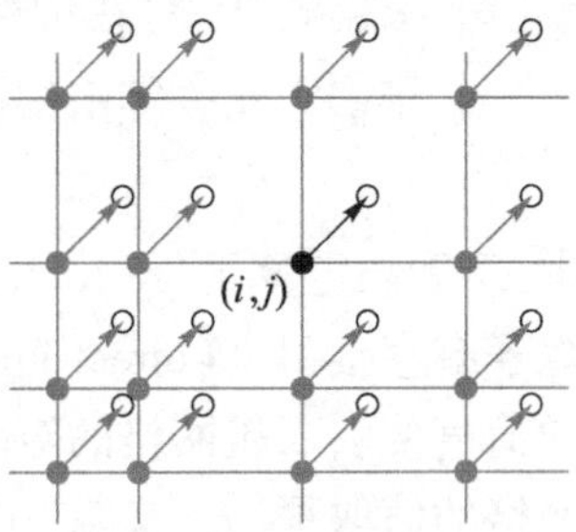

图 9.14 习题 9.1 附图

9-2 查阅相关参考文献可知,如 9.4 节所介绍的有限差分格子 Boltzmann 方法的运动黏性系数与松弛时间的关系一般表示为 $\nu=c_s^2\tau$,思考其与标准格子 Boltzmann 方法中 $\nu=c_s^2(\tau-0.5)\delta_t$ 为何不同,二者对于演化方程(9.2)右侧碰撞项的处理有何差别?

9-3 针对 9.5 节非结构化网格示意图 9.6,写出点 P 处有限容积公式(9.72)的具体表达式。查阅相关文献,分析有限容积格子 Boltzmann 方法中运动黏性与松弛时间 τ 之间的关系。

9-4 针对 9.7 节的多块网格,证明粗网格和细网格之间松弛时间的关系式(9.74);推导粗、细网格间分布函数的转换关系式(9.82)和式(9.83),注意流程图 9.10 中是在碰撞步骤之后采用上述关系式进行粗细网格转换。

参考文献

[1] Cao N Z, Chen S Y, Jin S, et al. Physical symmetry and lattice symmetry in the lattice Boltzmann method [J]. Physical Review E, 1997, 55(1): R21-R24.

[2] 郭照立,施保昌,王能超. 基于区域分裂的非均匀 Lattice Boltzmann 方法[J]. 计算物理,2001,18(2):181-184.

[3] Zheng H W, Shu C, Chew Y T, et al. A platform for developing new lattice Boltzmann models [J]. International Journal of Modern Physics C, 2005, 16(01): 61-84.

[4] Qu K, Shu C, Chew Y T. Alternative method to construct equilibrium distribution functions in lattice-Boltzmann method simulation of inviscid compressible flows at high Mach number[J]. Physical Review E, 2007, 75(3): 036706.

[5] He X Y, Luo L S, Dembo M. Some progress in lattice Boltzmann method. Part I. Nonuniform mesh grids [J]. Journal of Computational Physics, 1996, 129(2): 357-363.

[6] He X Y, Luo L S, Dembo M. Some progress in the lattice Boltzmann method: Reynolds number enhancement in simulations[J]. Physica A: Statistical Mechanics and its Applications, 1997, 239(1): 276-285.

[7] He X Y, Doolen G. Lattice Boltzmann method on curvilinear coordinates system: Flow around a circular cylinder[J]. Journal of Computational Physics, 1997, 134(2): 306-315.

[8] Dixit H N, Babu V. Simulation of high Rayleigh number natural convection in a square cavity using the lattice Boltzmann method[J]. International Journal of Heat and Mass Transfer, 2006, 49(3): 727-739.

[9] 王勇,何雅玲,童长青,等. 二维 Rayleigh-Bénard 对流的插值格子-Boltzmann 方法模拟研究[J]. 工程热物理学报,2007,28(2):313-315.

[10] 李明秀,陶文铨,何雅玲,等. 格子-Boltzmann 方法非均分网格的实施[J]. 工程热物理学报,2003,24(1):73-75.

[11] Shu C, Chew Y T, Niu X D. Least-squares-based lattice Boltzmann method: A meshless approach for simulation of flows with complex geometry[J]. Physical Review E, 2001, 64(4): 045701.

[12] Shu C, Niu X D, Chew Y T. Taylor-series expansion and least-squares-based lattice Boltzmann method: Two-dimensional formulation and its applications[J]. Physical Review E, 2002, 65(3): 036708.

[13] Chew Y T, Shu C, Niu X D. Simulation of unsteady incompressible flows by using Taylor series expandsion-and least squares-based lattice Boltzmann method[J]. International Journal of Modern Physics C, 2002, 13(06): 719-738.

[14] Shu C, Peng Y, Chew Y T. Simulation of natural convection in a square cavity by Taylor series expandsion-and least squares-based lattice Boltzmann method[J]. International Journal of Modern Physics C, 2002, 13(10): 1399-1414.

[15] Shu C, Niu X D, Chew Y T. Taylor series expandsion and least squares-based lattice Boltzmann method: Three-dimensional formulation and its applications[J]. International Journal of Modern Physics C, 2003, 14(07): 925-944.

[16] 吴杰. 格子 Boltzmann 方法及其应用研究[D]. 南京:南京航空航天大学,2005.

[17] Shu C, Peng Y, Zhou C F, et al. Application of Taylor series expansion and Least-squares-based lattice Boltzmann method to simulate turbulent flows[J]. Journal of Turbulence, 2006, 7: N38.

[18] Reider M B, Sterling J D. Accuracy of discrete-velocity BGK models for the simulation of the incompressible Navier-Stokes equations[J]. Computers & Fluids, 1995, 24(4): 459-467.

[19] Mei R W, Shyy W. On the finite difference-based lattice Boltzmann method in curvilinear coordinates [J]. Journal of Computational Physics, 1998, 143(2): 426-448.

[20] Guo Z L, Zhao T S. Explicit finite-difference lattice Boltzmann method for curvilinear coordinates [J]. Physical Review E, 2003, 67(6): 066709.

[21] Tölke J, Krafczyk M, Schulz M, et al. Implicit discretization and nonuniform mesh refinement approaches for FD discretizations of LBGK Models[J]. International Journal of Modern Physics C, 1998, 09(08): 1143-1157.

[22] Wang Y, He Y L, Zhao T S, et al. Implicit-explicit finite-difference lattice Boltzmann method for compressible flows[J]. International Journal of Modern Physics C, 2007, 18(12): 1961-1983.

[23] Pareschi L, Russo G. Implicit-explicit Runge-Kutta schemes and applications to hyperbolic systems with relaxation[J]. Journal of Scientific Computing, 2005, 25(1): 129-155.

[24] 傅德薰,马延文. 计算流体力学[M]. 北京:高等教育出版社,2002.

[25] Kataoka T, Tsutahara M. Lattice Boltzmann model for the compressible Navier-Stokes equations with flexible specific-heat ratio[J]. Physical Review E, 2004, 69(3): 035701.

[26] Zhang H X. Non-oscillatory and non-free-parameter dissipation difference scheme[J]. Acta Aerodynamica Sinica, 1988, 6(2): 143-165.

[27] Jiang G S, Shu C W. Efficient implementation of weighted ENO schemes[J]. Journal of Computational Physics, 1996, 126(1): 202-228.

[28] Watari M, Tsutahara M. Two-dimensional thermal model of the finite-difference lattice Boltzmann method with high spatial isotropy[J]. Physical Review E, 2003, 67(3): 036306.

[29] Ghiroldi G P, Gibelli L. A finite-difference lattice Boltzmann approach for gas microflows [J]. Communications in Computational Physics, 2015, 17(4): 1007-1018.

[30] Sofonea V, Lamura A, Gonnella G, et al. Finite-difference lattice Boltzmann model with flux limiters for liquid-vapor systems[J]. Physical Review E, 2004, 70(4): 046702.

[31] Hejranfar K, Ezzatneshan E. Simulation of two-phase liquid-vapor flows using a high-order compact finite-difference lattice Boltzmann method[J]. Physical Review E, 2015, 92(5): 053305.

[32] Li Q, He Y L, Wang Y, et al. Coupled double-distribution-function lattice Boltzmann method for the compressible Navier-Stokes equations[J]. Physical Review E, 2007, 76(5): 056705.

[33] Li Q, He Y L, Wang Y, et al. Three-dimensional non-free-parameter lattice-Boltzmann model and its application to inviscid compressible flows[J]. Physics Letters A, 2009, 373(25): 2101-2108.

[34] Chen F, Xu A G, Zhang G C, et al. Multiple-relaxation-time lattice Boltzmann approach to compressible flows with flexible specific-heat ratio and Prandtl number[J]. Europhysics Letters, 2010, 90(5): 54003.

[35] He Y L, Liu Q, Li Q. Three-dimensional finite-difference lattice Boltzmann model and its application to inviscid compressible flows with shock waves[J]. Physica A: Statistical Mechanics and its Applications, 2013, 392(20): 4884-4896.

[36] Xu A G. Finite-difference lattice-Boltzmann methods for binary fluids[J]. Physical Review E, 2005, 71 (6): 066706.

[37] Xu H, Dang Z. Finite difference lattice Boltzmann model based on the two-fluid theory for multicomponent fluids[J]. Numerical Heat Transfer, Part B: Fundamentals, 2017, 72(3): 250-267.

[38] Sakthivel M, Anupindi K. Axisymmetric compact finite-difference lattice Boltzmann method for blood flow

simulations[J]. Physical Review E,2019,100(4):043307.

[39] Aghakhani S,Pordanjani A H,Karimipour A,et al. Numerical investigation of heat transfer in a power-law non-Newtonian fluid in a C-shaped cavity with magnetic field effect using finite difference lattice Boltzmann method[J]. Computers & Fluids,2018,176:51-67.

[40] Sofonea V,Sekerka R F. Viscosity of finite difference lattice Boltzmann models[J]. Journal of Computational Physics,2003,184(2):422-434.

[41] Nannelli F,Succi S. The lattice Boltzmann equation on irregular lattices[J]. Journal of Statistical Physics, 1992,68(3/4):401-407.

[42] Succi S. Lattice Boltzmann Equation for Fluid Dynamics and Beyond[M]. Oxford:Clarendon Press,2001.

[43] Amati G, Succi S, Benzi R. Turbulent channel flow simulations using a coarse-grained extension of the lattice Boltzmann method[J]. Fluid Dynamics Research,1997,19:289-302.

[44] Peng G W,Xi H W,Duncan C,et al. Lattice Boltzmann method on irregular meshes[J]. Physical Review E,1998,58(4):R4124-R4127.

[45] Peng G W,Xi H W,Duncan C,et al. Finite volume scheme for the lattice Boltzmann method on unstructured meshes[J]. Physical Review E,1999,59(4):4675-4682.

[46] Xi H W,Peng G W,Chou S H. Finite-volume lattice Boltzmann method[J]. Physical Review E,1999,59(5):6202-6205.

[47] Ubertini S, Bella G, Succi S. Lattice Boltzmann method on unstructured grids: Further developments [J]. Physical Review E,2003,68(1):016701.

[48] Ubertini S, Succi S, Bella G. Lattice Boltzmann schemes without coordinates[J]. Philosophical Transactions of the Royal Society of London Series A: Mathematical, Physical and Engineering Sciences, 2004,362(1821):1763-1771.

[49] Rossi N, Ubertini S, Bella G, et al. Unstructured lattice Boltzmann method in three dimensions[J]. International Journal for Numerical Methods in Fluids,2005,49(6):619-633.

[50] Ubertini S,Bella G,Succi S. Unstructured lattice Boltzmann equation with memory[J]. Mathematics and Computers in Simulation,2006,72(2):237-241.

[51] Lee T, Lin C L. A characteristic Galerkin method for discrete Boltzmann equation[J]. Journal of Computational Physics,2001,171(1):336-356.

[52] Lee T,Lin C L. An Eulerian description of the streaming process in the lattice Boltzmann equation[J]. Journal of Computational Physics,2003,185(2):445-471.

[53] Li Y S, LeBoeuf E J, Basu P K. Least-squares finite-element lattice Boltzmann method[J]. Physical Review E,2004,69(6):065701.

[54] Li Y S,LeBoeuf E J,Basu P K. Least-squares finite-element scheme for the lattice Boltzmann method on an unstructured mesh[J]. Physical Review E,2005,72(4):046711.

[55] Shi X,Lin J Z,Yu Z S. Discontinuous Galerkin spectral element lattice Boltzmann method on triangular element[J]. International Journal for Numerical Methods in Fluids,2003,42(11):1249-1261.

[56] Düster A,Demkowicz L,Rank E. High-order finite elements applied to the discrete Boltzmann equation [J]. International Journal for Numerical Methods in Engineering,2006,67(8):1094-1121.

[57] Xiao S P. A lattice Boltzmann method for shock wave propagation in solids[J]. Communications in Numerical Methods in Engineering,2007,23(1):71-84.

[58] Filippova O,Hänel D. Grid refinement for lattice-BGK models[J]. Journal of Computational Physics, 1998,147(1):219-228.

[59] Yu D Z,Mei R W,Shyy W. A multi-block lattice Boltzmann method for viscous fluid flows[J]. International Journal for Numerical Methods in Fluids,2002,39(2):99-120.

[60] 王广超,施保昌,邓滨. 非均匀格子 Boltzmann 方法模拟方柱绕流[J]. 应用基础与工程科学学报,2003,11(4):335-344.

[61] 王广超,施保昌,邓滨. 嵌套边界的非均匀格子 Boltzmann 方法[J]. 水动力学研究与进展,A 辑,2004,19(1):19-25.

[62] 陶文铨. 数值传热学[M]. 2 版. 西安:西安交通大学出版社,2001.

[63] Tölke J, Krafczyk M, Rank E. A multigrid-solver for the discrete Boltzmann equation[J]. Journal of Statistical Physics,2002,107(1):573-591.

[64] Mavriplis D J. Multigrid solution of the steady-state lattice Boltzmann equation[J]. Computers & Fluids, 2006,35:793-804.

[65] 王兴勇,索丽生,程永光,等. 双重网格的 Lattice Boltzmann 方法[J]. 河海大学学报,2003,31(1):5-10.

[66] Wang X Y, Suo L S, Cheng Y G, et al. Lattice Boltzmann method with double meshes[J]. Journal of Hydrodynamics, Ser B,2003,(1):90-96.

第 10 章 格子Boltzmann方法的应用

在前面几章,我们介绍了格子 Boltzmann 方法的基础理论、单相流动与传热模型、多相流模型、多孔介质流动与传热模型、边界处理及网格技术。至此,可以基于前面的知识将格子 Boltzmann 方法应用于多种物理问题的数值模拟研究。实际上,经过 30 余年的发展,格子 Boltzmann 方法已在能源动力、石油化工、机械交通等多个领域都有着广泛的应用。

在本章,将给出格子 Boltzmann 方法的几个应用案例,分别就格子 Boltzmann 方法模拟封闭方腔自然对流、Rayleigh-Bénard 对流、激波、声波衰减、池沸腾中的气泡成核与生长以及泡沫金属内相变材料熔化问题等展开叙述。这些应用案例仅是我们将格子 Boltzmann 方法用于工程热物理领域的部分实践,希望能起到抛砖引玉的作用。此外,为了方便读者尽快掌握格子 Boltzmann 方法,在附录 D 和附录 E 分别给出了顶盖驱动流和相分离的格子 Boltzmann 方法程序,并对程序的实施和模拟结果做了适当的说明。

10.1 封闭方腔自然对流

封闭方腔自然对流是计算流体力学与计算传热学中的经典算例。在这一问题中,速度场与温度场耦合,且流动情况比较复杂,会随着瑞利数的不同而出现不同的流动结构,因此常用来作为考核模型或算法的例子。

10.1.1 物理模型

封闭方腔自然对流的物理模型如图 10.1 所示。方腔的高和宽均为 H,上、下壁面绝热。腔内充满着 $Pr=0.71$ 的均质均温空气。在初始瞬间,方腔左壁被加热至 T_0,右壁被冷却至 $T_1(T_1<T_0)$,并保持恒定。

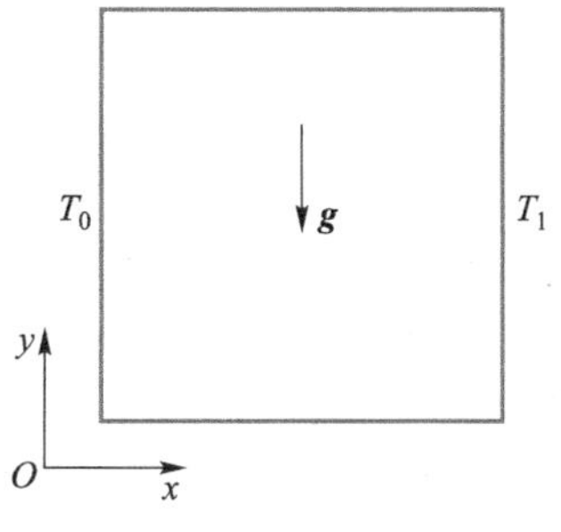

图 10.1 封闭方腔自然对流的物理模型

坐标系的原点取为方腔的左下角,并以水平方向为 x 方向,重力的反方向为 y 方向。流动的初始条件为 $\boldsymbol{u}=0$, $T_{\rm m}=(T_0+T_1)/2$;边界条件为

左壁:$\boldsymbol{u}=0, T=T_0$;

右壁:$\boldsymbol{u}=0, T=T_1$;

上壁和下壁:$\boldsymbol{u}=0, \partial T/\partial y=0$。

在许多流动中(如自然对流),常常可采用 Boussinesq 假设。该假设由三部分组成[1]:(1) 流动中的黏性热耗散忽略不计;(2) 除密度以外其他物性为常数;(3) 对密度仅考虑动量方程中与体积力有关的项，其余各项中的密度亦作常数处理，并且认为体积项中的密度具有如下形式：

$$\rho=\rho_0[1-\beta(T-T_{\mathrm{m}})] \tag{10.1}$$

式中：T_{m} 为参考温度；ρ_0 为与 T_{m} 相对应的流体密度；β 为体胀系数。

描述自然对流的两个最基本的量纲为 1 的参数是普朗特数和瑞利数，分别定义为

$$Pr=\frac{\nu}{\chi},\quad Ra=\frac{g\beta\Delta TH^3 Pr}{\nu^2} \tag{10.2}$$

式中：χ 为热扩散系数；ΔT 为高温壁和低温壁的温度差；g 为重力加速度。

瑞利数和格拉晓夫数

我们采用本书第 4 章提到的基于温度分布函数的热格子 Boltzmann 模型来模拟封闭方腔自然对流。由于体胀系数 β 是未知的，因此两个量纲为 1 的参数 Pr 数和 Ra 数通常并不足以确定松弛因子 τ_f 和 τ_T，此时必须利用另一个量纲为 1 的参数，即马赫数

$$Ma=\frac{u_{\mathrm{c}}}{c_{\mathrm{s}}} \tag{10.3}$$

式中：$u_{\mathrm{c}}=\sqrt{g\beta\Delta TH}$ 为自然对流的特征速度；$c_{\mathrm{s}}=c/\sqrt{3}$ 为格子声速。根据式(10.2)有[2]

$$Ma=\frac{\nu\sqrt{3Ra}}{cH\sqrt{Pr}} \tag{10.4}$$

式(10.4)表明，当 Pr 数和 Ra 数确定之后，马赫数与运动黏度成正比，与方腔尺寸 H 成反比。对格子 Boltzmann 方法，运动黏度为

$$\nu=\frac{1}{3}c^2\left(\tau_f-\frac{1}{2}\right)\delta_t \tag{10.5}$$

如果马赫数取定值，例如 $Ma=0.1$，则可通过如下式子计算松弛因子 τ_f：

$$\tau_f=\frac{1}{2}+\frac{MaH\sqrt{3Pr}}{c\delta_t\sqrt{Ra}} \tag{10.6}$$

而松弛因子 τ_T 则可以依据 Pr 数确定。

模拟中，平衡态温度分布函数由式(4.25)给出，流场采用如下形式的格子 Boltzmann 方程：

$$f_\alpha(\boldsymbol{r}+\boldsymbol{e}_\alpha\delta_t,t+\delta_t)-f_\alpha(\boldsymbol{r},t)=-\frac{1}{\tau_f}[f_\alpha(\boldsymbol{r},t)-f_\alpha^{\mathrm{eq}}(\boldsymbol{r},t)]+\delta_t\boldsymbol{G}\cdot\frac{(\boldsymbol{e}_\alpha-\boldsymbol{u})}{p}f_\alpha^{\mathrm{eq}} \tag{10.7}$$

式中，$\boldsymbol{G}=-\rho_0\beta(T-T_{\mathrm{m}})\boldsymbol{g}$。模拟中采用标准的迁移碰撞规则及格子单位，时间步长和网格步长均取单位 1，即 $\delta_t=\delta_x=\delta_y=1$，于是格子速度 $c=1$。

10.1.2 模拟结果及分析

对 $Ra=10^3\sim10^6$ 范围内的封闭方腔自然对流，文献中已有比较一致的数值计算结果[3-7]。其中，1983 年，De Vahl Davis[3]使用涡量-流函数法得到了一个基准解，使用的最细网格为 81×81。Hortmann 等[4]采用有限体积多重网格法提供了另一个基准解，最细的网格达到 640×640。为了与文献[4]的基准解进行对比，需测得如下数据：冷壁面的平均 Nusselt 数 Nu_{ave}，最大 Nusselt 数 $Nu_{\max}$ 及其位置 y_{Nu}；x 方向速度分量 u_x 在通过方腔中央的竖直线上的最大值 $u_{x,\max}$ 及其位置 $y_{\max}$；y 方向速度分量 u_y 在通过方腔中央的水平线上的最大值 $u_{y,\max}$ 及其位置 $x_{\max}$。

首先，需考核网格独立性以选取合适的网格大小。以 $Ra=10^4$ 为例，选取 64×64、

128×128、192×192 和 256×256 四种网格大小来考察网格对数值解的影响。从表 10.1 和图 10.2 可以看到,对 $Ra=10^4$ 来说,网格 192×192 下的数值解已与基准解吻合良好,而且当网格加密到 256×256 时,计算结果并没有大幅度的变化,因此可认为此时的结果即为 $Ra=10^4$ 的网格独立解。对 $Ra=10^3$、10^5 和10^6 做类似的分析可确定它们的计算网格分别为 128×128、256×256 和 256×256。

表 10.1 不同网格下的数值解与基准解的比较 ($Ra=10^4$)

物理量	64×64	128×128	192×192	256×256	文献[4]
$u_{x,\max}$	16.054 2	16.149 4	16.172 2	16.181 5	16.180 2
$y_{\max}$	0.828 1	0.820 3	0.822 9	0.824 2	0.826 5
$u_{y,\max}$	19.387 4	19.570 6	19.609 8	19.620 8	19.629 5
$x_{\max}$	0.125	0.117 2	0.119 8	0.117 2	0.119 3
$Nu_{\max}$	3.549 0	3.541 9	3.535 1	3.531 7	3.530 9
y_{Nu}	0.859 4	0.859 4	0.859 4	0.855 5	0.853 1
Nu_{ave}	2.243 8	2.246 4	2.245 2	2.244 3	2.244 8

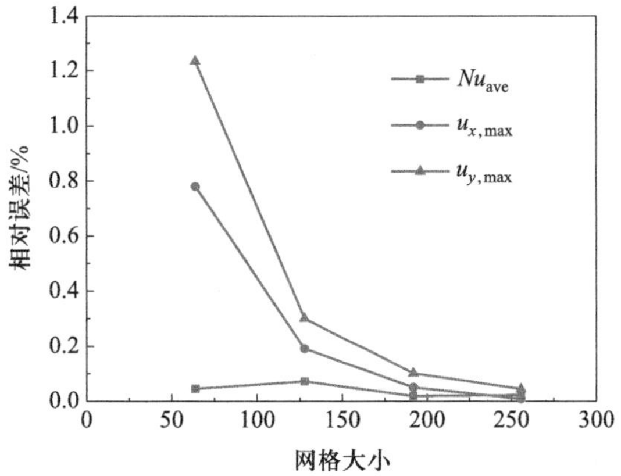

图 10.2 不同网格下的数值解的相对误差 ($Ra=10^4$)

图 10.3 和图 10.4 根据数值模拟的结果分别画出了不同瑞利数下流动稳定后的流线和等温线。事实上,封闭方腔自然对流是温差、压差和方腔有限空间限制三项因素综合作用的结果。从图 10.3 中可以看到,当 Ra 数较小时,流动的典型特点是在方腔中央出现一个近似圆形的涡。随着 Ra 数的增大,涡逐渐变成椭圆形。在 $Ra=10^5$ 时,由于浮升力的大幅度增大,竖壁边界层的作用也大幅度增强,因而涡的卷起作用更为强烈,在方腔中产生的旋涡偏离方腔中心,并分裂为两个涡。当 $Ra=10^6$ 时,这两个涡分别向左壁和右壁移动,并在方腔中央出现第三个涡。

从图 10.4 的等温线可以看出传热机制随 Ra 数增大的变化情况。当 Ra 数较小时,传热主要是由热壁和冷壁之间的热传导引起的,在这种情况下,等温线变化比较平缓。随着 Ra 数的增大,流场旋涡的椭圆化程度加大,从而方腔内的换热得到增强,传热机制逐渐由热传导占主导地位变为对流占主导地位。因此等温线在方腔中央逐渐变得水平,并且只在热壁和冷壁附近的薄边界层内保持竖直。以上观察到的现象与相关文献的报道是一致的。表 10.2 给出了数值解与基准解的定量对比,从中可以看到,绝大多数情况下的相对误差均小于 0.5%,部分数值解几乎与基准解完全一致。

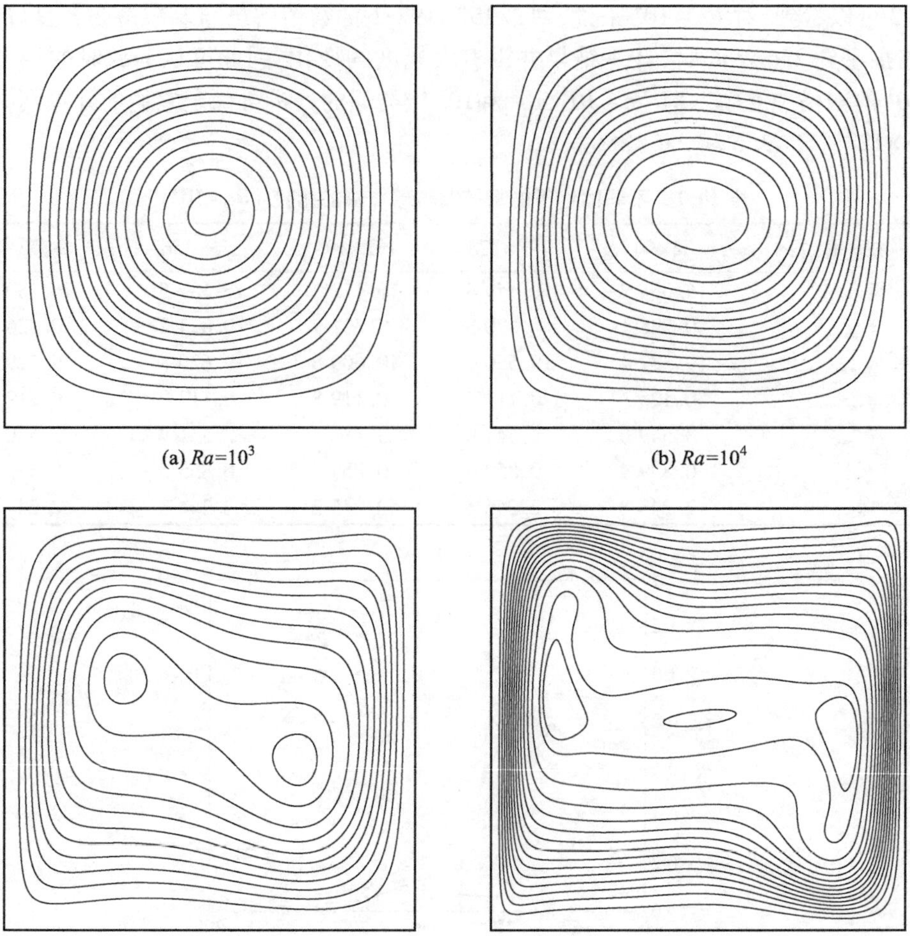

(a) $Ra=10^3$ (b) $Ra=10^4$

(c) $Ra=10^5$ (d) $Ra=10^6$

图 10.3 封闭方腔自然对流的流线

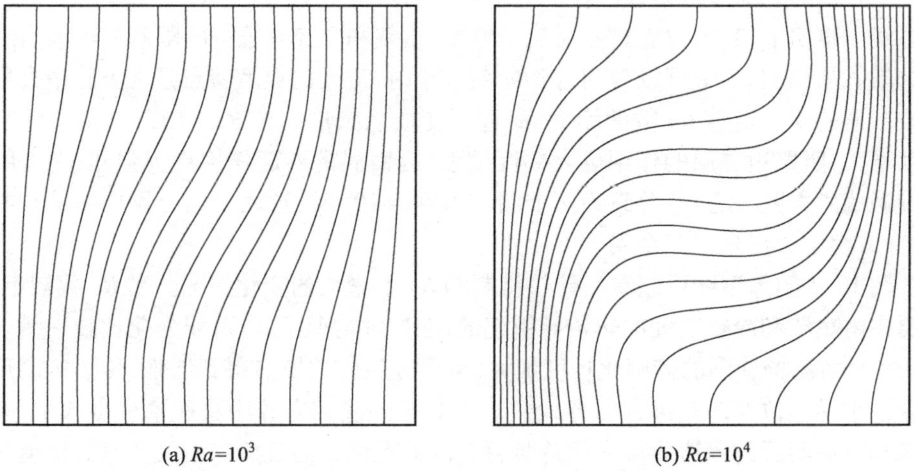

(a) $Ra=10^3$ (b) $Ra=10^4$

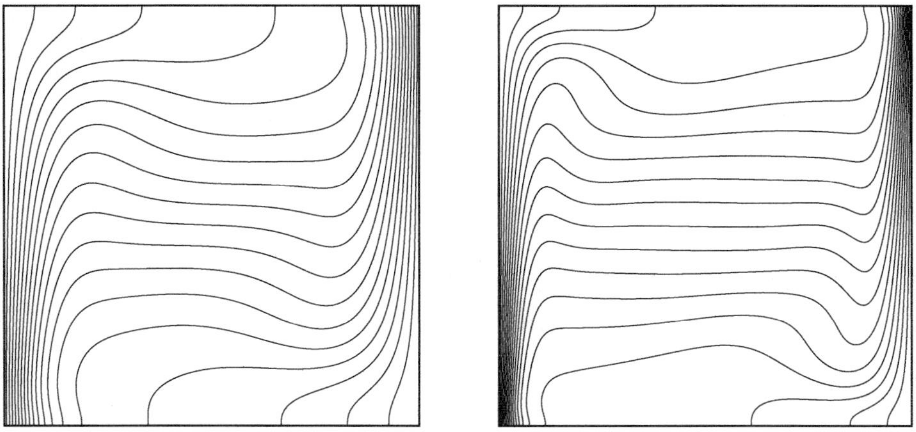

(c) $Ra=10^5$　　(d) $Ra=10^6$

图 10.4 封闭方腔自然对流的等温线

表 10.2 数值解与基准解的比较

Ra	比较	$u_{x,\max}$	$y_{\max}$	$u_{y,\max}$	$x_{\max}$	$Nu_{\max}$	y_{Nu}	Nu_{ave}
10^3	计算结果	3.649 3	0.812 5	3.701 0	0.179 7	1.501 7	0.906 3	1.116 9
	文献[3]	3.649	0.813	3.697	0.178	–	–	–
	相对误差/%	0.01	0.06	0.11	0.96	–	–	–
10^4	计算结果	16.172 2	0.822 9	19.609 8	0.119 8	3.535 1	0.859 4	2.245 2
	文献[4]	16.180 2	0.826 5	19.629 5	0.119 3	3.530 9	0.853 1	2.244 8
	相对误差/%	0.05	0.44	0.10	0.42	0.12	0.74	0.02
10^5	计算结果	34.736 2	0.855 5	68.495 4	0.066 4	7.750 0	0.921 9	4.521 9
	文献[4]	34.739 9	0.855 8	68.639 6	0.065 7	7.720 1	0.918 0	4.521 6
	相对误差/%	0.01	0.04	0.21	1.06	0.39	0.42	0.01
10^6	计算结果	64.868 7	0.851 6	219.334	0.039 1	17.614 0	0.964 8	8.792 6
	文献[4]	64.836 7	0.850 5	220.461	0.039 0	17.536	0.960 8	8.825 1
	相对误差/%	0.05	0.13	0.51	0.26	0.44	0.42	0.37

相关研究表明，在封闭方腔自然对流中，平均 Nu 数和 Ra 数之间存在着一定的关系，这一关系可以用一个幂律公式来描述：

$$Nu_{ave}=a\ (Ra)^b \tag{10.8}$$

式中，a 和 b 为常数。根据数值模拟结果，可得 $a=0.142\ 2$，$b=0.299\ 2$，这与文献[5]的结果（$a=0.142$，$b=0.299$）是一致的。

10.2 Rayleigh-Bénard 对流

从理论上讲，格子 Boltzmann 方法可以模拟从介观到宏观的一系列流动与传热问题。但对于宏观大尺度下（如高雷诺数、高瑞利数）的流动与传热问题，在程序稳定的情况下，标准格子 Boltzmann 方法需要大量的计算网格，由此，对计算资源提出了很高的要求。不

大涡模拟与亚格子

少学者致力于这方面的研究,例如杨帆等[8]在格子 Boltzmann 方法中引入 Smagorinsky 亚格子概念,并通过模拟高雷诺数下的顶盖驱动流证明了方法的有效性。我们也曾采用插值格子 Boltzmann 方法(参见 9.2 节),以二维 **Rayleigh-Bénard 对流**为例,验证其对高瑞利数问题模拟的可靠性[9]。在较大的瑞利数范围内,模拟结果与相关文献的结果吻合得很好。

10.2.1 物理模型

Rayleigh-Bénard 对流常作为不可压缩热格子 Boltzmann 模型的验证算例,其物理模型如图 10.5 所示,流体位于 x 方向上无限长的狭窄通道内,通道上壁面温度为 T_1,下壁面温度为 $T_0(T_0>T_1)$,通道高为 L_y。瑞利数定义为 $Ra=\beta(T_0-T_1)gL_y^3/(\nu\chi)$,其中 β 为体胀系数、g 为重力加速度、ν 和 χ 分别为运动黏度和热扩散系数。线性稳定性分析表明,在 Boussinesq 假设下的流体运动中,当流动 Ra 数大于临界值 $Ra_c=1\ 707.76$ 时,流动将丧失稳定性,传热机制将由导热转变为对流换热[10]。

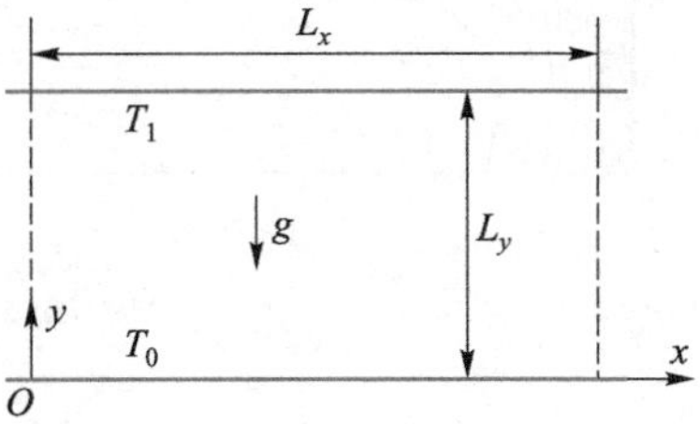

图 10.5 Rayleigh-Bénard 对流的物理模型

模拟中所使用的模型为 He 等的双分布函数热模型(见第 4 章)。考虑均匀正交网格,取 $L_x\times L_y=2L_y\times L_y=(rN_x)\times(rN_y)$,其中 N_x 和 N_y 分别为 x 和 y 方向上的网格数、r 为插值比。x 方向上采用周期性边界条件,y 方向上采用非平衡态外推格式。取 $Pr=0.71$,外力项 $\boldsymbol{G}=-\rho_0\beta(T-T_m)\boldsymbol{g}$,平均温度 $T_m=(T_0+T_1)/2$。此外,定义特征速度 $u_c=\sqrt{gL_y}$、特征时间 $t_c=L_y/v_c=\sqrt{L_y/g}$。模拟中,网格数固定为 $N_x\times N_y=200\times100$,通过改变 r 的大小来改变模拟区域的大小,各参数均采用格子单位。

10.2.2 模拟结果及分析

初始时刻,模拟区域内均给定一固定的温度和压强的小扰动[11]。此外,当 $Ra<10\ 000$ 时,r 取为 1,此时插值格子 Boltzmann 方法退化为标准的格子 Boltzmann 方法;当 $Ra\geqslant 10\ 000$ 时,r 随 Ra 数的增大而增大。具体地,当 $Ra=10\ 000$、$20\ 000$ 时,r 取为 2;当 $Ra=50\ 000$ 时,r 取为 4;当 $Ra=100\ 000$ 时,r 取为 8。

图 10.6 给出了不同 Ra 数下,模拟区域内 y 方向最大速度分量 $u_{y,\max}$ 随时间的变化规律。由图可知,当 Ra 数小于 Ra_c 时($Ra=1\ 000$),黏性力和扩散作用抑制了系统内的小扰动,$u_{y,\max}$ 保持零不变,系统处于静止状态;当 Ra 数大于 Ra_c 时($Ra=2\ 000\sim100\ 000$),初始时刻的扰动得到放大,$u_{y,\max}$ 迅速增大并最终保持为一个恒定值。当 $u_{y,\max}$ 保持不变时,系统达到稳定状态。此外,对比不同 Ra 数下的模拟结果可以发现,Ra 数越大,$u_{y,\max}$ 最终达到的恒定值也越大,即系统内的对流越强烈。

图 10.7 为不同 Ra 数下系统稳定后流线和等温线的分布情况。由图 10.7a 可见,当 Ra 数超过 Ra_c 时,上下壁面温差引起的浮升力的作用超过了黏性力和扩散作用,系统丧失稳定性,出现对流流动,y 方向上的温度不再呈线性分布,开始有涡的产生,且此时涡的形状近似为圆形。随着 Ra 数的增大(图 10.7b、c),壁面附近的温度变化剧烈,梯度逐渐增大;等温线分布变得复杂,局部区域出现冷热流体混合,涡的形状也逐渐变为扁平。

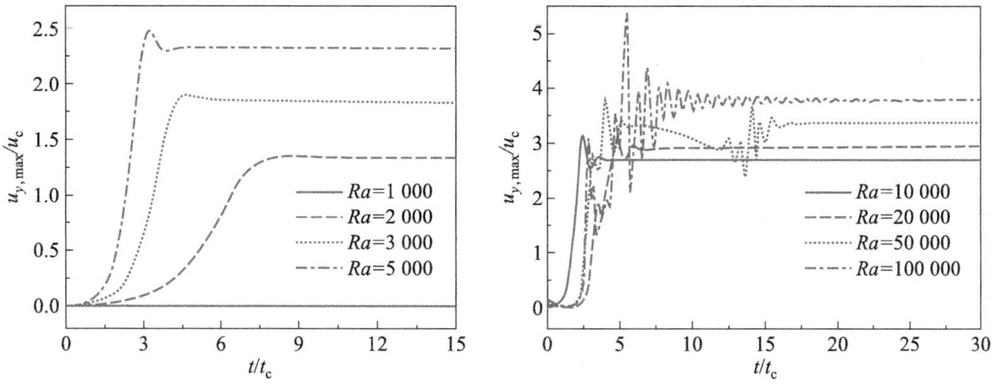

(a) $r=1$　　(b) $r>1$

图 10.6　y 方向最大速度分量随时间的变化规律

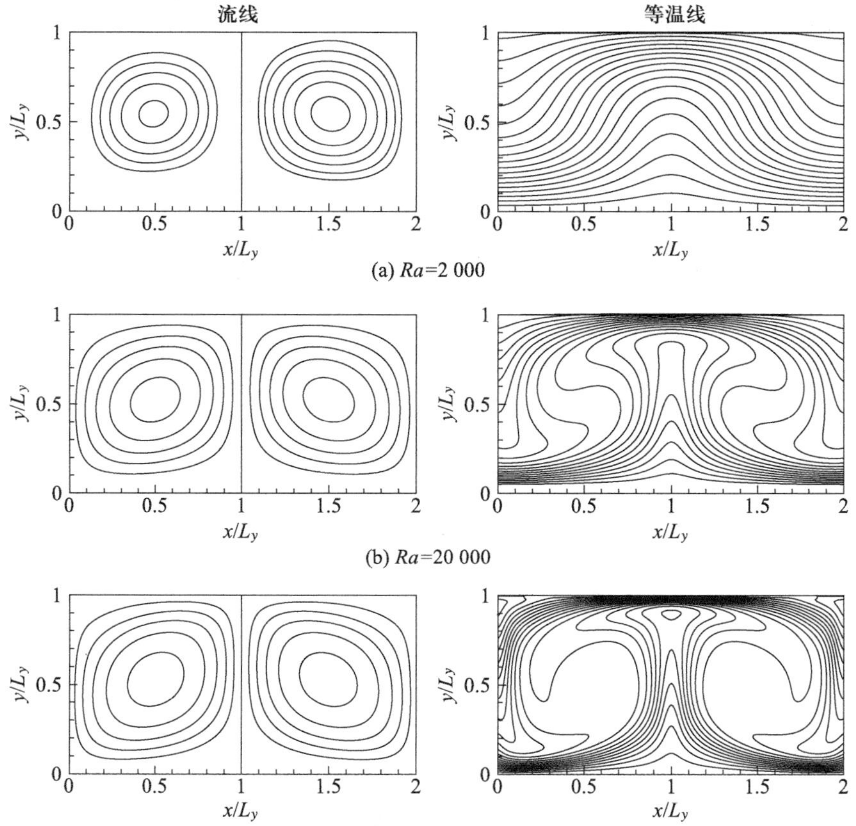

(a) $Ra=2\ 000$

(b) $Ra=20\ 000$

(c) $Ra=100\ 000$

图 10.7　不同 Ra 数下的流线和等温线分布

当系统达到稳定后，统计相应的 Nu 数，作为程序正确性的定量考核。Nu 数的定义为[11]

$$Nu=1+\frac{\langle u_y T\rangle L_y}{a(T_0-T_1)}$$

式中，$\langle\ \ \rangle$ 表示系统平均。

图 10.8 给出了 Nu 数随 Ra 数的变化规律，并与文献[12]采用 Galerkin 方法以及文献

[13]采用有限差分法所获得的结果做了对比。由图可见，当 Ra 数较小（$Ra \leqslant$ 2 000）时，结果介于文献[12]与[13]所获得的数据之间。总的说来，我们的模拟结果与文献吻合良好。

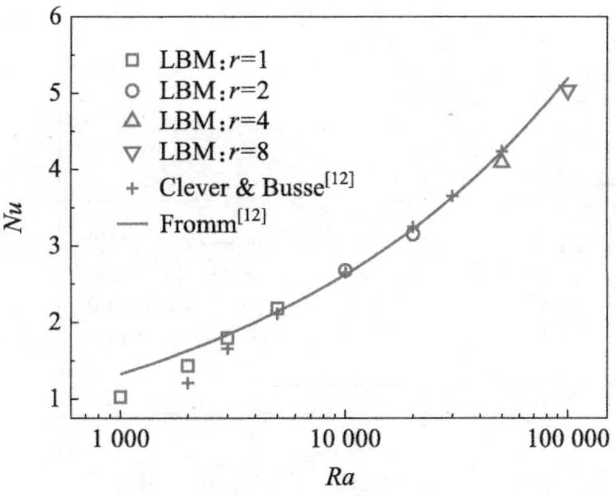

图 10.8　Nu 数随 Ra 数的变化规律

进一步，我们统计了流体的水平截面平均温度 $T_a(y)$。由图 10.9 可见，在下壁面附近，流体的温度梯度随 Ra 数的增大而增大，也即边界层厚度随着 Ra 数的增大而减小。因此，为了更详尽地分析壁面附近的物理场景，需要在壁面附近划分较细的网格。此外，当 Ra 数继续增大时，流体的运动呈现出三维特性，并逐渐向湍流过渡。

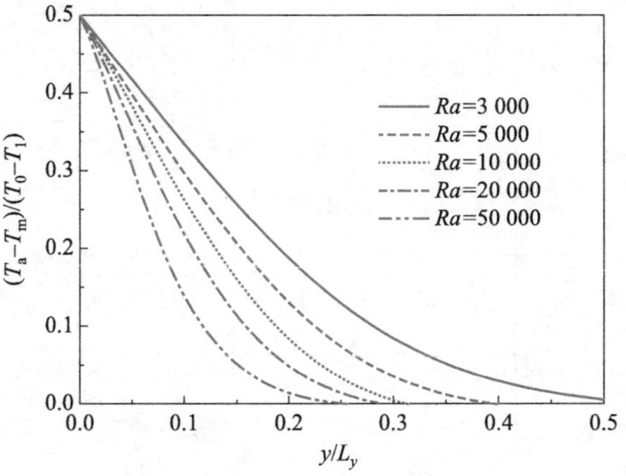

图 10.9　不同 Ra 数下截面平均温度随空间位置的变化

10.3　激 波 模 拟

气体中的微弱扰动是以当地声速向四周传播的，当飞行器以亚声速飞行时，扰动的传播速度比飞行器飞行速度大，所以扰动集中不起来，此时整个流场上流动参数（包括流速、压强等）的分布是连续的。而当飞行器以超声速飞行时，扰动来不及传到飞行器的前面去，结果前面的气体受到飞行器突跃式的压缩，形成集中的强扰动，这时会出现一个压缩过程的界面，就是**激波**。经过激波，气体的压强、密度和温度都会突然升高，流速则突然下降。压强的跃升会产生爆响：飞机在较低的空域中超声速飞行时，地面上的人就可以听见这种响声，即声爆。

模拟激波，首先要考虑流体的可压缩性，其次要考虑到流动是超声速的。反映在格子 Boltzmann 方法中，则是要采用适用于高马赫数可压缩流动的格子 Boltzmann 模型。在文献[14]中，我们利用耦合双分布函数格子 Boltzmann 模型（参见 4.5 节）对 **Riemann 问题**和**双马赫反射**（double Mach reflection）问题进行了数值模拟研究。这两类问题都属于无黏可压缩流动，因而在相应的模拟中，耦合双分布函数模型中的黏性项被当作误差来处理。从数值模拟的结果来看,这样处理对计算精度的影响不大。

10.3.1 Riemann 问题模型

Riemann 问题是计算流体力学的核心问题之一[15]。在计算流体力学和计算空气动力学中,几乎所有的高分辨率格式都是建立在求解 Riemann 问题的基础上。这是因为它不但包含了古典的光滑解,更是因为它是针对流体力学尤其是空气动力学中的各种间断问题,而这些间断问题的求解一直是计算流体力学发展的难点和核心问题。对复杂间断问题的求解,比光滑问题要难得多。

Riemann 的人物小传

1965 年,Glimm 提出了著名的 Riemann 问题——间断分解问题,Riemann 问题的命名是为了纪念德国数学家 Bernhard Riemann,因他首次尝试给出了该问题的解。后来,经过对 Riemann 问题的不断研究,不但得到了激波管非定常一维 Euler 方程的解析解,而且它还逐渐被称为现代计算流体力学的基石。

Riemann 问题中最典型的就是激波管问题。假设在一条均匀的直管中间有一个薄片,如图 10.10a 所示,薄片两边的气体参数是两种不同的常数,如果突然撤去薄片,管内所发展出的流态非常复杂,这类问题就是激波管问题。除了激波管问题外,Riemann 问题的实例还有很多,如两平面激波相对正面碰撞、平面激波冲击静止物体等[15]。

如图 10.10b 所示,设激波管在 $t=0$ 时刻,管内速度为零,隔膜左边为高压 p_4,右边为低压 p_1。若突然撤去隔膜,初始的压力间断将以非定常正激波的形式向右传播,其波速为 u_s,同时,一个等熵非定常中心稀疏波向左传播。如图 10.10c 所示,此时,管内流体被分为四个区,1 区和 4 区是未受扰动的区域,也就是波还没有传播到的区域;2 区是激波传播后的区域,其压力为正激波后压力 p_2;3 区是中心稀疏波传播后的区域,该区的压力 $p_3=p_2$,因为在 2 区和 3 区气体不能支持任何压力间断。2 区和 3 区的速度也相同,但由于 2 区是激波传播后的区,而 3 区是中心稀疏波传播后的区,因此,2 区和 3 区的熵、温度和密度是不同的,所以 2 区和 3 区以一个接触间断分开[15]。

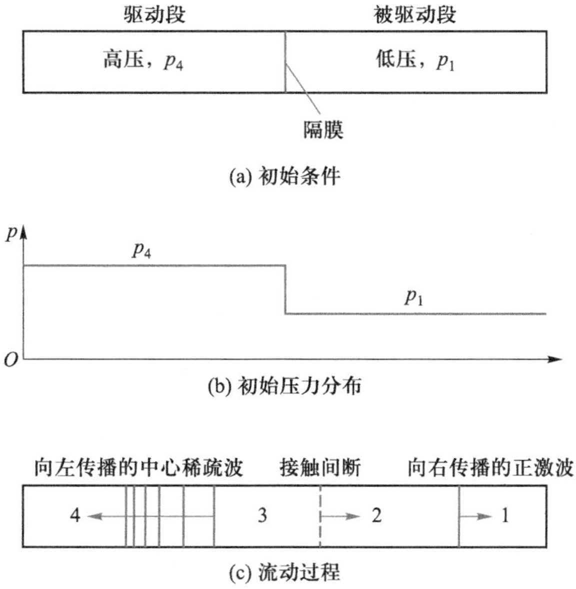

图 10.10 激波管流动示意图

激波管问题的精确解的推导可参考文献[16]的附录 B。

10.3.2 Riemann 问题的物理模型及模拟结果

Sod 问题,初始条件为

$$\begin{cases}(\rho/\rho_0,u_x/u_0,p/p_0)=(1,0,1), & 0\leqslant x/L_0\leqslant 1/2\\(\rho/\rho_0,u_x/u_0,p/p_0)=(0.125,0,0.1), & 1/2<x/L_0\leqslant 1\end{cases}\tag{10.9}$$

式中:$L_0=2$ m 为激波管的长度;ρ_0、$u_0=\sqrt{R_gT_0}$、$p_0=\rho_0R_gT_0$ 和 T_0 分别为参考密度、参考速度、参考压强和参考温度,取 $t_0=L_0/u_0$ 为系统的特征时间。对常温常压下的空气,有 $\rho_0=1.165\ \mathrm{kg/m^3}$,$R_g=287\ \mathrm{J/(kg\cdot K)}$,$T_0=303\ K$。与前面的算例不同,本算例采用实际物理单位。

Sod 问题是著名的一维激波管非定常流动,它包含了接触间断、激波、膨胀扇区等典型的流动现象。对含有激波的可压缩问题,为了捕捉激波或强间断,我们基于耦合双分布函数模型(参见 4.5 节),采用了有限差分格子 Boltzmann 方法(参见 9.4 节)。模拟中用到的时间离散格式为隐式-显式 Runge-Kutta 格式,空间离散格式为 WENO 格式,详细过程可参考文献[17-19]。

计算工质为空气,比热容比为 $\gamma=1.4$,普朗特数为 $Pr=0.71$。在这一算例中,我们选取的特征温度为 $T_c=2T_0$,时间步长 $\delta_t=30\,000\mu/p_0$,其中 $\mu=1.86\times10^{-5}\ \mathrm{kg/(m\cdot s)}$ 为动力黏度。取 $N_x\times N_y=400\times5$ 的均分网格,相应的网格步长为 $\delta_x=\delta_y=L_0/400$。$x$ 方向边界节点的宏观变量已知且恒定,因此密度和总能分布函数取其平衡态;在 y 方向上,为了用二维模型模拟一维物理问题,需采用周期性边界条件。

图 10.11 给出了基于圆函数分布的耦合双分布函数模型对 Sod 问题的数值模拟结果。实线为精确解,实心圆点代表模拟结果。从图中可以看到,除温度分布在间断处与精确解有稍许偏差外,其他宏观量的计算结果几乎与精确解完全吻合。

强激波问题,初始条件为

$$\begin{cases}(\rho/\rho_0,u_x/u_0,p/p_0)=(1,0,1\,000), & 0\leqslant x/L_0\leqslant 1/2\\(\rho/\rho_0,u_x/u_0,p/p_0)=(1,0,0.01), & 1/2<x/L_0\leqslant 1\end{cases}\tag{10.10}$$

可以看到,激波管隔膜两边的压力比为10^5,远大于 Sod 问题中的压力比。在这一问题中,将产生一个强激波。根据第 4 章介绍的特征温度的确定原则,在计算中,我们选取 $T_c=1\,000T_0$,其余参数与 Sod 问题的参数设置一样,数值结果如图 10.12 所示。实线为精确解,实心圆点代表模拟结果。整体看来,数值解与精确解吻合良好。不过在激波传播后的区域里,密度和温度的偏差比较明显。在计算流体力学中,这些偏差通常与空间离散格式有关。单从格子 Boltzmann 方法的角度来看,这一结果是令人满意的。

10.3.3 双马赫反射问题的物理模型及模拟结果

双马赫反射问题是描述一个强激波入射在与平面成30°角的斜坡上所发生的变化,来流是马赫数为 10 的强激波,它通常用来考核模型或算法对二维强激波可压缩流动的适应性。为了使问题尽可能简单,通常将入射激波倾斜,其物理模型如图 10.13 所示[14]。

计算区域为 3 m×1 m,反射面位于区域底面 $1/6<x\leqslant3$。在格子 Boltzmann 方法中,为了实施反射边界条件,可在反射面下附加两层虚拟格子,虚拟格子上的分布函数可根据镜

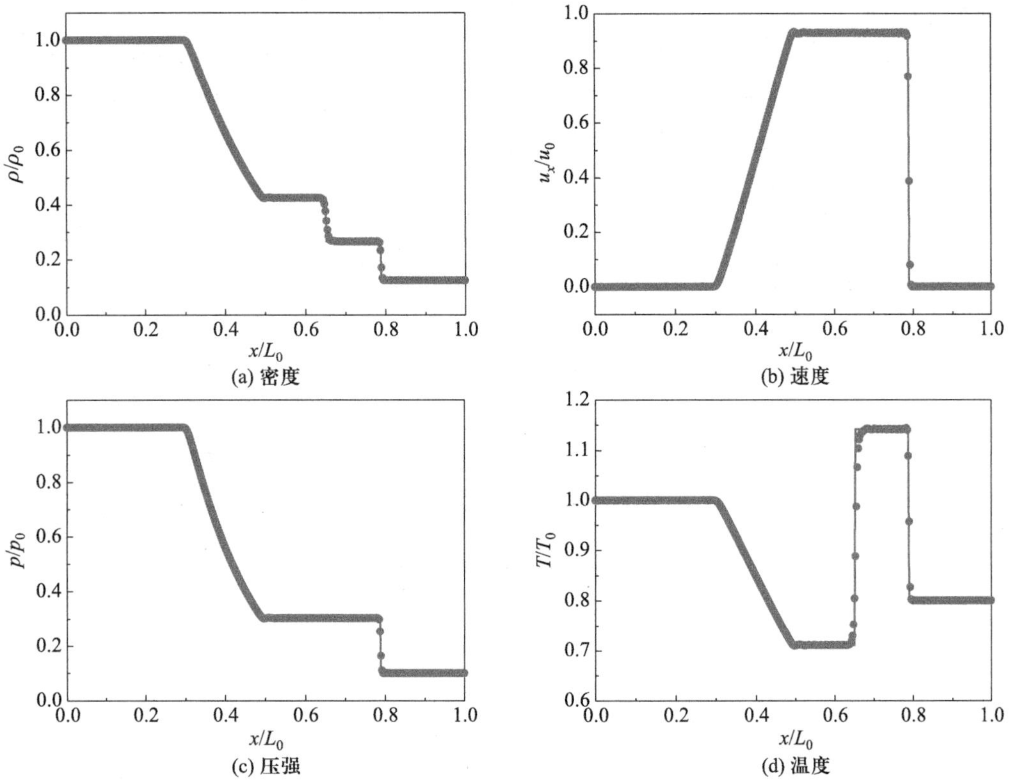

图 10.11　Sod 问题 $t=0.1644t_0$ 时刻数值解与理论解的对比

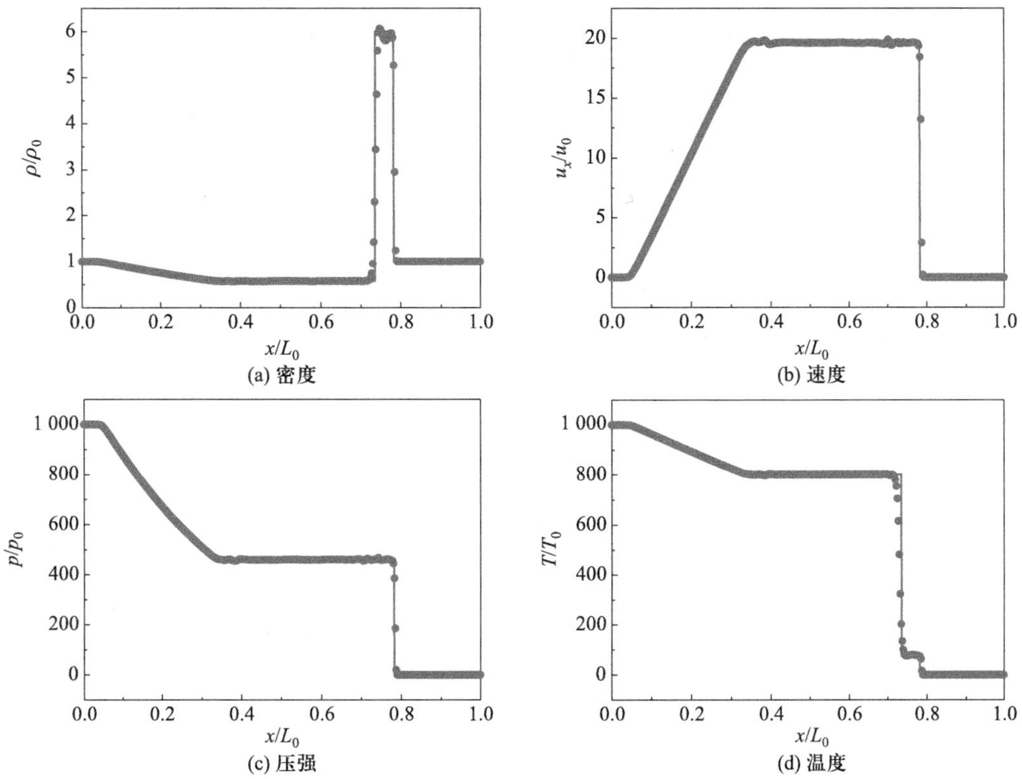

图 10.12　强激波问题 $t=0.012t_0$ 时刻数值解与理论解的对比

面边界条件确定。在底边 $0 \leqslant x \leqslant 1/6$，宏观参数为波后精确值（即来流参数）。上边界为马赫数为 10 的激波精确解，即以激波速度移动，波头左边赋以波后条件，波头右边赋以波前条件。对这些宏观量确定的边界，边界节点上的分布函数可直接赋以平衡态分布，即 $f_\alpha = f_\alpha^{eq}$ 及 $h_\alpha = h_\alpha^{eq}$。出口边界则可采用外推格式。模拟中，网格为 $N_x \times N_y = 360 \times 140$，$\gamma = 1.4$，$Pr = 0.71$，$T_c = 85T_0$，$\delta_t = 4\ 000\mu/p_0$。采用的时间离散格式为二阶显式隐式 Runge-Kutta 格式，采用的空间离散格式参见文献[17,18]。图 10.14 给出了 $t = 860\delta_t$ 时刻的等密度线和等压强线，这一结果与相关文献的实验和数值模拟结果吻合良好[19-20]。

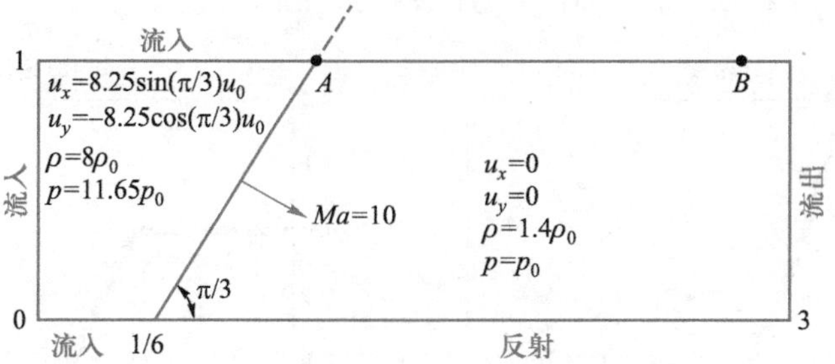

图 10.13 双马赫反射问题的物理模型

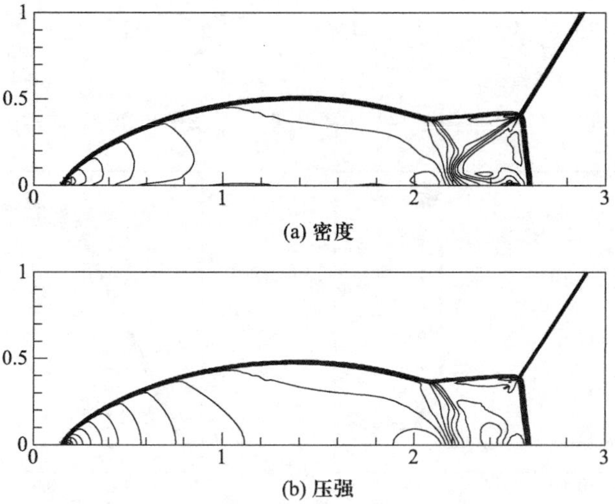

图 10.14 双马赫反射密度和压强等值线（$t = 860\delta_t$）

双马赫反射的 LBM 模拟

10.4 声波衰减

热致声自激振荡过程是热声发动机中典型的非线性现象，也是热声研究的难点[21-22]。起振是声波发展的物理过程，与之相反的则是声波衰减。通过对声波衰减的研究，有利于从相反角度探讨影响声波起振以及振动维持的因素，为开展热声现象的格子 Boltzmann 方法研究提供重要依据。在不考虑热问题的情况下，平面波的衰减主要是由黏性吸收引起的。

在声波模拟方面，Buick 等[23]采用 D2Q7 模型对声波衰减进行了数值研究，给出了声场介质的速度随时间的衰减情况，并证明模拟获得的相关频率及衰减系数与理论值吻合良好。Haydock 和 Yeomans[24]则采用格子 Boltzmann 方法模拟了行波声场中由于声波衰减而引起的声流,给出了一维情况下声场介质的速度分布以及压力梯度分布,但模拟中采用的入口边界条件需要对边界节点的密度分布函数进行额外计算,操作较繁复且物理意义不够明确。在文献[25]中,我们采用 D2Q9 模型分别模拟了一维及二维情况下平面波的衰减过程。模拟中采用了一种更为简单可行的边界条件,即入口处声源给定正弦变化的速度与密度扰动。模拟结果和理论分析解进行了比较,吻合得很好,证明了所开发的声波模拟程序的正确性。

10.4.1 物理模型

考虑平面声波在通道内沿 x 方向传播,声源位于 $x=0$ 处,通道的长和高分别为 L_x 和 L_y,如图 10.5 所示。声源的速度为 $u_{\text{in}}=U\sin(\omega t)$,其中 U 为振幅;$\omega=2\pi c_s/\Lambda$ 为角速度,Λ 为波长。相应地,声源处的密度为[24]

$$\rho=\rho_0+\delta_\rho\sin(\omega t)=\rho_0\left[1+\frac{U\sin(\omega t)}{c_s}\right] \tag{10.11}$$

式中:ρ_0 为平均密度;δ_ρ 为密度振幅。

模拟中采用 D2Q9 模型,均匀网格,格子数为 $N_x\times N_y=L_x\times L_y$。此外,$\rho_0=1.0$,$\delta_\rho=0.01$,则 $U=c_s\delta_\rho/\rho_0=0.005\ 77$;使用标准格子 Boltzmann 方法的碰撞迁移规则和格子单位。

10.4.2 模拟结果及分析

1. 一维声波衰减

对于一维声波衰减,y 方向上采用周期性边界条件。同时,x 方向上足够大,不考虑声波的反射,进口直接给定介质的速度及密度,出口则采用充分发展边界条件,网格数为 $N_x\times N_y=10\ 000\times 20$。速度及压力梯度的理论分析解为[26]

$$u_x(x,t)=U\exp(-a_s x)\sin\left(\omega t-\frac{2\pi}{\Lambda}x\right),\quad p'=\rho_0 a_s U^2\exp(-2a_s x) \tag{10.12}$$

式中,a_s 为单位长度上的声波衰减系数,定义为[23]

$$a_s=\frac{4\pi^2\nu}{c_s\Lambda^2} \tag{10.13}$$

模拟结果表明,随着声波的传播,声场中的介质在其平衡位置附近做简谐振动并表现出疏密相间,振动的方向与声波前进的方向一致,且沿声波方向密度振幅逐渐减小。图 10.15a 和图 10.15b 分别为 $\nu=0.1$、$\Lambda=50$ 和 100 时通道中心线上介质的瞬时速度分布。由图 10.15 可知,在介质黏性的作用下,声波沿着传播方向逐渐衰减,表现为速度振幅越来越小。且随着波长的增大,声波的衰减减缓。不同波长下模拟获得的速度分布与解析解均十分吻合。

由于黏性作用,声波速度振幅沿着传播方向逐渐减小,声场中存在非线性压力梯度。图 10.16 给出了 $\nu=0.1$、$\Lambda=50$ 和 100 时通道中心线上的压力梯度。模拟值经过了 100 个周期的时间平均。由图可知,沿声波传播方向,压力梯度呈负指数形式减小。且波长越大,声波衰减越慢(参见图 10.15),对应压力梯度越小,压力梯度减小也越慢。两种波长情况下,我们的模拟结果和理论值吻合得很好。

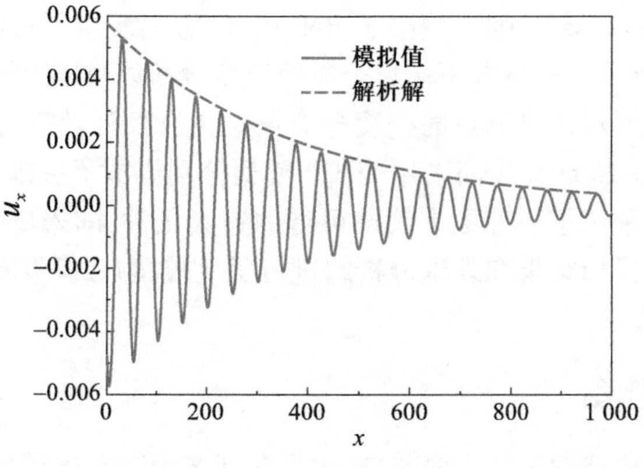

(a) Λ=50

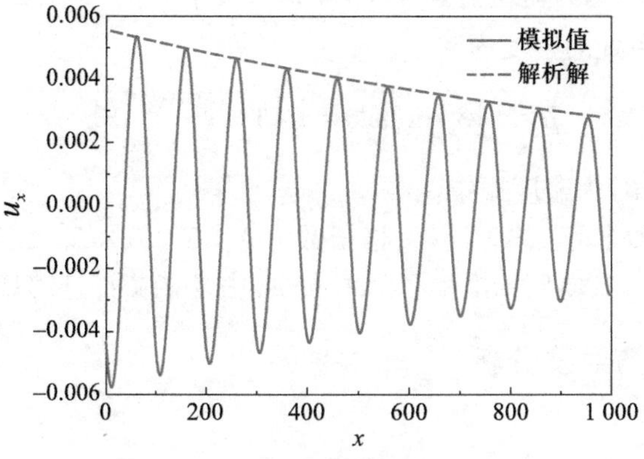

(b) Λ=100

图 10.15 一维声波衰减,两种波长下的瞬时速度分布

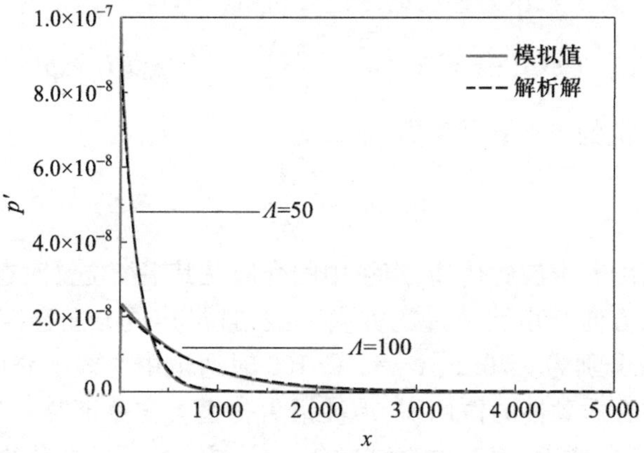

图 10.16 一维声波衰减,两种波长下的二阶压力梯度分布

2. 二维声波衰减

对于二维声波衰减,通道壁面上采用反滑移格式,模拟固体壁面对声场的影响。y 方向上宽度的选取需要满足 $\nu\Lambda/(2\pi c_s L_y^2) \ll 1$[23]。与一维问题相同,$x$ 方向上足够大,不考虑声波的反射,进口给定介质的速度及密度,出口采用充分发展边界条件。具体地,取 $N_x \times N_y = 3\ 000 \times 500$。由于受到壁面的摩擦,声波会在壁面附近形成厚度为 $(\nu\Lambda/(\pi c_s))^{1/2}$ 的边界层[26],而在远离壁面的通道中心区域,介质的速度以及压力梯度仍可采用式

(10.12)。与一维问题相比,二维物理模型下的声波衰减系数,除了要考虑介质黏性以外,还需要添加壁面的影响,定义如下[23]:

$$a_s=\frac{4\pi^2\nu}{c_s\Lambda^2}+\pi\left\{\frac{8\nu\Lambda}{6\pi c_s L_y^2}\cos^2\left\{\frac{1}{3}\cos^{-1}\left[-\left(\frac{27\nu\Lambda}{16\pi c_s L_y^2}\right)^{1/2}\right]\right\}\right\}^{1/2} \tag{10.14}$$

图 10.17 给出了 $\nu=0.2$、$\Lambda=25$ 和 50 时通道中心线上介质的瞬时速度分布。图 10.18 给出了 $\nu=0.2$、$\Lambda=25$ 和 50 时通道中心线上介质的压力梯度分布。分别对 100、200 以及 400 个周期内通道中心线上的压力梯度做了时间平均,确保结果不再随时间变化。由图可知,对于速度分布以及压力梯度,模拟获得的结果与理论值吻合良好。与一维物理模型类似,在二维问题中,波长越长,声波衰减越慢,对应的压力梯度越小,压力梯度减小也越慢。此外,当波长较大($\Lambda=50$)时,壁面影响加强,模拟结果与理论值略有偏差。

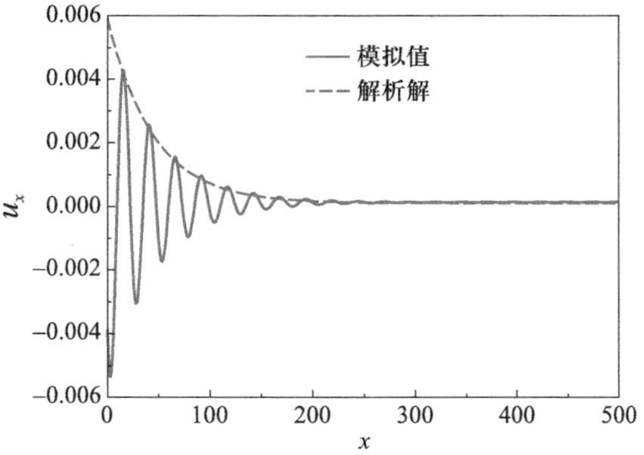

(a) Λ=25

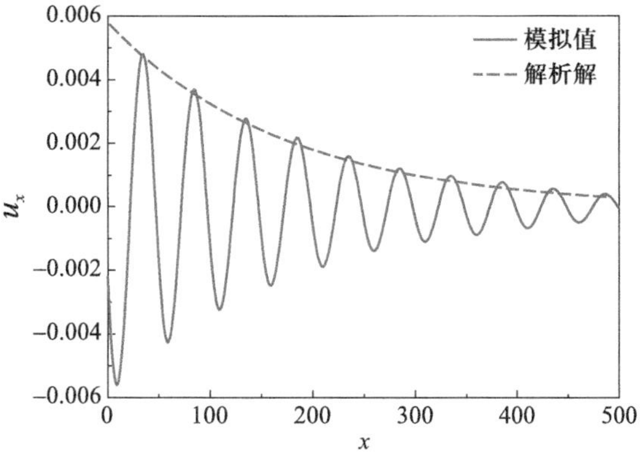

(b) Λ=50

图 10.17 二维声波衰减,两种波长下的瞬时速度分布

图 10.19 给出了二维声波衰减系数随黏度以及波长的变化规律。由图可知,在黏度一定的情况下,模拟获得的衰减系数随着波长的增大而减小,这与前面的结论是相吻合的。此外,在波长一定的情况下,模拟获得的衰减系数则随着黏度的增大而增大。在波长较小(频率较大)且黏度较大的时候,由于声波衰减非常剧烈,模拟结果和理论值稍有偏差。而多数情况下,模拟获得的衰减系数与理论值吻合得很好。

最后,计算声源处的激发声压级。定义声压级 $L_p=20\lg(p/p_{\mathrm{ref}})$,其中 p 为声波压力,$p_{\mathrm{ref}}=2\times10^{-5}$ Pa 为参考声压。考虑大气压力为 1×10^5 Pa,模拟中给定声源处声波压力的相

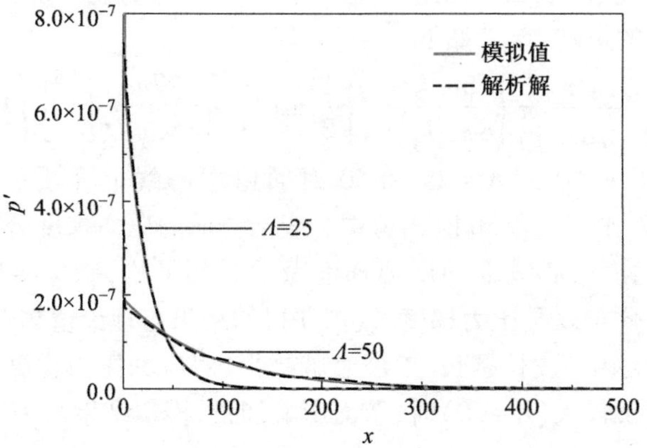

图 10.18 二维声波衰减,两种波长下的二阶压力梯度分布

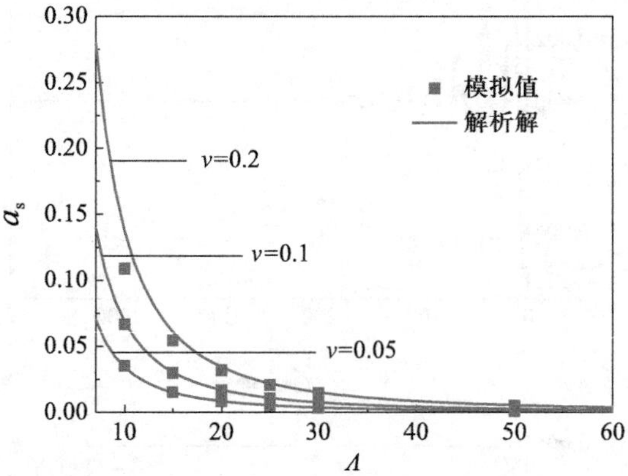

图 10.19 二维声波衰减,衰减系数随波长及黏度的变化

对变化为 1%,则声源处声波压力为 1 000 Pa,相应的激发声压级为 154 dB。当激发声压级过高时,声波的传播将表现出非线性。

10.5 池沸腾中的气泡成核与生长

相变传热以其高热流密度和小温差等特点和优势,成为传统工业与航空航天、电子信息、生物医学以及新能源等领域中优选的热质传递过程。与单相对流传热相比,相变对流传热更加复杂。以沸腾传热为例,该过程具有非常复杂的多相相互作用、非平衡、多尺度等特性,存在着众多的影响因素,如气泡成核、生长、融合、气-液-固三相相互作用等[27-29]。近年来,格子 Boltzmann 方法被应用于沸腾[30-41]、冷凝[42-49]、蒸发[50-52]等相变传热过程的研究,对探索相应的相变传热机理发挥了重要的作用,一些数值模拟研究的结论已被相关的实验研究所证实[53]。在沸腾的相关研究中,气泡成核一直是引起人们广泛关注的问题,因为它不仅是核化理论中的一个最基本的问题,而且关系到对核沸腾这一物理现象本质的认识以及对相关过程的调控。本节将结合热力学理论和格子 Boltzmann 模拟探讨池沸腾中的气泡成核与生长。

10.5.1 亚稳态

根据经典热力学，当液体达到所处压力下的饱和温度时，液体将发生沸腾，然而实际的观察却使人们发现，纯净液体在没有外界扰动的情况下沸腾温度会明显超过平衡热力学理论所预示的饱和温度[54]。当液体处在过热状态，尚未沸腾时，称为“过热液体”，其温度与相同压力下所对应的饱和温度的差值称为“过热度”。过热液体是一种处于亚稳平衡态(简称**亚稳态**，metastable)的液体，一旦有足够的扰动它将自发地由亚稳态转化为稳定的两相平衡态。我们所熟知的范德瓦耳斯状态方程可以描述这种亚稳态。如图 10.20 所示，图中横坐标为比体积，纵坐标为压力，点 C 为临界点，图中曲线 A-B-M-O-N-E-F 为温度低于临界温度时范德瓦耳斯状态方程所描述的某一等温线，其中点 B 和点 E 为两相共存点。

我们知道，范德瓦耳斯状态方程所对应的$(\partial p/\partial v)_T>0$ 的过程，即图 10.20 中的曲线 MON 并不对应实际的物理过程，但液相线在共存点以下的一段(即 BM 段)以及气相线在共存点以上的一段(即 EN 段)，它们并不违反热力学，分别对应“过热液体”和“过冷蒸气”的亚稳态[55-56]。亚稳态是实际存在的一种不稳定的平衡态，对于无穷小的扰动是稳定的，对有限大的扰动则是不稳定的。当液体被加热形成过热液时，液体将处于亚稳态，在受到足够的扰动时，过热液态就要消失，变为气液两相的混合物，即发生液气相变。由图 10.20 可以看到，点 M 和点 N 对应于$(\partial p/\partial v)_T=0$，它们是亚稳态的结束点。将各种温度下的亚稳态结束点分别连接起来，即可得到**亚稳边界线**，也称**旋节分解线**，如图 10.21 所示，它和共存曲线在临界点 C 相切。

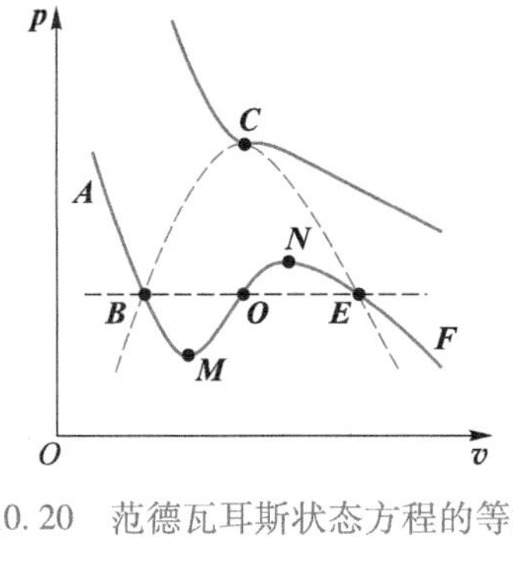

图 10.20 范德瓦耳斯状态方程的等温线

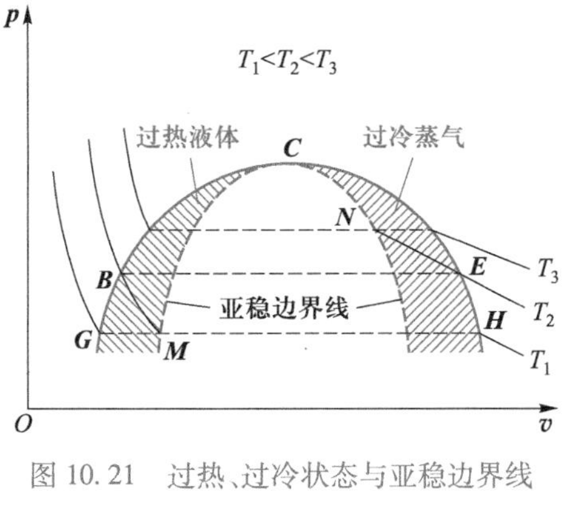

图 10.21 过热、过冷状态与亚稳边界线

关于亚稳态，一些学者认为[57]，可以将其想象为某种具有内部约束的状态，在内部约束存在时，它与稳定平衡态没有区别，一旦内部约束条件去除(在扰动作用下)，它将向较之更为稳定的状态运动。我国已故热力学专家曾丹苓曾撰文指出[58]，很难从热力学的角度去探究亚稳态向更稳定的平衡态转变的具体细节，特别是难以想象气泡是从半径 $r=0$ 变到 $r=r_c$ 的某种确定的渐变过程，因为 $r=0$ 或 r 接近于 0 时的蒸气不能视为一个热力学系统。故此，曾丹苓提出可以将核化过程看作是系统在扰动作用下产生的一个突变过程[58]。需要强调的是，基于状态方程通过$(\partial p/\partial v)_T=0$ 所确定的亚稳边界线并不是真实的亚稳边界线。由于亚稳态实验在实施上的困难，迄今为止所获得的实验数据仍十分有限。文献中较有共识的认识在于，当过热度较大时，过热液体偏离饱和液态较远，此时只

需要一个较小的扰动就可能破坏亚稳平衡态;相反地,当过热度较小时,则需要有更大尺度的扰动才能使亚稳平衡遭到破坏而引起移动。以图 10.21 为例,图中点 B 和点 E 为 T_2 温度下的两相共存点,点 G 和点 H 为 $T_1(T_1<T_2)$ 温度下的两相共存点,而点 M 和点 N 则为 T_2 温度下的亚稳边界点,假设点 M 的压力和点 G 的压力相等,则虚线 GM 为处在亚稳态区域的一段等压线;显然,GM 线段上的不同点代表不同的过热液体状态,且越往点 M 移动,过热度越大;对于离点 G 较远而离点 M 较近的点,可以认为其代表"非常深入"的亚稳态,对这种情况下的过热液体,只需要相对较小的扰动就可能引发亚稳态向稳定的两相平衡态的转变。

10.5.2 气液界面

关于气泡成核问题,已有不少学者进行过不同的论述,提出过不同的解释和设想,但许多理论研究或分析,由于缺乏相应的实验或数值模拟验证而让人们产生距离感。20 世纪 50 年代中期出现的分子动力学模拟方法对从微观尺度解释相变现象发挥了非常重要的作用,但其模拟过程所涉及的计算量巨大,从而限制了它所能应用的几何区域尺寸。以流体体积法(volume of fluid,VOF)、水平集法等方法为代表的界面捕捉或界面跟踪方法虽然也能应用于相变传热问题的研究,但由于这些方法并未纳入热力学理论从而难以模拟气泡或液滴的成核过程,往往需要预先布置种子气泡或种子液滴。

如前所述,范德瓦耳斯状态方程或其他形式的状态方程可以描述临界点和气液两相共存现象,并可以定性描述过热和过冷的亚稳态以及气液两相的转变,但是很长一段时期并未出现基于状态方程的气液相变数值模拟研究,这是因为仅仅依靠状态方程本身难以从数值模拟的角度来描述气液界面。事实上,无论是气泡还是液滴的成核,都存在着气液界面"从无到有"的过程。在热力学的教科书中,通常会提到范德瓦耳斯的两大重要贡献,即范德瓦耳斯状态方程和对比态原理。实际上,范德瓦耳斯还有一项非常重要的研究工作,即**毛细现象的热力学理论**(thermodynamic theory of capillarity)[59-60]。在这一理论中,他提出了一个重要假设,即气液两相界面可以看作是一个具有一定宽度且密度连续变化的过渡层,如图 10.22 所示。在此基础上,范德瓦耳斯引入了一个密度梯度项用以表征由于气液界面的存在所增加的系统自由能,即梯度自由能,参见第 5 章式(5.23)。

范德瓦耳斯的这一研究在后来被许多学者加以补充或完善。例如,Cahn 和 Hillard[61-62]将梯度自由能的概念从单组分两相系统推广到了两组分系统,提出了著名的 Cahn-Hillard 方程,奠定了相场方法(phase-field method)的基础。此外,范德瓦耳斯所指导的博士 Korteweg[63]研究了梯度自由能对流体力学方程的影响,构建了描述气液界面的 Korteweg 应力。这一应力的具体形式经过后来一些学者的修正形成了如下形式的压力张量[64]:

$$\boldsymbol{P}=\left(p_{\mathrm{EOS}}-\frac{\kappa}{2}\,|\nabla\rho|^2-\kappa\rho\nabla^2\rho\right)\boldsymbol{I}+\kappa\,\nabla\rho\nabla\rho \tag{10.15}$$

式中:p_{EOS}为范德瓦耳斯状态方程或其他形式的状态方程;κ 为与表面张力有关的系数。在第 5 章,我们曾多次提到这一压力张量,它与从平均场动理论所导出的宏观压力张量是一致的。事实上,结合上述压力张量以及 Navier-Stokes 方程和相应的状态方程就可以描述单组分等温气液两相流动,即

$$\partial_t\rho+\nabla\cdot(\rho\boldsymbol{u})=0 \tag{10.16}$$

$$\partial_t(\rho\boldsymbol{u})+\nabla\cdot(\rho\boldsymbol{uu})=-\nabla\cdot\boldsymbol{P}+\nabla\cdot\boldsymbol{\Pi}+\rho\boldsymbol{a} \tag{10.17}$$

式中：$\boldsymbol{a}$ 为外力加速度；$\boldsymbol{\Pi}$ 为黏性应力张量；$\boldsymbol{P}$ 为压力张量，由式(10.15)给出。式(10.15)~式(10.17)也正是扩散界面法(diffuse-interface method)[64]的基本控制方程组。

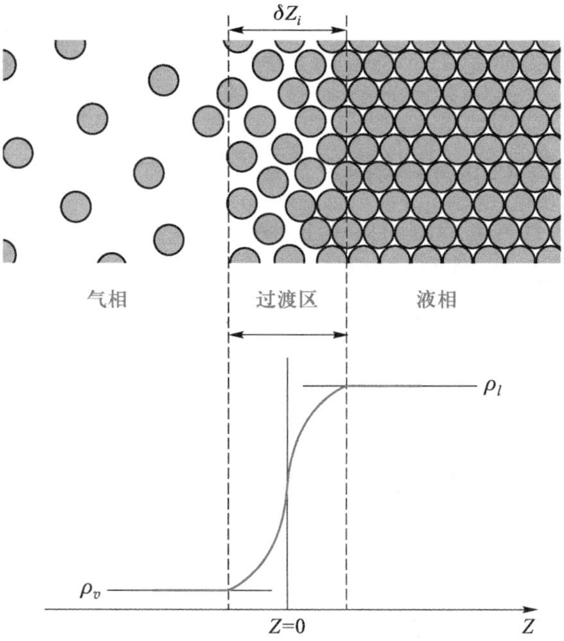

图 10.22　气液界面的描述

需要强调的是，尽管范德瓦耳斯在建立毛细现象的热力学理论时所采用的是一个假设[59]，但这一假设已被广泛证实为物理事实，即在微观层次气液界面确实存在一定的宽度。但真实气液界面的宽度可能在10^{-9} m 的量级(界面宽度在靠近临界点时会显著增大)[65]，如果模拟中的气液界面宽度设定在这样一个尺度，将严重限制扩散界面法所能应用的几何区域尺寸。事实上，这也是尖锐界面法(sharp-interface method)和扩散界面法争论的地方。从尖锐界面法的角度来看，既然真实界面的宽度在10^{-9}m 的量级，对许多气液两相流动来说，这样的界面宽度完全可以忽略，因此可以将气液界面看作是一个数学上的间断表面(mathematical surface of discontinuity)，也即所谓的尖锐界面。这一处理在文献中也得到了广泛的应用，它的主要缺点是在某些场合可能出现奇异性(singularity)。例如，采用尖锐界面法可能导致气-液-固三相接触点的应力奇异性，即“Huh-Scriven”佯谬[66]。相反，采用扩散界面法则可以有效地解决这一问题。此外，目前扩散界面法的应用已跳脱“真实界面”的概念，转而采用“数值界面”的概念，即模拟中的界面宽度并不代表真实的气液界面宽度，只需保证它远小于流动中最典型的特征尺度即可。

Huh-Scriven 佯谬

对于单组分非等温气液两相流动，可在连续方程和动量方程的基础上纳入相应的能量或温度方程。与等温系统相比，非等温系统的宏观方程要更复杂，这主要是因为对非等温系统来说，式(5.23)中的体相区自由能既是密度的函数也是温度的函数，从而导致相关的方程变得更加复杂。2002 年，He 和 Doolen[67]依据平均场动理论导出了单组分非等温气液两相流动的能量方程，即

$$
\begin{aligned}
&\partial_t(\rho e_{\mathrm{n}})+\nabla\cdot(\rho e_{\mathrm{n}}\boldsymbol{u}) \\
&=\nabla\cdot(\lambda\nabla T)+\boldsymbol{\Pi}:\nabla\boldsymbol{u}-\boldsymbol{P}:\nabla\boldsymbol{u}+\kappa\left[\nabla(\rho\nabla\rho)-\frac{1}{2}\nabla\cdot(\rho\nabla\rho)\boldsymbol{I}\right]:\nabla u
\end{aligned}
\tag{10.18}
$$

式中：λ 为导热系数；e_{n} 为系统的热力学能，包括内动能和内位能，形式为[41]

$$
\rho e_{\mathrm{n}}=\rho e_{\mathrm{n}}^{0}-\frac{1}{2}\kappa\rho\nabla^{2}\rho \tag{10.19}
$$

式中，e_{n}^{0} 为体相区的热力学能[68]。结合热力学关系式，可将式(10.18)转换为相应的温度方程，具体可参阅相关文献[41,69]。

10.5.3　格子 Boltzmann 模拟及分析

在第 5 章，我们对气液两相流的格子 Boltzmann 模型进行了详细的介绍，包括自由能模型、伪势模型、颜色梯度模型和相场模型。其中，伪势模型最突出的特点在于“自底向上”通过一个简单的伪势相互作用力反映了宏观层面相对复杂的压力张量，从而实现对相分离、气液界面和表面张力的描述。本节将采用伪势格子 Boltzmann 模型来模拟池沸腾中的气泡成核和生长过程。纯液体的核化通常可分为均相核化和异相核化，其中均相核化发生在液体的全部空间，是体积沸腾，而异相核化则发生在液体和加热表面的交界面[70]。我们在日常生活及实验中所观察到的水的沸腾通常都是异相核化。异相核化的发生是许多因素共同作用的结果，例如液体的性质、加热表面的形貌、表面湿润性等。许多实验研究均已证实，对于平表面，加热表面越疏水（接触角越大），越容易成核。此外，相对于凸起结构来说，加热表面的凹陷更有利于气泡成核。

本节将从格子 Boltzmann 模拟的角度，结合前述热力学理论，分析在格子 Boltzmann 模拟中沸腾现象是如何发生的。具体地，将讨论下面**两个问题**。

与已有实验研究的结论一致，对于平表面，格子 Boltzmann 模拟同样发现加热表面越疏水越容易形成沸腾，那么在其他条件完全相同的情况下，改变接触角的大小究竟在格子 Boltzmann 模拟中引起了什么样的变化而导致不同的起沸情况？

类似地，对于凸角和凹角，格子 Boltzmann 模拟发现凹角有利于气泡成核，这与经典成核结论是一致的。当结构化表面温度均匀且湿润性/接触角一致时，在格子 Boltzmann 模拟中，结构因素究竟产生了什么样的影响从而促使凹角有利于气泡成核？

1. 亲疏水复合湿润性平表面上的沸腾

首先，我们设计了如图 10.23 所示的二维算例，即亲水、疏水区域交替布置的复合湿润性平表面上的池沸腾，其中亲水区域的接触角为79°，疏水区域的接触角为121°；表面中心区域为疏水区域，以其为中心左右交替布置亲水、疏水区域，且在 x 方向采用周期性边界条件以模拟无限大平直表面。为了简化后面的理论分析，模拟中采用格子 Boltzmann-BGK 方程，即式(3.100)，相应的作用力项 $F_\alpha(\boldsymbol{r},t)$ 为

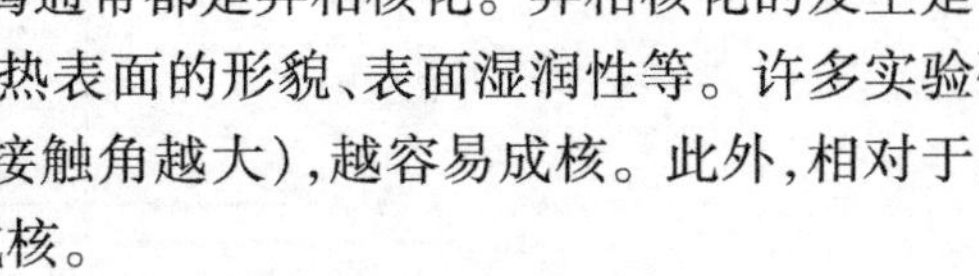

图 10.23　亲疏水复合湿润性平表面的示意图

$$
F_\alpha(\boldsymbol{r},t)=\omega_\alpha\left(1-\frac{1}{2\tau}\right)\left[\frac{(\boldsymbol{e}_\alpha-\boldsymbol{u})}{c_{\mathrm{s}}^{2}}+\frac{\boldsymbol{e}_\alpha\cdot\boldsymbol{u}}{c_{\mathrm{s}}^{4}}\boldsymbol{e}_\alpha\right]\cdot\boldsymbol{F} \tag{10.20}
$$

式中，$\boldsymbol{F}$ 为系统所受到的力，包括重力 $\boldsymbol{F}_{\mathrm{b}}$ 以及伪势相互作用力 $\boldsymbol{F}_{\mathrm{m}}$。

$$\boldsymbol{F}_{\mathrm{m}}(\boldsymbol{r},t)=-G\psi(\boldsymbol{r})\sum_{\alpha=0}^{8}w_{\alpha}\psi(\boldsymbol{r}+\boldsymbol{e}_{\alpha}\delta_{t})\boldsymbol{e}_{\alpha} \tag{10.21}$$

式中，伪势采用平方根形式，即 $\psi(\boldsymbol{r})=\sqrt{2(p_{\mathrm{EOS}}-\rho c_{\mathrm{s}}^{2})/Gc^{2}}$，其中 p_{EOS} 为状态方程。需要注意的是，采用平方根形式的伪势时，应当对外力项进行调整（参见 5.3 节），但为了便于后续的理论分析，这里仍然采用式（10.20）。

计算区域的网格为 $N_x \times N_y = 295\times300$，饱和温度为 $T_{\mathrm{s}}=0.86T_{\mathrm{c}}$（$T_{\mathrm{c}}$ 为临界温度），加热表面温度均匀且恒定，即 $T_{\mathrm{w}}=T_{\mathrm{s}}+\Delta T$，其中 ΔT 为过热度。初始时刻，除了加热表面以外，全场的温度均为饱和温度 T_{s}，且以 $y=2N_y/3$ 为界，其上方为饱和蒸汽，下方为饱和液体。计算区域在 x 方向采用周期性边界条件，加热表面的无滑移边界条件采用半步长反弹格式进行处理，其他模拟参数可参阅文献[38]。图 10.24 对比了三种不同过热度下该亲疏水复合湿润性平表面上的沸腾情况。从图中可以看到，当过热度 $\Delta T=0.04T_{\mathrm{c}}$ 时，加热表面上无论是亲水区域还是疏水区域均不起沸；当 $\Delta T=0.095T_{\mathrm{c}}$ 时，疏水区域上出现了气泡成核、生长及脱离现象，而亲水区域则一直无气泡产生；当过热度上升到 $\Delta T=0.135T_{\mathrm{c}}$ 时，亲水、疏水区域均产生了沸腾现象，从而形成了膜态沸腾。模拟结果清楚地展示了，随着加热表面过热度的上升，疏水区域会先于亲水区域起沸，即疏水区域沸腾起沸（onset of

(a) $\Delta T=0.04T_{\mathrm{c}}$

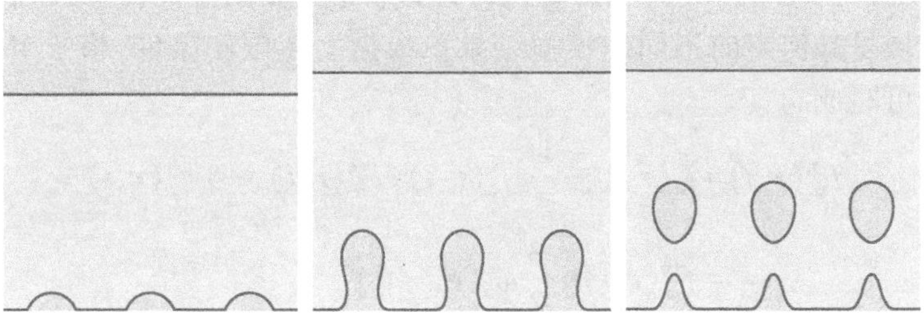

(b) $\Delta T=0.095T_{\mathrm{c}}$

(c) $\Delta T=0.135T_{\mathrm{c}}$

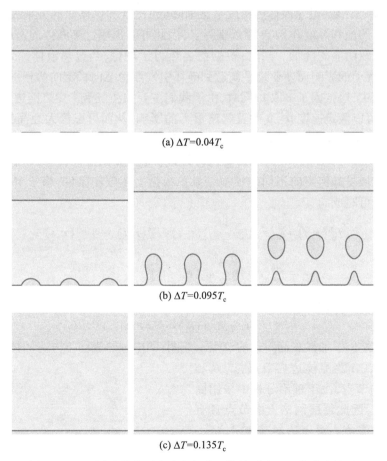

图 10.24 不同过热度下亲疏水复合湿润性平表面上的沸腾情况
（从左至右：$t=12\ 000\delta_t$、$36\ 000\delta_t$、$48\ 000\delta_t$）

nucleate boiling)的过热度小于亲水区域。如前所述，我们要讨论的重点是，改变接触角的大小究竟在格子 Boltzmann 模拟中引起了什么样的变化而导致不同的起沸情况。尽管格子 Boltzmann 方法模拟沸腾现象的研究最早可追溯至 2003 年[30]，但实际上文献中长期都未曾探讨过这一看似简单的问题。

要分析接触角的大小对沸腾现象的影响，首先应当清楚伪势格子 Boltzmann 模拟中接触角格式的本质。在第 5 章，我们对伪势模型的接触角格式进行了详细的介绍。实际上，无论是流固作用力格式还是虚拟密度(也即虚拟伪势)格式，它们的作用都是在靠近加热表面的流体边界节点上施加一个向上的流固作用力，且这一流固作用力越大，加热表面越疏水(即接触角越大)。如果将边界节点上的流体粒子想象成许多分子的团聚，则很明显，施加在每个分子上的竖直向上的力越大，分子就越具有离开表面的能力，继而形成更疏水的表面湿润性。

下面结合接触角/流固作用力的大小对上述沸腾模拟结果进行相应的理论分析。需要强调的是，沸腾过程非常复杂，对数值模拟进行全面细致的理论分析十分困难并且也不现实。因此我们做了一定程度的简化处理，且主要聚焦于流体边界节点上的密度变化，即重点关注接触角的大小对流体边界节点密度的影响。这一考虑的出发点在于，从图 10.21 可以看到，无论是沿等温线 BM 还是等压线 GM，亦或是沿着其他线段从饱和液态逐渐“深入”亚稳态区域，过热液体的密度会逐渐降低(密度和比体积的关系为 $\rho=1/v$)。由于模拟中选取的饱和温度 $0.86T_c$ 靠近临界点，因此可以预期过热液体的密度会随着过热度的增大有比较明显的降低。同时，根据前述热力学理论，当过热液体深入亚稳态区域时，极易受到扰动的影响而使亚稳平衡遭到破坏引发亚稳态向稳定的两相平衡态的转变。

理论分析中，松弛因子 τ 取 1；同时，由于我们主要关注气泡产生之前边界节点上的密度变化，因此可以假定速度 $\boldsymbol{u}=\boldsymbol{0}$ 并且忽略重力的影响，从而可以极大地简化相应的理论分析。此外，由于初始时刻，即 $t=0$ 时刻亲水区域和疏水区域的密度、速度、温度等都完全相同，因此从 $t=0$ 时刻到 $t=\delta_t$ 时刻的变化将具有非常重要的意义，它可以启示我们，亲水区域和疏水区域是如何走向不同的模拟结果。根据上述限定条件，格子 Boltzmann-BGK 方程的碰撞步可写成

$$\begin{aligned} f_\alpha^*(\boldsymbol{r},t) &= f_\alpha(\boldsymbol{r},t)-\frac{1}{\tau}[f_\alpha(\boldsymbol{r},t)-f_\alpha^{\mathrm{eq}}(\boldsymbol{r},t)]+\delta_t F_\alpha(\boldsymbol{r},t) \\ &= f_\alpha^{\mathrm{eq}}(\boldsymbol{r},t)+\frac{3}{2}\omega_\alpha(\boldsymbol{e}_\alpha\cdot\boldsymbol{F}) \end{aligned} \tag{10.22}$$

式中，$\boldsymbol{e}_\alpha\cdot\boldsymbol{F}=e_{\alpha x}F_x+e_{\alpha y}F_y$。相应的迁移步则为

$$f_\alpha(\boldsymbol{r}+\boldsymbol{e}_\alpha\delta_t,t+\delta_t)=f_\alpha^*(\boldsymbol{r},t) \tag{10.23}$$

图 10.25 给出了加热表面附近的节点示意图，图中加热表面下方为固体节点，上方为流体节点，且点 A 为流体边界节点，点 B 和点 C 为流体内节点，由于理论分析中采用周期性边界条件，因此每层网格上的节点具有相同的密度和温度。对于边界节点 A，每一新时层 $t+\delta_t$ 时刻离散速度 $\boldsymbol{e}_2$、$\boldsymbol{e}_5$ 和 $\boldsymbol{e}_6$ 方向的未知分布函数可由半步长反弹格式确定，即 $f_{\bar{\alpha}}(A,t+\delta_t)=f_\alpha^*(A,t)$，于是有

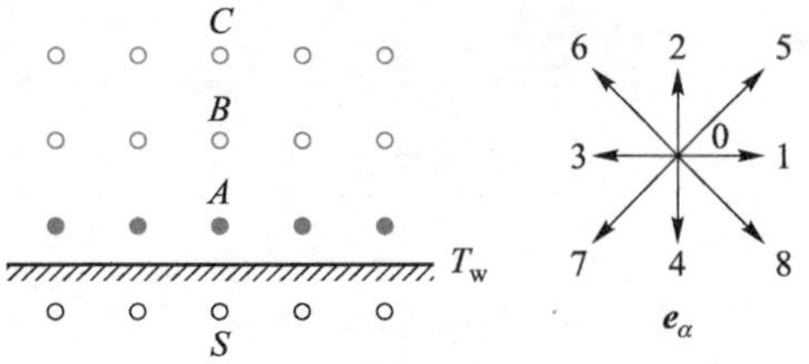

图 10.25 加热表面附近的节点示意图

$$f_2(A,t+\delta_t)=f_4^*(A,t)=f_4^{eq}(A,t)+\frac{3}{2}\omega_4[-F_y(A,t)] \tag{10.24}$$

$$f_5(A,t+\delta_t)=f_7^*(A,t)=f_7^{eq}(A,t)+\frac{3}{2}\omega_7[-F_x(A,t)-F_y(A,t)] \tag{10.25}$$

$$f_6(A,t+\delta_t)=f_8^*(A,t)=f_8^{eq}(A,t)+\frac{3}{2}\omega_8[F_x(A,t)-F_y(A,t)] \tag{10.26}$$

其他离散速度方向的分布函数则由迁移步式(10.23)确定,即

$$f_0(A,t+\delta_t)=f_0^*(A,t)=f_0^{eq}(A,t) \tag{10.27}$$

$$f_1(A,t+\delta_t)=f_1^*(A,t)=f_1^{eq}(A,t)+\frac{3}{2}\omega_1[F_x(A,t)] \tag{10.28}$$

$$f_3(A,t+\delta_t)=f_3^*(A,t)=f_3^{eq}(A,t)+\frac{3}{2}\omega_3[-F_x(A,t)] \tag{10.29}$$

$$f_4(A,t+\delta_t)=f_4^*(B,t)=f_4^{eq}(B,t)+\frac{3}{2}\omega_4[-F_y(B,t)] \tag{10.30}$$

$$f_7(A,t+\delta_t)=f_7^*(B,t)=f_7^{eq}(B,t)+\frac{3}{2}\omega_7[-F_x(B,t)-F_y(B,t)] \tag{10.31}$$

$$f_8(A,t+\delta_t)=f_8^*(B,t)=f_8^{eq}(B,t)+\frac{3}{2}\omega_8[F_x(B,t)-F_y(B,t)] \tag{10.32}$$

联立式(10.24)~式(10.32)可得

$$\begin{aligned}\rho(A,t+\delta_t) &= \sum_\alpha f_\alpha(A,t+\delta_t)\\ &= f_0^{eq}(A,t)+f_1^{eq}(A,t)+f_3^{eq}(A,t)+f_4^{eq}(A,t)+f_7^{eq}(A,t)+f_8^{eq}(A,t)+\\ &\quad f_4^{eq}(B,t)+f_7^{eq}(B,t)+f_8^{eq}(B,t)-\frac{3}{2}(\omega_4+\omega_7+\omega_8)[F_y(A,t)+F_y(B,t)]\end{aligned} \tag{10.33}$$

将平衡态分布函数的表达式以及相应的权系数 ω_α 代入式(10.33)可得

$$\rho(A,t+\delta_t)=\frac{5}{6}\rho(A,t)+\frac{1}{6}\rho(B,t)-\frac{1}{4}[F_y(A,t)+F_y(B,t)] \tag{10.34}$$

式(10.34)即为平表面上边界节点密度的通用表达式,它对亲水表面和疏水表面都是适用的。由式(10.34)可知,边界节点 A 在 $t+\delta_t$ 时刻的密度与 t 时刻点 A 和点 B 的密度以及它们所受到的 y 方向的力有关。基于式(10.34)可以分析从 $t=0$ 时刻到 $t=\delta_t$ 时刻边界节点密度的变化。在 $t=0$ 时刻有 $\rho(B)=\rho(A)=\rho_L$ 且 $F_y(B,0)=0$,这里 ρ_L 为饱和液体的密度,于是有

$$\rho(A,\delta_t)=\rho_L-\frac{1}{4}F_y(A,0) \tag{10.35}$$

由于我们主要关注气泡产生之前的密度变化,因而重力的影响可以忽略,此时仅考虑伪势相互作用力。结合式(10.21)和式(10.35)可得

$$\rho(A,\delta_t)=\rho_L-\frac{1}{8}\psi(A)[\psi(B)-\psi(S)] \tag{10.36}$$

式中,$\psi(A)$、$\psi(B)$ 和 $\psi(S)$ 分别为边界节点 A、流体内节点 B 和固体节点 S 的伪势。采用平方根形式的伪势 $\psi(\boldsymbol{r})=\sqrt{2(p_{\mathrm{EOS}}-\rho c_s^2)/Gc^2}$ 时,伪势大于零。为了实现不同的接触角,

固体节点的虚拟伪势 $\psi(S)<\psi(B)$，且随着接触角的增大，$\psi(S)$ 减小。因此根据式(10.36)可知，由于边界节点和固体节点之间流固作用力的存在，边界节点 $t=\delta_t$ 时刻的密度将小于初始时刻设定的饱和液体的密度 ρ_L，也即边界节点的流体温度升高且密度下降从而进入过热液态。

此外，前面曾提到，伪势格子 Boltzmann 模型中接触角格式的本质作用是在流体边界节点上施加一个向上的流固作用力，且这一流固作用力越大，加热表面的接触角越大。结合式(10.36)可得疏水表面上流体节点的密度 $\rho_{\text{pho}}(A,\delta_t)$ 与亲水表面上流体节点的密度 $\rho_{\text{phi}}(A,\delta_t)$ 之间的关系为

$$\rho_{\text{pho}}(A,\delta_t)-\rho_{\text{phi}}(A,\delta_t)=\frac{1}{8}\psi(A)\left[\psi(S)_{\text{pho}}-\psi(S)_{\text{phi}}\right] \tag{10.37}$$

由于 $\psi(A)>0$ 且 $\psi(S)_{\text{pho}}<\psi(S)_{\text{phi}}$，因此可得 $\rho_{\text{pho}}(A,\delta_t)<\rho_{\text{phi}}(A,\delta_t)$，也即在其他条件相同的情况下，疏水表面上流体节点的密度比亲水表面上流体节点的密度下降得更快。这里仅给出了 $t=0$ 时刻到 $t=\delta_t$ 时刻的典型变化，后续的变化可基于式(10.34)进行类似的分析，并可结合数值模拟输出的密度场予以印证。综上，疏水表面由于具有更大的流固作用力，导致流体边界节点在相同过热度下更深入亚稳态区域(密度更靠近亚稳边界点的密度)；相应地，疏水表面上沸腾起沸所需要的过热度也就更小，这也就不难理解为什么图 10.24b 显示在过热度 $\Delta T=0.095T_c$ 时疏水区域已经起沸而亲水区域仍无气泡产生。至此，我们也就回答了第一个问题(在其他条件完全相同的情况下，改变接触角的大小究竟在格子 Boltzmann 模拟中引起了什么样的变化而导致不同的起沸情况)。

2. 二维结构化表面上的沸腾

接下来考虑二维结构化表面上的气泡成核与生长。本算例采用的计算网格为 $N_x\times N_y=151\times300$，加热表面布置在计算区域的底部，其高度为 30 个格子单位(不考虑固体区域与流体区域的共轭传热，整个表面设置为均匀温度)，在加热表面的中心有一个宽度为 45 个格子单位、深度为 25 个格子单位的凹坑。加热表面不再设置亲疏水交替区域，而是整个表面具有均匀的湿润性(接触角为79°)。其他模拟参数及边界处理格式与前一个算例基本一致，具体可参阅文献[38]。图 10.26 对比了不同过热度下结构化表面上的沸腾情况。从图中可以看到，当过热度 $\Delta T=0.04T_c$ 时，整个表面一直无气泡产生；当过热度上升到 $\Delta T=0.095T_c$ 时，在凹坑的两个凹角处出现了气泡成核，随着时间推移，两个气泡逐渐长大，并且融合充满整个凹坑，之后可以观察到气泡脱离现象。值得注意的是，本算例中加热表面的接触角为79°，与前一个算例亲水区域的接触角是一致的。对比图 10.26b 与图 10.24b 可以发现，当过热度和接触角相同时，图 10.24b 中平直亲水区域并无气泡产生，而图 10.26b 中的结构化表面上则出现了气泡成核，这也就涉及前面所提到的第二个问题，即当加热表面温度均匀且湿润性/接触角一致时，在格子 Boltzmann 模拟中，表面结构因素究竟产生了什么样的影响从而促使凹角有利于气泡成核?

为了厘清第二个问题，可进行类似的分析。如图 10.27 所示，以凹坑的左下角为例，点 A 是离凹角最近的流体边界节点，点 B_1 和点 B_2 是与其相邻的两个流体边界节点，点 C 是流体内节点，点 S 是固体节点。由于加热表面温度均匀且湿润性一致，因此所有固体节点具有相同的虚拟伪势 $\psi(S)$。对于边界节点 A 点，$t+\delta_t$ 时刻离散速度 $\boldsymbol{e}_1$、$\boldsymbol{e}_2$、$\boldsymbol{e}_5$、$\boldsymbol{e}_6$ 和 $\boldsymbol{e}_8$ 方向的未知分布函数可由半步长反弹格式确定，即

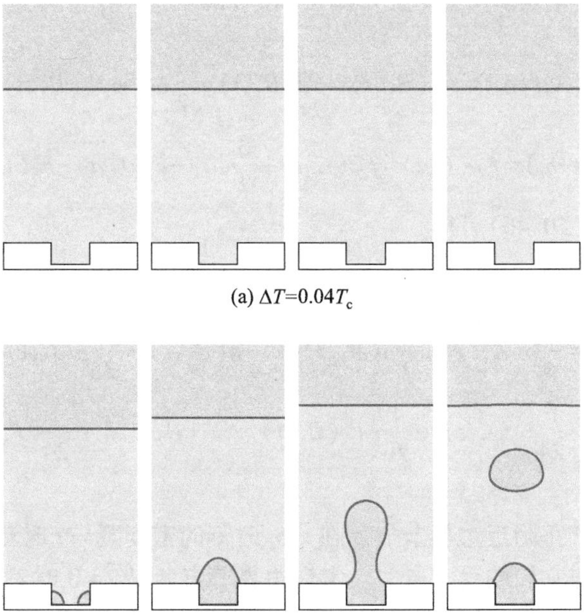

图 10.26 不同过热度下均匀湿润性结构化表面上的沸腾情况
（从左至右：$t=5\times10^3\delta_t$、$10^4\delta_t$、$2.5\times10^4\delta_t$、$3.2\times10^4\delta_t$）

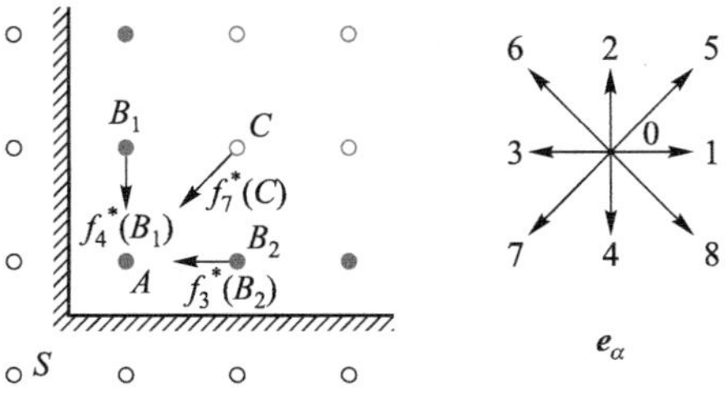

图 10.27 凹角附近的节点示意图

$$f_1(A,t+\delta_t)=f_3^*(A,t)=f_3^{eq}(A,t)+\frac{3}{2}\omega_3[-F_x(A,t)] \tag{10.38}$$

$$f_2(A,t+\delta_t)=f_4^*(A,t)=f_4^{eq}(A,t)+\frac{3}{2}\omega_4[-F_y(A,t)] \tag{10.39}$$

$$f_5(A,t+\delta_t)=f_7^*(A,t)=f_7^{eq}(A,t)+\frac{3}{2}\omega_7[-F_x(A,t)-F_y(A,t)] \tag{10.40}$$

$$f_6(A,t+\delta_t)=f_8^*(A,t)=f_8^{eq}(A,t)+\frac{3}{2}\omega_8[F_x(A,t)-F_y(A,t)] \tag{10.41}$$

$$f_8(A,t+\delta_t)=f_6^*(A,t)=f_6^{eq}(A,t)+\frac{3}{2}\omega_6[-F_x(A,t)+F_y(A,t)] \tag{10.42}$$

其他离散速度方向的分布函数由迁移步式(10.23)，即

$$f_0(A,t+\delta_t)=f_0^*(A,t)=f_0^{eq}(A,t) \tag{10.43}$$

$$f_3(A,t+\delta_t)=f_3^*(B_2,t)=f_3^{eq}(B_2,t)+\frac{3}{2}\omega_3[-F_x(B_2,t)] \tag{10.44}$$

$$f_4(A,t+\delta_t)=f_4^*(B_1,t)=f_4^{eq}(B_1,t)+\frac{3}{2}\omega_4[-F_y(B_1,t)] \tag{10.45}$$

$$f_7(A,t+\delta_t)=f_7^*(C,t)=f_7^{eq}(C,t)+\frac{3}{2}\omega_7[-F_x(C,t)-F_y(C,t)] \tag{10.46}$$

联立式(10.38)~式(10.46)可得

$$\begin{aligned}\rho(A,t+\delta_t) &= \sum_\alpha f_\alpha(A,t+\delta_t)\\ &= \frac{3}{4}\rho(A,t)+\frac{1}{9}\rho(B_1,t)+\frac{1}{9}\rho(B_2,t)+\frac{1}{36}\rho(C,t)-\frac{5}{24}F_x(A,t)-\\ &\quad \frac{5}{24}F_y(A,t)-\frac{1}{6}[F_y(B_1,t)+F_x(B_2,t)]-\frac{1}{24}[F_x(C,t)+F_y(C,t)]\end{aligned} \tag{10.47}$$

由式(10.47)可知,凹角附近边界节点 A 在 $t+\delta_t$ 时刻的密度既与 t 时刻点 A、点 B_1、点 B_2 和点 C 的密度有关也与它们的受力有关。考察边界节点 A 从 $t=0$ 时刻到 $t=\delta_t$ 时刻的密度变化,初始时刻有 $\rho(A)=\rho(B_1)=\rho(B_2)=\rho(C)=\rho_L$ 且 $F_x(B_2)=F_y(B_1)=F_x(C)=F_y(C)=0$,于是可得

$$\rho(A,\delta_t)=\rho_L-\frac{5}{24}F_x(A,0)-\frac{5}{24}F_y(A,0) \tag{10.48}$$

依据伪势相互作用力的表达式可将式(10.48)转化为

$$\rho(A,\delta_t)=\rho_L-\frac{25}{144}\psi(A)[\psi(B)-\psi(S)] \tag{10.49}$$

与前一个例子类似,由于平方根形式的伪势 ψ 大于零,且有 $\psi(S)<\psi(B)$,因此根据式(10.49)可知,凹角附近边界节点 A 在 $t=\delta_t$ 时刻的密度将小于初始时刻设定的饱和液体的密度 ρ_L,从而进入过热液态区域。

特别地,对比式(10.49)和式(10.36)可以发现,凹角附近边界节点的密度和平直表面附近边界节点的密度存在如下关系:

$$\rho_{cor}(A,\delta_t)=\rho_{flat}(A,\delta_t)-\frac{7}{144}\psi(A)[\psi(B)-\psi(S)] \tag{10.50}$$

式中:下标“cor”代表凹角附近的边界节点;下标“flat”代表平直表面附近的边界节点。由式(10.48)和式(10.50)可知,与平直表面附近的边界节点相比,凹角附近的边界节点既受到 x 方向的流固作用力又受到 y 方向的流固作用力,导致其密度下降得更快,即 $\rho_{cor}(A,\delta_t)<\rho_{flat}(A,\delta_t)$。这也就不难理解为什么在同样的过热度和同样的接触角下,凹角处会先产生气泡,也即由于凹角的存在降低了其附近流体边界节点沸腾起沸所需要的过热度。事实上,可对凸角附近的边界节点进行类似的分析,并且可以发现与凹角及平直表面相比,凸角的存在将不利于气泡成核。对于 $t=\delta_t$ 时刻之后的密度变化,可基于式(10.47)进行相应的分析。图10.28给出了凹角附近流体节点 B_1、B_2 和点 C 伪势相互作用力 $F_y(B_1,t)$、$F_x(B_2,t)$、$F_x(C,t)$ 和 $F_y(C,t)$ 随时间的变化,由于这些相互作用力大于零,从而会使得 $\rho(A,t+\delta_t)$ 进一步下降,继而促进气泡成核。

综上,结合热力学理论,从格子 Boltzmann 模拟的角度探讨了池沸腾中的气泡成核与

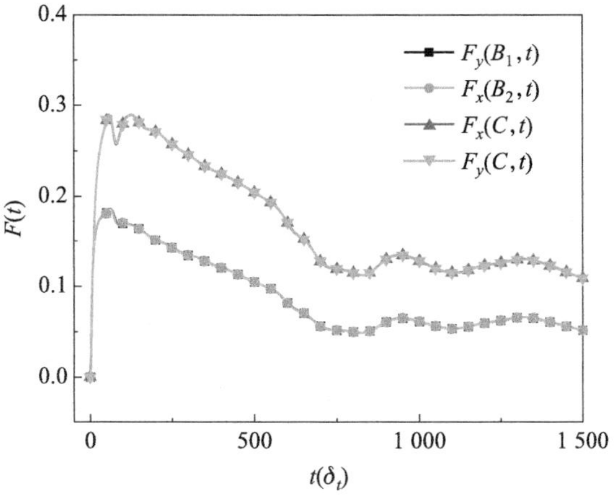

图 10.28 凹角附近流体节点伪势相互作用力随时间的变化

生长，分析了接触角和表面结构对沸腾起沸的影响。通过碰撞迁移规则及反弹格式，可以比较方便地直接确定边界节点处的密度变化，这是介观格子方法（格子气自动机方法和格子 Boltzmann 方法）的独特优势，也给我们从密度变化的角度来分析过热液态的发展提供了方便。如前所述，一些针对气泡成核的分析往往是从自由能或者能量变化的角度展开，缺乏直观的感受，让读者产生距离感。实际上，基于格子 Boltzmann 模拟，结合密度和温度的变化，同样可以分析相应的自由能或化学势的变化。以笔者的研究经验，在格子 Boltzmann 模拟中，尽管气泡产生之前需要过热液态发展一段时间，但气泡从无到有产生的瞬态过程更接近一个突变过程，这也印证了前述曾丹苓所提出的观点。

沸腾与冷凝传热由于涉及核化，实际上与热力学密切相关，丰富和完善热力学理论对促进相变传热学科的发展大有裨益。于渌等[56]在相关著作中曾指出：“对于古老的气液相变，我们今天的知识并不比范德瓦耳斯增加了很多。建立包括过冷、过热和两相共存现象在内的、严格的气液相变统计理论，仍然是尚未全部解决的难题。”

10.6 泡沫金属内相变材料熔化问题

泡沫金属内相变材料熔化问题的物理模型如图 10.29 所示。二维方腔中填充泡沫金属，泡沫金属的孔隙内则充满着相变材料。方腔的左壁面保持恒定的高温 T_h，右壁面保持恒定的低温 T_c，上下壁面绝热。相变材料和泡沫金属的初始温度均为 T_i（小于熔化温度 T_{melt}，即初始时刻相变材料处于固态）。当 $t>0$ 时，左壁面保持恒定的温度 T_h（高于熔化温度 T_{melt}），相变材料开始熔化。为了与文献[71]的数值模拟结果进行对比，选取 $T_h=1$，$T_{melt}=0.3$，$T_c=T_i=T_0=0$。

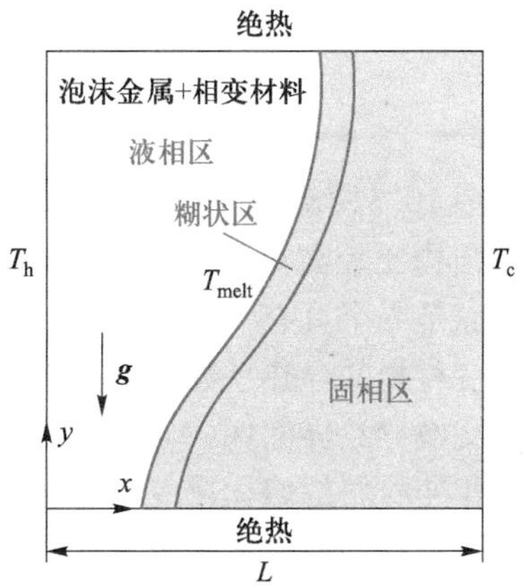

图 10.29 泡沫金属内相变材料熔化问题的物理模型

采用的 REV 尺度多孔介质固液相变格子 Boltzmann 模型参见文献[72]，模拟中的达西数 Da、

瑞利数 Ra 和傅里叶数 Fo 分别定义如下：

$$Da=\frac{K}{L^2},\quad Ra=\frac{g\beta\Delta TL^3}{\nu_f\alpha_f},\quad Fo=\frac{t\alpha_f}{L^2}\tag{10.51}$$

式中：K 为泡沫金属的渗透率；L 为计算区域的特征尺寸；$\Delta T=T_h-T_c$；t 为时间步数；ν_f 为液态相变材料的运动黏度；α_f 为相变材料的等效热扩散系数，其定义为 $\alpha_f=\alpha_l f_l+\alpha_s(1-f_l)$，其中 f_l 为液相分数，α_l 和 α_s 分别为液态和固态相变材料的热扩散系数。

当 $Ra=10^6$ 时，计算网格选取为 $N_x\times N_y=150\times150$ 的均匀网格；当 $Ra=10^8$ 时，计算网格选取为 $N_x\times N_y=300\times300$ 的均匀网格。图 10.30 给出了 $Ra=10^6$ 和10^8 时不同 Fo 数下的固液相界面的位置（为了与文献[71]进行对比，固液相界面对应的液相分数为 $f_l=0.5$）。从图中可以看到，本文的模拟结果与文献[71]的数值模拟结果吻合得很好。当 $Ra=10^6$ 时，由于泡沫金属与相变材料的导热系数之比非常大，方腔内液相区的传热方式主要是热传导，因此固液相界面基本是竖直的。当 $Ra=10^8$ 时，在熔化的开始阶段（$Fo\leqslant0.000\,4$），方腔内液相区的传热方式主要是热传导，对流作用非常弱；随着熔化过程的进行，受到逐渐增强的对流作用的影响，固液相界面开始变得弯曲，这表明逐渐增强的对流作用加快了方腔上半部分相变材料的熔化速度。需要注意的是，泡沫金属与液相相变材料之间的内部热交换会对液相区的自然对流强度起到较强的抑制作用。

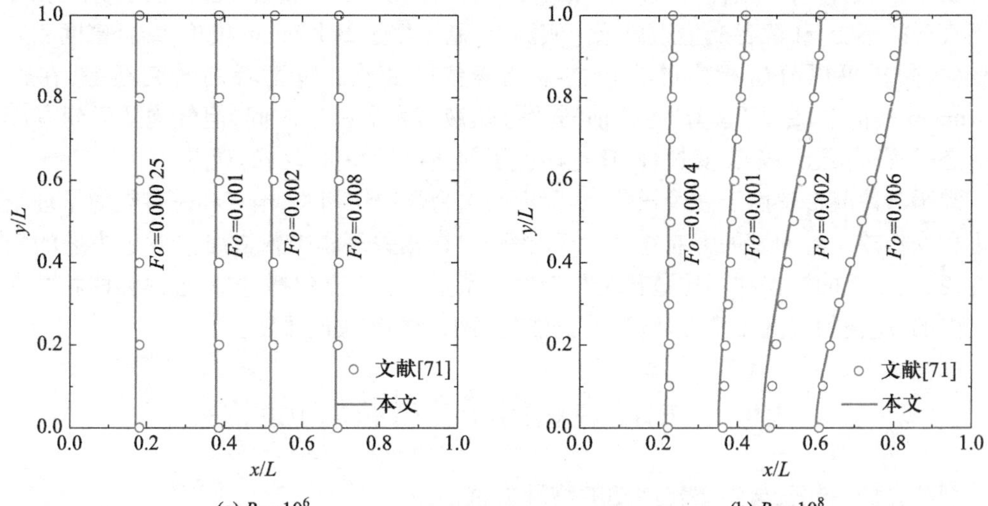

图 10.30 不同 Fo 数下的固液相界面位置

图 10.31 给出了 $Ra=10^6$ 时不同 Fo 数下的流线和固液相界面。从图中可以看到，流场的结构比较简单，在方腔的液相区只存在一个椭圆形的大涡。此外，还可以看到，固液相界面是一个具有一定宽度的区域。在熔化的开始阶段（$Fo\leqslant0.000\,25$），由于泡沫金属与相变材料的导热系数之比非常大，泡沫金属与相变材料之间存在较大的温差，两者之间的热交换导致界面区（$0.05<f_l<0.95$）占据的网格超过了 10 个。同时，这种内部热交换加快了相变材料的熔化速度；随着熔化过程的进行，泡沫金属与相变材料之间的温差变小，界面区占据的网格数逐渐减少；当 $Fo=0.008$ 时，方腔内液相区的流动与传热开始进入稳态，界面区占据的网格数不超过 3 个。图 10.32 给出了 $Ra=10^8$ 时不同 Fo 数下的流线和固液相界面。从图中可以看到，随着瑞利数的增大，对流作用大幅度增大，从而加快了方

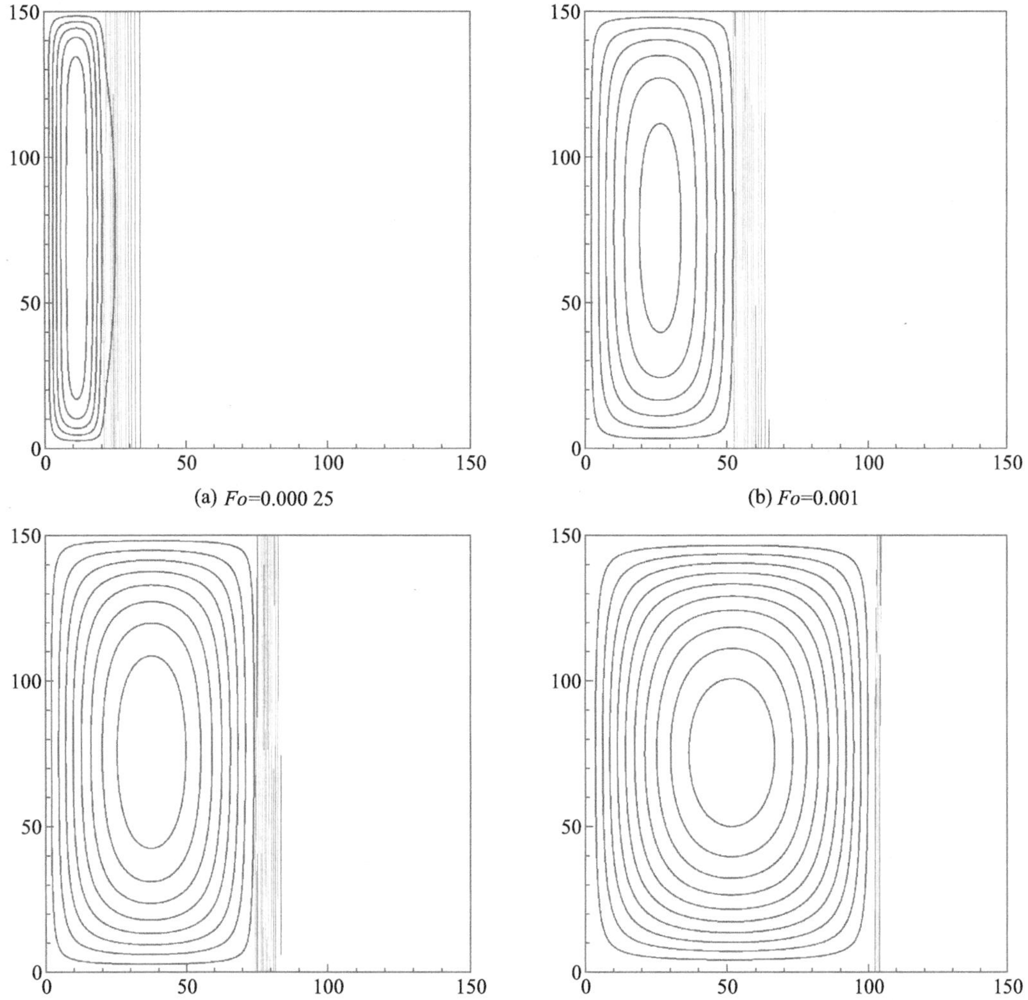

(a) Fo=0.000 25　(b) Fo=0.001

(c) Fo=0.002　(d) Fo=0.008

图 10.31　不同 Fo 数下的流线和固液相界面($Ra=10^6$)

腔上半部分相变材料的熔化速度。但由于泡沫金属的存在,液相区的自然对流强度受到了抑制,流场的结构仍然比较简单。

图 10.33 给出了 $Ra=10^6$ 和 10^8 时不同 Fo 数下方腔中央水平线上($y/L=0.5$)的温度分布。图中 T_m 和 T_f 分别为泡沫金属和相变材料的温度分布。泡沫金属与相变材料之间的温度差异是由两者之间的内部热交换决定的。如果内部热交换的热响应时间慢于泡沫金属和相变材料的热响应时间,泡沫金属和相变材料温度场的演化将是独立的,并且到达稳态的时间由热响应时间慢的相变材料决定;反之,泡沫金属和相变材料温度场的演化将耦合在一起。在本文考虑的工况下,内部热交换的热响应时间快于泡沫金属和相变材料的热响应时间,泡沫金属和相变材料温度场的演化耦合在一起。图 10.33 表现出了非常明显的局部非热平衡效应特征。在熔化的开始阶段,泡沫金属与相变材料之间的温差非常大(同一时刻泡沫金属与相变材料温差最大的区域是相变区);随着熔化过程的进行,泡沫金属与相变材料之间的温差逐渐变小,在稳态时两者之间的温差趋于零,局部非热平衡效应消失。图中可见,本文的结果与文献[71]的数值模拟结果吻合得很好。

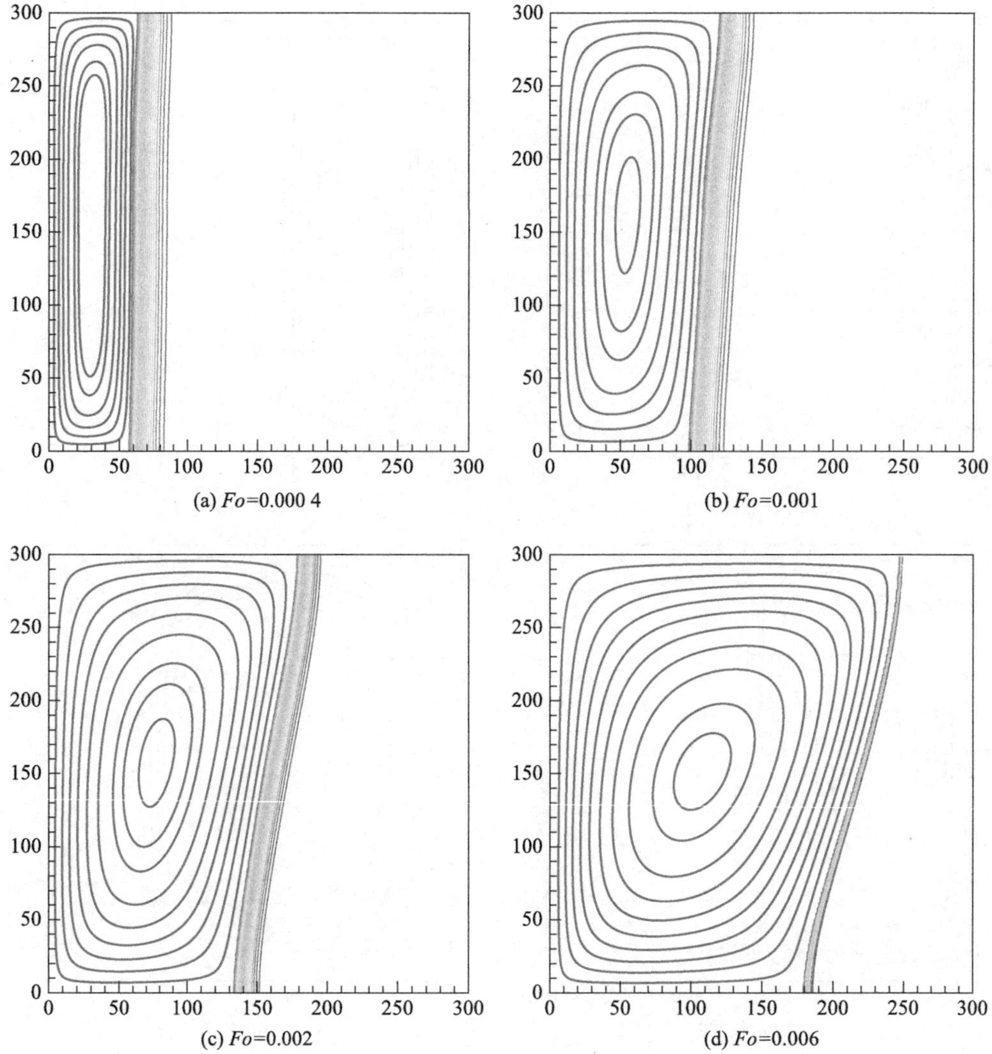

图 10.32 不同 Fo 数下的流线和固液相界面($Ra=10^8$)

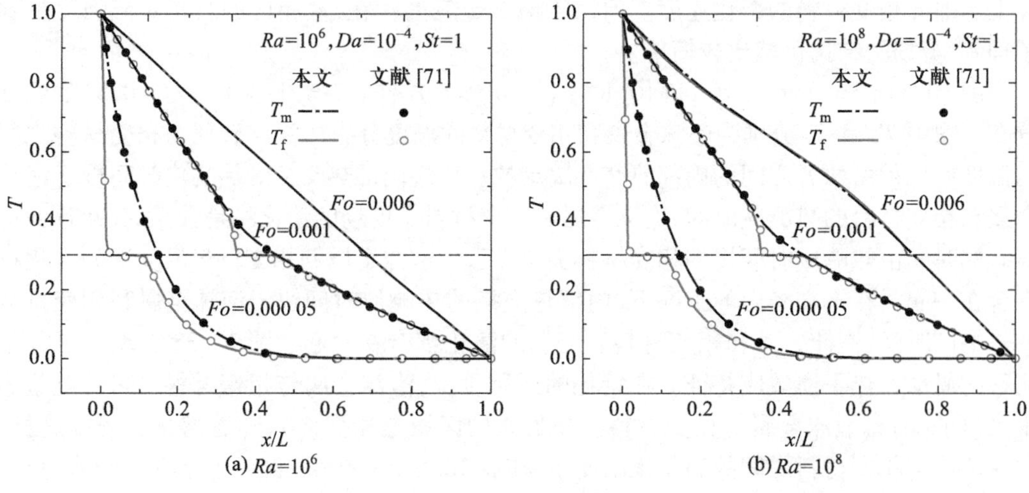

图 10.33 不同 Fo 数下方腔中央水平线上($y/L=0.5$)的温度分布

通常情况下,在相变材料中填充高导热率泡沫金属可以大幅度提高其蓄热效率[72]。但需要注意的是,虽然泡沫金属的加入使得复合相变材料的整体蓄热效率得到了大幅度提升,但也会导致相变材料所占比例降低,从而减少复合相变材料的蓄热量。因此,在设计泡沫金属相变蓄热装置时,需要对蓄热量和蓄热效率这两个因素进行综合考虑。在实际应用中,通常采用高渗透率和高孔隙率的泡沫金属来改善相变材料的蓄热性能。

10.7 小　　结

随着其理论的逐步完善,格子 Boltzmann 方法在许多领域都得到了越来越多的关注。本章所列举的 6 个应用既包括格子 Boltzmann 方法应用于封闭方腔自然对流、Rayleigh-Bénard 对流、Riemann 问题等计算流体力学与计算传热学中的经典问题,也包括格子 Boltzmann 方法在高马赫数可压缩流动、气液相变与沸腾、固液相变等复杂流动与传热问题中的应用。从中可以看到,格子 Boltzmann 方法已逐渐成为一种非常有竞争力的实用数值模拟手段。关于格子 Boltzmann 方法在其他应用领域的研究,读者可参阅第 1 章的介绍及相关文献。为了帮助初学者更好地理解格子 Boltzmann 方法"简单算法模拟复杂问题"的特性并掌握相应的程序实施,书后的附录 D 和附录 E 分别提供了格子 Boltzmann 方法模拟顶盖驱动流和相分离的程序代码。

参考文献

[1] 陶文铨. 数值传热学[M]. 2 版. 西安:西安交通大学出版社,2001.

[2] Li Q, He Y L, Wang Y, et al. An improved thermal lattice Boltzmann model for flows without viscous heat dissipation and compression work[J]. International Journal of Modern Physics C, 2008, 19(1): 125-150.

[3] De Vahl Davis G. Natural convection of air in a square cavity: A benchmark numerical solution[J]. International Journal for Numerical Methods in Fluids, 1983, 3(3): 249-264.

[4] Hortmann M, Perić M, Scheuerer G. Finite volume multigrid prediction of laminar natural convection: Bench-mark solutions[J]. International Journal for Numerical Methods in Fluids, 1990, 11(2): 189-207.

[5] Markatos N C, Pericleous K A. Laminar and turbulent natural convection in an enclosed cavity[J]. International Journal of Heat and Mass Transfer, 1984, 27(5): 755-772.

[6] Peng Y, Shu C, Chew Y T. Simplified thermal lattice Boltzmann model for incompressible thermal flows [J]. Physical Review E, 2003, 68(2): 026701.

[7] Dixit H N, Babu V. Simulation of high Rayleigh number natural convection in a square cavity using the lattice Boltzmann method[J]. International Journal of Heat and Mass Transfer, 2006, 49: 727-739.

[8] 杨帆,刘树红,唐学林,等. 格子 Boltzmann 亚格子模型的研究[J]. 工程热物理学报,2004(1): 43-46.

[9] 王勇,何雅玲,童长青,等. 二维 Rayleigh-Bénard 对流的插值格子-Boltzmann 方法模拟研究[J]. 工程热物理学报,2007,28(2):313-315.

[10] Shan X W. Simulation of Rayleigh-Bénard convection using a lattice Boltzmann method[J]. Physical Review E, 1997, 55(3): 2780-2788.

[11] He X Y, Chen S Y, Doolen G D. A novel thermal model for the lattice Boltzmann method in incompressible limit[J]. Journal of Computational Physics, 1998, 146(1): 282-300.

[12] Clever R M, Busse F H. Transition to time-dependent convection[J]. Journal of Fluid Mechanics, 1974, 65

(4):625-645.
[13] Fromm J E. Numerical solutions of the nonlinear equations for a heated fluid layer[J]. Physics of Fluids, 1965,8(10):1757-1769.
[14] Li Q, He Y L, Wang Y, et al. Coupled double-distribution-function lattice Boltzmann method for the compressible Navier-Stokes equations[J]. Physical Review E,2007,76(5):056705.
[15] 阎超. 计算流体力学方法及其应用[M]. 北京:北京航空航天大学出版社,2006.
[16] 赵海洋. 高精度数值计算方法研究及应用[D]. 长沙:国防科学技术大学,2002.
[17] Wang Y, He Y L, Zhao T S, et al. Implicit-explicit finite-difference lattice Boltzmann method for compressible flows[J]. International Journal of Modern Physics C,2007,18(12):1961-1983.
[18] 张涵信. 无波动、无自由参数的耗散差分格式[J]. 空气动力学学报,1988,6:143-165.
[19] Woodward P, Colella P. The numerical simulation of two-dimensional fluid flow with strong shocks [J]. Journal of Computational Physics,1984,54(1):115-173.
[20] Sun C H. Simulations of compressible flows with strong shocks by an adaptive lattice Boltzmann model [J]. Journal of Computational Physics,2000,161(1):70-84.
[21] Swift G W. Thermoacoustic engines[J]. The Journal of the Acoustical Society of America,1988,84(4): 1145-1180.
[22] Chen Y, Liu X, Zhang X Q, et al. Thermoacoustic simulation with lattice gas automata[J]. Journal of Applied Physics,2004,95(8):4497-4499.
[23] Buick J M, Greated C A, Campbell D M. Lattice BGK simulation of sound waves[J]. Europhysics Letters, 1998,43(3):235-240.
[24] Haydock D, Yeomans J M. Lattice Boltzmann simulations of attenuation-driven acoustic streaming[J]. Journal of Physics A:Mathematical and General,2003,36(20):5683-5694.
[25] 王勇,何雅玲,刘迎文,等. 声波衰减的格子-Boltzmann 方法模拟[J]. 西安交通大学学报,2007,41 (1):5-8.
[26] Nyborg W L M. 11-Acoustic Streaming[M]//Mason W P. Physical Acoustics. Academic Press. 1965: 265-331.
[27] Dhir K. Boiling heat transfer[J]. Annual Review of Fluid Mechanics,1998,30:365-401.
[28] Dhir V K. Numerical simulations of pool-boiling heat transfer[J]. Aiche Journal,2001,47(4):813-834.
[29] Koizumi Y, Shoji M, Monde M, et al. Boiling: Research and Advances[M]. Elsevier,2017.
[30] Zhang R Y, Chen H D. Lattice Boltzmann method for simulations of liquid-vapor thermal flows[J]. Physical Review E,2003,67(6):066711.
[31] Gong S, Cheng P. A lattice Boltzmann method for simulation of liquid-vapor phase-change heat transfer [J]. International Journal of Heat and Mass Transfer,2012,55(17-18):4923-4927.
[32] Li Q, Kang Q J, Francois M M, et al. Lattice Boltzmann modeling of boiling heat transfer: The boiling curve and the effects of wettability[J]. International Journal of Heat and Mass Transfer,2015,85:787-796.
[33] Gong S, Cheng P. Lattice Boltzmann simulations for surface wettability effects in saturated pool boiling heat transfer[J]. International Journal of Heat and Mass Transfer,2015,85:635-646.
[34] Li Q, Yu Y, Zhou P, et al. Enhancement of boiling heat transfer using hydrophilic-hydrophobic mixed surfaces: A lattice Boltzmann study[J]. Applied Thermal Engineering,2018,132:490-499.
[35] Ma X J, Cheng P, Quan X J. Simulations of saturated boiling heat transfer on bio-inspired two-phase heat sinks by a phase-change lattice Boltzmann method[J]. International Journal of Heat and Mass Transfer, 2018,127:1013-1024.
[36] Ma X J, Cheng P. Dry spot dynamics and wet area fractions in pool boiling on micro-pillar and micro-cavity hydrophilic heaters: A 3D lattice Boltzmann phase-change study[J]. International Journal of Heat and

Mass Transfer,2019,141:407-418.

[37] Li W X,Li Q,Yu Y,et al. Enhancement of nucleate boiling by combining the effects of surface structure and mixed wettability:A lattice Boltzmann study[J]. Applied Thermal Engineering,2020,180:115849.

[38] Li Q,Yu Y,Wen Z X. How does boiling occur in lattice Boltzmann simulations? [J]. Physics of Fluids, 2020,32(9):093306.

[39] Li W X, Li Q, Yu Y, et al. Nucleate boiling enhancement by structured surfaces with distributed wettability-modified regions: A lattice Boltzmann study [J]. Applied Thermal Engineering, 2021, 194:117130.

[40] Deng Z L, Liu X D, Wu S C, et al. Pool boiling heat transfer enhancement by bi-conductive surfaces [J]. International Journal of Thermal Sciences,2021,167:107041.

[41] Li Q,Yu Y,Luo K H. Improved three-dimensional thermal multiphase lattice Boltzmann model for liquid-vapor phase change[J]. Physical Review E,2022,105(2):025308.

[42] Liu X L,Cheng P. Lattice Boltzmann simulation of steady laminar film condensation on a vertical hydrophilic subcooled flat plate[J]. International Journal of Heat and Mass Transfer,2013,62:507-514.

[43] Zhang Q Y, Sun D K, Zhang Y F, et al. Lattice Boltzmann modeling of droplet condensation on superhydrophobic nanoarrays[J]. Langmuir,2014,30(42):12559-12569.

[44] Li X P,Cheng P. Lattice Boltzmann simulations for transition from dropwise to filmwise condensation on hydrophobic surfaces with hydrophilic spots[J]. International Journal of Heat and Mass Transfer,2017, 110:710-722.

[45] Li M J, Huber C, Mu Y T, et al. Lattice Boltzmann simulation of condensation in the presence of noncondensable gas[J]. International Journal of Heat and Mass Transfer,2017,109:1004-1013.

[46] Wu S C,Yu C,Yu F W,et al. Lattice Boltzmann simulation of co-existing boiling and condensation phase changes in a confined micro-space[J]. International Journal of Heat and Mass Transfer, 2018, 126: 773-782.

[47] Zheng S F,Eimann F,Fieback T,et al. Numerical investigation of convective dropwise condensation flow by a hybrid thermal lattice Boltzmann method[J]. Applied Thermal Engineering,2018,145:590-602.

[48] Shen L Y,Tang G H,Li Q,et al. Hybrid wettability-induced heat transfer enhancement for condensation with noncondensable gas[J]. Langmuir,2019,35(29):9430-9440.

[49] Tang S, Li Q, Yu Y, et al. Enhancing dropwise condensation on downward-facing surfaces through the synergistic effects of surface structure and mixed wettability [J]. Physics of Fluids, 2021, 33 (8):083301.

[50] Li Q,Zhou P,Yan H J. Pinning-depinning mechanism of the contact line during evaporation on chemically patterned surfaces:A lattice Boltzmann study[J]. Langmuir,2016,32:9389-9396.

[51] Yu Y,Li Q,Zhou C Q,et al. Investigation of droplet evaporation on heterogeneous surfaces using a three-dimensional thermal multiphase lattice Boltzmann model[J]. Applied Thermal Engineering,2017,127: 1346-1354.

[52] Yuan W Z,Zhang L Z. Pinning-depinning mechanisms of the contact line during evaporation of micro-droplets on rough surfaces:A lattice Boltzmann simulation[J]. Langmuir,2018,34:7906-7915.

[53] Sun X Z,Li Q,Li W X,et al. Enhanced pool boiling on microstructured surfaces with spatially-controlled mixed wettability[J]. International Journal of Heat and Mass Transfer,2022,183:122164.

[54] 曾丹苓,敬成君．利用涨落理论确定液体的极限过热度[J]．中国科学(A 辑),1995,25(10): 1075-1081.

[55] 王竹溪．热力学[M]．2 版．北京:北京大学出版社,2004.

[56] 于渌,郝柏林,陈晓松．边缘奇迹:相变和临界现象[M]．北京:科学出版社,2005.

[57] Kestin J. A Course In Thermodynamics, vol. 1[M]. Hemisphere publishing Corportion, 1979.

[58] 曾丹苓. 汽-液相变中成核的热力学理论[J]. 重庆大学学报, 1995, 18(2): 1-8.

[59] Rowlinson J S. Translation of J. D. van der Waals "The thermodynamik theory of capillarity under the hypothesis of a continuous variation of density" [J]. Journal of Statistical Physics, 1979, 20(2): 197-200.

[60] Rowlinson J S, Widom B. Molecular Theory of Capillarity[M]. Oxford: Clarendon Press, 1982.

[61] Cahn J W, Hilliard J E. Free energy of a nonuniform system. I. Interfacial free energy[J]. The Journal of Chemical Physics, 1958, 28(2): 258-267.

[62] Cahn J W, Hilliard J E. Free energy of a nonuniform system. Ⅲ. Nucleation in a two-component incompressible fluid[J]. The Journal of Chemical Physics, 1959, 31(3): 688-699.

[63] Korteweg D J. Sur la forme que prennent les équations du mouvements des fluides si l'on tient compte des forces capillaires causées par des variations de densité considérables mais continues et sur la théorie de la capillarité dans l'hypothèse d'une variation continue de la densité [J]. Archives Neerlandaises des Sciences Exactes et Naturelles, Série II, 1901, 6: 1-24.

[64] Anderson D M, McFadden G B, Wheeler A A. Diffuse-interface methods in fluid mechanics[J]. Annual Review of Fluid Mechanics, 1998, 30(1): 139-165.

[65] Carey V P. Liquid-Vapor Phase-Change Phenomena[M]. Third ed. New York: CRC Press, 2020.

[66] 赵亚溥. 表面与界面物理力学[M]. 北京:科学出版社, 2012.

[67] He X Y, Doolen G D. Thermodynamic foundations of kinetic theory and lattice Boltzmann models for multiphase flows[J]. Journal of Statistical Physics, 2002, 107(1-2): 309-328.

[68] Giovangigli V. Kinetic derivation of diffuse-interface fluid models [J]. Physical Review E, 2020, 102: 012110.

[69] Li Q, Zhou P, Yan H J. Improved thermal lattice Boltzmann model for simulation of liquid-vapor phase change[J]. Physical Review E, 2017, 96: 063303.

[70] 吕俊复, 吴玉新, 李舟航, 等. 气液两相流动与沸腾传热[M]. 北京:科学出版社, 2017.

[71] Krishnan S, Murthy J Y, Garimella S V. A two-temperature model for solid-liquid phase change in metal foams[J]. Journal of Heat Transfer, 2004, 127(9): 995-1004.

[72] Liu Q, He Y L, Li Q. Enthalpy-based multiple-relaxation-time lattice Boltzmann method for solid-liquid phase-change heat transfer in metal foams[J]. Physical Review E, 2017, 96(2): 023303.

附录

附录 A　笛卡儿张量的基本知识

附录 A 中，我们归纳了格子 Boltzmann 方法中用到的部分张量知识，以便读者阅读本书。这些张量知识主要用在模型构建和宏观方程推导等方面。需要注意的是，这些并不是张量运算的全部，有兴趣的读者可参阅相关的张量以及流体力学书籍[1-3]。

A.1　标量、矢量与张量

流体力学中所处理的各种物理量，按其维数划分可分为**标量**(scalar)、**矢量**(vector)和**张量**(tensor)。

标量是一维的量，它只需 1 个数来表示，例如流体的密度、温度等都是标量。

矢量不仅有数量的大小，而且有指定的方向，它必须由某一空间坐标系的 3 个坐标轴方向的分量来表示，因此矢量是三维的量。常用黑体 $\boldsymbol{x}$ 或 $\boldsymbol{r}$ 来表示空间坐标位置矢量。对笛卡儿坐标，$\boldsymbol{x}$ 的 3 个分量为 x_1、x_2 和 x_3。3 个坐标方向的单位矢量(基矢量)分别用 $\boldsymbol{e}_1$、$\boldsymbol{e}_2$ 和 $\boldsymbol{e}_3$ 表示。因此位置矢量可写为

$$\boldsymbol{x}=x_1\boldsymbol{e}_1+x_2\boldsymbol{e}_2+x_3\boldsymbol{e}_3 \tag{A1}$$

三维空间中的二阶张量则是一个九维的量，必须用 9 个分量才能完整地表示，例如应力、应变率等都是二阶张量。在三维空间中，n 阶张量由 3^n 个分量组成，所以标量和矢量均可作为低阶张量而纳入张量的范畴。标量为零阶张量，而矢量则为一阶张量。

A.2　张量的表示法及爱因斯坦求和约定

为简化张量的书写，约定在同一项中如出现两个相同下标，就意味着对这个指标求和，而求和符号 $\sum$ 并不写出，该指标称为**哑指标**(dummy index)，这就是**爱因斯坦求和约定**(Einstein's summation convention)。利用这一约定，矢量 $\boldsymbol{x}$ 可表示为

$$\boldsymbol{x}=x_i\boldsymbol{e}_i \tag{A2}$$

又如

$$a_ib_i=a_1b_1+a_2b_2+a_3b_3 \tag{A3}$$

由于哑指标在求和运算后就消失了，因此改变哑指标的字母并不会改变表达式的内容，如 $a_ib_i=a_jb_j$。在张量运算中，合理利用这一性质，将会带来很大便利。

在方程同一项中只出现一次的指标称为**自由指标**(free index)，在同一方程的所有项中出现的自由指标必须相同。为避免混淆，求和约定要求在方程同一项中同一指标出现

的次数不能多于两次。

一个二阶张量可以看作是两个矢量 $\boldsymbol{a}$ 和 $\boldsymbol{b}$ 的并矢,即

$$\boldsymbol{P}=\boldsymbol{ab}=a_i b_j \boldsymbol{e}_i \boldsymbol{e}_j \tag{A4}$$

并矢运算不同于两个矢量的点乘运算和叉乘运算:点乘运算的结果是一个标量,叉乘运算的结果是一个矢量,而并矢运算的结果则是一个二阶张量。上述表达式中的 $\boldsymbol{e}_i\boldsymbol{e}_j$ 的顺序不能交换,因为 $a_i b_j \boldsymbol{e}_j \boldsymbol{e}_i$ 对应的是 $\boldsymbol{ba}$。如果把张量 $\boldsymbol{P}$ 的分量表示为 P_{ij},则张量 $\boldsymbol{P}$ 可以表示为 $\boldsymbol{P}=P_{ij}\boldsymbol{e}_i\boldsymbol{e}_j$。可以看到,由两个基矢量伴随一个二阶张量,在三维空间这样的基矢量对有 9 个,称为并基。三维空间中 n 阶张量由 3^n 个分量组成,其每一个分量有 n 个基矢量相伴随。二阶张量的分量可以写成矩阵的形式,即

$$\boldsymbol{P}=\begin{bmatrix} P_{11} & P_{12} & P_{13} \\ P_{21} & P_{22} & P_{23} \\ P_{31} & P_{32} & P_{33} \end{bmatrix} \tag{A5}$$

A.3 克罗内克符号

单位张量**克罗内克符号**(Kronecker symbol)δ_{ij}的定义为

$$\delta_{ij}=\begin{cases} 1 & i=j \\ 0 & i\neq j \end{cases} \tag{A6}$$

写成矩阵形式为

$$\boldsymbol{\delta}=\begin{bmatrix} 1 & 0 & 0 \\ 0 & 1 & 0 \\ 0 & 0 & 1 \end{bmatrix} \tag{A7}$$

δ_{ij}的基本性质有

$$\delta_{ij}=\delta_{ji}, \quad a_i\delta_{ij}=a_j, \quad P_{ik}\delta_{kj}=P_{ij} \tag{A8}$$

两个基矢量 $\boldsymbol{e}_i$ 和 $\boldsymbol{e}_j$ 的点积

$$\boldsymbol{e}_i \cdot \boldsymbol{e}_j=\delta_{ij} \tag{A9}$$

由定义可知,δ_{ij}是各向同性张量,即当坐标系转动后,张量的分量不变:$\delta'_{ij}=\delta_{ij}$。

A.4 对称张量与反对称张量

以二阶张量为例。如果一个二阶张量的下标 i 和 j 互换后所代表的分量仍不变,即 $S_{ij}=S_{ji}$,则称之为二阶**对称张量**(symmetric tensor)。也即二阶张量的矩阵表示形式中各元素关于主对角线对称,从而只有六个独立元素,可表示为

$$\boldsymbol{S}=S_{ij}=\begin{bmatrix} S_{11} & S_{12} & S_{13} \\ S_{12} & S_{22} & S_{23} \\ S_{13} & S_{23} & S_{33} \end{bmatrix} \tag{A10}$$

类似地,对于二阶**反对称张量**(anti-symmetric tensor),则有 $A_{ij}=-A_{ji}$,其主对角线元素为零,从而只有三个独立元素,即

$$\boldsymbol{A}=A_{ij}=\begin{bmatrix} 0 & A_{12} & A_{13} \\ -A_{12} & 0 & A_{23} \\ -A_{13} & -A_{23} & 0 \end{bmatrix} \tag{A11}$$

任何一个二阶张量 P_{ij} 都可以唯一地分解为一个二阶对称张量和一个二阶反对称张量的和，即

$$P_{ij}=\frac{1}{2}(P_{ij}+P_{ji})+\frac{1}{2}(P_{ij}-P_{ji}) \tag{A12}$$

当变换 i、j 后，右侧第一项不变，第二项改变符号。因而，$(P_{ij}+P_{ji})/2$ 为张量 P_{ij} 的对称部分，而 $(P_{ij}-P_{ji})/2$ 则为其反对称部分。

A.5 二阶张量的代数运算

与标量一样，张量也有加减以及求积等代数运算。

1）张量相等

两个张量相等则各分量一一对应相等。若张量 $\boldsymbol{A}=A_{ij}\boldsymbol{e}_i\boldsymbol{e}_j$ 与张量 $\boldsymbol{B}=B_{ij}\boldsymbol{e}_i\boldsymbol{e}_j$ 相等，则 $A_{ij}=B_{ij}$。

2）张量加减

张量加减为其同一坐标系下对应元素相加减。两张量必须同阶才能加减，即

$$\boldsymbol{A}\pm\boldsymbol{B}=A_{ij}\pm B_{ij} \tag{A13}$$

3）二阶张量的点积与双点积

以二阶张量为例，两个张量的**点积**（dot product，又名点乘、内积、数量积等）定义为

$$\begin{aligned}\boldsymbol{A}\cdot\boldsymbol{B}&=(A_{ij}\boldsymbol{e}_i\boldsymbol{e}_j)\cdot(B_{mn}\boldsymbol{e}_m\boldsymbol{e}_n)\\&=A_{ij}B_{mn}\delta_{jm}\boldsymbol{e}_i\boldsymbol{e}_n=A_{ij}B_{jn}\boldsymbol{e}_i\boldsymbol{e}_n\end{aligned} \tag{A14}$$

显然，二阶张量的点积即为并矢中相邻的两个单位矢量做点积，得到一个新的张量。

二阶张量与矢量的点积则定义为

$$\boldsymbol{a}\cdot\boldsymbol{B}=(a_i\boldsymbol{e}_i)\cdot(B_{mn}\boldsymbol{e}_m\boldsymbol{e}_n)=a_iB_{mn}\delta_{im}\boldsymbol{e}_n=a_iB_{in}\boldsymbol{e}_n \tag{A15}$$

$$\boldsymbol{B}\cdot\boldsymbol{a}=(B_{mn}\boldsymbol{e}_m\boldsymbol{e}_n)\cdot(a_i\boldsymbol{e}_i)=B_{mn}a_i\delta_{in}\boldsymbol{e}_m=B_{mn}a_n\boldsymbol{e}_m \tag{A16}$$

由上述两式可知，二阶张量与矢量的点积为一个新的矢量，且通常有 $\boldsymbol{a}\cdot\boldsymbol{B}\neq\boldsymbol{B}\cdot\boldsymbol{a}$。但单位二阶张量 $\boldsymbol{I}$ 与任意矢量的点积均给出原矢量，即

$$\boldsymbol{a}\cdot\boldsymbol{I}=\boldsymbol{I}\cdot\boldsymbol{a}=\boldsymbol{a} \tag{A17}$$

二阶张量的**双点积**（double product，又名双点乘）定义为

$$\boldsymbol{A}:\boldsymbol{B}=(A_{ij}\boldsymbol{e}_i\boldsymbol{e}_j):(B_{mn}\boldsymbol{e}_m\boldsymbol{e}_n)=A_{ij}B_{mn}(\boldsymbol{e}_i\cdot\boldsymbol{e}_m)(\boldsymbol{e}_j\cdot\boldsymbol{e}_n)=A_{ij}B_{ij} \tag{A18}$$

显然，两个二阶张量的双点积结果为一个新的标量。

A.6 张量的微分和积分运算

在流体力学的相关公式中，常涉及张量的微分和积分运算，其中最常见的是梯度、散度、拉普拉斯算子、高斯公式等。

A.6.1 梯度

如果**哈密顿算符**（Hamiltonian operator）用张量指标法表示为 $\nabla=\boldsymbol{e}_i\partial/\partial x_i$，则标量 ϕ 的**梯度**（gradient）为

$$\nabla\phi=\left(\boldsymbol{e}_i\frac{\partial}{\partial x_i}\right)\phi=\boldsymbol{e}_i\frac{\partial\phi}{\partial x_i} \tag{A19}$$

利用哈密顿算符进行运算时，先进行微分运算，后进行张量运算。对矢量 $\boldsymbol{u}=u_i\boldsymbol{e}_i$，其梯度为

$$\nabla \boldsymbol{u}=\left(\boldsymbol{e}_i \frac{\partial}{\partial x_i}\right)(u_j\boldsymbol{e}_j)=\frac{\partial u_j}{\partial x_i}\boldsymbol{e}_i\boldsymbol{e}_j \tag{A20}$$

可见,标量和矢量的梯度分别是矢量和二阶张量,因而一个 n 阶张量的梯度是 $n+1$ 阶张量。

当上面的 $\boldsymbol{u}$ 表示流体力学中的速度矢量时,$\nabla \boldsymbol{u}$ 即为速度梯度张量,而$(\nabla \boldsymbol{u})^{\mathrm{T}}$ 则为速度梯度张量的转置

$$(\nabla \boldsymbol{u})^{\mathrm{T}}=\left(\frac{\partial u_j}{\partial x_i}\boldsymbol{e}_i\boldsymbol{e}_j\right)^{\mathrm{T}}=\frac{\partial u_j}{\partial x_i}\boldsymbol{e}_j\boldsymbol{e}_i \tag{A21}$$

交换 i、j 可得

$$(\nabla \boldsymbol{u})^{\mathrm{T}}=\frac{\partial u_i}{\partial x_j}\boldsymbol{e}_i\boldsymbol{e}_j \tag{A22}$$

因而流体力学中的应变率张量为

$$\boldsymbol{S}=\frac{1}{2}[\nabla \boldsymbol{u}+(\nabla \boldsymbol{u})^{\mathrm{T}}]=\frac{1}{2}\left(\frac{\partial u_j}{\partial x_i}+\frac{\partial u_i}{\partial x_j}\right)\boldsymbol{e}_i\boldsymbol{e}_j \tag{A23}$$

此外,速度梯度张量与单位二阶张量双点积为

$$\nabla \boldsymbol{u}:\boldsymbol{I}=(\nabla \boldsymbol{u})^{\mathrm{T}}:\boldsymbol{I}=\nabla\cdot \boldsymbol{u} \tag{A24}$$

即一个张量与单位张量做双点积,得到该张量对角线元素的和,称为张量的**迹**(trace)。

A.6.2 散度

张量的**散度**(divergence)相当于哈密顿算符与张量的点积。对矢量 $\boldsymbol{a}=a_i\boldsymbol{e}_i$,其散度为

$$\nabla\cdot \boldsymbol{a}=\left(\boldsymbol{e}_i \frac{\partial}{\partial x_i}\right)\cdot(a_j\boldsymbol{e}_j)=\frac{\partial a_j}{\partial x_i}(\boldsymbol{e}_i\cdot\boldsymbol{e}_j)=\frac{\partial a_i}{\partial x_i} \tag{A25}$$

对二阶张量 $\boldsymbol{P}=P_{ij}\boldsymbol{e}_i\boldsymbol{e}_j$,其散度为

$$\nabla\cdot \boldsymbol{P}=\left(\boldsymbol{e}_i \frac{\partial}{\partial x_i}\right)\cdot(P_{jk}\boldsymbol{e}_j\boldsymbol{e}_k)=\frac{\partial P_{jk}}{\partial x_i}(\boldsymbol{e}_i\cdot\boldsymbol{e}_j)\boldsymbol{e}_k=\frac{\partial P_{ik}}{\partial x_i}\boldsymbol{e}_k \tag{A26}$$

可见,矢量和二阶张量的散度分别为标量和矢量,因而一个 n 阶张量的散度是 $n-1$ 阶张量。

A.6.3 拉普拉斯算符

拉普拉斯算符(Laplace operator,或 Laplacian)可表示为先求梯度,然后再求散度的运算,如

$$\nabla\cdot\nabla\phi=\nabla^2\phi=\left(\boldsymbol{e}_i \frac{\partial}{\partial x_i}\right)\cdot\left(\boldsymbol{e}_j \frac{\partial}{\partial x_j}\right)\phi=\frac{\partial}{\partial x_i}\left(\frac{\partial \phi}{\partial x_i}\right) \tag{A27}$$

拉普拉斯算符作用于矢量 $\boldsymbol{a}=a_i\boldsymbol{e}_i$,则有

$$\nabla\cdot\nabla \boldsymbol{a}=\nabla^2\boldsymbol{a}=\frac{\partial}{\partial x_i}\left(\frac{\partial a_j}{\partial x_i}\boldsymbol{e}_j\right)=\boldsymbol{e}_j\frac{\partial}{\partial x_i}\left(\frac{\partial a_j}{\partial x_i}\right) \tag{A28}$$

A.6.4 高斯公式

设 $\boldsymbol{P}$ 是二阶张量,则高斯公式为

$$\int_A \boldsymbol{n}\cdot \boldsymbol{P}\mathrm{d}A=\int_V \nabla\cdot \boldsymbol{P}\mathrm{d}V \tag{A29}$$

附录B 格子张量

B.1 正多边形格子张量的性质及基本粒子速度模型

在格子 Boltzmann 模型的构造过程中,经常会遇到粒子速度的 n 阶张量

$$\boldsymbol{E}_{i_1i_2\cdots i_n}^{(n)} = \sum_{\alpha} (\boldsymbol{e}_{\alpha})_{i_1}\cdots(\boldsymbol{e}_{\alpha})_{i_n} \tag{B1}$$

式中,$\boldsymbol{e}_{\alpha}(\alpha=1,2,\cdots,M)$为一矢量序列,类似于格子 Boltzmann 模型中的粒子速度。对于正 M 边形网格,$\boldsymbol{e}_{\alpha}$ 的配置为

$$\boldsymbol{e}_{\alpha}=e\left(\cos\frac{2\pi\alpha}{M},\sin\frac{2\pi\alpha}{M}\right) \tag{B2}$$

式中,$e=|\boldsymbol{e}_{\alpha}|$。由于 $\boldsymbol{e}_{\alpha}$ 具有**各向同性**(isotropic)的特点,其对应的张量 $\boldsymbol{E}^{(n)}$ 亦满足各向同性,具有如下性质[4-5]:

(1) 奇数阶$(2n+1,n\geqslant0)$的各向同性张量为0;

(2) 偶数阶$(2n,n\geqslant1)$的各向同性张量与 $\Delta_{i_1i_2\cdots i_{2n}}^{(2n)}$ 成比例。

对于性质(1),一阶张量 $\boldsymbol{E}^{(1)}$ 对应于矢量,由于 $\boldsymbol{e}_{\alpha}$ 具有各向同性,则其相对于坐标轴是对称的,将 $\boldsymbol{e}_{\alpha}$ 投射到坐标轴上并求矢量和,根据对称性其和 $\boldsymbol{E}^{(1)}$ 必然为0。依次类推,其他奇数阶张量 $\boldsymbol{E}^{(n)}$ 同样为0。性质(1)(2)相应的数学表达为

$$\boldsymbol{E}^{(2n+1)}=0 \tag{B3}$$

$$\boldsymbol{E}^{(2n)}=\frac{M}{d(d+2)\cdots(d+2n-2)}e^{2n}\Delta^{(2n)} \tag{B4}$$

式中,张量算子 $\Delta^{(2n)}$ 为

$$\Delta_{ij}^{(2)}=\delta_{ij} \tag{B5}$$

$$\Delta_{ijkl}^{(4)}=\delta_{ij}\delta_{kl}+\delta_{ik}\delta_{jl}+\delta_{il}\delta_{jk} \tag{B6}$$

$$\Delta_{ijklpq}^{(6)}=\delta_{ij}\Delta_{klpq}^{(4)}+\delta_{ik}\Delta_{jlpq}^{(4)}+\delta_{il}\Delta_{jkpq}^{(4)}+\delta_{ip}\Delta_{jklq}^{(4)}+\delta_{iq}\Delta_{jklp}^{(4)} \tag{B7}$$

并存在如下的递归关系:

$$\Delta_{i_1i_2\cdots i_{2n}}^{(2n)}=\sum_{j=2}^{2n}\delta_{i_1i_j}\Delta_{i_2\cdots i_{j-1}i_{j+1}\cdots i_{2n}}^{(2n-2)} \tag{B8}$$

文献[5]对正多边形格子进行了相关的张量分析,得出了各向同性 $\boldsymbol{E}^{(n)}$ 的阶数与正多边形的边数之间的对应关系,详见表 B.1。

表 B.1 各向同性张量的阶数与正多边形边数 M 对应表

$\boldsymbol{E}^{(2)}$	$M>2$
$\boldsymbol{E}^{(3)}$	$M\geqslant2,M\neq3$
$\boldsymbol{E}^{(4)}$	$M>2,M\neq4$
$\boldsymbol{E}^{(5)}$	$M\geqslant2,M\neq3$、5
$\boldsymbol{E}^{(6)}$	$M>4,M\neq6$
$\boldsymbol{E}^{(7)}$	$M\geqslant2,M\neq3$、5、7

从表 B. 1 可以看到,当 M 不等于 $n,n-2,n-4,\cdots$ 时,$\boldsymbol{E}^{(n)}$ 是各向同性的。同时,对于足够大的 M,$M>n$,其所对应的张量 $\boldsymbol{E}^{(n)}$ 一定是各向同性的。例如,正四边形最多能保证三阶格子张量是各向同性的,而其四阶张量则是各向异性的

$$E_{ijkl}^{(4)}\mid_{M=4}=2\delta_{ijkl} \tag{B9}$$

式中,δ_{ijkl} 为非对称张量,当且仅当 $i=j=k=l$ 时,$\delta_{ijkl}=1$,否则为 0。

正方形格子中两种基本的粒子速度模型如图 B. 1 所示。其中,图 B. 1 中模型(a)对应正四边形;而图 B. 1 中模型(b)则是从正八边形而来,其对应的各阶张量为图 B. 2 所示正八边形的格子张量减去图中虚线对应的正四边形的格子张量。两种基本模型的格子张量详见表 B. 2。许多二维多速模型都是由这两种基本粒子速度模型组合而成。

对于三维模型,其对应的为多面体,因此要比二维模型复杂得多,感兴趣的读者可参阅相关文献[5]。

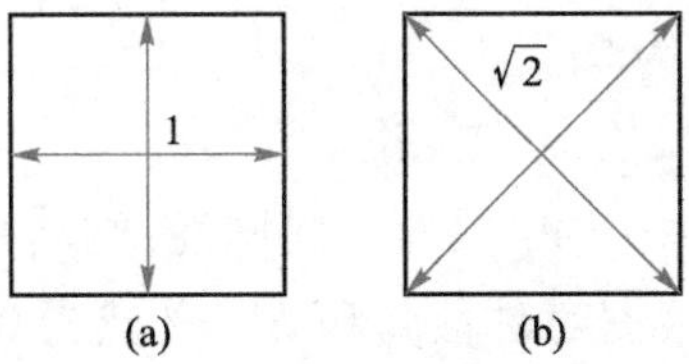

图 B. 1 正方形格子中两种基本的粒子速度模型

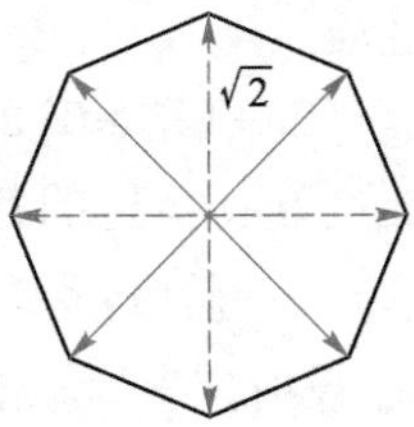

图 B. 2 正八边形

表 B. 2 两种基本模型的张量计算

模型	e_α	M	$E_{ij}^{(2)}$	$E_{ijkl}^{(4)}$	$E_{ijklpq}^{(6)}$
(a)	cyc:(±1,0)	4	$2\delta_{ij}$	$2\delta_{ijkl}$	$2\delta_{ijklpq}$
(b)	(±1,±1)	8	$4\delta_{ij}$	$4\Delta_{ijkl}^{(4)}-8\delta_{ijkl}$	$4\Delta_{ijklpq}^{(6)}/3-16\delta_{ijklpq}$

注:cyc 表示循环排列,即(±1,0),(0,±1)。

B. 2 常用离散速度模型及其张量计算

我们整理了常用离散速度模型的速度配置及其张量计算,详见表 B. 3。

表 B. 3 常用离散速度模型的速度配置及其相应的张量

离散速度模型	速度配置	$E_{ij}^{(2)}$	$E_{ijkl}^{(4)}$	$E_{ijklpq}^{(6)}$
D2Q9	(0,0)	0	0	0
	cyc:(±1,0)	$2\delta_{ij}$	$2\delta_{ijkl}$	$2\delta_{ijklpq}$
	(±1,±1)	$4\delta_{ij}$	$4\Delta_{ijkl}^{(4)}-8\delta_{ijkl}$	$4\Delta_{ijklpq}^{(6)}/3-16\delta_{ijklpq}$
D3Q15	(0,0,0)	0	0	0
	cyc:(±1,0,0)	$2\delta_{ij}$	$2\delta_{ijkl}$	$2\delta_{ijklpq}$
	(±1,±1,±1)	$8\delta_{ij}$	$8\Delta_{ijkl}^{(4)}-16\delta_{ijkl}$	$8(\Delta_{ijklpq}^{(6)}-2\Delta_{ijklpq}^{(4,2)}+16\delta_{ijklpq})$

续表

离散速度模型	速度配置	$E_{ij}^{(2)}$	$E_{ijkl}^{(4)}$	$E_{ijklpq}^{(6)}$
D3Q19	(0,0,0)	0	0	0
	cyc:(±1,0,0)	$2\delta_{ij}$	$2\delta_{ijkl}$	$2\delta_{ijklpq}$
	cyc:(±1,±1,0)	$8\delta_{ij}$	$4\Delta_{ijkl}^{(4)}-4\delta_{ijkl}$	$4(\Delta_{ijklpq}^{(4,2)}-13\delta_{ijklpq})$

表中,δ_{ijklpq}、$\Delta_{ijklpq}^{(4,2)}$ 均为非对称张量,且有

$$\Delta_{ijklpq}^{(4,2)}=\delta_{ij}\delta_{klpq}+\delta_{ik}\delta_{jlpq}+\delta_{il}\delta_{jkpq}+\delta_{ip}\delta_{jklq}+\delta_{iq}\delta_{jklp}+\delta_{jk}\delta_{ilpq}+\delta_{jl}\delta_{ikpq}+\delta_{jp}\delta_{iklq}+\delta_{jq}\delta_{iklp}+\delta_{kl}\delta_{ijpq}+\delta_{kp}\delta_{ijlq}+\delta_{kq}\delta_{ijlp}+\delta_{lp}\delta_{ijkq}+\delta_{lq}\delta_{ijkp}+\delta_{pq}\delta_{ijkl} \tag{B10}$$

附录C 单位转换

对于实际物理问题的数值模拟,通常可以采取两种思路编写相关程序:一种是在程序代码中,所有物理量均采用实际**物理单位**(physical unit);另一种是将所有的物理参数都做无量纲处理,并确保无量纲化前后的流动与传热准则数一致。类似地,在格子Boltzmann方法的相关模拟中,我们可以直接采用物理单位,也可以采用**格子单位**(lattice unit)。因而,这就涉及物理单位与格子单位之间的转换问题。

尽管早在2001年,Succi[6]就在他的著作中提到了格子Boltzmann方法的单位转换问题,但我们发现有不少初学者对这一问题较为困惑。本节,基于我们的理解,对该转换问题予以阐述。

C.1 参考量与单位转换

以D2Q9模型为例,模拟中最基本的参数为长度L、密度ρ、时间t、速度u以及运动黏度ν,假设这些量都为格子单位。相应地,实际物理单位下的参数为长度L'、密度ρ'、时间t'、速度u'以及运动黏度ν'。为了实现格子单位与实际物理单位之间的转换,需要确定对应的参考量,即参考长度L_r、参考密度ρ_r、参考时间t_r、参考速度u_r和参考运动黏度ν_r,其定义如下:

$$L_r=\frac{L'}{L},\quad \rho_r=\frac{\rho'}{\rho},\quad t_r=\frac{t'}{t},\quad u_r=\frac{u'}{u},\quad \nu_r=\frac{\nu'}{\nu} \tag{C1}$$

以上物理量只包含质量、长度和时间3个基本物理量,且由于质量只包含在密度中,因此,L_r、t_r、u_r和ν_r只有2个是独立的,另2个可以通过计算确定。

对于一个物理问题,实际的L'、ρ'、u'以及ν'通常是已知的。单相不可压缩流动不受密度取值的影响,ρ_r可任意取值,因此可以令$\rho=1.0$。而通常格子单位下采用$\delta_x=\delta_y=1$,$\delta_t=1$,当网格划分完成之后,根据特征长度所划分的网格数,L_r可以被确定。考虑到格子Boltzmann计算的稳定性,可以选取模型的运动黏度ν为一合理数值,从而确定ν_r。于是,速度和时间的参考量可通过L_r和ν_r计算

$$u_r=\frac{\nu_r}{L_r},\quad t_r=\frac{L_r^2}{\nu_r} \tag{C2}$$

至此,就可完成格子单位与实际物理单位之间的转换,换算成物理单位为

$$\delta_x' = \delta_y' = L_r, \quad \delta_t' = t_r = \frac{L_r^2}{\nu_r} = \frac{L_r}{u_r} \tag{C3}$$

需要说明的是，由于 L_r、t_r、u_r 和 ν_r 互相约束，在格子 Boltzmann 计算模型的建立及单位换算过程中可能需要调整网格数、运动黏度和速度的取值，使其在合理范围。

C.2 算例

以一个实际中可能遇到的换热管圆柱绕流为例，假设流体为 20℃ 的空气，其运动黏度 $\nu' = 1.5\times10^{-5}\text{m}^2/\text{s}$，入口流速 $u' = 5\text{m/s}$ 且为不可压缩流动，圆柱直径 $D' = 38$ mm。若模拟中取格子单位下的参数为 $\delta_x = \delta_y = 1$、$\delta_t = 1$、$\nu = 0.005$，且将圆柱直径划分为 $N_x = 380$ 个网格，即 $D = N_x\delta_x = 380$，则参考长度 $L_r = D'/D = 10^{-4}\text{m}$，参考运动黏度 $\nu_r = \nu'/\nu = 3\times10^{-3}\text{m}^2/\text{s}$。于是，速度和时间的参考量可通过 L_r 和 ν_r 计算：

$$u_r = \frac{\nu_r}{L_r} = 30 \text{ m/s} \tag{C4}$$

$$t_r = \frac{L_r^2}{\nu_r} = \frac{L_r}{u_r} = 3.33\times10^{-6} \text{ s} \tag{C5}$$

则格子 Boltzmann 模型计算中的入口速度为 $u = u'/u_r = 0.167$。

分析上述单位转换结果，首先，$\delta_t' = \delta_t t_r = 3.33\times10^{-6}$ s 在量级上较小，对于实际时间尺度的模拟需要较多的计算步骤。其次，$\nu = 0.005$ 对于单松弛格子 Boltzmann 模型已经过小，很容易导致计算发散，因此需要采用多松弛模型。同时，考虑到 $u \ll c_s = 1/\sqrt{3}$ 的不可压缩流动条件，$u = 0.167$ 的速度也已经较大。从式(C4)可以发现，ν 和 u 的取值存在一定矛盾，为了增加 ν 的数值以提高计算稳定性，需要减小参考值 ν_r，在 L_r 一定的条件下，会同时减小 u_r，导致格子 Boltzmann 计算中的 u 增大。例如，若取 $\nu = 0.01$，则有 $u = 0.333$。解决以上矛盾的方式只有同时减小 L_r，即进一步增加所划分的网格数，从而增大计算量。

可见，格子 Boltzmann 方法作为一种介观模拟方法，如果被用来处理宏观问题，即较大的空间时间尺度，就会对计算资源要求很高。这也在一定程度上解释了为什么高雷诺数或高瑞利数问题的模拟是格子 Boltzmann 方法领域的一大难题。为了克服这一困难，可采用非均匀网格，并在时间上采用一些加速措施，如第 9 章提到的隐式-显式有限差分格子 Boltzmann 方法。当然，也可基于区域分裂，采用并行计算，这正是格子 Boltzmann 方法的一大优势。

最后，需要指出的是，在采用格子单位做计算时，还可撇开物理单位的概念，只需保证模拟的关键准则数与实际物理问题的相关准则数相同即可。事实上，这一思路在计算流体力学与计算传热学中使用较广。当然，本节所示的单位转换方法，同样能够保证相关准则数相同。

附录 D 顶盖驱动流的格子 Boltzmann 模拟

D.1 物理模型

顶盖驱动流(lid-driven flow)是计算流体与计算传热学中的一个经典问题(图 D.1)，

常用作等温不可压缩算法的校核算例，同时也是一个很好的格子 Boltzmann 方法入门算例。在顶盖驱动流中，方腔的上边界以一个恒定速度水平右移，而其他三个边界则保持静止不动。其基本特征是：流动稳定后，方腔的中央有一个一级大涡出现，而在左下角和右下角会分别出现一个二级涡，当雷诺数超过一临界值后，在方腔的左上角还会出现一个涡。这些涡的中心位置是雷诺数的函数。雷诺数的定义为 $Re=LU/\nu$，其中，L 是方腔的高度，U 是顶盖的移动速度，ν 是运动黏度。

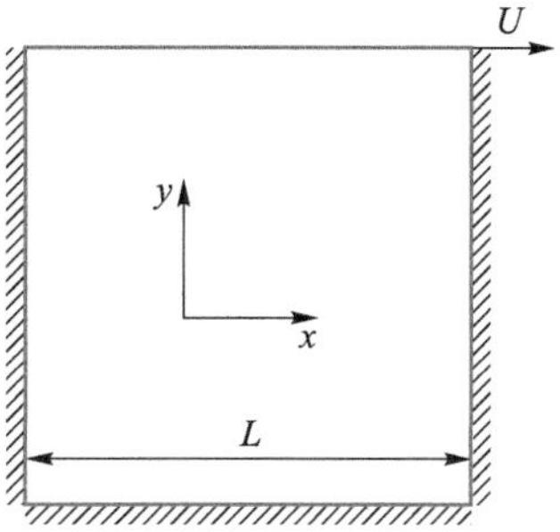

图 D.1 顶盖驱动流示意图

在本附录，我们提供一个顶盖驱动流的格子 Boltzmann 模拟程序，采用 D2Q9 模型以及标准的格子 Boltzmann 方程，边界处理采用非平衡态外推格式，程序收敛判据如下：

$$E_r=\frac{\sqrt{\sum_{i,j}\{[u_x(i,j,t+\delta_t)-u_x(i,j,t)]^2+[u_y(i,j,t+\delta_t)-u_y(i,j,t)]^2\}}}{\sqrt{\sum_{i,j}[u_x(i,j,t+\delta_t)^2+u_y(i,j,t+\delta_t)^2]}}<10^{-6}$$

模拟中，流场初始密度 $\rho_0=1$，顶盖驱动速度 $U=0.1$，计算区域的大小为 $L_x\times L_y=256\times256$，也即 $L=256$；采用湿节点式布置方式（参见 8.1 节），运动黏度由雷诺数定义式反算得到。各参数均为格子单位。

虽然该程序是针对顶盖驱动流而编写的，读者可通过改变初始条件、边界条件等，来计算其他物理问题。程序由 C++语言编写，支持的环境为 Microsoft Visual Studio 2010。

D.2 程序变量表及源程序

变量表

变量名	变量含义
Q	离散速度的总个数
Lx,Ly	x, y 方向的长度
NX+1,NY+1	x,y 方向的节点数
dx,dy	x, y 方向的网格步长
dt	时间步长
c	格子速度
n	演化次数
k	离散速度方向 α
Re	Re 数
niu	运动黏度 ν
tau_f	量纲为 1 的松弛时间 τ
U	顶盖速度
e[Q][2]	离散速度 $\boldsymbol{e}_\alpha$
w[Q]	权系数 ω_α
rho[NX+1][NY+1]	密度 ρ
u0[NX+1][NY+1][2]	n 时层的速度

续表

变量名	变量含义
u[NX+1][NY+1][2]	n+1 时层的速度
f[NX+1][NY+1][Q]	演化前的分布函数
F[NX+1][NY+1][Q]	演化后的分布函数
rho0	流场初始密度 ρ_0
error	两个相邻时层速度的最大相对误差

源程序

```
#include" StdAfx. h"
#include<iostream>
#include<cmath>
#include<cstdlib>
#include<iomanip>
#include<fstream>
#include<sstream>
#include<string>

using namespace std;
const int Q = 9;    //D2Q9 模型
const int NX = 256;
const int NY = 256;
const double U = 0. 1;

int  e[Q][2] = {{0,0},{1,0},{0,1},{-1,0},{0,-1},{1,1},{-1,1},{-1,-1},{1,-1}};
double  w[Q] = {4. 0/9,1. 0/9,1. 0/9,1. 0/9,1. 0/9,1. 0/36,1. 0/36,1. 0/36,1. 0/36};
double rho[NX+1][NY+1],u[NX+1][NY+1][2],u0[NX+1][NY+1][2],f[NX+1][NY+1][Q],
  F[NX+1][NY+1][Q];
int i,j,k,ip,jp,n;
double c,Re,dx,dy,Lx,Ly,dt,rho0,P0,tau_f,niu,error;

void init();
double feq(int k,double rho,double u[2]);
void evolution();
void output(int m);
void Error();

int main()
{
using namespace std;
init();
for(n = 0;  ; n++)
{
```

```
    evolution( );
    if( n%500 = = 0)
    {
      Error( );
    cout<<"The "<<n<<"th computation result:"<<endl<<"The u,v of point(NX/2,NY/2) is:"
      <<setprecision(6)<<u[NX/2][NY/2][0]<<","<<u[NX/2][NY/2][1]<<endl;
    cout<<"The max relative error of uv is:"<<setiosflags(ios::scientific)<<error<<endl;
     if(n >=2000)
     {
      if(n%2000 = = 0) output(n);
      if(error<1.0e-6) break;
      }
    }
  }
  return 0;
}

void init( )
{
    dx = 1.0;
    dy = 1.0;
    Lx = dx * double(NY);
    Ly = dy * double(NX);
    dt = dx;
    c = dx/dt;
    rho0 = 1.0;
    Re = 1000;
    niu = U * Lx/Re;
    tau_f = 3.0 * niu+0.5;
    std::cout<<"tau_f = " <<tau_f<<endl;

    for(i=0; i<=NX; i++)        //初始化
      for(j=0; j<=NY; j++)
    {
      u[i][j][0] = 0;
      u[i][j][1] = 0;
      rho[i][j] = rho0;
      u[i][NY][0] = U;
      for(k=0; k<Q; k++)
    {
        f[i][j][k] = feq(k,rho[i][j],u[i][j]);
    }
  }
}
```

```
double feq(int k,double rho,double u[2])    //计算平衡态分布函数
{
    double eu,uv,feq;
    eu = (e[k][0] * u[0]+e[k][1] * u[1]);
    uv = (u[0] * u[0]+u[1] * u[1]);
    feq = w[k] * rho * (1.0+3.0 * eu+4.5 * eu * eu-1.5 * uv);
    return feq;
}

void evolution()
{
for(i=1; i<NX; i++) //演化
  for(j=1; j<NY; j++)
    for(k=0; k<Q; k++)
    {
      ip = i-e[k][0];
      jp = j-e[k][1];
      F[i][j][k] = f[ip][jp][k] + (feq(k,rho[ip][jp],u[ip][jp])-f[ip][jp][k])/tau_f;
    }

for(i=1; i<NX; i++)    //计算宏观量
    for(j=1; j<NY; j++)
    {
      u0[i][j][0] = u[i][j][0];
      u0[i][j][1] = u[i][j][1];
      rho[i][j] = 0;
      u[i][j][0] = 0;
      u[i][j][1] = 0;
      for(k=0; k<Q; k++)
      {
        f[i][j][k] = F[i][j][k];
        rho[i][j] += f[i][j][k];
        u[i][j][0] += e[k][0] * f[i][j][k];
        u[i][j][1] += e[k][1] * f[i][j][k];
      }
      u[i][j][0] /= rho[i][j];
      u[i][j][1] /= rho[i][j];
    }

//边界处理
    for(j=1; j<NY; j++)        //左右边界
    for(k=0; k<Q; k++)
    {
      rho[NX][j] = rho[NX-1][j];
```

```
      f[NX][j][k] = feq(k,rho[NX][j],u[NX][j]) + f[NX-1][j][k]-feq(k,rho[NX-1][j],
      u[NX-1][j]);
      rho[0][j] = rho[1][j];
      f[0][j][k] = feq(k,rho[0][j],u[0][j]) + f[1][j][k]-feq(k,rho[1][j],u[1][j]);
    }
  for(i=0; i<=NX; i++)      //上下边界
    for(k=0; k<Q; k++)
    {
      rho[i][0] = rho[i][1];
      f[i][0][k] = feq(k,rho[i][0],u[i][0]) + f[i][1][k]-feq(k,rho[i][1],u[i][1]);

      rho[i][NY] = rho[i][NY-1];
      u[i][NY][0] = U;
      f[i][NY][k] = feq(k,rho[i][NY],u[i][NY]) + f[i][NY-1][k]-feq(k,rho[i][NY-1],
    u[i][NY-1]);
    }
}

void output(int m)   //输出
{
    ostringstream name;
    name<<"cavity_"<<m<<".dat";
    ofstream out(name.str().c_str());
    out<<"Title =\"LBM Lid Driven Flow\"\n"<<"VARIABLES = \"X\",\"Y\",\"U\",\"V\"\n"
      <<"ZONE T=\"BOX\",I="<<NX+1<<",J="<<NY+1<<",F=POINT"<<endl;
    for(j=0; j<=NY; j++)
      for(i=0; i<=NX; i++)
      {
        out<<double(i)/Lx<<" "<<double(j)/Ly<<" "<<u[i][j][0]<<" "<<u[i][j][1]<<endl;
        }
}

void Error()
{
    double temp1,temp2;
    temp1 = 0;
    temp2 = 0;
    for(i=1; i<NX; i++)
      for(j=1; j<NY; j++)
      {
        temp1 +=
      ((u[i][j][0]-u0[i][j][0]) * (u[i][j][0]-u0[i][j][0])+(u[i][j][1]-u0[i][j][1]) *
        (u[i][j][1]-u0[i][j][1]) );
      temp2 += (u[i][j][0] * u[i][j][0]+u[i][j][1] * u[i][j][1]);
```

```
    }
    temp1 = sqrt(temp1);
    temp2 = sqrt(temp2);
    error = temp1/(temp2+1e-30);
}
```

D.3 数值模拟结果

图 D.2 给出了不同雷诺数下顶盖驱动流的等流函数线，从图中可以清楚地看到雷诺数对流动模式的影响。当雷诺数较小时（$Re \leqslant 1\ 000$），方腔中只出现三个涡：一个位于方腔中央的一级涡和一对位于左下角和右下角附近的二级涡。当 $Re=2\ 000$ 时，在左上角出现第三个二级涡。当雷诺数上升至 5 000 时，在右下角出现了一个三级涡。从图中还可以看到，随着雷诺数的增加，一级涡的中心向方腔的中央位置移动[7]。

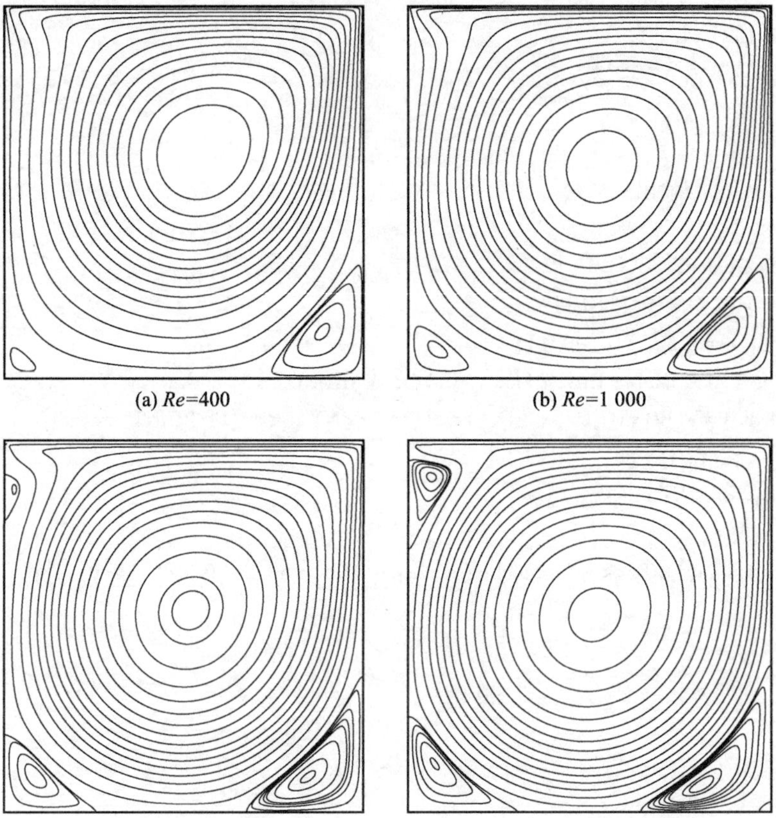

(a) Re=400　(b) Re=1 000　(c) Re=2 000　(d) Re=5 000

图 D.2　顶盖驱动流的等流函数线

为了量化上述结果，我们测量了方腔中央的一级涡以及左右下角附近的两个二级涡的中心位置，结果列于表 D.1。表中，a、b、c 和 d 分别代表文献[8]、[9]、[10]和本程序的模拟结果。可以看到，本程序的模拟结果与其他文献的结果吻合得很好。

表 D.1　顶盖驱动流的涡的位置

Re		一级涡		左下涡		右下涡	
		x	*y*	*x*	*y*	*x*	*y*
400	a	0.556 3	0.600 0	0.050 0	0.050 0	0.887 5	0.118 8
	b	0.554 7	0.605 5	0.050 8	0.046 9	0.890 6	0.125 0
	c	0.560 8	0.607 8	0.054 9	0.051 0	0.890 2	0.125 5
	d	0.566 8	0.606 9	0.047 1	0.048 2	0.885 7	0.123 1
1 000	a	0.543 8	0.562 5	0.075 0	0.081 3	0.862 5	0.106 3
	b	0.531 3	0.562 5	0.085 9	0.078 1	0.859 4	0.109 4
	c	0.533 3	0.564 7	0.090 2	0.078 4	0.866 7	0.113 7
	d	0.533 6	0.567 5	0.082 0	0.073 1	0.864 4	0.112 8
2 000	a	0.525 0	0.550 0	0.087 5	0.106 3	0.837 5	0.093 8
	c	0.525 5	0.549 0	0.902	0.105 9	0.847 1	0.098 0
	d	0.522 6	0.548 2	0.086 3	0.101 6	0.843 9	0.097 7
5 000	a	0.512 5	0.531 3	0.062 5	0.156 3	0.850 0	0.081 3
	b	0.511 7	0.535 2	0.070 3	0.136 7	0.808 6	0.074 2
	c	0.517 6	0.537 3	0.078 4	0.137 3	0.807 8	0.074 5
	d	0.515 8	0.535 8	0.074 2	0.133 7	0.805 5	0.074 5

附录 E　基于伪势多相模型的相分离格子 Boltzmann 模拟

E.1　问题描述

与亚稳态区域基于成核生长而发生的相变不同，**相分离**(phase separation)通常是指发生在体系的不稳定区域的**旋节分解**(spinodal decomposition)。在不稳定区域，任何微小的涨落或扰动都可能导致自发的相分离。以淬火过程为例，当体系淬火到不稳定区域时，会立即发生旋节分解，即相分离。类似地，对气液两相系统，假设初始时刻的密度或比体积处在图 10.20(参见 10.5.1 节)中曲线 *MON* 所对应的热力学不稳定区域，只要施加微小的扰动就可能促使相分离的发生。

在本附录，我们提供了一个基于伪势模型的相分离格子 Boltzmann 模拟程序。具体地，采用 D2Q9 离散速度模型以及相应的格子 Boltzmann-BGK 方程，即式(3.100)。作用力项和伪势相互作用力参见式(10.20)和式(10.21)。式(10.21)中的权系数 w_α 与平衡态分布函数中的权系数 ω_α 存在 $w_\alpha=\omega_\alpha/c_s^2=3\omega_\alpha$ 的关系。在本算例中，不考虑重力，并且采用指数形式的伪势(参见 5.3.1 节)，即 $\psi(\rho)=\psi_0\exp(-\rho_0/\rho)$ [11-12]。相关的模拟参数为 $\psi_0=1$、$\rho_0=1$、$G=-10/3$。根据 Maxwell 等面积法则，对平直气液界面，这些模拟参数对应的气液共存密度为 $\rho_l\approx2.783$ 和 $\rho_g\approx0.367\ 5$；对曲线界面，气液共存密度与平直表面的结果稍有不同。

计算区域的大小为 $L_x \times L_y = 150\times150$（格子单位），采用湿节点式布置方式（参见 8.1 节），因此相应的网格节点数为 $N_x \times N_y = 151\times151$。$x$ 方向和 y 方向均采用周期性边界条件，为此须在计算区域的四周各布置一层虚拟层，即 $x=0$ 和 $x=N_x+1$ 为 x 方向布置的虚拟层，而 $y=0$ 和 $y=N_y+1$ 则为 y 方向布置的虚拟层。对于这样一个二维问题的周期性边界条件，可以采取一种较为简便的处理：假定虚拟层的节点也参与碰撞迁移，因此在迁移步之后，$t+\delta_t$ 时刻整个真实物理区域的分布函数都是已知的，于是可直接将真实物理层 $x=N_x$ 上的分布函数赋予虚拟层 $x=0$ 的节点，同时将真实物理层 $x=1$ 上的分布函数赋予虚拟层 $x=N_y+1$ 的节点，y 方向也可进行类似的处理。这样处理符合周期性边界条件的物理本质。程序由 C++语言编写，支持的环境为 Microsoft Visual Studio 2010。初始时刻，给定一个随机的密度扰动，由 C++语言中的函数 rand( ) 产生随机数。虽然该程序是基于指数形式的伪势，读者可在此基础上采用平方根形式的伪势[13]及相应的作用力格式[14-15]，并将程序拓展至多松弛碰撞算子[15]。此外，还可在该程序的基础上加入壁面边界及相应的接触角格式[16-19]。

E.2 程序变量表及源程序

变量表

变量名	变量含义
Q	离散速度的总个数
Nx,Ny	x, y 方向的节点数
Lx,Ly	x, y 方向的长度
i,j	节点坐标
dx,dy	x,y 方向的网格步长
dt	时间步长
m	演化次数
k	离散速度方向 α
niu	运动黏度 ν
tau_f	量纲为 1 的松弛时间 τ
e[Q][2]	离散速度 $\boldsymbol{e}_\alpha$
w[Q]	权系数 ω_α
totalmass0	系统初始总质量
totalmass	系统瞬态总质量
rho[Nx+2][Ny+2]	密度
u[Nx+2][Ny+2][2]	速度
fai[Nx+2][Ny+2]	伪势
Fm[Nx+2][Ny+2][2]	伪势相互作用力
F[Nx+2][Ny+2][Q]	碰撞后的分布函数

源程序

```
#include "stdafx.h"
#include<iostream>
#include<cmath>
#include<cstdlib>
#include<iomanip>
#include<fstream>
#include<sstream>
#include<string>

using namespace std;

const int Q = 9;    //D2Q9 模型
const int Nx = 151;
const int Ny = 151;

int       e[Q][2] = {{0,0},{1,0},{0,1},{-1,0},{0,-1},{1,1},{-1,1},{-1,-1},{1,-1}};
double    w[Q] = {4.0/9,1.0/9,1.0/9,1.0/9,1.0/9,1.0/36,1.0/36,1.0/36,1.0/36};
double    rho[Nx+2][Ny+2],u[Nx+2][Ny+2][2],f[Nx+2][Ny+2][Q],F[Nx+2][Ny+2][Q];
double    tau_f,dx,dy,dt,Lx,Ly,niu,G,fai[Nx+2][Ny+2],Fm[Nx+2][Ny+2][2],totalmass0,totalmass;
int       i,j,k,ip,jp,m;

void SetPrameter()
{
  dx = 1.0;
  dy = 1.0;
  Lx = dx * double(Nx-1);
  Ly = dy * double(Ny-1);
  dt = dx;
  tau_f = 0.8;
  niu = (tau_f-0.5)/3.0;
  std::cout <<" niu = "<<niu<<endl;
}

double feq(int k,double rho,double u[2]) //计算平衡态分布函数
{
    double eu,uv,feq;
    eu = (e[k][0] * u[0]+e[k][1] * u[1]);
    uv = (u[0] * u[0]+u[1] * u[1]);
    feq = w[k] * rho * (1.0+3.0 * eu+4.5 * eu * eu-1.5 * uv);
    return feq;
}

void init()
```

```
{
    totalmass0 = 0;
    for(i=0; i<=Nx+1; i++)   //流场及密度场初始化
      for(j=0; j<=Ny+1; j++)
      {
        u[i][j][1] = 0.0;   u[i][j][0] = 0.0;
        int range = 10;
        int min =0;
        int r = rand()&range + min;   //生成随机数
        rho[i][j] = 1.0-0.01 * double(r)/50.0;   //初始时刻密度施加随机扰动
        totalmass0 += rho[i][j];
      }
}

void compute_interaction_force() //计算伪势相互作用力
{
      for(i=0; i<=Nx+1; i++)
        for(j=0; j<=Ny+1; j++)
        {
          fai[i][j] = 1.0 * exp(-1.0/rho[i][j]);
        }
        for(i=1; i<=Nx; i++)
      for(j=1; j<=Ny; j++)
      {
        G =-10.0/3.0;
          Fm[i][j][0] = 0.0;
          Fm[i][j][1] = 0.0;
          for(k=1; k<Q; k++)
          {
            Fm[i][j][0] += (-G) * fai[i][j] * 3.0 * w[k] * fai[i+e[k][0]][j+e[k][1]] * e[k][0];
            Fm[i][j][1] += (-G) * fai[i][j] * 3.0 * w[k] * fai[i+e[k][0]][j+e[k][1]] * e[k][1];
          }
        }

        for(j=1; j<=Ny; j++)
        {
          Fm[0][j][0] = Fm[Nx][j][0];
          Fm[0][j][1] = Fm[Nx][j][1];
        Fm[Nx+1][j][0] = Fm[1][j][0];
          Fm[Nx+1][j][1] = Fm[1][j][1];
        }
        for(i=0; i<=Nx+1; i++)
        {
          Fm[i][0][0] = Fm[i][Ny][0];
```

```
        Fm[i][0][1] = Fm[i][Ny][1];
      Fm[i][Ny+1][0] = Fm[i][1][0];
        Fm[i][Ny+1][1] = Fm[i][1][1];
      }
}

void init_micro() //分布函数初始化
{
    for(i=0; i<=Nx+1; i++)
      for(j=0; j<=Ny+1; j++)
        for(k=0; k<Q; k++)
        {
          f[i][j][k] = feq(k,rho[i][j],u[i][j]);
        }
}

void evolution()
{
    for(i=0; i<=Nx+1; i++) //碰撞步
      for(j=0; j<=Ny+1; j++)
        for(k=0; k<Q; k++)
        {
          double Force[Q],ux,uy,Fx,Fy;
        ux = u[i][j][0];
          uy = u[i][j][1];
        Fx = Fm[i][j][0];
          Fy = Fm[i][j][1];
        Force[k] = (1.0-0.5/tau_f) * w[k] * 3.0 * ( (e[k][0]-ux) * Fx + (e[k][1]-uy) * Fy
          + 3.0 * (e[k][0] * ux + e[k][1] * uy) * ( e[k][0] * Fx + e[k][1] * Fy) );
        F[i][j][k] = f[i][j][k]-(f[i][j][k]-feq(k,rho[i][j],u[i][j]))/tau_f + Force[k];
      }

for(i=0; i<=Nx+1; i++) //迁移步
    for(j=0; j<=Ny+1; j++)
      for(k=0; k<Q; k++)
      {
        ip = i-e[k][0];
        jp = j-e[k][1];
      if( ip>= 0 && ip<=Nx+1 && jp>=0 && jp<=Ny+1)
      {
        f[i][j][k] = F[ip][jp][k];
      }
    }
```

```
//边界处理
  for(j=1; j<=Ny; j++)
    for(k=0; k<Q; k++)
    {
        f[0][j][k] = f[Nx][j][k];
        f[Nx+1][j][k] = f[1][j][k];
    }
  for(i=0; i<=Nx+1; i++)
    for(k=0; k<Q; k++)
    {
        f[i][0][k] = f[i][Ny][k];
        f[i][Ny+1][k] = f[i][1][k];
    }

totalmass = 0;
for(i=0; i<=Nx+1; i++) //计算宏观密度
  for(j=0; j<=Ny+1; j++)
  {
      rho[i][j] = 0;
      for(k=0; k<Q; k++)
      {
          rho[i][j] += f[i][j][k];
      }
      totalmass += rho[i][j];
  }

compute_interaction_force(); //更新伪势相互作用力

for(i=0; i<=Nx+1; i++) //计算宏观速度
  for(j=0; j<=Ny+1; j++)
  {
    double Utemp1,Utemp2;
          Utemp1 = 0;   Utemp2 = 0;
      for(k=0; k<Q; k++)
      {
          Utemp1 += e[k][0] * f[i][j][k];
          Utemp2 += e[k][1] * f[i][j][k];
      }
      u[i][j][0] = (Utemp1 + 0.5 * (Fm[i][j][0] ))/rho[i][j];
      u[i][j][1] = (Utemp2 + 0.5 * (Fm[i][j][1] ))/rho[i][j];
    }
}

void output(int m) //输出
```

```
{
    ostringstream name;
    if(m==0)
    {
      name<<"Phase_sepa0000"<<m<<".dat";
    }
    if(m>1 && m<100)
    {
      name<<"Phase_sepa000"<<m<<".dat";
    }
    if(m>=100 && m<1000)
    {
      name<<"Phase_sepa00"<<m<<".dat";
    }
    if(m>=1000 && m<10000)
    {
      name<<"Phase_sepa0"<<m<<".dat";
    }
    if(m>=10000 && m<100000)
    {
      name<<"Phase_sepa"<<m<<".dat";
    }
  ofstream out(name.str().c_str());
    out<<"Title =\"LBM Phase_sepa\"\n"<<"VARIABLES = \"X\",\"Y\",\"rho\"\n"
      <<"ZONE T=\"BOX\",I="<<Nx+1<<",J="<<Ny+1<<",F=POINT"<<endl;
    for(j=1; j<=Ny+1; j++)
      for(i=1; i<=Nx+1; i++)
      {
        out<<double(i-1)/Lx<<"  "<<double(j-1)/Ly<<"  "<<rho[i][j]<<endl;
      }
}
int main()
{
    using namespace std;
    SetPrameter();
    init();
    compute_interaction_force();
    init_micro();
    for(m=0;m<=80000; m++)
    {
      evolution();
      if(m%200 == 0)
  {
      cout<<"The "<<m<<"th computation result:"<<endl<<"The total mass ratio is:" <<setprecision(7)
```

```
        <<totalmass/totalmass0<<endl;
        output(m);
      }
    }
  return 0;
}
```

E.3 数值模拟结果

图 E.1 给出了相分离模拟的动态演化过程,对应的时刻分别为 $t=100\delta_t$、$400\delta_t$、$2\,000\delta_t$和 $50\,000\delta_t$。由于初始时刻的密度处在热力学不稳定区域且施加了一个微小扰动,因此很快就会发生相分离,在 $t=100\delta_t$ 时刻可以看到许多大小不一且形状不规则的液滴。随着时间的推移,一些液滴会与其周围的液滴融合而形成更大的液滴。此外,在系统达到稳定之前,会一直发生气液两相的转变(与蒸发及冷凝现象类似),因此可以看到一些小液滴的消失以及一些大液滴的长大,直至系统达到稳定平衡,形成一个大液滴(参见 $t=50\,000\delta_t$ 时刻)。程序中输出了系统瞬态总质量与初始时刻系统总质量的比值,由于模拟中的气液界面为扩散界面,而初始时刻并不存在界面,因此这一比值会在 1.0 附近有非常微小的波动。当系统稳定时,比值同样趋于稳定,约为 0.999 28。

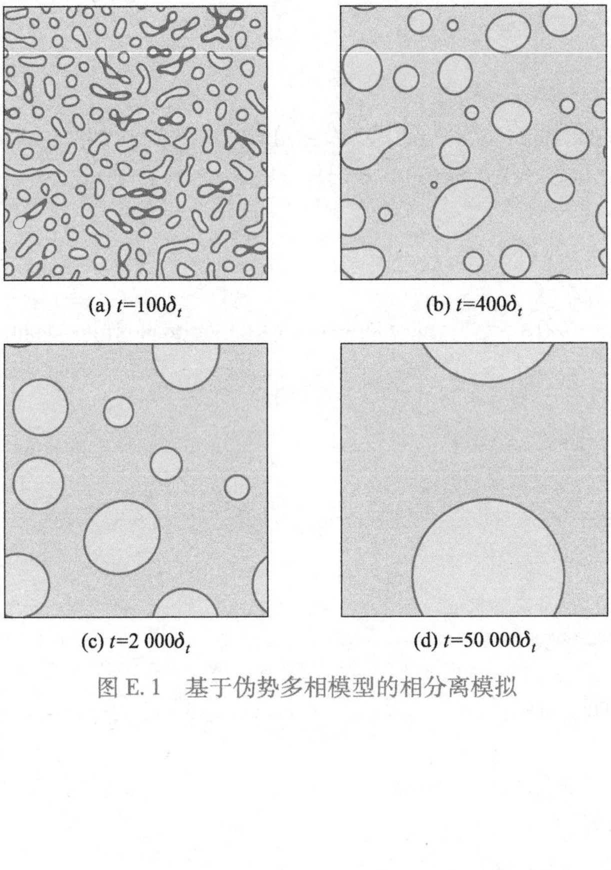

(a) $t=100\delta_t$　(b) $t=400\delta_t$　(c) $t=2\,000\delta_t$　(d) $t=50\,000\delta_t$

图 E.1 基于伪势多相模型的相分离模拟

附录 F 主题索引(中英文对照)

A

爱因斯坦求和约定(Einstein's summation convention) 261

B

Boltzmann-BGK 方程(Boltzmann-BGK equation) 44
Boltzmann H 定理(Boltzmann H-theorem) 40
Boltzmann 方程(Boltzmann equation) 37
Boussinesq 假设(Boussinesq approximation) 227
半步长反弹格式(halfway bounce-back scheme) 186
本构方程(constitutive equation) 25
被动标量模型(passive scalar model) 75
边界处理方法(boundary condition method) 183
边界处理格式(boundary condition scheme) 183
标量(scalar) 261
标准格子 Boltzmann 方法(standard lattice Boltzmann method) 46,255
表面张力(surface tension) 103
表征体元(representative elementary volume) 163,164

C

Cahn-Hilliard 方程(Cahn-Hilliard equation) 127
Chapman-Enskog 展开(Chapman-Enskog expansion) 49
插值格子 Boltzmann 方法(interpolation-supplemented lattice Boltzmann method) 202
充分发展(fully developed) 186
稠密流体(dense fluid) 101

D

D2G9 73
D2Q9 47
D3Q15 49
D3Q19 49
DdQm 47
单松弛碰撞算子(single-relaxation-time) 5
动力学格式(hydrodynamic scheme) 188
动量交换法(momentum-exchange method) 146
点积(dot product) 263
顶盖驱动流(lid-driven flow) 268
对称边界(symmetric boundary) 185
对称张量(symmetric tensor) 262
多尺度展开(multiscale expansion) 49
多重网格(multi-grid) 221
多块网格(multi-block grid) 217
多粒子碰撞模型(multi-particle collision model) 5
多松弛碰撞算子(multiple-relaxation-time) 6,52
多组分伪势模型(multicomponent pseudopotential model) 118

E

Euler 方程(Euler equation) 2

F

FHP 模型(FHP model) 4
反对称张量(anti-symmetric tensor) 262
反滑移热能(counter-slip thermal energy) 189
反滑移速度(counter-slip velocity) 189
非结构化网格(unstructured grid) 215
非平衡反弹格式(non-equilibrium bounce-back scheme) 188
非平衡态外推格式(non-equilibrium extrapolation scheme) 191
分配函数(assigning function) 97
封闭方腔自然对流(natural convection in a square cavity) 227
傅里叶定律(Fourier's law) 28

G

刚性问题(stiff problem) 210
高斯公式(Gauss theorem) 264
各向同性(isotropic) 265
格子(lattice) 46
格子 BGK 模型(lattice-BGK model) 5
格子 Boltzmann-BGK 方程(lattice Boltzmann-BGK equation) 46
格子 Boltzmann 方法(lattice Boltzmann method) 1

格子单位(lattice unit) 202,267
光滑因子(smoothness indicator) 213

H

焓法模型(enthalpy-based model) 151
He-Luo 模型(He-Luo model) 72
HPP 模型(HPP model) 4
H 定理(*H* theorem) 40
H 函数(*H* function) 39
哈密顿算符(Hamiltonian operator) 263
宏观(macroscopic) 1,2
宏观流动速度(macroscopic flow velocity) 38
混合模型(hybrid model) 74

J

迹(trace) 264
激波(shock wave) 88
计算流体力学(computational fluid dynamics) 2
伽利略不变性(Galilean invariance) 124
接触角(contact angle) 108
截断形式(truncated form) 5
结构化网格(structured grid) 214
介观(mesoscopic) 1,2
镜面反射格式(specular reflection scheme) 187
就近平均(averaging nearby values) 221
局部粒子分布函数(local particle distribution function) 5
局部非热平衡(local thermal non-equilibrium) 165
局部热平衡(local thermal equilibrium) 165
矩方程(moment equation) 40
矩阵模型(matrix model) 5

K

克罗内克符号(Kronecker symbol) 262
孔隙尺度(pore scale) 163
控制体(control volume) 22

L

拉格朗日法(Lagrangian method) 22
拉普拉斯算符(Laplace operator 或 Laplacian) 264
雷诺第二输运方程(second Reynolds' transport equation) 24
雷诺输运方程(Reynolds's transport equation) 23
离散 Boltzmann-BGK 方程(discrete Boltzmann-BGK equation) 201
离散速度模型(discrete velocity model) 46
连续 Boltzmann 方程(continuous Boltzmann equation) 201
连续性方程(continuity equation) 24
连续介质假设(continuous hypothesis) 21
流固两相流(fluid-solid two-phase flow) 145
流体质点(fluid particle) 22

M

Maxwell 分布(Maxwell distribution) 37
Maxwell 输运方程(Maxwell transport equation) 40
密度分布函数(density distribution function) 77

N

纳维-斯托克斯方程(Navier-Stokes equations) 27
牛顿流体(Newtonian fluid) 25

O

耦合双分布函数格子 Boltzmann 模型(coupled double distribution function lattice Boltzmann model) 89
欧拉法(Euler method) 22

P

碰撞不变量(collision invariant) 37
碰撞项(collision term) 44
平均场动理论(kinetic mean-field theory) 101

Q

启发式格式(heuristic scheme) 184
气体动理论(gas kinetic theory) 33
气液界面(gas-liquid interface) 244
强激波问题(strong shock wave problem) 236

R

Rayleigh-Bénard 自然对流(Rayleigh-Bénard convection) 231
热力学第一定律(first law of thermodynamics) 28
热力学能分布函数(internal energy distribution function) 77

热力学一致性(thermodynamic consistency) 106,108
REV 尺度(REV scale) 163
Riemann 问题(Riemann problem) 235

S

Sod 问题(Sod problem) 236
散度(divergence) 264
声波衰减(attenuation of sound waves) 238
矢量(vector) 261
守恒形式的动量方程(conservation form of momentum equation) 25
数值传热学(numerical heat transfer) 2
双重网格(double meshes) 221
双点积(double product) 263
双二次插值(bi-quadratic interpolation) 222
双分布函数模型(double distribution function model) 74
双马赫反射(double Mach reflection) 236
双线性插值(bilinear interpolation) 222
松弛时间(relaxation time) 45
速度分布函数(velocity distribution function) 44
随机速度(random velocity) 38

T

泰勒展开和最小二乘格子 Boltzmann 方法(Taylor series expansion-and least square-based lattice Boltzmann method) 205
梯度(gradient) 263
特异速度(peculiar velocity) 95
通量限制器(flux limiter) 213

W

温度分布函数(temperature distribution function) 74
WENO(weighted essentially non oscillatory) 213
外推格式(extrapolation scheme) 190
微分形式的动量方程(differential form of momentum equation) 25
微分形式的总能方程(differential form of total energy equation) 28
微观(microscopic) 1,2
伪势(pseudo-potential) 111
伪势模型(pseudo-potential model) 111
物理单位(physical unit) 202,267

X

系统(system) 22
系综理论(ensemble theory) 34
限定(restriction) 221
限定算子(restriction operator) 221
相场(phase-field) 127
相场模型(phase-field model) 127
序参数(order parameter) 127
虚拟平衡态分布函数(virtual equilibrium distribution function) 193

Y

雅可比行列式(Jacobian determinant) 23
哑指标(dummy index) 261
亚稳态(metastable state) 243
演化方程(evolution equation) 4
颜色梯度模型(color-gradient model) 120
延拓(prolongation) 221
隐式-显式有限差分格子 Boltzmann 方法(implicit-explicit finite-difference lattice Boltzmann method) 210
有限差分格子 Boltzmann 方法(finite-difference lattice Boltzmann method) 209
有限容积格子 Boltzmann 方法(finite-volume lattice Boltzmann method) 214
有限元格子 Boltzmann 方法(finite-element lattice Boltzmann method) 217
圆函数(circle function) 93

Z

Zou-Hou 模型(Zou-Hou model) 71
张量(tensor) 261
直接引入(direct inject) 221
质量守恒(mass conservation) 24
质量修正格式(mass modified scheme) 189
周期性边界条件(periodic boundary) 184
自由能(free-energy) 104
自由能模型(free-energy model) 104
自由指标(free index) 261
总能分布函数模型(total energy distribution function model) 84

参考文献

[1] 章梓雄,董曾南. 粘性流体力学[M]. 2版. 北京:清华大学出版社,2011.

[2] 董曾南,章梓雄. 非粘性流体力学[M]. 北京:清华大学出版社,2003.

[3] 李开泰,黄艾香. 张量理论与应用[M]. 北京:科学出版社,2004.

[4] 钟霖浩. 风生环流机制模式的格子 Boltzmann 数值模拟[D]. 青岛:中国海洋大学,2005.

[5] Wolfram S. Cellular automaton fluids 1: Basic theory[J]. Journal of Statistical Physics, 1986, 45(3): 471-526.

[6] Succi S. The lattice Boltzmann equation for fluid dynamics and beyond[M]. Oxford: Clarendon Press, 2001.

[7] 郭照立,郑楚光,李青,等. 流体动力学的格子 Boltzmann 方法[M]. 武汉:湖北科学技术出版社,2002.

[8] Vanka S P, Leaf G K. Block-implicit multigrid solution of Navier-Stokes equations in primitive variables [J]. Journal of Computational Physics, 1986, 65(1): 138-158.

[9] Ghia U, Ghia K N, Shin C T. High-Re solutions for incompressible flow using the Navier-Stokes equations and a multigrid method[J]. Journal of Computational Physics, 1982, 48(3): 387-411.

[10] Hou S L, Zou Q S, Chen S Y, et al. Simulation of cavity flow by the lattice Boltzmann method[J]. Journal of Computational Physics, 1995, 118(2): 329-347.

[11] Shan X W. Pressure tensor calculation in a class of nonideal gas lattice Boltzmann models[J]. Physical Review E, 2008, 77(6): 066702.

[12] Li Q, Zhou P, Yan H J. Revised Chapman-Enskog analysis for a class of forcing schemes in the lattice Boltzmann method[J]. Physical Review E, 2016, 94: 043313.

[13] Yuan P, Schaefer L. Equations of state in a lattice Boltzmann model[J]. Physics of Fluids, 2006, 18: 042101.

[14] Li Q, Luo K H, Li X J. Forcing scheme in pseudopotential lattice Boltzmann model for multiphase flows [J]. Physical Review E, 2012, 86(1): 016709.

[15] Li Q, Luo K H, Li X J. Lattice Boltzmann modeling of multiphase flows at large density ratio with an improved pseudopotential model[J]. Physical Review E, 2013, 87: 053301.

[16] Benzi R, Biferale L, Sbragaglia M, et al. Mesoscopic modeling of a two-phase flow in the presence of boundaries: the contact angle[J]. Physical Review E, 2006, 74(2): 021509.

[17] Ding H, Spelt P D M. Wetting condition in diffuse interface simulations of contact line motion[J]. Physical Review E, 2007, 75(4): 046708.

[18] Li Q, Luo K H, Kang Q J, et al. Contact angles in the pseudopotential lattice Boltzmann modeling of wetting[J]. Physical Review E, 2014, 90(5): 053301.

[19] Li Q, Yu Y, Luo K H. Implementation of contact angles in pseudopotential lattice Boltzmann simulations with curved boundaries[J]. Physical Review E, 2019, 100(5): 053313.